Susan Schorr, Claudia Weidenthaler (Eds.)
Crystallography in Materials Science

Also of interest

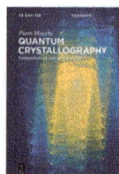

Quantum Crystallography.
Fundamentals and Applications
Macchi, 2021
ISBN 978-3-11-060710-9, e-ISBN 978-3-11-060712-3

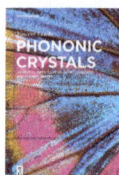

Phononic Crystals.
Artificial Crystals for Sonic, Acoustic, and Elastic Waves
Laude, 2020
ISBN 978-3-11-063728-1, e-ISBN 978-3-11-064118-9

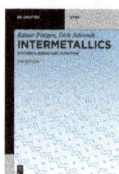

Intermetallics.
Synthesis, Structure, Function
Pöttgen, Johrendt, 2019
ISBN 978-3-11-063580-5, e-ISBN 978-3-11-063672-7

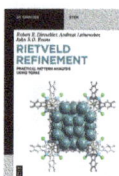

Rietveld Refinement.
Practical Powder Diffraction Pattern Analysis using TOPAS
Dinnebier, Leineweber, Evans, 2019
ISBN 978-3-11-045621-9, e-ISBN 978-3-11-046138-1

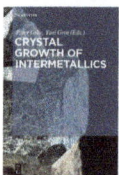

Crystal Growth of Intermetallics
Gille, Grin, (Eds.) 2019
ISBN 978-3-11-049584-3, e-ISBN 978-3-11-049678-9

Crystallography in Materials Science

From Structure-Property Relationships to Engineering

Edited by
Susan Schorr, Claudia Weidenthaler

DE GRUYTER

Editors

Prof. Dr. Susan Schorr
Helmholtz-Zentrum Berlin
für Materialien u. Energie GmbH
Department Structure and Dynamics of Energy Materials
Hahn-Meitner-Platz 1
14109 Berlin
Germany
susan.schorr@helmholtz-berlin.de

Freie Universität Berlin
Institute of Geological Sciences
Malteserstr. 74–100
12249 Berlin
Germany

Dr. Claudia Weidenthaler
Max-Planck-Institut
für Kohlenforschung
Kaiser-Wilhelm-Platz 1
45470 Mülheim
Germany
claudia.weidenthaler@mpi-mail.mpg.de

ISBN 978-3-11-067485-9
e-ISBN (PDF) 978-3-11-067491-0
e-ISBN (EPUB) 978-3-11-067504-7

Library of Congress Control Number: 2021932030

Bibliographic information published by the Deutsche Nationalbibliothek
The Deutsche Nationalbibliothek lists this publication in the Deutsche Nationalbibliografie;
detailed bibliographic data are available on the Internet at http://dnb.dnb.de.

© 2021 Walter de Gruyter GmbH, Berlin/Boston
Cover image: BlackJack3D/E+/Getty Images
Typesetting: Integra Software Services Pvt. Ltd.
Printing and binding: CPI books GmbH, Leck

www.degruyter.com

Foreword

Modern developments in science and technology require the production of new materials and designs. Knowing the material's properties is mandatory to develop and manufacture tailor-made materials for technical applications.

Crystallography is a profoundly interdisciplinary science and focuses nowadays on the investigation of the spatial arrangement of matter in all different types of materials and their changes under external influences like temperature or pressure. The crystal structure itself is very complex: it can be seen as an interplay of static and dynamic structures, it contains imperfections and structural disorder, such as point defects or dislocations, and it has to be distinguished between average and local structures. The crystal structure plays a crucial role in determining the physical properties of a material. A material of certain chemical composition can have different structural parameters and thus have different properties.

But crystallography not only applies to the atomic structure of materials, it is also crucial for the understanding of disordered materials such as glasses and collective quantum phenomena such as superconductivity. Thus, the knowledge about a material's structural characteristics and its correlation with technologically relevant properties is the prerequisite for designing new materials, with tailored properties central to technological needs, and is at the forefront of materials research.

This book shows that crystallography provides the foundation for the understanding of structure-function relationships and bridges the fundamental understanding of materials with applications.

The purpose of this book is twofold: (1) to present an overview on crystallographic techniques and methods to study complex structural material properties and (2) to present exemplarily investigations of structure-function relations in various technologically relevant materials like those for solar energy conversion and for solar-thermal water splitting or batteries. In the book, it is shown that crystallography can also be extremely useful in research areas that are normally not using structure studies such as corrosion research. Finally, the book also pays attention to recent advances in structure modeling as well as structure-property investigations with the help of density functional theory calculations. We have collected chapters written by renowned experts in the different research fields, giving insights into their scientific areas as well as reflecting modern aspects of crystallographic materials research.

This book is addressed to a wide audience of scientists in the field of physics, chemistry, mineralogy, and – of course – crystallography and materials science. It will be also interesting for graduate students in these disciplines. The book is intended to

https://doi.org/10.1515/9783110674910-202

pique the curiosity of younger students about crystallography and its fascinating insights into materials.

We thank all authors for their valuable contributions which made *Crystallography in Materials Science* possible.

Susan Schorr und Claudia Weidenthaler

Contents

List of contributors

Claudia Weidenthaler
Max-Planck-Institut für Kohlenforschung
Kaiser-Wilhelm-Platz 1
45470 Mülheim
Germany
claudia.weidenthaler@mpi-mail.mpg.de

Robert Dinnebier
Max-Planck Institute for Solid State Research
Heisenbergstrasse 1
70569 Stuttgart
Germany

Sebastian Bette
Max-Planck Institute for Solid State Research
Heisenbergstrasse 1
70569 Stuttgart
Germany

Alain Lafond
Université de Nantes
Nantes
France

Catherine Guillot-Deudon
Université de Nantes
Nantes
France

Michaël Paris
Université de Nantes
Nantes
France

Maria Teresa Caldes
Université de Nantes
Nantes
France

Stéphane Jobic
Université de Nantes
Nantes
France

Susan Schorr
Helmholtz-Zentrum Berlin für Materialien
u. Energie GmbH
Department Structure and Dynamics of
Energy Materials
Hahn-Meitner-Platz 1
14109 Berlin
Germany
susan.schorr@helmholtz-berlin.de

Galina Gurieva
Helmholtz-Zentrum Berlin für Materialien
u. Energie GmbH
Department Structure and Dynamics of
Energy Materials
Hahn-Meitner-Platz 1
14109 Berlin
Germany

Joachim Breternitz
Helmholtz-Zentrum Berlin für Materialien
und Energie
Department Structure and Dynamics of
Energy Materials
Hahn-Meitner-Platz 1
14109 Berlin
Germany

Martin Schmücker
Deutsches Zentrum für Luft- und Raumfahrt
(DLR)
Linder Höhe
51147 Köln
Germany

Nicole Knoblauch
Deutsches Zentrum für Luft- und Raumfahrt
(DLR)
Linder Höhe
51147 Köln
Germany

https://doi.org/10.1515/9783110674910-204

Vanessa Peterson
ANSTO - Sydney
New Illawarra Rd
Lucas Heights
NSW 2234
Australia

Christophe Didier
ANSTO - Sydney
New Illawarra Rd
Lucas Heights
NSW 2234
Australia

Wei Kong Pang
ANSTO - Sydney
New Illawarra Rd
Lucas Heights
NSW 2234
Australia

Arndt Remhof
Empa
Überlandstrasse 129
CH-8600 Dübendorf
Switzerland

Radovan Černý
University of Geneva
24 Quai Ernest Ansermet
1211 Genève 4
Switzerland

Christiane Stephan
Bundesanstalt für Materialforschung
und -prüfung (BAM)
Berlin
Germany

Daniel Fritsch
Helmholtz-Zentrum Berlin für Materialien
und Energie
Department Structure and Dynamics of
Energy Materials
Hahn-Meitner-Platz 1
14109 Berlin
Germany

Bernd Hinrichsen
BASF SE
Ludwigshafen
Germany

Claudia Weidenthaler

1 In situ tools for the exploration of structure–property relationships

Abstract: For understanding structure-property relationships, investigations under non-ambient conditions or process conditions are indispensable. Today, a large number of different spectroscopic, microscopic, and diffraction methods are available to monitor how materials change under reaction conditions. Based on the results, it is possible to draw conclusions about structure-property interactions under operation. Depending on the specific application and its respective requirements, materials can be modified and optimized.

Keywords: in situ, structure-property relationship, diffraction, spectroscopy

1.1 Introduction

Nowadays, crystal structures are with many different physical techniques, highlighting different aspects of structures. High experimental accuracy on structural details is achieved, providing deep insights into comprehensive structural aspects. This is important since the physical and chemical properties of materials result from their individual crystal structures. The detailed understanding of how crystal structures influence the physical and chemical properties of the related materials is essential for tailor-made preparation or modification of materials used in specific applications. The term "structure" in this context stands not only for the averaged crystal structure of a compound. Of course, it is important to know the averaged crystal structure of a material. However, for many applications, such as catalysis, the presence of structural defects or amorphous phases is in many cases the decisive factor for good or bad performance. For this reason, the local atomic structure arrangements in materials with a lack of structural long-range order are gaining more interest. Finally, microstructure properties such as finite-size effects, and defects such as dislocation, stacking faults, or twins are important parameters for many applications and need to be investigated. Figure 1.2 summarizes structures on different length scales starting with the averaged crystal structure of Co_3O_4 evaluated by conventional powder diffraction or single crystal methods (Fig. 1.1a). The atomic structure arrangement representing only the short-range structural order is shown in Fig. 1.1b. The first nearest atomic pair correlations of disordered silicon oxide were studied by total scattering experiments and atomic pair

Claudia Weidenthaler, Max-Planck-Institut für Kohlenforschung, Kaiser-Wilhelm-Platz 1, 45470 Mülheim, Germany, weidenthaler@mpi-muelheim.mpg.de

https://doi.org/10.1515/9783110674910-001

(a) crystal structure

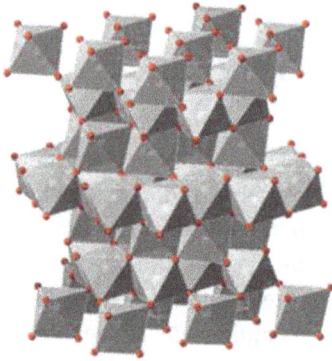

(b) local structure

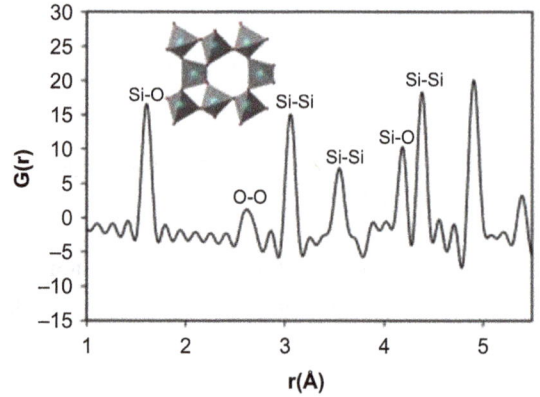

(c) microstructure

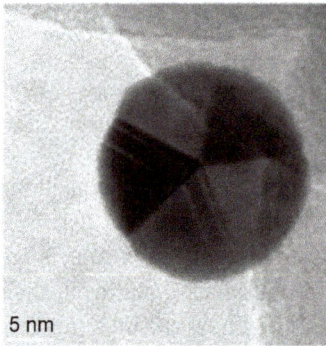

Fig. 1.1: The term structure used in this chapter covers (a) averaged crystal structures (example shows Co$_3$O$_4$), (b) local atomic structures evaluated by total scattering experiments and atomic pair distribution function analysis (example shows PDF data of a disordered silicon dioxide), and (c) microstructure reflecting finite-size effects or defects such as stacking faults or twinning as shown for a gold nanoparticle.

distribution function (PDF) analysis. The microstructure of a twinned nanosized gold particle (Fig. 1.1c) represents structure details also on the atomic level but visualized by microscopy techniques [1].

Not only the knowledge about the crystal structures or microstructures under ambient conditions is of great importance but also following structure changes during treatment under non-ambient conditions and relating this to the performance of materials became more and more interesting during the last decades. In situ studies provide information about synthesis and crystallization reactions, reaction pathways, formation of potential metastable intermediates, structural phase transformations, compositional changes taking place during reactions, or microstructure changes. These so-called in situ studies are performed under various physical or chemical conditions using different analytical tools. These tools may focus on the different types of reactions or monitor different material properties. A selection of possibilities of how experimental conditions can be varied, which analytical methods can be applied in situ/*operando*, and what kind of reactions can be monitored is summarized in Tab. 1.1. There are a large number of review articles on in situ/*operando* studies related to structure–property relationships of which only a few recently published ones can be mentioned in this chapter as examples [2–7].

To describe structure–property relations comprehensively, it is important to consider that it may not be sufficient to apply only one experimental method. Adequate combinations of different tools probing structures on different length scales are mandatory to get a picture of structure–property relationships. Even one step further, modern analytical tools nowadays enable structure studies under real operating conditions (*operando*). *Operando* studies evaluate materials under realistic process conditions by simultaneous analysis of structural or electronic features and their performances such as reaction rates of catalysts or redox reactions that take place in batteries [8, 9]. While

Tab. 1.1: The main variable experimental parameters, analytical methods, and reactions used for in situ/*operando* studies of the structure–property relationship of functional materials.

Variation of physical and/or chemical conditions	Temperature
	Pressure (gas pressure or hydrostatic pressure)
	Gas atmosphere
	Humidity
	Electric field
	Mechanical energy (ball-milling)
	Light illumination
	Stress field
	Hydrothermal/solvothermal
Analytical methods	Diffraction of periodic atomic structures with X-rays, neutrons, and electrons
	Scattering on amorphous matter with X-rays, neutrons, and electrons
	Microscopy
	Spectroscopy methods such as Raman, infrared, ultraviolet–visible, nuclear magnetic resonance, extended X-ray absorption fine structure), and neutron spectroscopy
Reactions monitored under in situ conditions	Synthesis: nucleation and crystallization
	Growth of particles
	Alloying, segregation
	Electrochemical reactions
	Catalytic reactions
	Structural phase transformation
	Magnetic structure transformation

in situ studies evaluate structure–property relationships, *operando* studies study structure–performance relations. From *operando* data, information on the kinetics of reactions can be derived. Such types of investigations became indispensable for many different materials used in large-scale industrial applications as well as for fundamental research. The broad variety ranges from energy-related materials (batteries, fuel cells, solar cells, solid hydrogen storage materials, and supercapacitors), materials for electronic devices, to catalysts for heterogeneous catalysis, or functional materials performing under high-stress loads to name but a few. Nowadays, in situ or *operando* studies are mandatory and will become state of the art if structure properties under reaction conditions will be really understood.

Many examples are available, which show that materials characterized "*ex situ*" (before and after reaction) may not necessarily represent the state of the material during a reaction or under performance conditions. Exposure of materials to the atmosphere result in forming oxides or hydroxides in case the compounds are sensitive to humidity. Structural phase transformations into the metastable state may take place during non-ambient conditions. Such intermediate phases may influence the performance of the desired material under process conditions but might go unnoticed when the compound is characterized after the reaction. This again stresses the importance of in situ studies. As one example, Fig. 1.2 shows in situ X-ray powder diffraction patterns of $LaCO_3$, a perovskite tested as a catalyst for electrochemical applications. During the reduction of $LaCoO_3$ (Fig. 1.2a) with an appropriate amount of H_2 at defined temperatures, a layered Ruddlesden–Popper phase, La_2CoO_4, is obtained (Fig. 1.2b). Exposure to air leads to the reoxidation of the Ruddlesden–Popper phase back to $LaCoO_3$ (Fig. 1.2c). Measuring *ex situ* X-ray powder patterns of

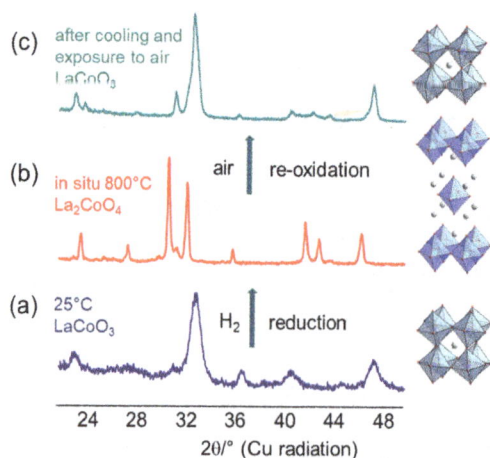

Fig. 1.2: In situ powder diffraction patterns of (a) $LaCoO_3$, (b) collected at 800 °C after partial reduction with H_2 to La_2CoO_4 Ruddlesden–Popper phase, and (c) sample after exposure to air and reoxidation to $LaCoO_3$.

the sample before reduction and after reduction but exposure to air would have simply missed the formation of the Ruddlesden–Popper phase. Based on the observations under in situ conditions, La_2CoO_4 could be synthesized by the targeted reduction of $LaCoO_3$ [10].

Crystal structures are obtained either from single crystals or from polycrystalline powder materials. Structure solution and refinement are the first steps if crystallography is related to material properties. Whenever single crystals of a certain size are available, the crystal structures of new compounds are solved relatively straightforward in most cases. Crystal structure solution from powder diffraction data can be very complicated or even impossible, especially for inorganic or nanosized compounds. Nevertheless, due to improved algorithms, the number of structures solved from powder diffraction data is growing year by year. Powders can contain small crystals of one crystalline compound (single phase) or of several crystalline phases (phase mixtures). Besides the averaged atomic crystal structure, microstructure properties due to infinite crystallite size effects and/or defects may have a strong influence on material properties. To go one step further down on the length scale of structure analysis, local structures of partially amorphous or disordered compounds are investigated by total scattering experiments. Even though the measured powder patterns of such samples do not provide information on the averaged crystal structure, useful information on local ordering and hence physical and chemical properties become accessible.

The spectrum of experimental techniques available for in situ or *operando* studies has become remarkably broad over the two last decades. Besides, new generation large research facilities offer the possibility for high-quality studies on very short experimental timescales. Already accessible but also future directing are combined experiments that focus on different length scales of structures or different content of information. In this chapter, the emphasis will be on different techniques available for in situ studies using in-house instrumentation as well as large research facilities such as neutron or synchrotron radiation sources.

Complementary methods such as solid-state nuclear magnetic resonance (NMR) spectroscopy, infrared (IR) or Raman spectroscopy, or X-ray absorption techniques such as extended X-ray absorption fine structure (EXAFS) can complement diffraction or scattering techniques. Also microscopy such as transmission electron microscopy or X-ray photoemission microscopy provides very useful structural information or even allow structure solution of nanosized materials.

In principle, studies on structure–property relationships are based on two different approaches as discussed in a review article by Sharma et al. for battery materials (Fig. 1.3) [11]. Experiments can follow behavior of materials either (a) in equilibrium or nonequilibrium state, or they can study materials as (b) a single component or in an operating device working under real conditions. Experiments in nonequilibrium states are most difficult to realize, and data evaluation is often supported by equilibrium single component studies, which provide information on structures under well-defined conditions.

equilibrium		non-equilibrium	
single component (ex *situ*)	operating device (*in situ*)	single component (*in situ*/ operando)	operating device (operando)
electrode structure at defined temperature or composition	electrode structure within a battery as function of charge	electrode structure during reaction with different electrodes	electrode structure within a battery during charge/discharge

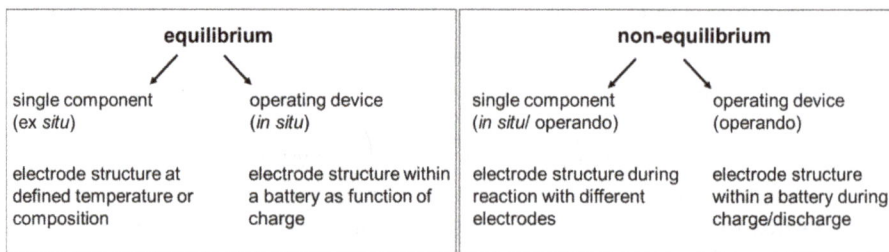

Fig. 1.3: In situ studies on structure–property relationships can either follow behavior of materials (a) in equilibrium state or (b) in the nonequilibrium state.

Before designing any in situ experiment, several basic questions need to be addressed:

(a) What kind of structural information is required? Is the aim of investigations to obtain information about the local structure on atomic scale or about the average crystal structure? If the magnetic properties of a material are studied, then the magnetic structure of the compound should be known.

(b) What are the sample properties that might limit analysis? There are many factors that can influence the experimental results. To name but a few is the presence of light elements together with heavier elements, the presence of absorbing elements, or hydrogen if X-ray techniques will be used. Another challenge is the handling of air-sensitive samples, which need to be under a protective atmosphere not only during sample preparation but also during the entire experiment.

(c) Which analytical technique can provide the required structure information?

(d) What kind of sample environment and technical setup is required?

(e) Which excitation source is suitable to solve questions addressed in (a–d)?

This chapter gives an overview of the diversity of scientific questions related to structure–property relationships, which can be addressed by in situ/*operando* studies. A selection of analytical tools is introduced, using different excitation sources and different sample environments. The chapter is by no means a complete account of in situ tools for structure–property studies, but rather provides examples of how in situ/*operando* studies can beneficially contribute to the understanding of structure–property relationships.

1.2 Probe sources for in situ studies

Depending on the scientific problem and sample properties, the appropriate technique for the experiment needs to be chosen. Structure investigations are performed in principle with one of the following tools:

(a) Electromagnetic radiation
(b) Neutron radiation
(c) Electrons

All radiation sources can be used for various types of scattering and spectroscopy experiments. It is beneficial to use them as complementary methods. The most common spectroscopic methods probing local structure properties of solid materials are EXAFS, Raman, IR, ultraviolet–visible, and solid-state NMR spectroscopy. However, the focus of this chapter is the investigation of structure–property interactions using diffraction methods. For this reason, spectroscopic methods are only briefly touched upon in the following.

1.2.1 Electromagnetic radiation

The electromagnetic spectrum covers radiation with a wavelength from radio waves (10^3 m) to gamma rays (10^{-12} m) (Fig. 1.4). Depending on the length scale on which structures are studied, the light of the appropriate wavelength can be chosen for a specific experiment. For example, radio waves excite nuclei in an external magnetic field into NMR spectroscopy producing signals that are detected with sensitive radio receivers. This technique is most useful for determining the structure of molecules in solution but it is also widely applied to solid compounds. Solid-state NMR spectroscopy is an important technique, for instance, in the structural characterization of framework structures, such as zeolites, since it provides information about the connectivity between $[SiO_4]^{4-}$ tetrahedra or replacement of Si by Al in the framework.

In situ solid-state NMR spectroscopy has been applied to study the structure formation of microporous aluminosilicates (AlPO) during hydrothermal synthesis starting from the precursor gels [12]. From liquid-state NMR studies of the initial precursor, the nature of individual reactive species prior to heating could be identified. ^{31}P, ^{27}Al, and ^{19}F NMR spectra showed octahedral complexes of mixed fluoroaluminophosphate accompanied by tetrahedrally coordinated phosphate anions and octahedrally coordinated aluminum cations (Fig. 1.5). These elementary units are considered as primary building units (PBU) from which more complex structures can be formed.

After the identification of PBUs, high-temperature NMR experiments under hydrothermal conditions show the temperature-dependent evolution of the structure units, where octahedrally coordinated Al changes to penta-coordinated species, so-called prenucleation building units (PNBUs) (Fig. 1.6). During temperature increase, the penta-coordinated PBUs change their Al coordination number before they start to dimerize and form rings.

An important conclusion from the in situ NMR studies is that the structure of PNBUs differs from the secondary building units known from solid alumosilicate

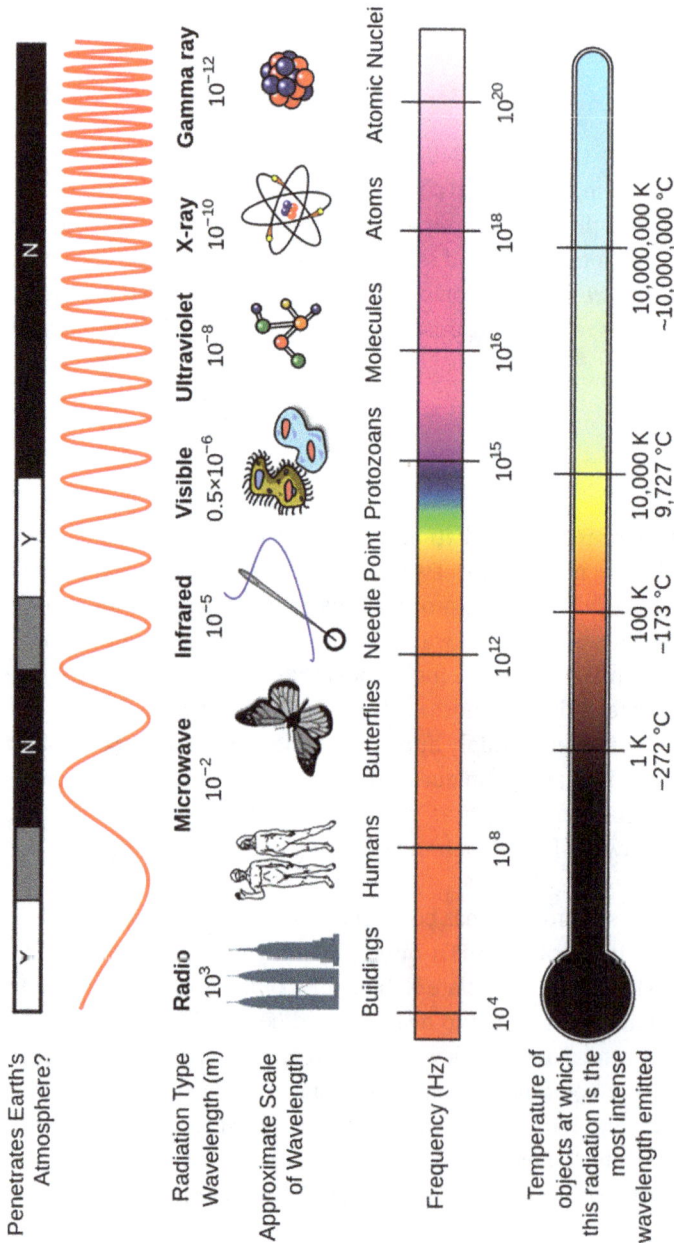

Fig. 1.4: Spectrum of electromagnetic radiation (all text and articles published by Sun.org are licensed under a Creative Commons Attribution-ShareAlike 4.0 International License).

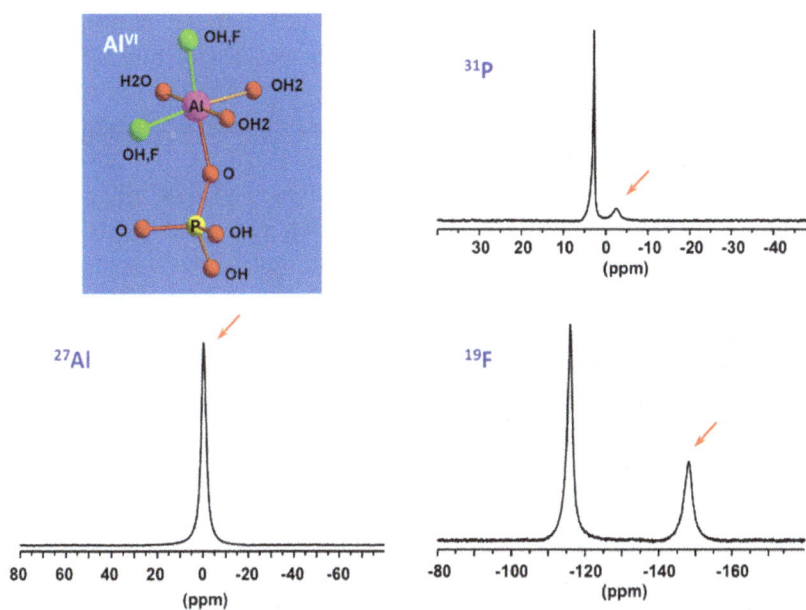

Fig. 1.5: ^{27}Al, ^{31}P, and ^{19}F NMR spectra of starting precursor for AlPO$_4$ prior to heating. The red arrows show signatures of the octahedral mixed fluoroaluminophosphates, the primary building unit (PBU) shown in the upper left image (reprinted with permission from Ref. [12]. Copyright © 2018, MDPI).

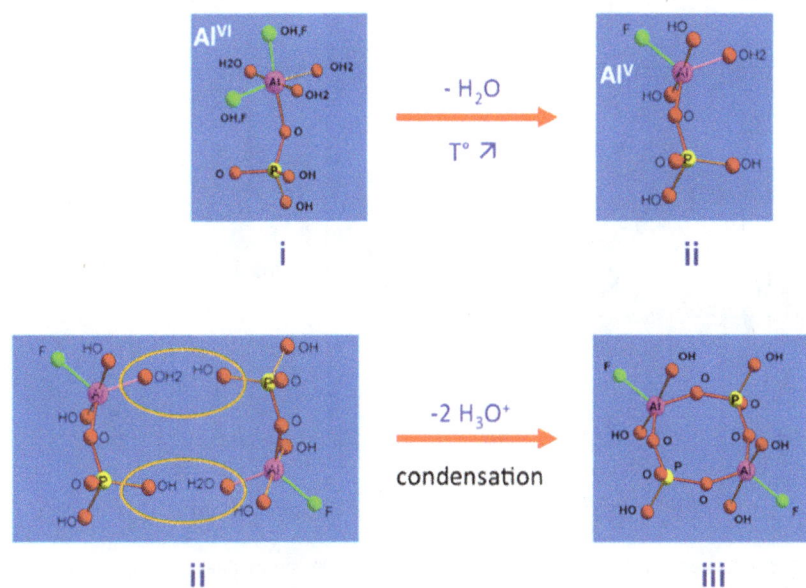

Fig. 1.6: Proposed sequential pathway of PNBU formation from (i) PBUs (octahedral complexes of mixed fluoroaluminophosphate) to (ii) metastable penta-coordinated species, which subsequently transform to (iii) PNBUs (reprinted with permission from Ref. [12]. Copyright © 2018, MDPI).

structures. The prenucleation species are tetrameric units composed of two Al and two phosphate units. The secondary building units of the crystal structure of the alumosilicate show comparable tetrameric units but the coordination of Al is different. Both the structures of PNBU and SBU are related and the PNBU can be transformed to the SBUs by simple conformational rearrangement [12].

When the wavelength is decreased to the IR and visible range, IR spectroscopy and Raman spectroscopy can support the evaluation of structure properties on a more local length scale. Both techniques are vibrational spectroscopies with complementary information. Raman spectroscopy provides reliable information about crystalline structure as well as on structural and chemical inhomogeneity. As a local probe, Raman spectroscopy is appropriate for the identification of structure defects that can influence material properties significantly. In situ Raman spectroscopy is nowadays well established either as a single probe or in combination with other methods such as X-ray diffraction (XRD) experiments. An example of in situ Raman studies is the investigations of intercalation reactions into graphitic carbon electrodes of ion batteries during cycled charging (Fig. 1.7a). Cohn et al. studied the intercalation of solvated K^+ ions into graphitic electrodes [13]. Prior to co-intercalation, a strong Raman signal G_{UC} is observed corresponding to pristine graphene material. When K^+ ions start to intercalate with decreasing potential (stage 4), a second Raman (G_C) band appears at ~1,600 cm^{-1} (Fig. 1.7b). This signal indicates negatively charged graphene layers, which surround filled galleries of positively charged K^+ ions. The disappearance of the G_{UC} band indicates that there exist no uncharged graphene layers

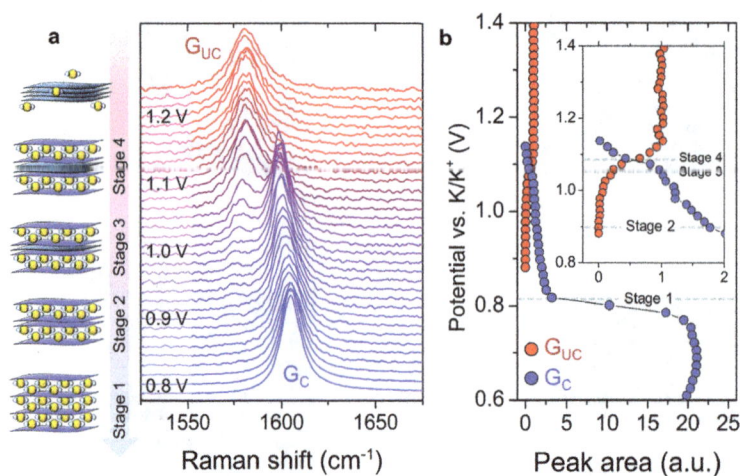

Fig. 1.7: (a) In situ Raman spectra (normalized) acquired every 30 s during linear voltage sweep at 0.5 mV/s with corresponding cell voltages. (b) Analysis of the evolving G peak doublet with the component peak areas plotted with respect to cell potential. Inset: expanded portion showing initial staging between 1.4 and 0.8 V (reprinted with permission from Ref. [13]. Copyright © 2016, Royal Society of Chemistry).

anymore. Further insertion of K^+ ions leads to an additional blueshift of the G_C peak. The graphene layers accept more electrons to balance the insertion of K^+ ions and the electrolyte into every interlayer gallery (stage 1). This example shows that in situ Raman spectroscopy is a tool to study sequential processes not accessible by simple *ex situ* measurements.

A further decrease of the wavelength into the range of X-rays facilitates the detailed study of crystal structures on the atomic level because the wavelength corresponds to the range of interatomic distances of crystalline materials. Crystal structures and crystal structure changes are usually studied with X-rays either on in-house diffractometers or, if required, at synchrotron sources. For kinetic studies of processes proceeding on very short timescales, in-house diffractometers may not enable data collection fast enough for monitoring such rapid changes. Modern beamlines at synchrotron facilities provide not only fast detectors but also a high brilliance. Tunable wavelengths/energies of the radiation and space for large sample environment setups are also important for in situ/*operando* studies.

Crystal structure analysis with X-rays requires crystallite sizes large enough to perform either single crystal or powder diffraction studies on polycrystalline materials. The lower size limit for single-crystal studies with laboratory instruments nowadays is about 30–300 µm depending on the chemical composition of the compound and the instrumental setup. With synchrotron radiation, sizes can be smaller. For crystal structure refinement of polycrystalline powders, the sizes of the individual crystals should be smaller than 40 µm but not smaller than about 100 nm. If crystals get smaller than 100–120 nm, additional influences of the microstructure need to be considered. Reflections become broader which provides useful information but crystal structure analysis is restricted due to limited resolution.

1.2.2 Neutrons

Neutron radiation is applied for elastic as well as inelastic scattering techniques. Elastic neutron scattering measures the scattered intensity with varying scattering angles. The main types of instruments are diffractometers (single crystal, powder diffraction, or diffuse scattering from amorphous materials), reflectometers, and small-angle neutron scattering instruments. While elastic scattering techniques are used for structure analysis, inelastic neutron scattering probes atomic vibrations and is used for spectroscopy. Neutrons offer another radiation source for structure–property studies. Compared to X-ray radiation, neutrons interact differently with the matter. This provides advantages, which support the complementary usage of neutron radiation. Neutrons are uncharged but carry a magnetic moment and therefore interact with magnetic moments of electrons in the sample. Therefore, magnetic structures can be obtained from neutron diffraction studies. Another advantage of neutrons is that they interact with the nucleus instead of electrons and the contribution to the diffracted intensity is different for each isotope. Light atoms (low Z) such as hydrogen, lithium, carbon, or oxygen may contribute strongly to diffracted intensities even in the presence of atoms with high Z, which is interesting for the study of Li-ion batteries. Since the interaction of the neutron beam is mostly with the small nuclei of the atoms, the scattering power of an atom does not fall off with the scattering angle as it does for X-rays due to the spatial distribution of the electrons. No atomic form factor is required to describe the shape of the electron cloud of the atom. Since intensities do not fall off with the diffraction angle, reflections with suitable high intensities appear at higher angles. This provides additional useful structural information. Hydrogen in structures is almost invisible for X-rays, while deuterium, an isotope of hydrogen, is a very strong scatterer for neutrons. The exchange of hydrogen by deuterium in crystal structures enables the localization of the isotope in crystal structures. For example, crystal structures of complex metal aluminum hydrides, considered for solid-state hydrogen storage material, as a heat storage material, or as solid electrolytes in batteries can be determined from neutron diffraction data including the deuterium/hydrogen positions. The location of hydrogen in the crystal structures is essential because this defines the orientation of the tetrahedra or octahedra in the crystal structures. Only if the position of hydrogen/deuterium is determined, not only the orientation of the polyhedra but also their connectivity can be determined. From X-ray powder diffraction studies of a new complex metal aluminum hydride, $CsAlH_4$, the positions of Cs and Al cations could be determined while the hydrogen positions remained unclear [14]. Theoretical density functional theory (DFT) calculations predicted hypothetical hydrogen positions. However, neutron diffraction data revealed deuterium positions, which showed that the space group $I4_1/amd$ predicted from DFT calculations is not correct (Fig. 1.8a). The reason for this is a disorder of the deuterium coordination. However, this can only be determined from neutron diffraction data. The disorder results in the real space group $I4_1/a$ (Fig. 1.8b).

Neutrons are, for example, used for time-resolved deuteration studies to trace phase formations [15]. They studied deuteration of Li_3N by in situ neutron diffraction studies upon heating the nitride at 1.0 MPa of deuterium gas pressure (Fig. 1.9a). After about 50 min of deuteration, $LiND_2$ (lithium amide) starts to form in a one-step reaction together with LiD at a temperature of about 470 K. Unfortunately, also lithium oxide (Li_2O) is formed, which results from the reaction of LiD with some lithium hydroxide impurities in the starting material. From quantitative Rietveld refinements, the time-dependent change of the phase composition during deuteration is obtained (Fig. 1.9b).

Another example of the advantage of using neutron radiation is structure studies of elements in the periodic table. Due to the similar number of electrons and thus similar scattering power, XRD cannot distinguish between Si and Al, or Co and Ni. For instance, it is important to determine the cation distribution in spinels. From the cation distribution thermodynamic and physical properties can be derived [16]. For example, the powder patterns of $CoAl_2O_4$ and $Co_{0.5}Ni_{0.5}Al_2O_4$ obtained by X-ray experiments are

(a)

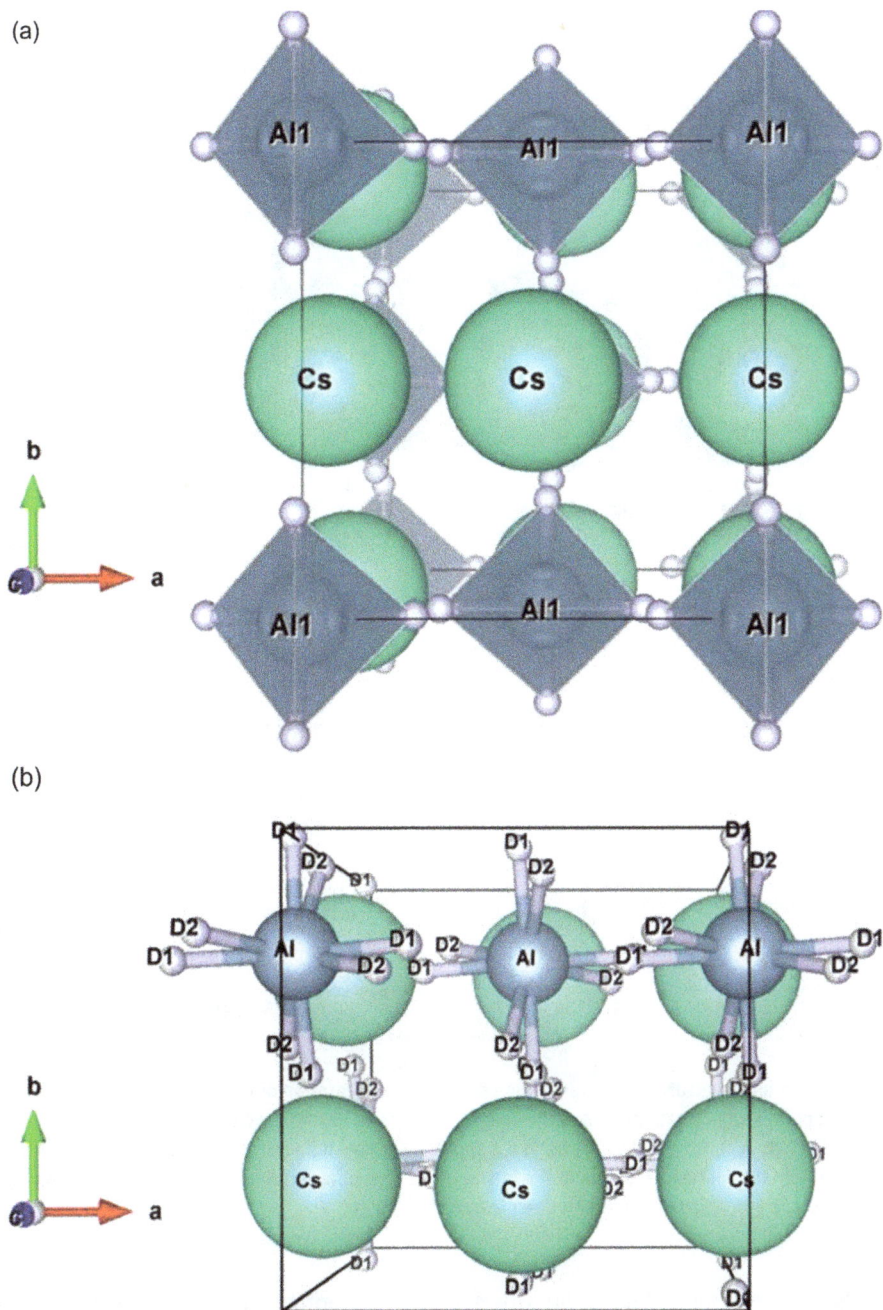

(b)

Fig. 1.8: Crystal structure of CsAlD$_4$ in (a) space group $I4_1/amd$ as predicted by DFT calculations with an ordered coordination of deuterium around aluminum. (b) Crystal structure in space group $I4_1/a$ as determined from neutron diffraction with disordered coordination of deuterium (reprinted with permission from Ref. [14]. Copyright © 2015, Wiley-VCH Verlag GmbH & Co. KGaA, Weinheim).

(a)

(b)

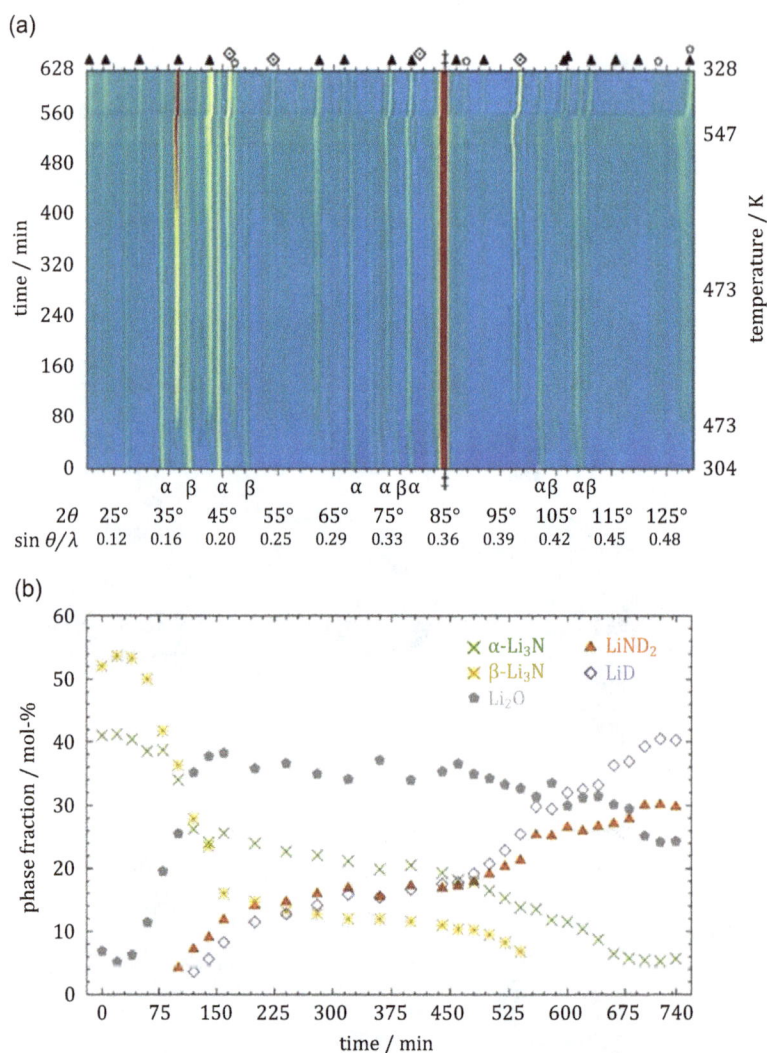

Fig. 1.9: (a) In situ neutron powder diffraction data of the deuteration of Li_3N measured in a single-crystal sapphire gas cell at 1.0 MPa deuterium gas pressures and varying temperatures. Phases: α, Li_3N; β, Li_3N; ▲, $LiND_2$; ▢, LiD; ⬠, Li_2O; ‡, sapphire. (b) Phase fractions of different phases obtained from the Rietveld analysis of the in situ diffraction data (reprinted with permission from Ref. [15]. Copyright © 2016, American Chemical Society).

hardly distinguishable (Fig. 1.10a). However, the scattering contributions of Ni and Co are significantly different from neutron diffraction experiments as shown in Fig. 1.10b and c.

One of the most important advantages of neutrons for in situ/*operando* studies is their penetrating power. The neutron absorption cross sections are small; therefore,

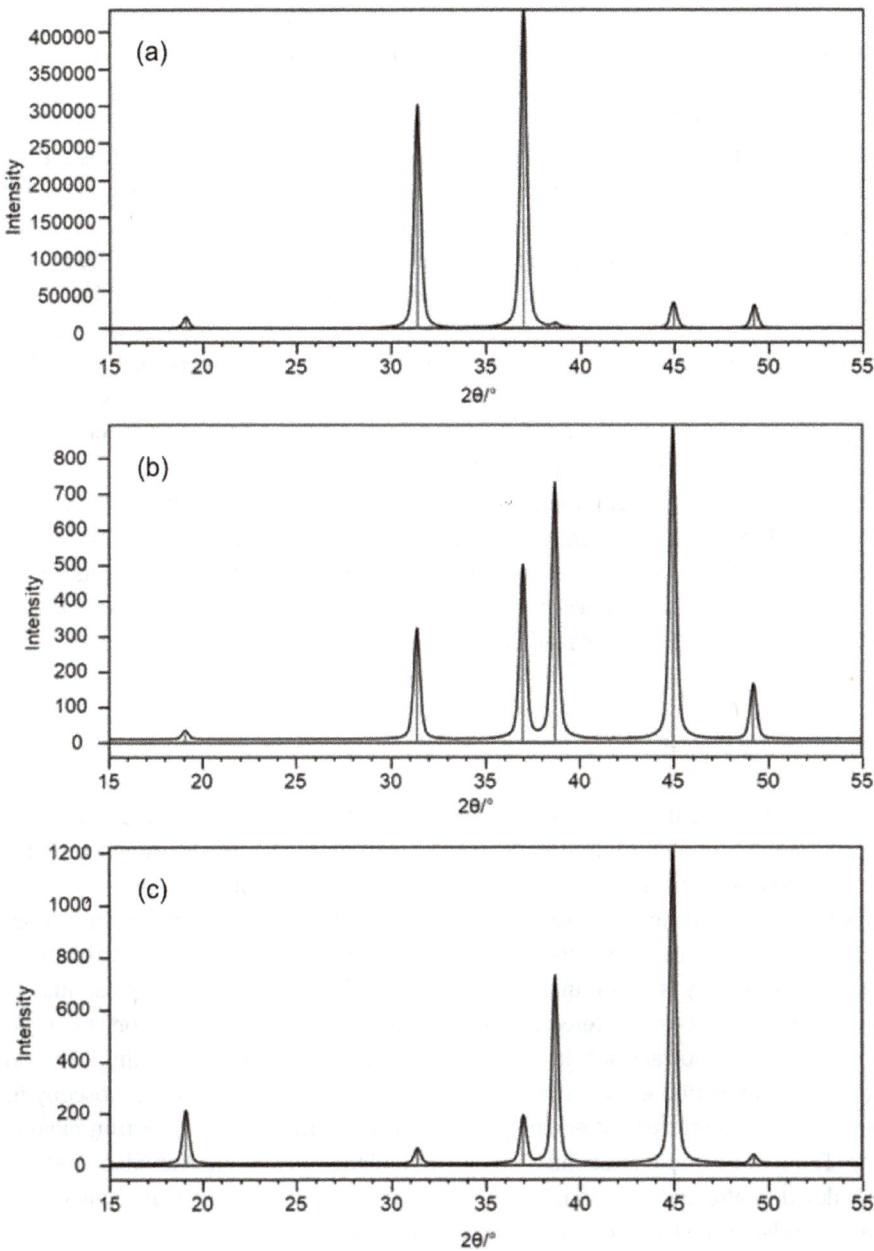

Fig. 1.10: (a) Simulated X-ray powder patterns of $CoAl_2O_4$ and $Co_{0.5}Ni_{0.5}Al_2O_4$, and simulated neutron diffraction patterns of (b) $Co_{0.5}Ni_{0.5}Al_2O_4$ and (c) $CoAl_2O_4$. Structure simulation was based on [9].

thermal neutrons in general can penetrate matter easily. While for X-rays only a limited number of sample cell materials are available which are transparent for radiation, neutrons can penetrate through several millimeters of cell walls depending on the used materials. This is highly valuable when sample environments such as furnaces, cooling devices, reaction cells, pressure cells, or electronic devices are required. Typical examples of sample environments used for in situ neutron powder diffraction experiments are summarized in [17], notably to examine materials for lithium-ion batteries, magnetic or hard materials, high-strength alloys, hydrogen storage materials, ferroelectric or superconducting materials, and solid catalysts.

One disadvantage of neutron sources is that they cannot compete with X-rays (synchrotron radiation) in terms of available flux. Low fluxes in general are limiting time resolution of a measurement. Another limitation might be the required amount of samples for powder studies or relatively large crystals for single-crystal studies. Depending on the neutron facility and the beamline, powder samples of volume $1-2\ cm^3$ are required for analysis. A typical sample size recommended for single-crystal neutron diffraction is ~100 mm^3 while for synchrotron experiments a size of ~0.01 mm^3 is more typical [18]. Neutron studies provide many different types of experiments, starting from crystallization studies from amorphous precursors, elemental ordering of alloy structures, monitoring of reaction mechanism, or electrochemical reactions.

1.2.3 Electrons

Electrons are elementary particles that carry a negative charge, possess an intrinsic angular momentum (spin), and have a very small mass (1/1,836 mass of a proton). They interact more strongly with matter compared to X-rays or neutrons. Since electrons as X-rays and neutrons exhibit properties of both waves and particles, they can be used for both diffraction and imaging experiments. One disadvantage of electron microscopy is that it may cause radiation damage to the samples similar to X-rays. This has to be considered when experiments are planned. Electrons are used for either bulk or surface-sensitive microscopy methods such as scanning electron microscopy, reflection electron microscopy, and scanning tunneling microscopy in imaging sample surfaces. When compared with these, transmission electron microscopy (TEM) and scanning transmission electron microscopy (STEM) can display structures down to atomic resolution. Since this chapter focusses on structure-sensitive in situ methods, only TEM and STEM will be considered further. Fundamental developments in environmental electron microscopy instrumentation provide the option to visualize and understand structure–property relations.

1.3 Cell designs: advantages and problems

The appropriate design of sample cells for a specific in situ/*operando* experiment has to consider many technical aspects. The cells should be as similar as possible to the reaction vessels normally used for a reaction or as close as possible to the real systems. Testing new materials under reaction conditions requires usually running several experimental series including the need for experimental time. Some reactions are very fast and to follow them, fast detection systems are required. Even though nowadays X-ray detection systems for in-house instruments became much faster, data collection times cannot compete with the speed of 2D detectors used at synchrotron facilities. Another challenge is finding a material for cell construction, which complies with the requirements of the reaction and is at the same time transparent for the incoming and outgoing radiation. Several reviews provide excellent overviews on the development of sample cells suitable for X-ray and neutron scattering experiments [19–23]. A very recent review by Liu et al. [7] summarizes sample cells suitable for in situ microscopy and diffraction of battery materials. Literature research has revealed numerous publications on a large variety of in situ cells for *operando* studies of battery materials. However, the development of specific sample environments was rather challenging for also other topics. In this chapter, only a small selection of sample environments and applications will be presented.

Berlinghof et al. designed a cell for real-time grazing-incidence wide- and small-angle scattering and X-ray reflectivity (Fig. 1.11) [24]. The flexible setup allows in situ and real-time studies of drying kinetics after doctor blading of thin films, of solvent vapor annealing of thin films, and hydration of samples. The automated and portable setup is applicable at beamlines and laboratory instruments.

Many different electrochemical cell designs have been developed for in situ and *operando* studies. They differ in their geometries: some of them are cylindrical, while others are flat. Leriche et al. [25] presented a cylindrical electrochemical cell, which enables fast investigation of insertion materials under high current density conditions (up to 5 C, 850 mA/g; Fig. 1.12). The cell can be used at different beamlines for diffraction and spectroscopy. The top part of the cell contains the X-ray transparent beryllium window. The electrode material is directly placed behind the window. Parts A, B, and C belong to the cylindrical plunger.

A design consisting of layered components was used for building a planar cell for neutron diffraction experiments [26]. From the special design, the cell is applicable to study phase transitions in both electrodes (anode/cathode) with simultaneous use of minimum electrode volumes (Fig. 1.13). Single-crystal Si sheets are used as casing and window materials. This reduces the amount of background scattering. Besides, the cell closes tightly to avoid oxidation of the electrolyte.

Other examples for in situ diffraction experiments requiring specific sample environments are gas reactions. Until a few years ago, hydrogenation reactions of solids at elevated temperatures and gas pressure were quite challenging. This becomes

a)

b)

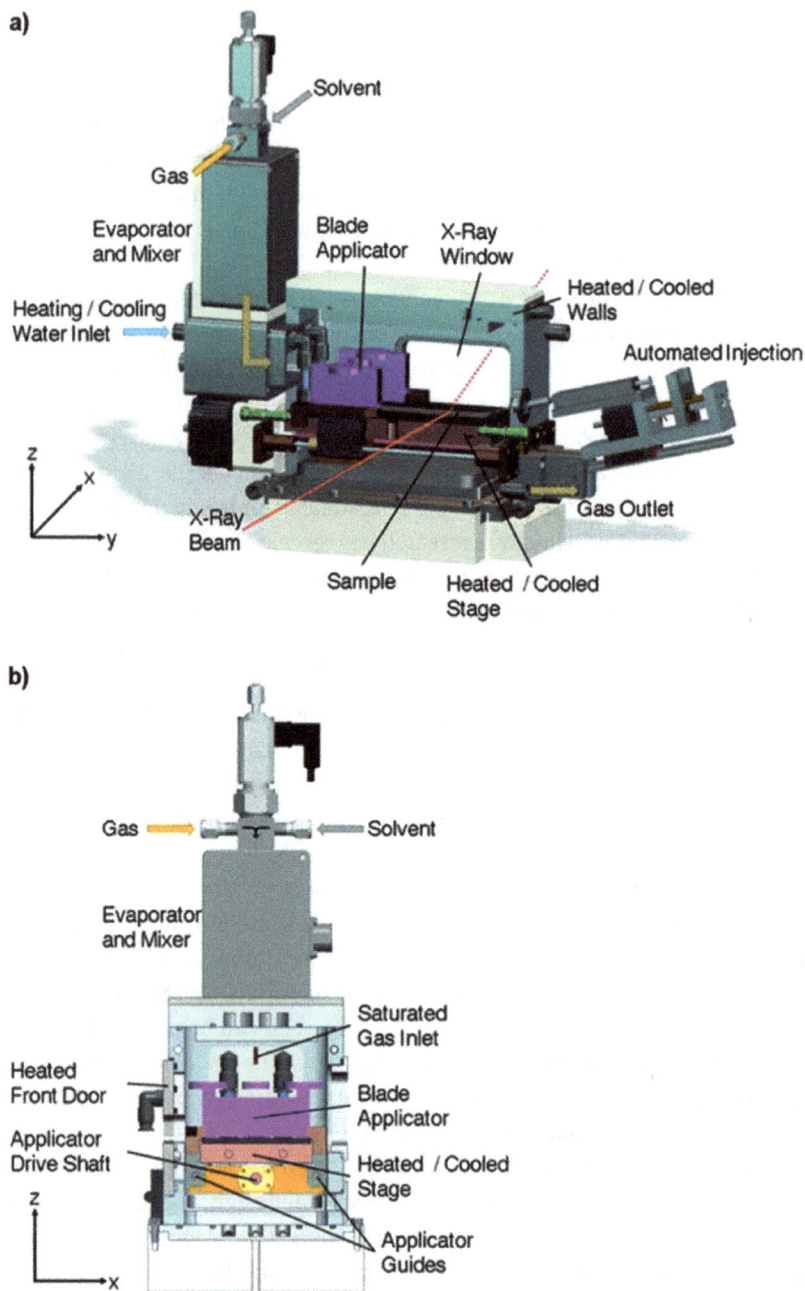

Fig. 1.11: Technical drawings of the automated in situ cell suitable for grazing-incidence wide- and small-angle scattering and X-ray reflectivity (reprinted with permission from Ref. [24]. Copyright © 2018, International Union of Crystallography).

(a)

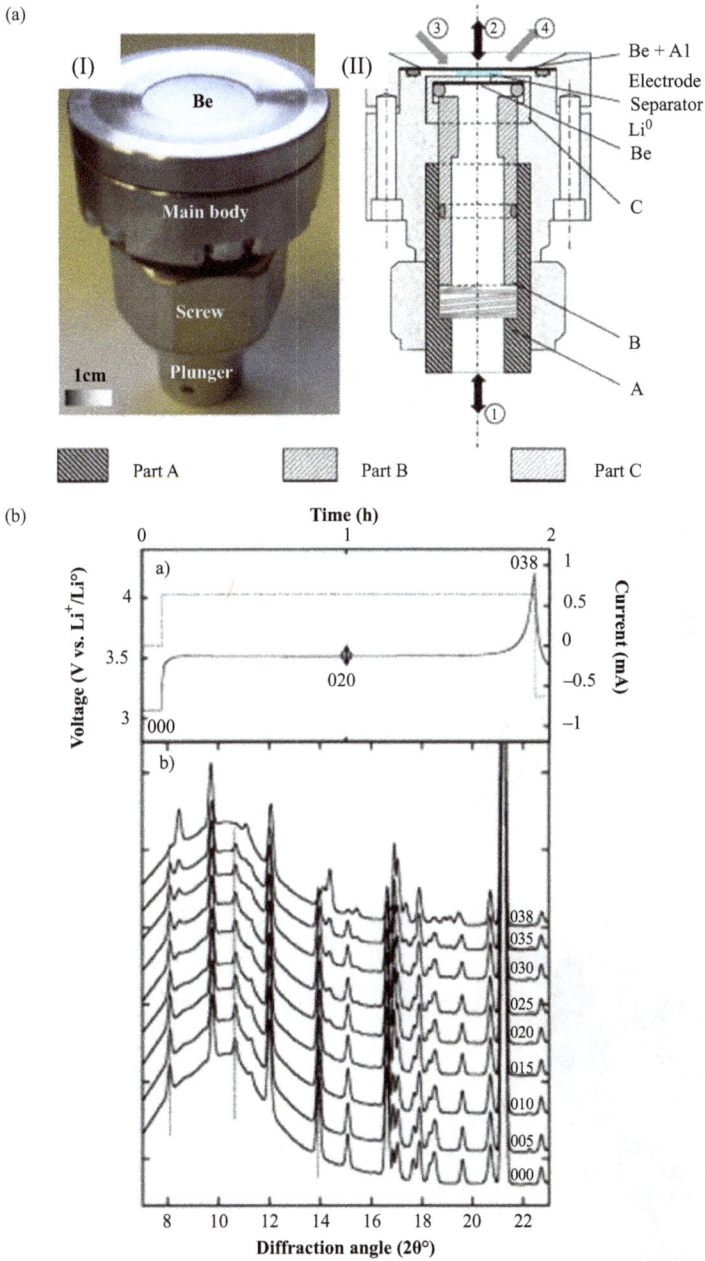

Fig. 1.12: Electrochemical cell for operando measurements. (a) (I) Photo and (II) detailed view of the cell with incoming and outcoming beam paths in transmission (1 + 2) and reflection geometry (3 + 4). (b) Upper part: electrochemical data, lower part XRD patterns recorded *operando* during charging LiFePO$_4$ (reprinted with permission from Ref. [25]. Copyright © 2010, The Electrochemical Society).

(a)

(b)

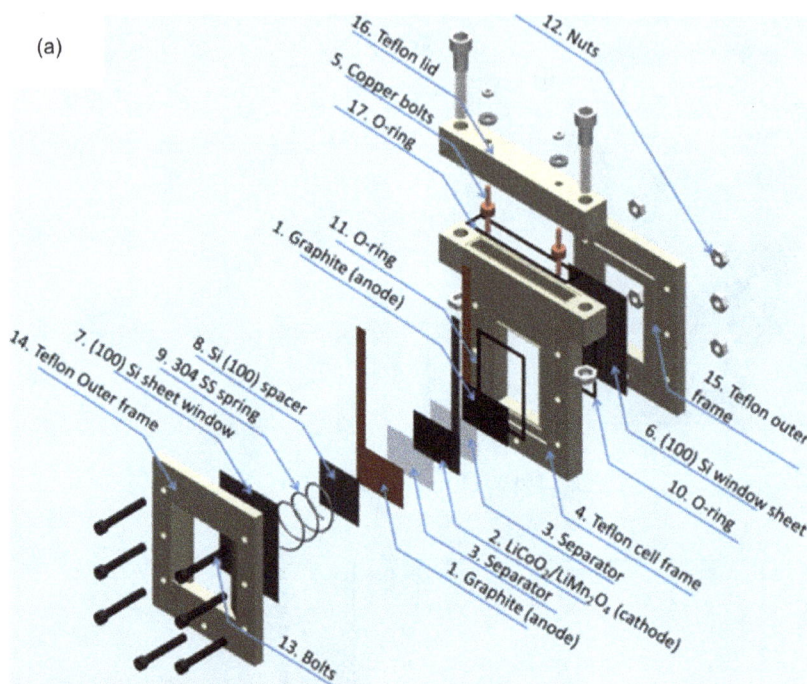

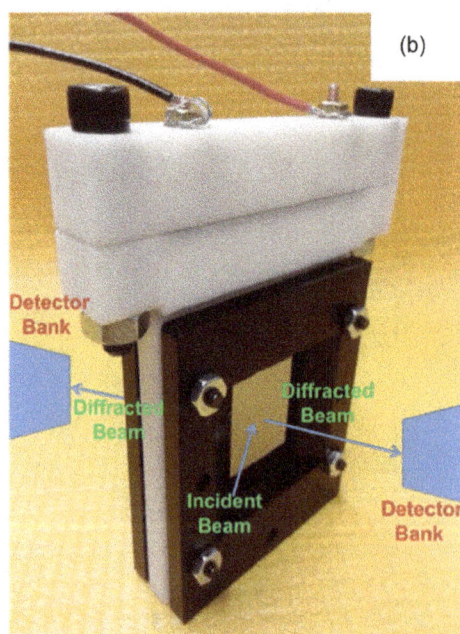

Fig. 1.13: (a) Schematic view of the electrochemical cell showing the individual components. (b) Assembled electrochemical cell (reprinted with permission from Ref. [26]. Copyright © 2014, The Electrochemical Society).

even more complicated if the samples are highly air sensitive and need protection during the experiments. Cells for synchrotron and in-house hydrogenation experiments were developed, which could overcome these problems. Conventional glass or quartz glass capillaries break if gas pressures inside the closed capillaries become too high. An alternative is sapphire capillaries which can be heated and withstand hydrogen pressures of up to 1,000 bar [21, 27]. Either the capillaries are sealed at one end or both ends are open allowing for flow-through experiments (Fig. 1.14a). However, the use of single-crystal sapphire capillaries requires high-energy synchrotron radiation to assure penetration. To follow hydrogenation reactions with laboratory diffractometers, a special cell made of a polymer material (polyetheretherketone) was developed (Fig. 1.14b) [28]. The cell can be operated to hydrogen pressures of 300 bar at room temperature or 150 bar and 100 °C, which makes it possible to observe hydrogenation processes qualitatively.

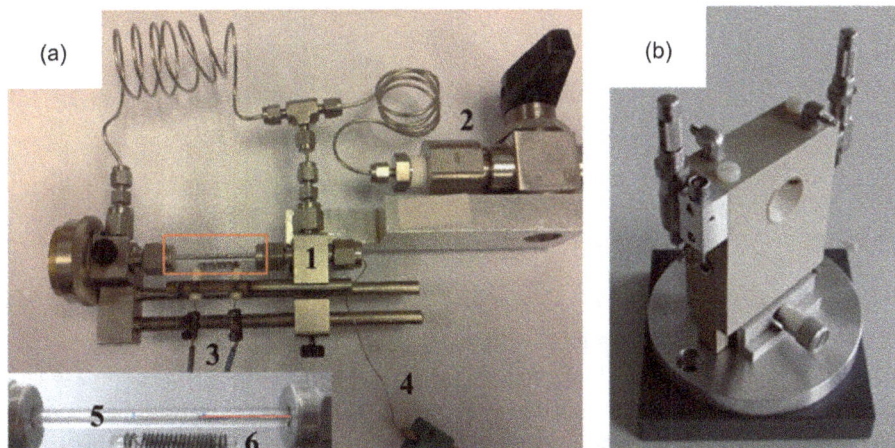

Fig. 1.14: High-pressure cells for in situ diffraction studies obtained during hydrogenation studies (a) in a cell made of single-crystal sapphire and (b) a cell made of X-ray transparent polymer (polyetheretherketone) suitable for in-house diffraction experiments (reprinted with permission from Refs. [27, 28]. Copyright © 2015, International Union of Crystallography).

The cell shown in Fig. 1.15 was constructed for in situ synchrotron diffraction studies performed during the hydrothermal synthesis of microporous materials [29]. The apparatus was introduced already in 1995 by Evens et al. [30] for the study of kinetics and mechanisms of hydrothermal synthesis reactions. The original cell was operating at temperatures of 230 °C and 27 bar. An outer stainless steel tube allows working at elevated pressures. An inner Teflon liner limits the contact of the reactants with the cell walls. A magnetic stirrer below the Teflon liner allows continuous stirring of the reaction solutions. Due to the high intensity of the synchrotron radiation, diffraction data could be measured very fast although 2-mm-thick steel walls had to be penetrated.

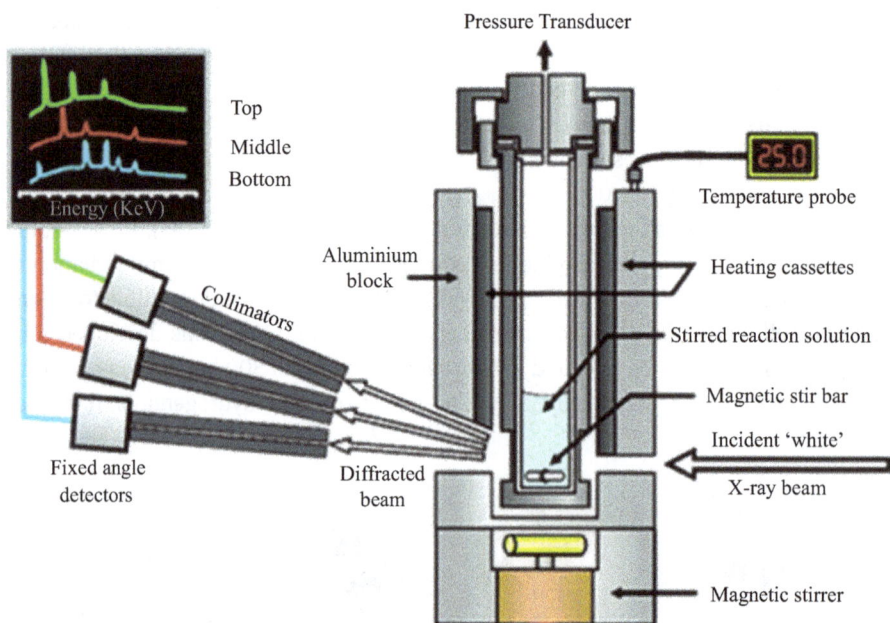

Fig. 1.15: Drawing of the experimental apparatus used for in situ diffraction studies performed during the synthesis of microporous phosphates at the Daresbury Laboratory (reprinted with permission from Ref. [29]. Copyright © 2004, American Chemical Society).

A sample cell for in situ single-crystal diffraction enables gas dosing into a quartz glass capillary containing a loop with a single crystal (Fig. 1.16a). The cell is connected via polymer tubing to a gas-dosing manifold (Fig. 1.16b). The capillary can be evacuated with a turbomolecular pump and heated by an external heat source. The authors studied the gas uptake into metal-organic framework (MOF) structures. This requires the removal of solvent molecules from the pores before the adsorption of guest molecules can take place [31].

Another possible cell design for gas–solid reaction is shown in Fig. 1.17a [32]. At first glance, the cell looks like many other cells of this type. Samples are filled into quartz glass capillaries with 1 mm diameter. However, the flow-through capillary is not static but can be rotated by 200° providing uniform heating and better particle statistics. The sample can be heated by external heaters such as heat blowers or cryostream systems, and product gases are analyzed by mass spectrometry or gas chromatography. The cell is mounted on an in-house diffractometer equipped with a position-sensitive detector (Fig. 1.17b).

a

3.0 mm quartz capillary ————————•   crystal on a
                                    MiTeGen Loop

cap with _____
Viton® O-ring

                                    Beswick ball valves
                                    ——— for gas-dosing
goniometer _____   and exhaust
head

                                    quick-connect
                                    ——— ports to
                                    PTFE tubing

b

                    Swagelok® Ultra-Torr   1.0 mm borosilicate
                    vacuum fitting   glass capillary
                    with Viton O-ring   with crystal on
                                            a glass fiber

Cajon® VCO   HiP taper
fitting   seal valve

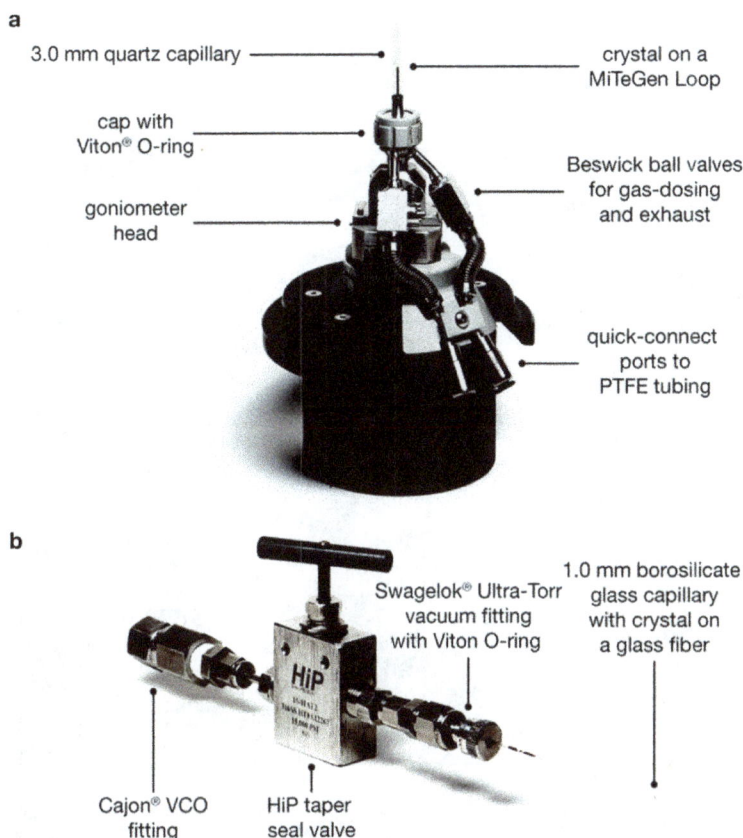

Fig. 1.16: (a) Gas cell for single-crystal studies during gas dosing designed and built at Advanced Light Source Beamline 11.3.1. (b) Gas-dosing assembly (reprinted with permission from Ref. [31]. Copyright © 2017, The Royal Society of Chemistry).

1.4 Crystallization studies

In situ diffraction techniques have been already used in the 1990s to study the crystallization of porous materials. One of the first studies was published in 1992 on in situ time-resolved synchrotron energy-dispersive diffraction experiments monitoring the early stages of the autoclave synthesis of VPI5, an aluminophosphate molecular sieve with one-dimensional channel structure [33]. Sample tubes made of 0.5–1.0 mm aluminum alloy were covered with an inner 0.2 mm polytetrafluoroethylene sleeve to protect the metal from the hot corrosive alkaline solutions and to provide an inert environment for the synthesis. Hard X-rays (white beam) as provided by the synchrotron at Daresbury Laboratory could transmit through the sample cell and the diffracted radiation was measured by an energy-dispersive detector. The sample tube containing

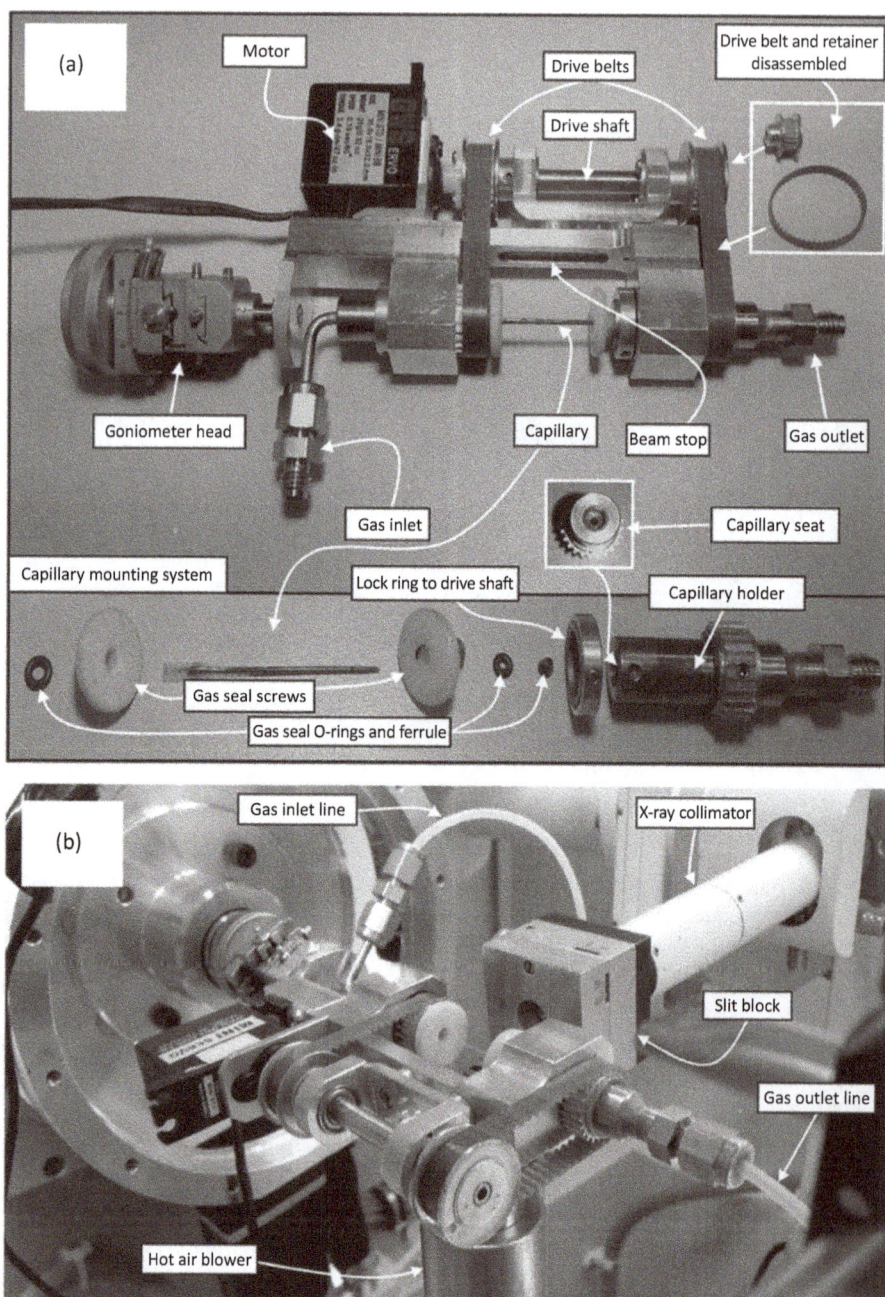

Fig. 1.17: (a) Assembled capillary flow cell and disassembled capillary mounting ring (bottom) and (b) capillary flow cell mounted on an Inel 3000 diffractometer (reprinted with permission from Ref. [32]. Copyright © 2004, American Chemical Society).

the precursor gel was placed inside a copper heating block. The diffraction data collected during heating in intervals of 60 s are shown in Fig. 1.18. The energy-dispersive diffraction experiment reveals reflections at about 22 keV (16.4 Å) and 45.6 keV (8.2 Å). Both reflections belong to the forming VPI-5 phase. The additional peak of 74.7 keV (5.0 Å) belongs to pseudo-boehmite, which is consumed as the amount of VPI-5 increases. The studies provide two important pieces of information: (a) VPI-5 forms rapidly already after 13 min and (b) formation starts already at low temperatures of 112 °C.

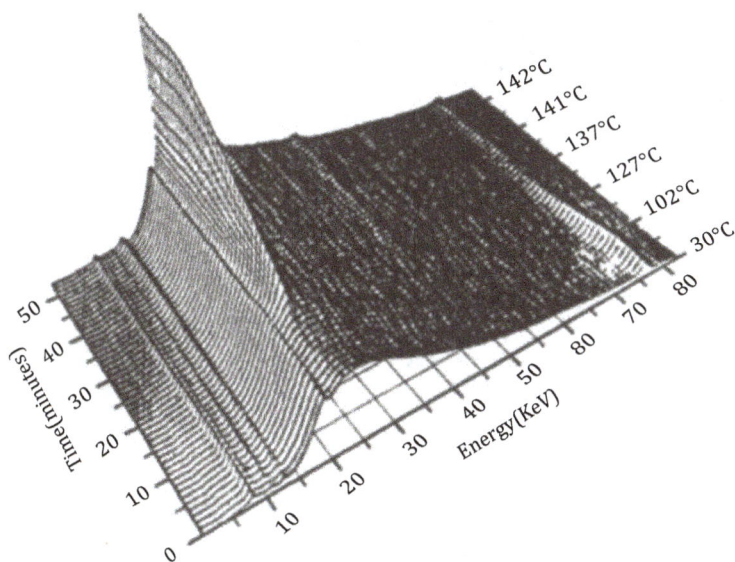

Fig. 1.18: In situ time-resolved synchrotron diffraction patterns collected during the synthesis of VPl-5 (reprinted with permission from Ref. [33]. Copyright © 1992, Elsevier Science Publishers B.V.).

Christensen et al. studied metal-substituted microporous alumosilicates by in situ synchrotron radiation experiments [34]. They did not only watch the temperature and time-dependent crystallization of a Zn-substituted ZnAPO-47, they could also determine the apparent activation energies for the hydrothermal nucleation (Fig. 1.19).

Chalcogenide glasses are attractive materials because they possess large ionic conductivities to three orders of magnitude larger than that of the oxide glasses with the same mobile ion content. Photodeposition of Ag in a chalcogenide layer is used to produce diffraction gratings, microlenses, or optical memories [35]. For Ag–Ge–Se glasses, it was investigated whether phase separation and the formation of Ag-rich domains take place which could explain reversible changes of the conductivity of films by several orders of magnitudes when weak voltages were applied. Neutron thermodiffractometry was applied to study the crystallization of Ag–Ge–Se

(a)

(b)

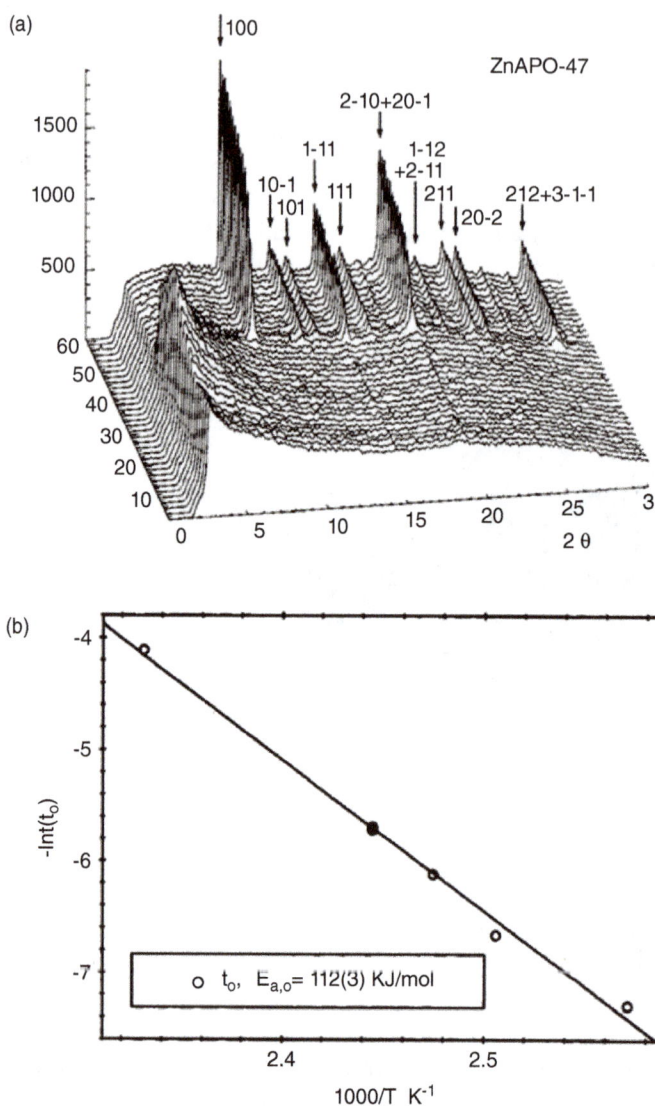

Fig. 1.19: (a) Stack of powder patterns obtained upon heating Zn^{2+}-substituted precursor gels from room temperature to 200 °C. The reaction product is the microporous alumosilicate ZnAPO-47. (b) The Arrhenius plot for determination of the activation energy for the nucleation process in the crystallization of ZnAPO-47. Activation energies are reproduced from (reprinted with permission from Ref. [34]. Copyright © 1998, American Chemical Society).

glasses upon heating and under isothermal conditions (Fig. 1.20). The evaluation of the in situ data of three different glass compositions revealed for all samples the formation of two stable phases Ag_8GeSe_6 and $GeSe_2$ as main products. However, the crystallization process of Ag-rich glasses was more complex. Unstable Ag_2GeSe_3

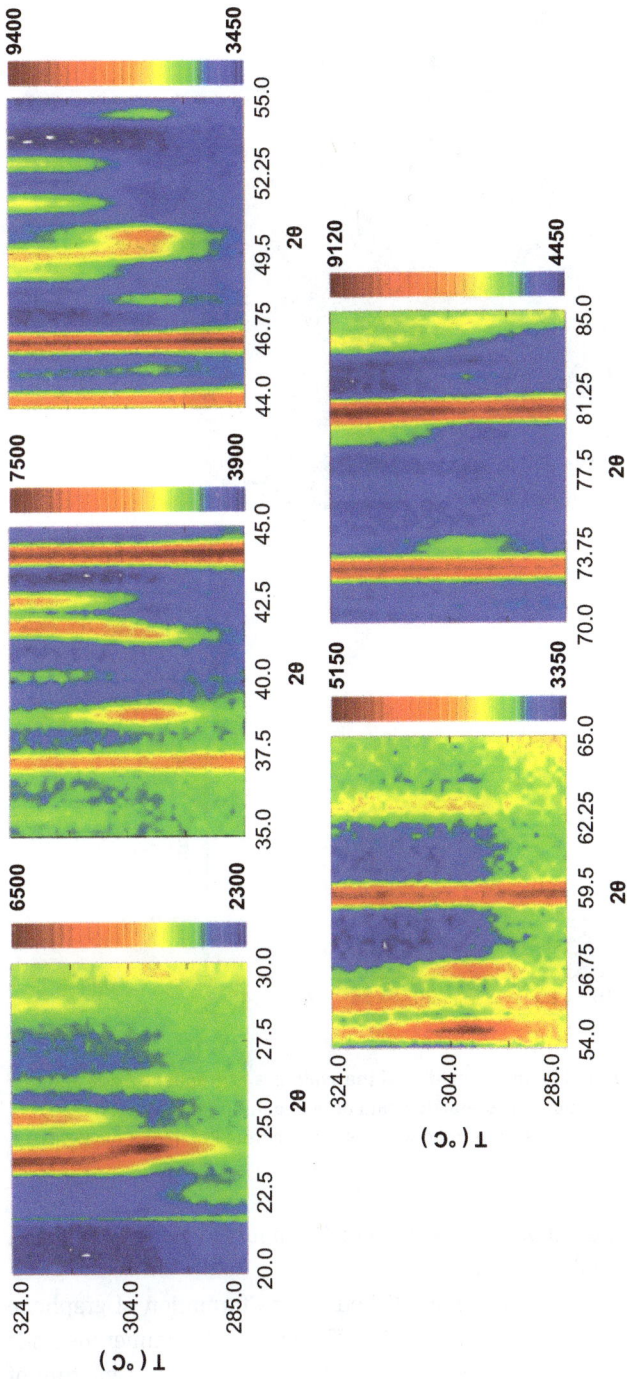

Fig. 1.20: Neutron thermodiffractograms of Ag–Ge–Se glass collected during heating from 285 °C up to 324 °C ($\lambda = 2.5295(1)\,\text{Å}$) (reprinted with permission from Ref. [35]. Copyright © 2008, IOP Publishing).

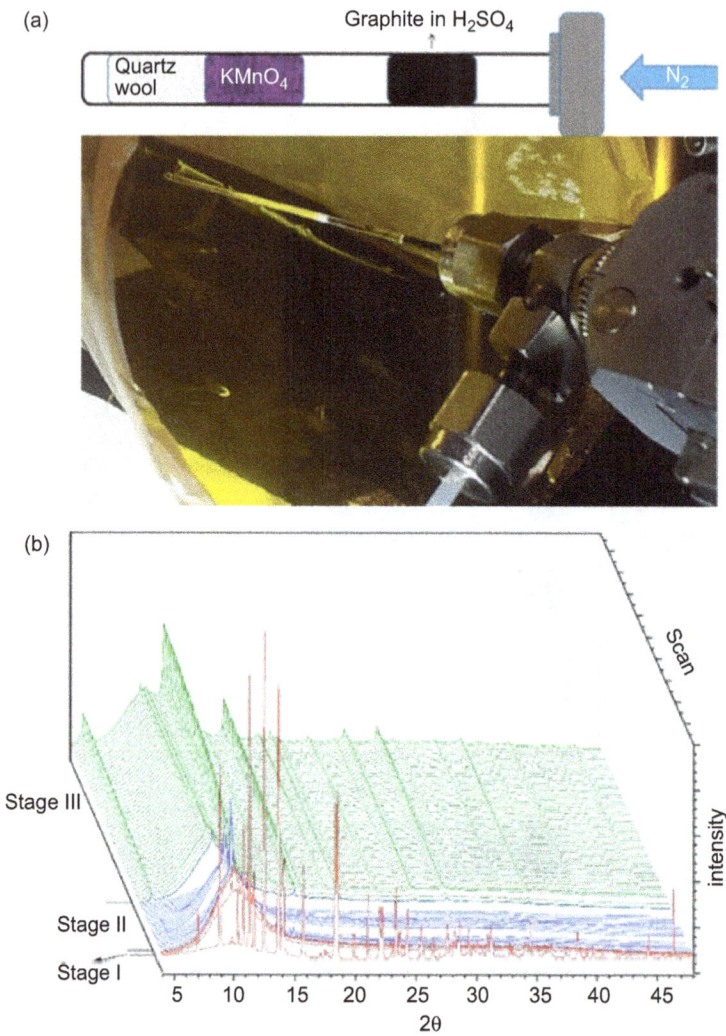

Fig. 1.21: (a) Experimental setup used for in situ studies of the synthesis of graphene oxide and reduced graphene oxide and (b) in situ X-ray diffraction data obtained during the synthesis displaying three different synthesis stages (reprinted with permission from Ref. [36]. Copyright © 2016, Elsevier Inc.).

appeared at high temperature but decomposed after a few hours to give a new phase, $Ag_{10}Ge_3Se_{11}$, along with the stable $GeSe_2$.

Another interesting in situ study was published on the formation of graphene oxide (GO) and reduced graphene oxide (rGO) [36]. The modified Hummer methods are often used for the synthesis of GO. Graphite oxidizes partially in a mixture of concentrated H_2SO_4 and $KMnO_4$ and graphene layers with oxygen functionalities and defects formed. Subsequent thermal reduction produces rGO. The aim of the in situ studies performed in a capillary cell shown in Fig. 1.21a was to clarify the reaction mechanism. The in situ synchrotron powder diffraction experiment revealed a three-step synthesis of GO (Fig. 1.21b). In stage I, the dissolution of $KMnO_4$ and the intercalation of the synthesis mixture into graphite (stage II) are taking place. Stage III shows the formation of additional crystalline material. GO forms during intercalation but also during the crystallization period. It was shown that thermal reduction of GO to rGO proceeds via an intermediate disordered stage before ordered rGO appears.

1.5 In situ studies using single crystals

Single crystals can be analyzed by diffraction methods, spectroscopy, or microscopy. While single-crystal diffraction gives a picture of the averaged crystal structure, spectroscopy of a single crystal provides additional information on molecular vibrations. In addition, microscopy tools provide complementary insights to local structures such as structural disorder or defects in a single crystal, which are not accessible via single-crystal diffraction methods. A large number of in situ single-crystal experiments can be performed, which document structural changes at non-ambient pressures or temperatures. There are many interesting topics for single crystal studies such as gas uptake and release in porous structures as well as crystallization studies or intercalation processes in electrode materials. In situ single-crystal diffraction at neutron or synchrotron scattering facilities can be performed depending on temperature, under high pressures, with varied magnetic (neutron) and electric fields, or different gas atmospheres. Compared to experiments at large research facilities, single-crystal X-ray experiments on in-house diffractometers are restricted with respect to the lower limit of the sample size and dynamic studies.

   As an example for detailed in situ studies of structure–property relationships, Jaffe et al. [37] reported the first high-pressure single-crystal structures of 3D halide hybrid perovskites of the type $(MA)PbX_3$ (with $MA = CH_3NH_3^+$, $X = Br^-$ or I^-). Single-crystal measurements were performed at a synchrotron source in a diamond anvil cell (DAC) at pressures <2 GPa, confirming the structures of starting materials and high-pressure polymorphs. In addition, high-pressure experiments of powder samples were conducted in a DAC up to 51 GPa. The study aimed at understanding how compression influences the structures and thereby the optoelectronic properties such as photoluminescence, optical bandgaps, and electronic conductivity. Figure 1.22 displays the crystal structures of two different lead-halide hybrid perovskites obtained at different pressures. Both structures undergo a pressure-induced phase transformation to a cubic structure with space group $Im\bar{3}$. High-pressure studies of powder samples reveal that the materials start to become amorphous above pressures of 2.7 GPa with some long-range order be maintained up to pressures of about 50 GPa.

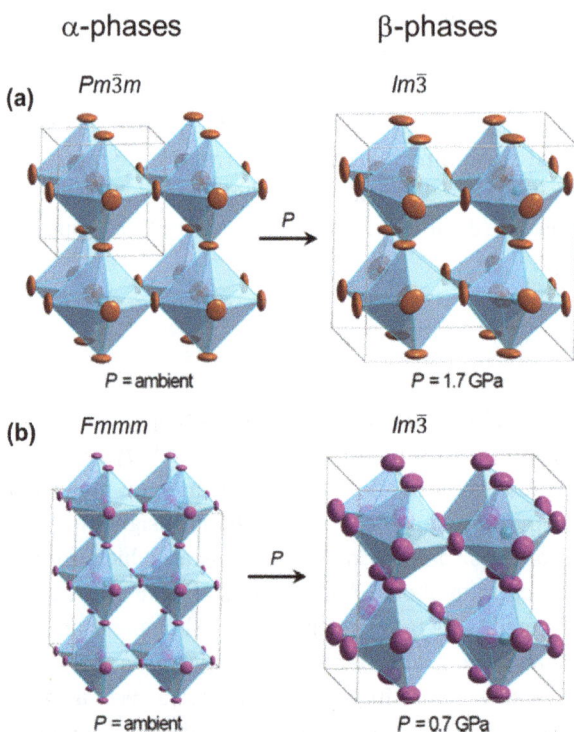

Fig. 1.22: Crystal structure changes from the α-phases to β-phases as obtained from high-pressure single-crystal studies observed for (a) $CH_3NH^+PbBr_3$ during pressure increase to 1.7 GPa and of (b) $CH_3NH^+PbI_3$ at 0.7 GPa. Adapted from Ref. [29] (reprinted with permission from Ref. [37]. Copyright © 2016, American Chemical Society).

In addition to structural changes, also pressure-related conductivity and photoluminescence measurements were performed. Low pressures lead to lead-halide bond compressions and therefore a volume reduction. This is associated with a redshift in photoluminescence and a decrease in the optical bandgap of $CH_3NH^+PbI_3$ at higher pressures. The α–β phase transformation is connected with a tilt of the octahedra in the structures causing an abrupt blueshift in the photoluminescence energy.

Porous framework materials such as zeolites or MOFs are suitable candidates for gas storage, gas separation, or gas sensing. Halder and Kepert applied in situ single-crystal XRD experiments using an in-house diffractometer to follow adsorption and desorption processes of different guest species into a porous coordination framework as a function of temperature and vapor pressure [38]. The crystal structure of the host materials enables flexibility of the framework that guest molecules can homogeneously adsorb into and desorb from the porous structure. Gate-opening processes and breathing phenomena were investigated for flexible MOF structures by the evaluation of single-crystal data obtained during CO_2 adsorption at synchrotron sources (Fig. 1.23) [39].

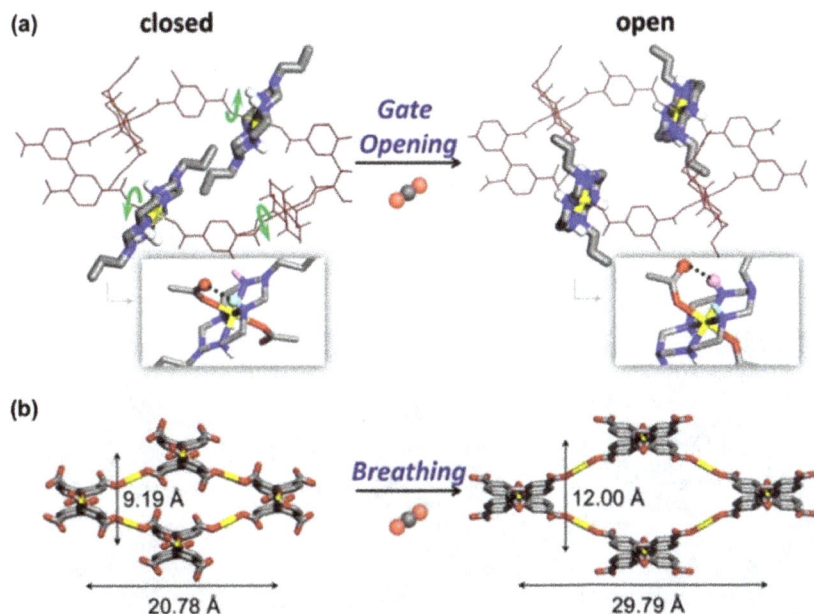

Fig. 1.23: Comparison of structures obtained by single-crystal studies of p-flexMOF and as-flexMOF representing (a) the gate-opening and (b) breathing phenomena upon CO_2 adsorption (reprinted with permission from Ref. [39]. Copyright © 2016, American Chemical Society).

Another example showing the potential of single-crystal X-ray analysis following solid-state chemical reactions under *operando* conditions was carried out in a specially designed electrochemical cell. The electrochemical oxidation of $SrFeO_{2.5}$ to $SrFeO_3$ was followed for a 70 μm single crystal at the European Synchrotron Radiation Facility (beamline BM01A) [40]. The single crystal was placed in an electrochemical cell consisting of two telescoped quartz-glass capillaries mounted on a goniometer head. The electrolyte was pumped in through the open-end inner capillary and pumped out through the outer capillary and also flushed away oxygen formed during electrolysis. Fast data collection of 0.3 s per frame resulted in full-sphere data collection within 18 min. The authors were able to investigate microstructural aspects, which could not be studied by powder diffraction experiments before. The in situ studies showed that brownmillerite-type $SrFeO_{2.5}$ oxidizes to perovskite-type $SrFeO_3$ via an electro-chemical oxygen intercalation reaction forming two intermediate structures. The observed appearance of the phase sequence $SrFeO_{2.5}$/$SrFeO_{2.75}$/$SrFeO_{2.875}$/$SrFeO_3$ is accompanied by long-range oxygen vacancy ordering (Fig. 1.24). The authors

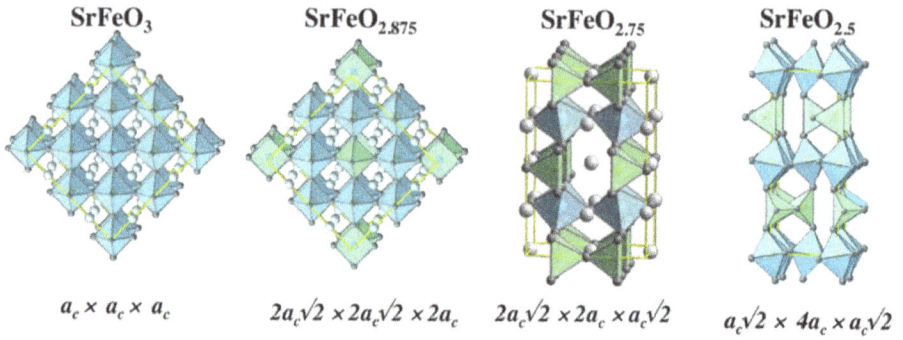

Fig. 1.24: Phase sequence of SrFeO$_{3-x}$ structures showing superstructures due to long-range oxygen vacancies (reprinted with permission from Ref. [40]. Copyright © 2015, IOP Publishing).

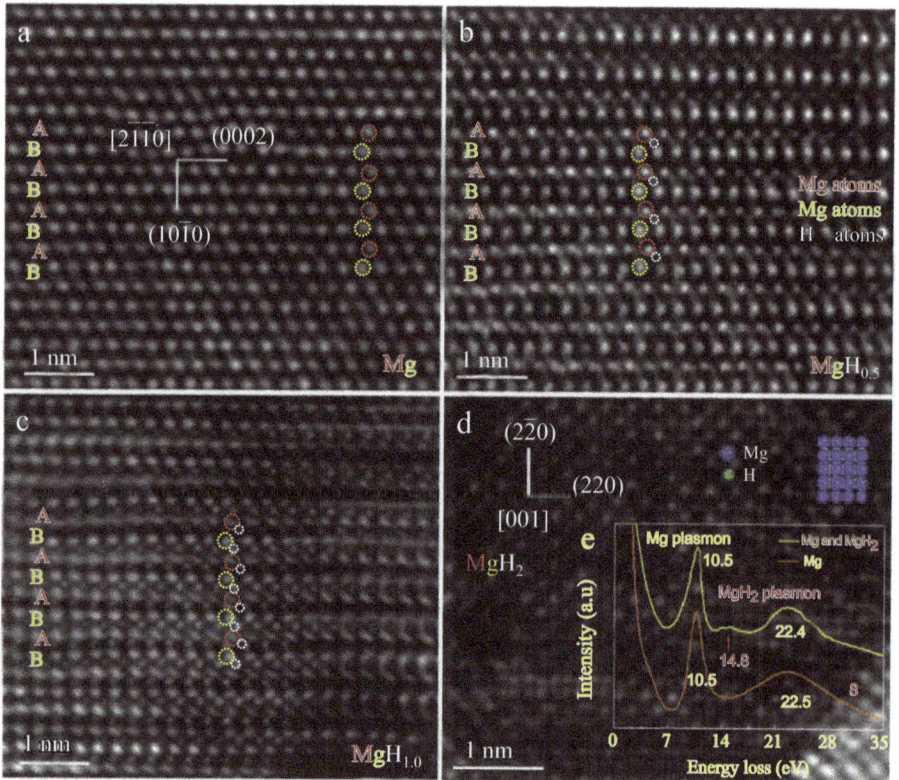

Fig. 1.25: TEM images of structures of (a) hexagonal closed packed Mg, (b) metastable MgH$_{0.5}$, (c) metastable MgH$_{1.0}$, (d) stable MgH$_2$. Inset atomic model: unit cells of MgH$_2$ along the [001] with blue (Mg atoms) and green spheres (H atoms). (e) EELS maps of Mg (~10.5 and ~ 22.5 eV) and MgH$_2$ (~14.8 eV) (reprinted with permission from Ref. [41]. Copyright © 2020, Acta Materialia Inc. Published by Elsevier Ltd).

could show that due to the synthesis procedure, starting $SrFeO_{2.5}$ crystals are twinned. From changes of the twin domain structure, a topotactic reaction mechanism from $SrFeO_{2.5}$ to $SrFeO_{2.75}$ was proven. The next step to $SrFeO_{2.875}$ goes along with changes of the microstructure by adding a third twin domain.

Sun et al. [41] studied the hydrogenation of single-crystal magnesium by in situ environmental TEM using a FEI Titan ETEM G2 microscope at 300 kV. The sample was heated to 200 °C in 10^{-3} Pa H_2 atmosphere. The authors could resolve the intercalation of hydrogen atoms into the tetrahedral sites of the hexagonal close-packed Mg structure along the $[2\bar{1}\bar{1}0]$ direction. Incipient hydrogenation of magnesium results in the incorporation of hydrogen into the bulk. The experimental observations were combined with DFT studies. Figure 1.25 shows the atomic-scale TEM images of (a) metallic Mg, the metastable intercalated phases (b) $MgH_{0.5}$ and (c) $MgH_{1.0}$, and finally (d) stable MgH_2.

1.6 In situ studies on polycrystalline powder samples

Polycrystalline samples can consist of one crystalline phase or mixtures of several compounds. While single-crystal studies only provide information about the specific crystal, in situ diffraction studies on powder samples consider the structure–property relationships including all components contributing to the diffraction pattern. While in situ XRD has been used for many decades, in situ neutron studies have been increasingly popular in recent years, with in situ experiments of catalytic processes a good example of this [42, 43]. Kandemir et al. reported neutron powder diffraction as a method for in situ analysis of inorganic catalysts [44]. $Cu/ZnO/Al_2O_3$ catalyst for methanol synthesis was studied under industrially relevant conditions at 523 K and 6 MPa with a syngas mixture ($CO_2/CO/D_2/Ar$) in a flow cell (Fig. 1.26a). The cell was connected to a gas supply for the feed gas and online gas analytics of the product gases. The experiments were performed on the instrument Echidna at the ANSTO facility, which has addressed well the technical challenges associated with in situ studies. One crucial issue is finding a cell material that is transparent for neutrons but does not contribute to much own scattering to the data. In addition, the material has to be resistant to high gas pressures at elevated temperatures. Further, since hydrogen is used at high pressures, a material with long-term resistance against hydrogen embrittlement and corrosion is required. The authors decided on $AlMg_3$ with an effective wall thickness of 4.52 mm [45]. First, the catalyst was reduced in a gas feed of pure D_2 in the temperature region from 301 to 523 K, while diffraction data of single reflections of CuO and reduced Cu were collected with an acquisition time of 5 min per pattern. The normalized integrated peak intensities are correlated with the catalyst temperature and the fluent gas

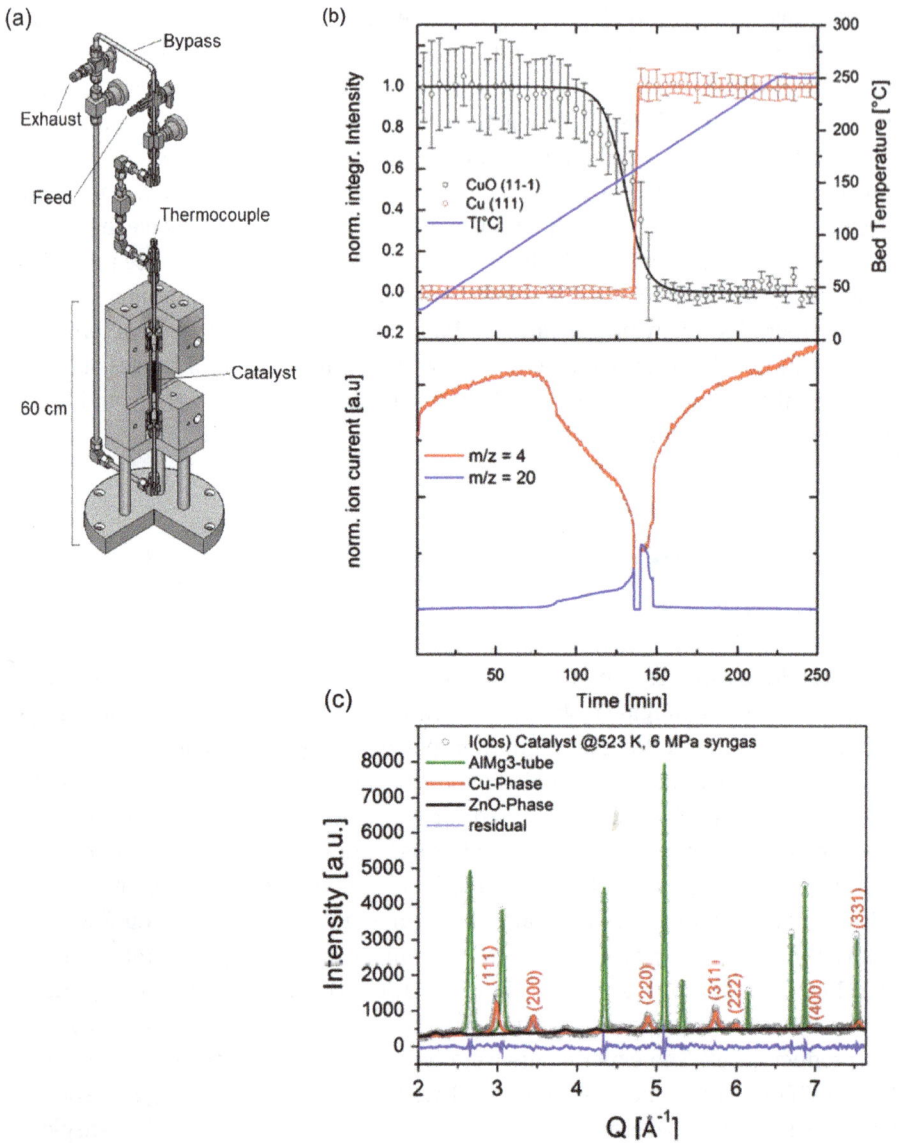

Fig. 1.26: (a) Assembled flow cell and (b) reduction of commercial Cu/ZnO/Al$_2$O$_3$ catalyst. Normalized integrated intensities of CuO($1\bar{1}1$) and Cu(111) peaks correlated with the catalyst bed temperature (top) and product gas composition (bottom) during isobar reduction from 301 to 523 K in D$_2$ feed. (c) The Rietveld refined neutron powder diffraction pattern under syngas at 523 K and 6 MPa (reprinted with permission from Ref. [45]. Copyright © 2012, Elsevier B.V.).

composition (Fig. 1.26b, c). After reaching 523 K, the feed gas was switched to syngas and diffraction data were collected with acquisition times of 1 or 2 h.

Another example of in situ neutron diffraction studies is the reversible hydrogenation of Zintl phases BaGe and BaSn for the mechanistic understanding of hydrogen uptake and release [46]. The experiments were performed in sapphire single-crystal cells with a 6 mm inner diameter. The cell was connected to a gas supply system providing pressurized D_2. The in situ diffraction data show that BaGe forms already at room temperature and low pressures (1–2 MPa) an orthorhombic β-BaGeD$_y$ (Fig. 1.27). Reaching 5 MPa D_2 pressure, heating was started while pressure was kept constant. At about 350 K, γ-BaGeD$_y$ starts to form, and this reaction is finished at 425 K.

A very good example of this, how helpful in situ neutron diffraction studies are for the understanding of the behavior of functional materials, can be temperature-dependent studies on high-strength steel used in the automotive industry [47]. The authors showed that the heat treatment design of novel metallic materials can be informed by in situ neutron diffraction studies. Transformation-induced plasticity (TRIP) steels combine strength, ductility, and work hardening capacity, which provides formability and a high-energy absorption capacity. The mechanical properties are based on the microstructure of the composite consisting of hard retained austenite island embedded in a soft matrix. The sample of cooled rolled C–Mn–Si steel was mounted on a diffractometer and heated with an induction coil (Fig. 1.28). With this setup, large assembly units can be analyzed without destruction. In situ neutron diffraction studies monitored the temperature behavior of the composite during high-temperature treatment with respect to phase transformation of austenite to ferrite and secondly the transformation that takes place upon cooling (Fig. 1.29a). The analysis of the lattice parameter obtained during heating and cooling provides information on thermal expansion, carbon diffusion, and internal stress. As shown in Fig. 1.29b, the starting material consists of a mixture of bcc (ferrite, bainite, and martensite) and fcc (retained austenite) phases. With increasing temperature, the reflections belonging to retained austenite disappear at 600 °C while the intensities of the bcc phases increase. This indicates the transformation of retained austenite to ferrite. However, during further heating retained austenite reappears and both phases coexist at high temperatures. Upon cooling, retained austenite gradually disappears and is barely observable below 400 °C, indicating the transformation of austenite to ferrite. During the second annealing cycle, the material behaves differently. Austenite is the only crystalline phase stable up to 1,000 °C, whereas after cooling austenite is completely transformed to ferrite. From the in situ neutron diffraction studies, the high-temperature annealing of TRIP sheet steel could be monitored with respect to structural phase transformations and stress/strain properties.

Rechargeable batteries are widely used as energy storage materials for consumer electronics such as mobile phones or laptops. In addition, there is a growing market for Li-ion batteries for mobile applications in cars. In order to increase the storage capacity, modified batteries are developed and these batteries are also intensively analyzed with

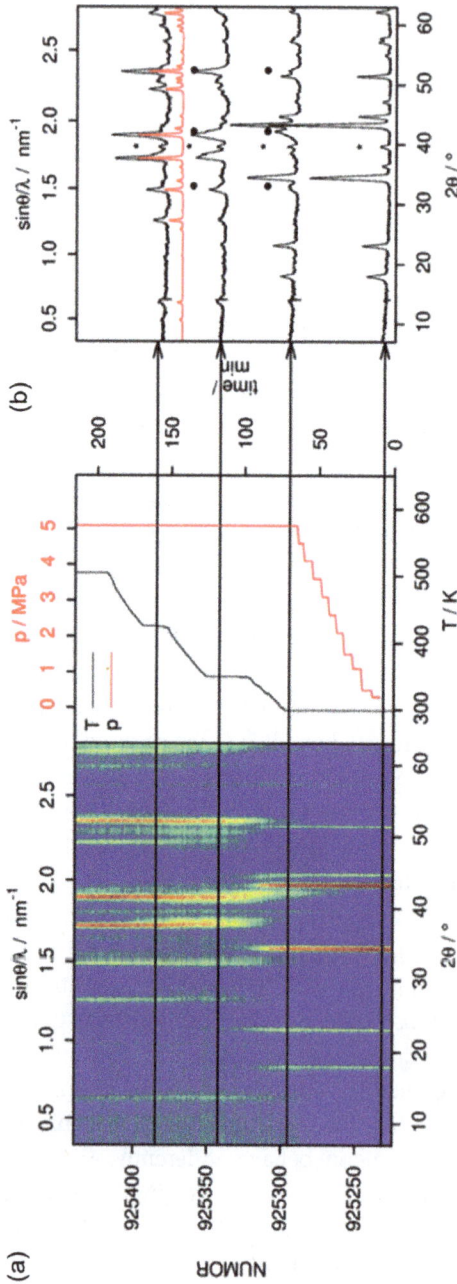

Fig. 1.27: (a) Two-dimensional plot of in situ neutron powder diffraction collected from BaGe under D_2 pressure. (b) Selected diffraction patterns. The diffraction patterns correspond to BaGe (bottom), partial formation of β-BaGeDy (middle), and γ-BaGeHy (top, a simulation is given as red curve). (*) marks BaO (right) (reprinted with permission from Ref. [46]. Copyright © 2018, De Gruyter).

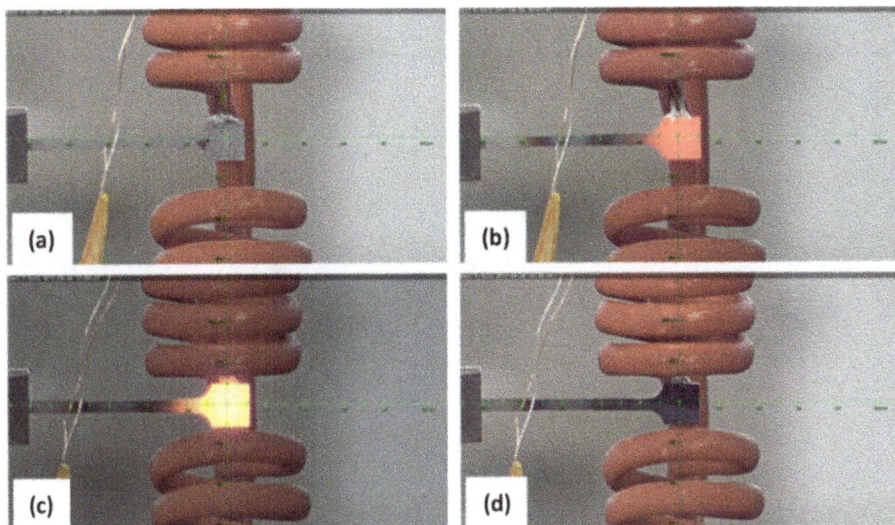

Fig. 1.28: Photographs of the sample taken at different temperatures: (a) 60 °C in cycle 1 before heating up, (b) isothermal holding at 850 °C in cycle 1, (c) isothermal holding at 1,000 °C in cycle 2, and (d) 60 °C in cycle 2 after cooling down (reprinted with permission from Ref. [47]. Copyright © 2018, MDPI).

respect to their structural alteration during operation. Due to the ability of neutrons to pass easily through matter, neutron diffraction is an appropriate tool for in situ/*operando* studies of batteries during discharging/charging cycles. The (de)lithiation of the electrodes during cycling may lead to structure damages. To avoid such damages, understanding of all processes and influence on the structures is required. An electrochemical cell with a cylindrical design enables the collection of high-quality *operando* neutron diffraction data suitable for precise Rietveld refinement analyses [48]. Changes of the 222 reflection of the corresponding spinel cathode material $LiNi_{0.5}Mn_{1.5}O_4$ (LNMO) are given in Fig. 1.30 together with the galvanostatic curves. From the galvanostatic curves, three potential plateaus that are assigned to the stepwise oxidation of Ni^{2+} to Ni^{4+} can be identified. The first oxidation step Ni^{2+}–Ni^{3+} is associated with the decrease of Li^+ in the LNMO structure accompanied by a shift of the 222 reflections caused by a decrease of the lattice parameter. During charging, spinels with different Li contents are formed and Ni is further oxidized to Ni^{4+}. Structural changes observed during charging are mostly reversible during discharging. However, from the *operando* studies, it could be concluded that the lithium content in the LNMO structure after the first cycle was lower, indicating that some lithium was consumed by the formation of a solid electrolyte interface and/or by electrolyte side reactions.

The structural evolution of $LaNi_5$-type metal hydride electrodes at high charge/discharge rates was studied by a combination of neutron powder diffraction and electron microscopy [49]. In $LaNi_5$-type metal electrodes, both La and Ni can be substituted

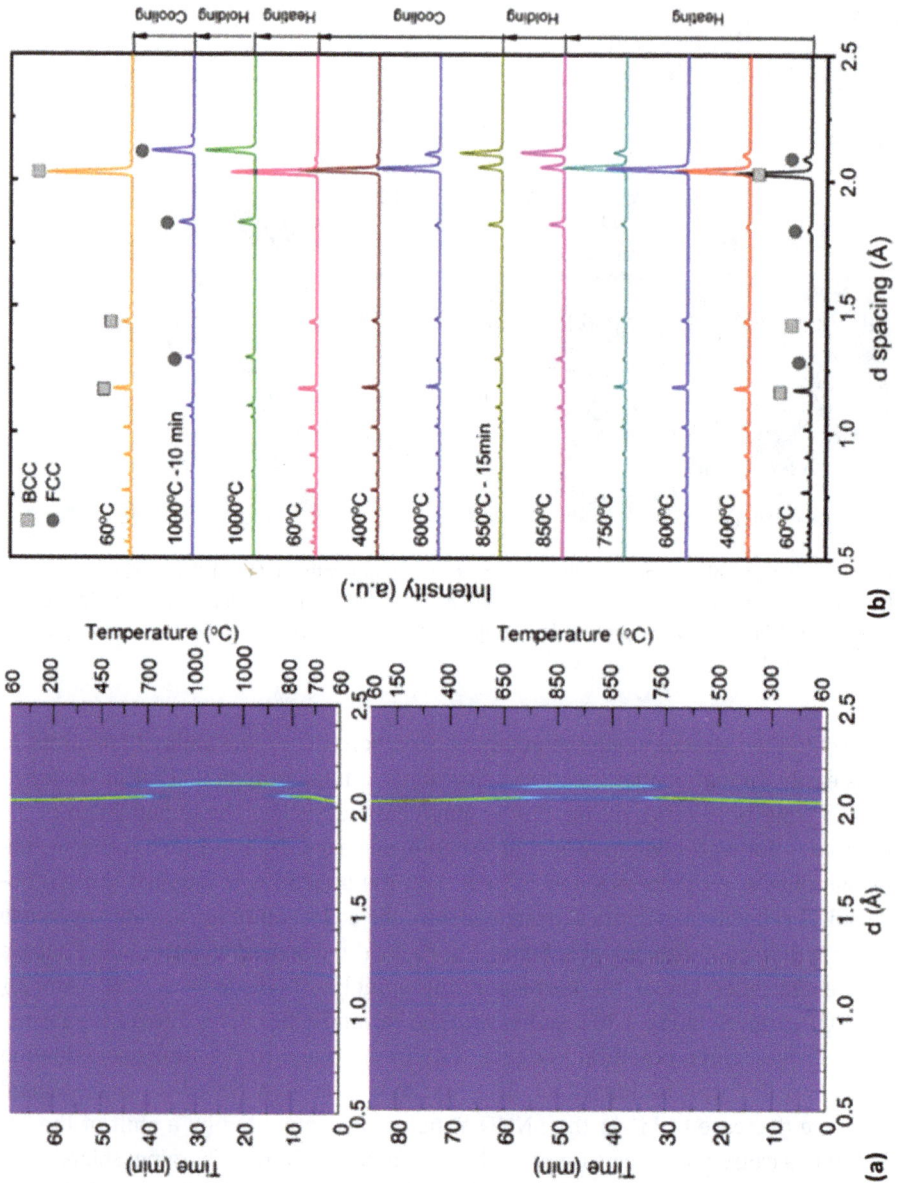

Fig. 1.29: (a) Contour plots of in situ neutron diffraction data as a function of time/temperature for cycle 1 (upper) and cycle 2 (lower image), and (b) selected diffraction patterns collected during annealing (data collection time per scan: 1 min) (reprinted with permission from Ref. [47]. Copyright © 2018, MDPI).

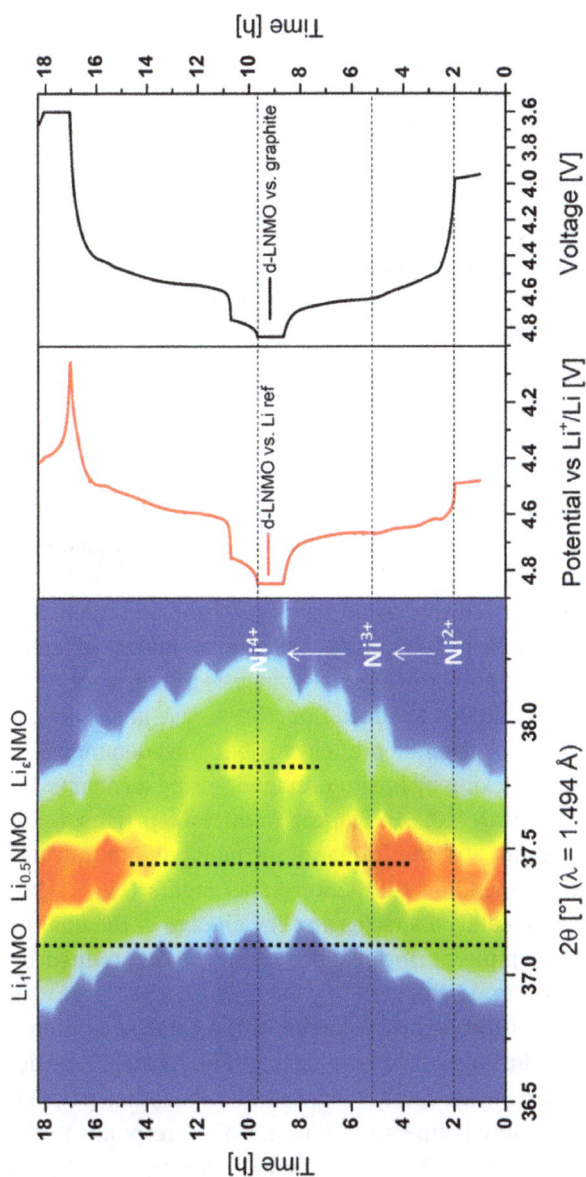

Fig. 1.30: Contour plot of neutron powder patterns collected *operando* during the first cycle of a cylindrical electrochemical cell consisting of a Li spinel versus graphite cylindrical cell: (left) change of (222) spinel reflection over time (right) the galvanostatic cycle of d-LNMO versus Li reference or versus graphite (reprinted with permission from Ref. [48]. Copyright © 2017, The Royal Society of Chemistry).

by other rare-earth and transition metal cations and during the charging process, hydrides are formed. During charging of the battery, hydrogenation occurs at the hydrogen storage alloy used as the negative electrode material. The hydrides can be described by two different hydrogenation states, the dilute solid-solution, α-LaNi$_5$H$_x$, $x \leq$ 0.5, and concentrated solid solution, β-LaNi$_5$H$_x$, $x \geq 6$.

For the in situ neutron powder diffraction experiments, the electrochemical cell setup shown in Fig. 1.31a was used. The setup enables the simultaneous measurement of the diffraction patterns and the discharge capacity. The diffraction patterns obtained as a function of time reveal different regions where reactions take place (Fig. 1.31b). During discharge, only the β-phase is visible for 1.7 h. The peak shift to higher 2θ angles indicates a continuous decrease of the unit cell volume. Between reaction times of 1.7 and 3.2 h, a two-phase region is observed, where the transformation of the β- to the α-phase takes place and an increased discharge capacity was detected. Finally, at reaction times >3.2 h, only the α-phase solid solution is present. The phase change over time during discharging as given in Fig. 1.30c shows that even in the fully charged state, 10% of the α-phase (dilute solid solution) remains in the electrode.

The last example of in situ diffraction studies on the polycrystalline sample has been performed with laboratory equipment for running solid–gas reactions. Different transition metal-based compounds are currently tested as catalysts for the cracking of ammonia: NH$_3$ into N$_2$ and H$_2$. The hydrogen obtained from the storage molecule ammonia is then used in fuel cells for the generation of electricity. Molybdenum-based catalysts were studied during NH$_3$ cracking by in situ diffraction using a commercial reaction chamber (XRK900, Anton Paar) [50, 51]. Starting from MoO$_3$ precursor, the successive reduction of molybdenum and the formation of molybdenum nitrides could be traced by temperature-dependent diffraction experiments (Fig. 1.32a). Mo$_x$N$_y$ could be identified as active catalyst species and the change of the quantitative phase composition of the sample during reaction could be determined by quantitative Rietveld refinements (Fig. 1.32b).

Furthermore, detailed studies of the cycling behavior of the catalyst exhibited systematic shifts of the reflections during heating and cooling (Fig. 1.33a) [51]. Structure refinement of the molybdenum nitride catalyst showed that the lattice parameter decrease obtained during increasing temperature (Fig. 1.33b) corresponds to the release of nitrogen from the structure (Fig. 1.33c). In situ studies during several catalytic cycles confirmed the formation of nitrides with temperature-dependent nitrogen content.

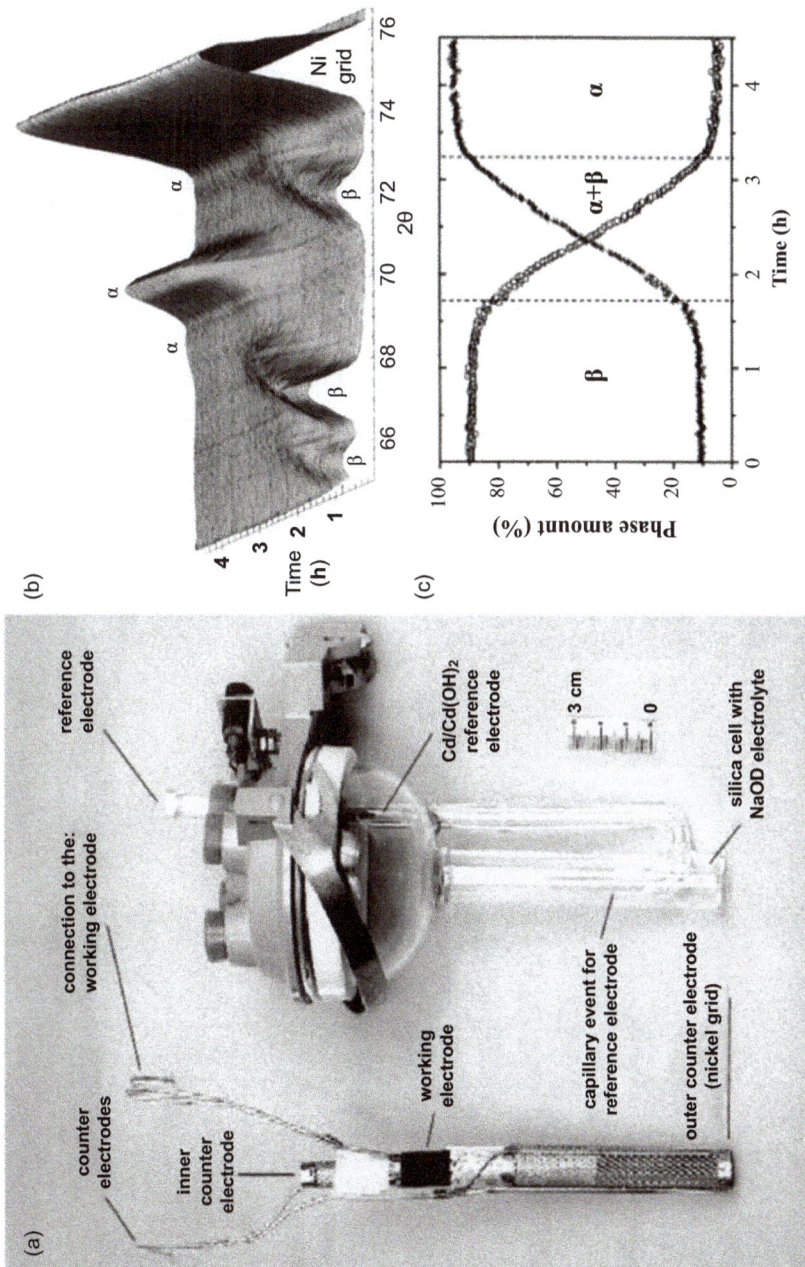

Fig. 1.31: (a) Cell assembly used for in situ neutron powder diffraction experiments. (b) Three-dimensional plot of the neutron powder diffraction patterns collected during potentiodynamic discharge of the LaNi$_5$-type electrode. (c) Time-dependent phase change was observed during discharging (reprinted with permission from Ref. [49]. Copyright © 2002, Elsevier Science B.V.).

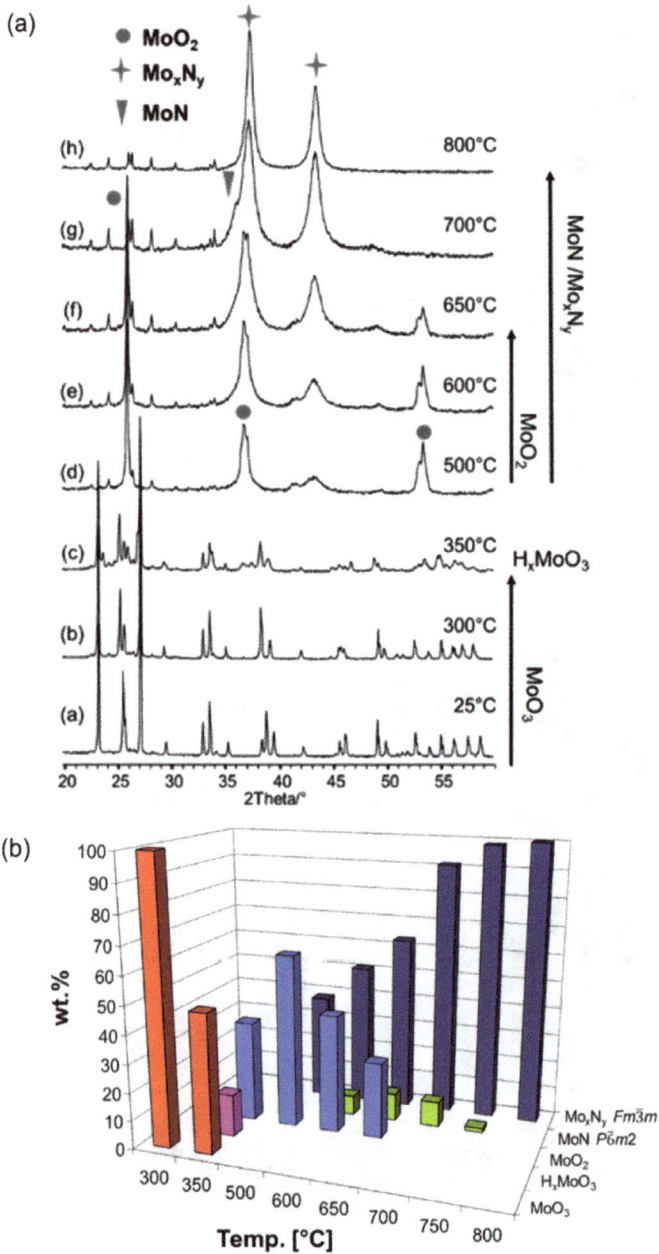

Fig. 1.32: (a) Selected powder diffraction patterns collected in situ during NH_3 decomposition starting from MoO_3 precursor in a flow of 100% NH_3 (space velocity of 15,000 mL/g sample). (b) Quantitative phase analysis from 300 to 800 °C starting from MoO_3. The quantitative analysis was obtained by Rietveld refinements (reprinted with permission from Ref. [50]. Copyright © 2013, Elsevier Science B.V.).

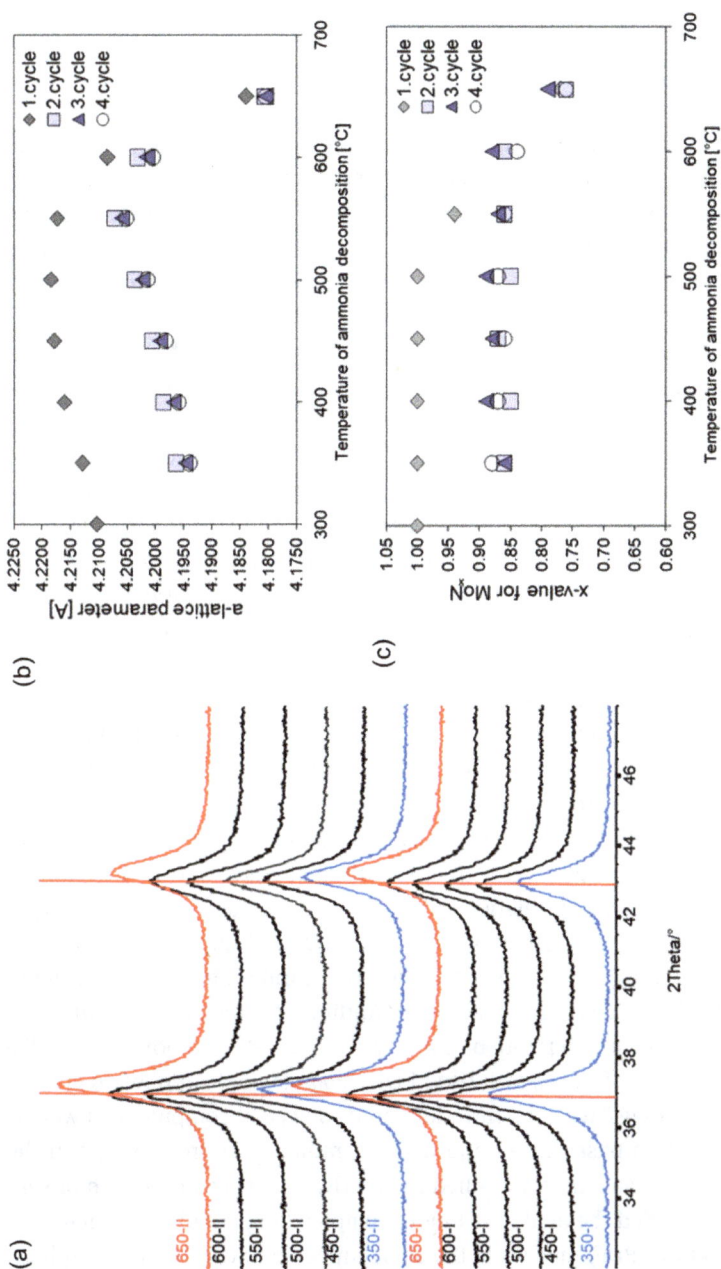

Fig. 1.33: (a) Powder patterns measured during two cycles of NH$_3$ cracking on MoN$_x$ catalysts showing temperature-dependent peak shifts. (b) Temperature dependence of the a-lattice parameter of MoN$_x$ and (c) the x-value in MoN$_x$ (representing the composition of the cubic molybdenum nitride during cycling of MoN$_x$ under NH$_3$ (reprinted with permission from Ref. [51]. Copyright © 2014, The Royal Society of Chemistry).

1.7 Local structure analysis by total scattering experiments

As has been discussed in the previous chapters, average crystal structure information can be deduced either from single crystals or from powder diffraction experiments. Both methods require crystals of specific minimum crystal dimensions. The resulting averaged crystal structures are based on the evaluation of Bragg scattering and consider long-range order only. However, many materials show structural disorder, defects, or they are partially or fully amorphous. Most information about disorder in a crystal can be obtained via analyzing single-crystal diffuse scattering. Downsizing compounds to the nanometer scale is important for the successful use of energy applications or in catalysis. Conventional structure analysis of nanosized materials, on the other hand, is limited due to the lack of long-range order. The implementation of total scattering experiments on nanosized powders allows the analysis of structures on the local scale. PDF analysis uses not only the Bragg reflections but also diffuse scattering to extract structural information on the atomic scale. The book of Egami and Billinge *Underneath the Bragg Peaks, Structural Analysis of Complex Materials* [52] is an excellent summary of how the PDF method has been applied for the structure analysis of amorphous materials since the early stages of development. The PDF method is a well-established technique for structural studies of glasses and liquids since the beginning of the last century. The method is based on the approach of Debye in 1915 who took a three-dimensional average of the sample structure amplitude to obtain the scattering expected from isotropic samples such as amorphous materials (e.g., glass) or liquids [53]. Later, the formalism of correlation functions was introduced by Zernicke and Prins, which shows the relationship between the atom pair correlation function, and the isotropically averaged scattering function [54]. The introduction of the distribution function for the mutual distance of two molecules enables the derivation of a formula for the scattering of X-ray in liquids. PDF provides the probability of finding an atom at a given distance *r* from another atom and can therefore be considered as a weighted bond length distribution. Until the 1960s of the last century, PDF was not widely applied to crystalline materials that changed with the introduction of powerful computers as well as the setup of synchrotron-based X-ray and neutron radiation sources. While in the early year PDF analysis was mainly applied to liquids, the usage of neutron or synchrotron radiation enabled the analysis of more complex nonperiodic solid materials [55, 56]. The strength of PDF analysis is the short-range atomic order in amorphous materials but also of nanostructured materials with structural features lying between the two extremes: amorphous and crystalline materials [57]. By PDF analysis, structure properties on different length scales from internal molecular to nanometer and even larger scales can be studied. In situ total scattering experiments and PDF analysis can be used to study nucleation and crystallization of solids on a local structure scale. Even though the conventional XRD pattern would only show an amorphous scattering contribution due to lack of long-range order, PDF analysis will provide interatomic correlations, which gives insights into the structural ordering on the atomic length scale. White et al. performed a series of in situ neutron and synchrotron PDF analysis on geopolymer binders obtained from high-purity metakaolin by activation with sodium-based alkaline solutions [58, 59]. Local structure changes occurring during geopolymerization provide information on molecular processes that are responsible for the performance of the materials. Metakaolin was first obtained by calcination of kaolinite and in a second step mixed with an alkaline activator solution. The activator consists of a solution of KOD in D_2O mixed with amorphous fumed silica. Due to the low scattering power in case of X-rays, O–O distances beside strong scatterers are hardly visible as well as H–O bonds. In the case of neutrons, O–O can be detected easily as well as D–O bonds become prominent due to the scattering length of deuterium. The obtained gels displayed only small local structural changes over the initial 17 h as shown by in situ PDF studies (Fig. 1.34a) [58]. Significant changes occur during 90 days as the gel gradually evolves to a more ordered structure. PDF analysis also revealed that the dissolution of metakaolin is strongly dependent on

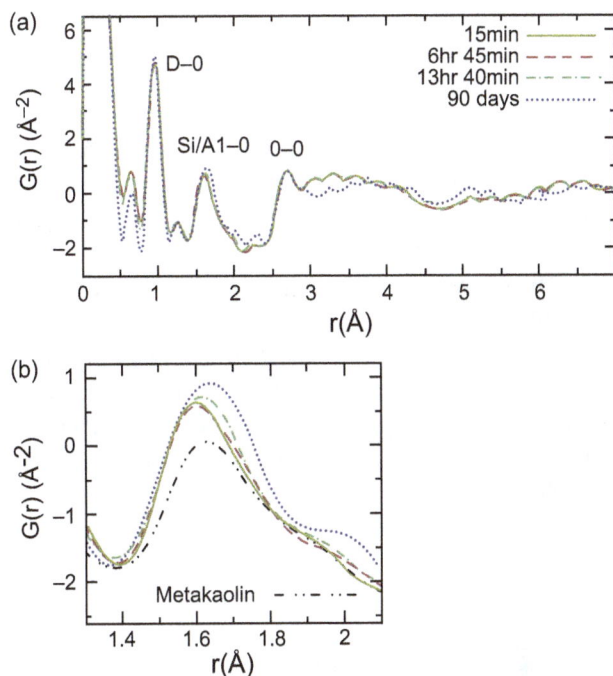

Fig. 1.34: Neutron pair distribution functions (PDFs) obtained during geopolymerization. (a) Correlations up to 7 Å; (b) (Si,Al)–O correlation and comparison with the PDF obtained for pure metakaolin (reprinted with permission from Ref. [58]. Copyright © 2011, The American Ceramic Society).

the pH of the activating solution. At higher pH values, dissolution is faster. The correlation in the measured PDFs at about 1.6 Å is assigned to Si–O and Al–O bonds in metakaolin with the cations in IV coordination (Fig. 1.34b). After 90 days, the correlation at ~1.9 Å is becoming a more intense indication of the dissolution of metakaolin. This distance is attributed to Al–O bonds with aluminum in higher VI-fold coordination states.

In situ PDF together with small-angle X-ray scattering or small-angle neutron scattering experiments are the major tools for the evaluation of nanoparticle formation. However, small-angle scattering does not focus on atomic local structures, whereas PDF provides useful insights into the early stages of atomic structure arrangement in nanomaterials. Iversen and coworkers published a large number of PDF studies on the synthesis of different nanomaterials. The studies aim to how precursor solutions react under solvothermal conditions to form solid precipitates, and later transform into nanocrystals. One example is the combination of time-resolved in situ total scattering experiments and PDF analysis to follow the formation of WO_3 under solvothermal conditions [60]. As electrochromic materials, WO_3 has found applications in solar cells, smart windows, and gas sensors. WO_3 crystallizes in different stable crystal structures as well as a metastable modification. An aqueous precursor solution of $(NH_4)_6H_2W_{12}O_{40}\cdot xH_2O)$ was heated in a high-pressure in situ capillary cell to selected temperatures while total scattering data were collected. Time-resolved PDF data obtained during isothermal treatment at 250 °C are presented in Fig. 1.35a and b. As no changes of PDFs are observed, the precursor structure in solution remains almost unchanged after heating is started. A representative PDF obtained after 25 s is given in Fig. 1.35c together with a calculated PDF based on a modeled cluster structure (Fig. 1.35d). Corner- and edge-sharing WO_6 octahedra form a cluster represented by two peaks at 1.87 and 2.22 Å corresponding to W–O correlations. The distances at 3.28 Å belong to W–W pairs from edge-sharing WO_6 octahedra, while the W–W distances at 3.68 Å arise from corner-sharing octahedra. The absence of any correlations beyond a distance of 7 Å indicates the presence of local ordering only. First, changes in the local ordering occur after 7–8 min of heating indicating that the edge-sharing octahedra transform into corner-sharing octahedra. After 10 min of heating at 250 °C, the precursor nuclei in the solution abruptly form crystalline particles with less than 3 min.

1.8 In situ studies combining different experimental tools

Although in situ diffraction experiments already provide a lot of information on structure–property relationships compared to *ex situ* studies, the additional combination of different in situ techniques can be much more enlightening. Fischer–Tropsch

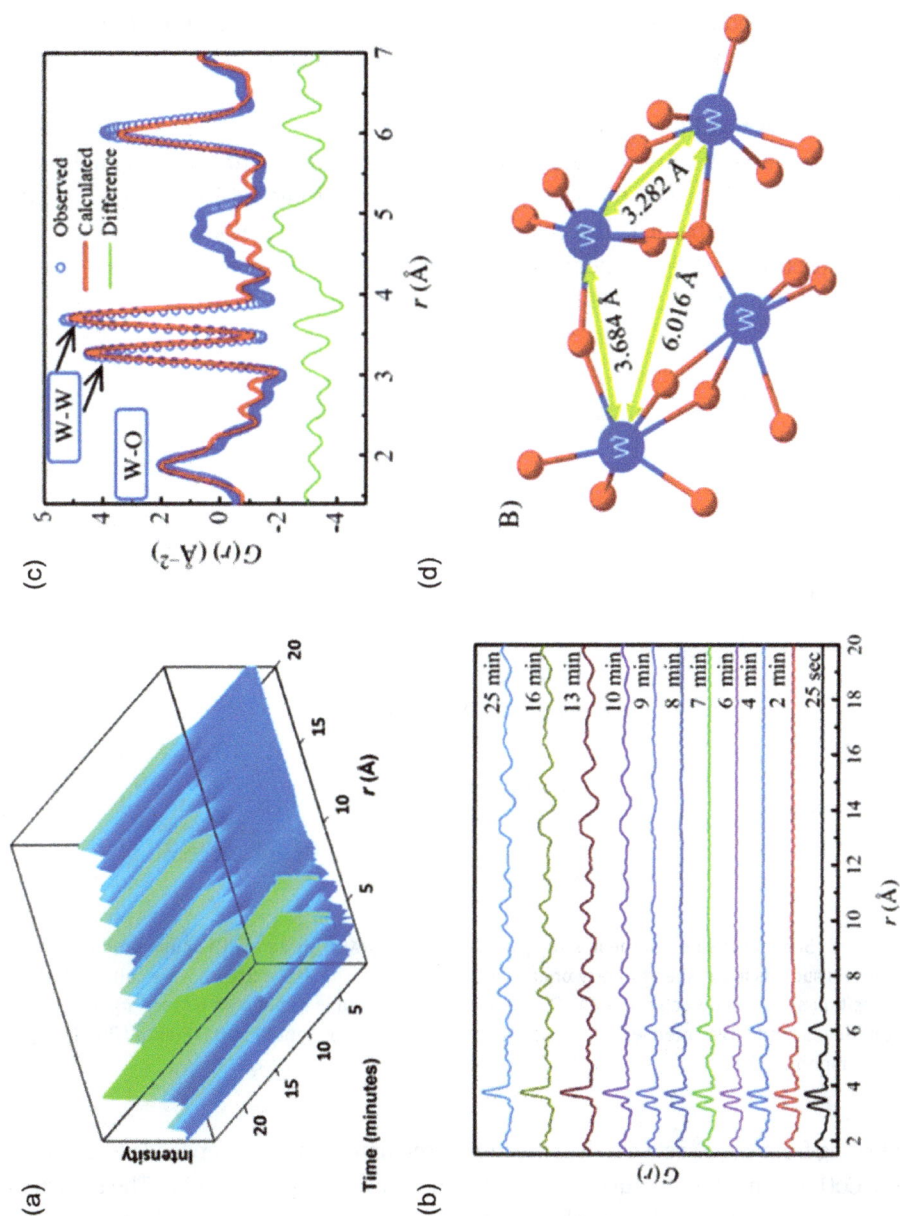

Fig. 1.35: (a) Time-dependent evolution of PDF collected at 250 °C. (b) Selected PDFs demonstrating structure changes at 250 °C with time. (c) PDF for the precursor solution obtained after 25 s of heating. (d) Molecular cluster present in the precursor solution derived from the measured PDF (reprinted with permission from Ref. [60]. Copyright © 2014, Wiley-VCH Verlag GmbH & Co. KGaA, Weinheim).

catalysts applied for the production of synthetic fuels from a mixture of CO and H_2 were studied under industrially relevant conditions. Synchrotron XRD in combination with X-ray absorption near-edge structure (XANES) was used to reveal information on crystallite sizes and the oxidation states of the active Co-based component [61]. The behavior of a laboratory plug flow reactor was simulated by an experimental setup similar to that displayed in Fig. 1.36 with the catalyst packed in a 1-mm diameter quartz capillary tube [62]. The catalytic conversion was monitored using an online mass spectrometer. The experiments were performed at a reaction temperature of 483 K and 10 bar respective 18 bar gas pressures.

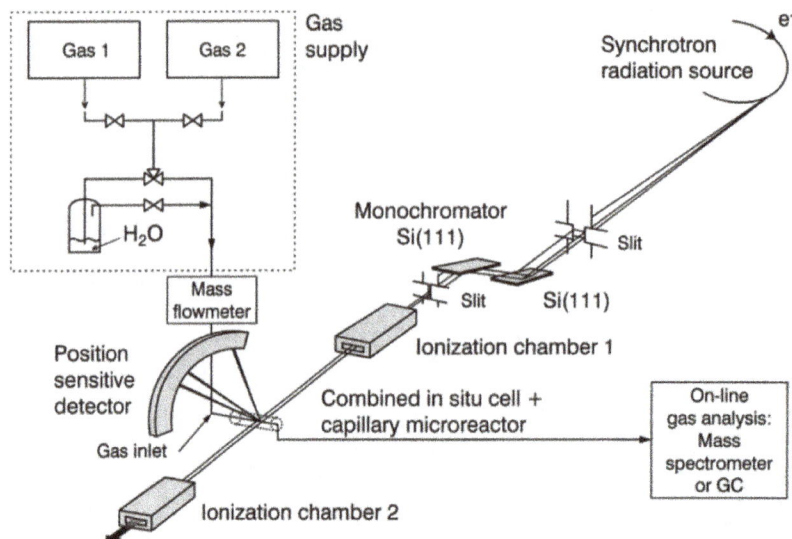

Fig. 1.36: Schematic setup of the in situ synchrotron XANES/XRD studies with online product analysis such as mass spectrometry or gas chromatography. The XRD patterns were collected with a position-sensitive detector and XANES spectra were obtained in either a step scanning or a continuous mode (reprinted with permission from Ref. [62]. Copyright © 2002, Plenum Publishing Corporation).

Starting from Co_3O_4 as a catalyst precursor, the first reduction step leads to metastable CoO which finally reduces to metallic Co^0 (fcc and hcp; Fig. 1.37a). The reduction process followed by XANES confirmed the XRD results. However, it is evident from the XANES data that about 10% of the cobalt oxide remains in the nonmetallic phase after reduction for 4 h at 673 K in H_2 (Fig. 1.37b). This is additional information inaccessible by XRD analysis. Line broadening analysis of the XRD data reveals that the crystallize size of the resulting metallic Co remains quite small (10 nm as determined by the Scherrer method). During the first 2 h of reaction at 483 K no significant sintering

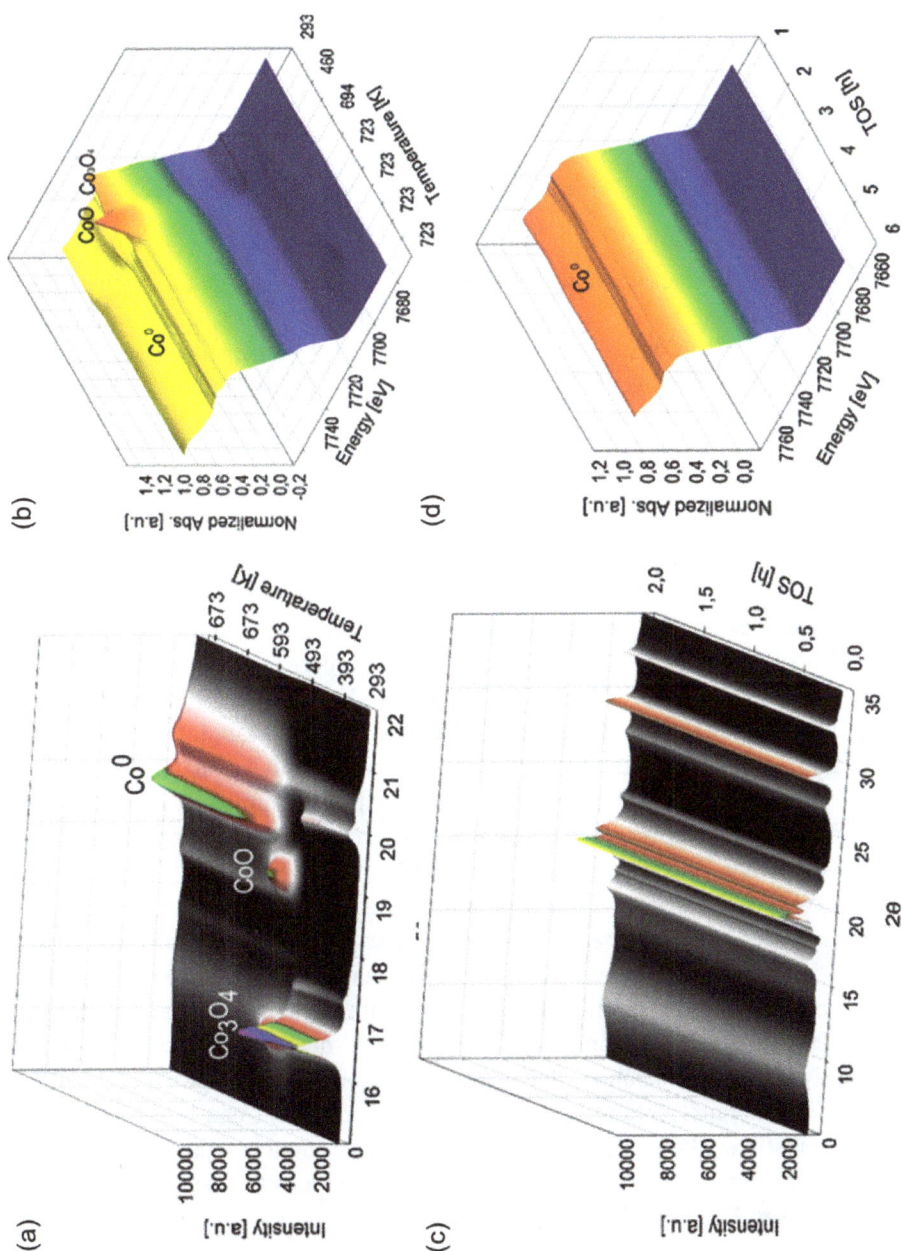

Fig. 1.37: (a) In situ XRD patterns collected during the reduction of Co_3O_4 via CoO to metallic Co in pure H_2 atmosphere ($\lambda = 0.70417$ Å). (b) XANES profiles collected during reduction. Scans were collected in intervals of approximately 30 min. (c) In situ XRD patterns collected during the first 2 h on stream at 493 K and 10 bar gas pressure. (d) XANES profiles collected during Fischer–Tropsch reaction at 18 bar and 483 K (reprinted with permission from Ref. [61]. Copyright © 2009, Elsevier B.V.).

occurs (Fig. 1.37c) and also the oxidation state remains stable (Fig. 1.37d). Further experiments showed that running the reaction at 673 K for 1 h initiates the sintering of the Co crystallites.

1.9 Outlook

Comprehensive characterization of structure properties is essential for the fundamental understanding of structure–property relationships. The investigation of materials under working conditions, either in situ or *operando*, is becoming increasingly important. This is especially relevant when crystal structures change under reaction conditions in such a way that the resulting properties also change significantly. As has been illustrated in this chapter, structure analysis can be performed on different length scales, from local atomic ordering to averaged crystal structures. A large number of experimental tools are available ranging from microscopy over spectroscopy to scattering methods. Due to great effort in the development of powerful large research facilities such as synchrotron or neutron sources, most types of materials could be studied under reaction conditions far beyond simple heat treatment. This is not only because of the installation of more brilliant radiation sources or microscopes with high atomic resolution but also the development of fast and quantum-efficient detectors is essential. Moreover, modern laboratory instruments have also become more suitable for in situ/*operando* studies of materials.

Nevertheless, scientific questions related to structure properties are becoming more and more complex and the desire to get even more accurate insight into reactions is growing. The simultaneous combination of experimental techniques providing complementary information about materials is becoming more and more important. One challenge hereby is to adapt all experimental methods to the same timescale. Another challenge is to find suitable materials for sample environments that can withstand the reaction conditions without falsifying the measurement results. There is a great demand for the development of sample cells, which enable the study of structure changes during real operation conditions. Another question arises on how representative a specific in situ experiment can be for a material under realistic working conditions. The use of large amounts of material in a commercial process may lead to different experimental results than a small sample quantity used in an in situ setup analyzed at a synchrotron source with a very small sample volume investigated. Experimental conditions for specific applications should be comparable and processing should be standardized. Otherwise, experimental data obtained for the same material by different groups cannot be discussed comparatively. In situ/*operando* experiments produce large amounts of data ("big data"). These data need treatment in such a way that the results are reliable and representative. Intelligent experimental strategies are required for the large variety of experiments nowadays available for in situ or *operando* studies of structure–property relationships. In the end, what should not be overlooked is a critical examination of the experimental data and careful interpretation of the results.

References

[1] Dangwal Pandey, A., Güttel, R., Leoni, M., Schüth, F., Weidenthaler, C. Influence of the microstructure of gold-zirconia yolk-shell catalysts on the CO oxidation activity. J. Phys. Chem. C 2010, 114, 19386–19394. Doi: 10.1021/jp106436h.

[2] Banares, M.A. Operando methodology: Combination of in situ spectroscopy and simultaneous activity measurements under catalytic reaction conditions. Catal. Today 2005, 100, 71–77. Doi: 10.1016/j.cattod.2004.12.017.

[3] Goonetilleke, D., Sharma, N. In situ neutron powder diffraction studies. Curr. Opin. Electrochem. 2019, 15, 18–26. Doi: 10.1515/psr-2018-0155.

[4] Peterson, V.K., Auckett, J.E., Pang, W.K. Real-time powder diffraction studies of energy materials under non-equilibrium conditions. IUCrJ. 2017, 4, 540–554. Doi: 10.1107/s2052252517010363.

[5] Liu, Y.S., Glans, P.-A., Chuang, C.H., Kapilashrami, M., Guo, J.H. Perspectives of in situ/ operando resonant inelastic X-ray scattering in catalytic energy materials science. J. Electron Spectrosc. Relat. Phenom. 2015, 200, 282–292. Doi: 10.1016/j.elspec.2015.07.004.

[6] Wu, X., Li, S., Yang, B., Wang, C. In Situ transmission electron microscopy studies of electrochemical reaction mechanisms in rechargeable batteries. J. Electrochem. Energy Rev. 2019, 3, 467–491. Doi: 10.1007/s41918-019-00046-2.

[7] Liu, D., Shadike, Z., Lin, R., Qian, K., Li, H., Li, K., Wang, S., Yu, Q., Liu, M., Ganapathy, S., Qin, X., Yang, Q.H., Wagemaker, M., Kang, F., Yang, X.Q., Li, B. Review of recent development of in situ/operando characterization techniques for lithium battery research. Adv. Mater. 2019, 31, 1806620. Doi: 10.1002/adma.201806620.

[8] Topsøe, H. Developments in operando studies and in situ characterization of heterogeneous catalysts. J.Catal. 2003, 216, 155–164. Doi: 10.1016/s0021-9517(02)00133-1.

[9] Liu, Y.S., Jeong, S., White, J.L., Feng, X.F., Cho, E.S., Stavila, V., Allendorf, M.D., Urban, J.J., Guo, J.H. In-situ/operando X-ray characterization of metal hydrides. Chem. Sus. Chem. 2019, 20(10), 1261–1271. Doi: 10.1002/cphc.201801185.

[10] Ortatatlı, S., Ternieden, J., Weidenthaler, C. Low temperature formation of Ruddlesden–Popper-type layered $La_2CoO^{4\pm\delta}$ Perovskite monitored via in situ X-ray powder diffraction. Eur. J. Inorg. Chem. 2018, 5. Doi: 10.1002/ejic.201801162.

[11] Sharma, N., Pang, W.K., Guo, Z.P., Peterson, V.K. In situ powder diffraction studies of electrode materials in rechargeable batteries. Chem. Sus. Chem 2015, 8(17), 2826–2853. Doi: 10.1002/cssc.201500152.

[12] Haouas, M. Nuclear magnetic resonance spectroscopy for in situ monitoring of porous materials formation under hydrothermal conditions. Materials 2018, 11, 1416–1434. Doi: 10.3390/ma11081416.

[13] Cohn, A.P., Muralidharan, N., Carter, R., Share, K., Oakes, L., Pint, C.L. Durable potassium ion battery electrodes from high-rate cointercalation into graphitic carbons. J. Mater. Chem. A 2016, 4(39), 14954–14959. Doi: 10.1039/C6TA06797B.

[14] Bernert, T., Krech, D., Kockelmann, W., Felderhoff, M., Frankcombe, T.J., Weidenthaler, C. Crystal structure relation between tetragonal and orthorhombic $CsAlD_4$: DFT and time-of-flight

neutron powder diffraction studies. Eur. J. Inorg. Chem. 2015, 33, 5545–5550. Doi: 10.1002/ejic.201500841.

[15] Behrendt, G., Reichert, C., Kohlmann, H. Hydrogenation reaction pathways in the systems Li_3N-H_2, $Li_3N-Mg-H_2$, and $Li_3N-MgH_2-H_2$ by in situ X-ray diffraction, in situ neutron diffraction, and in situ thermal analysis. J. Phys. Chem. C. 2016, 120(25), 13450–13455. Doi: 10.1021/acs.jpcc.6b04902.

[16] O'Neill, H.S.C. Temperature dependence of the cation distribution in $CoAl_2O_4$ spinel. Eur. J. Mineral. 1994, 6, 603–609. Doi: 10.1127/ejm/6/5/0603.

[17] Hansen, T.C., Kohlmann, H. Chemical reactions followed by in situ neutron powder diffraction. Z. Anorg. Allg. Chem. 2014, 640(15), 3044–3063. Doi: 10.1002/zaac.201400359.

[18] Welberry, R., Whitfield, R. Singe crystal diffuse neutron scattering. Quantum Beam Sci. 2018, 2, 2. Doi: 10.3390/qubs2010002.

[19] Bailey, I.F. A review of sample environments in neutron scattering. Z. Kristall 2003, 218(2), 84–95. Doi: 10.1524/zkri.218.2.84.20671.

[20] Norby, P. In-situ XRD as a tool to understanding zeolite crystallization. Curr. Opin. Colloid Interface Sci. 2006, 11(2), 118–125. Doi: 10.1016/j.cocis.2005.11.003.

[21] Isnard, O. A review of in situ and/or time resolved neutron scattering. C. R. Physique 2007, 8, 789–805. Doi: 10.1016/j.crhy.2007.10.002.

[22] Jensen, T.R., Nielsen, T.K., Filinchuk, Y., Jorgensen, J.E., Cerenius, Y., Gray, E.M., Webb, C.J. Versatile in situ powder X-ray diffraction cells for solid-gas investigations. J. Appl. Crystallogr. 2010, 43, 1456–1463. Doi: 10.1107/s0021889810038148.

[23] Møller, K.T., Hansen, B.R.S., Dippel, A.C., Jorgensen, J.E., Jensen, T.R. Characterization of gas-solid reactions using in situ powder X-ray diffraction. Z. Anorg. Allg. Chem. 2014, 640(15), 3029–3043. Doi: 10.1002/zaac.201400262.

[24] Berlinghof, M., Bar, C., Haas, D., Bertram, F., Langner, S., Osvet, A., Chumakov, A., Will, J., Schindler, T., Zech, T., Brabec, C.J., Unruh, T. Flexible sample cell for real-time GISAXS, GIWAXS and XRR: Design and construction. J. Synchrotron Rad. 2018, 25, 1664–1672. Doi: 10.1107/S1600577518013218.

[25] Leriche, J.B., Hamelet, S., Shu, J., Morcrette, M., Masquelier, C., Ouvrard, G., Zerrouki, M., Soudan, P., Belin, S., Elkaim, E., Baudelet, F. An electrochemical cell for operando study of lithium batteries using synchrotron radiation. J. Electrochem. Soc. 2010, 157(5), A606–A610. Doi: 10.1149/1.3355977.

[26] Vadlamani, B., An, K., Jagannathan, M., Chandran, R.K.S. An In-situ electrochemical cell for neutron diffraction studies of phase transitions in small volume electrodes of Li-Ion batteries. J. Electrochem. Soc. 2014, 161, A1731-A1741. Doi: 10.1149/2.0951410jes.

[27] Hansen, B.R.S., Møller, K.T., Paskevicius, M., Dippel, A.C., Walter, P., Webb, C.J., Pistidda, C., Bergemann, N., Dornheim, M., Klassen, T., Jorgensen, J.E., Jensen, T.R. In situ X-ray diffraction environments for high-pressure reactions. J. Appl. Crystallogr. 2015, 48, 1234–1241. Doi: 10.1107/s1600576715011735.

[28] Moury, R., Hauschild, K., Kersten, W., Ternieden, J., Felderhoff, M., Weidenthaler, C. An in situ powder diffraction cell for high-pressure hydrogenation experiments using laboratory X-ray diffractometers. J. Appl. Crystallogr 2015, 48, 79–84. Doi: 10.1107/s1600576714025692.

[29] Norquist, A.J., O'Hare, D. Kinetic and mechanistic investigations of hydrothermal transformations in zinc phosphates. J. Am. Chem. Soc. 2004, 126, 6673–6679. Doi: 10.1021/ja049860w.

[30] Evans, S.O., Francis, R.J., O'Hare, D., Price, S.J., Clark, S.M., Flaherty, J., Gordon, J., Nield, A., Tang, C.C. An apparatus for the study of the kinetics and mechanism of hydrothermal reactions by in situ energy dispersive x-ray diffraction. Rev. Sci. Instrum. 1995, 66(3), 2442–2445. Doi: 10.1063/1.1146451.

[31] Gonzalez, M.I., Mason, J.A., Bloch, E.D., Teat, S.J., Gagnon, K.J., Morrison, G.Y., Queen, W.L., Long, J.R. Structural characterization of framework–gas interactions in the metal–organic framework Co2(dobdc) by in situ single-crystal X-ray diffraction. Chem. Sci. 2017, 8, 4387–4398. Doi: 10.1039/c7sc00449d.

[32] Scarlett, N.V.Y., Hewish, D., Pattel, R., Webster, N.A.S. A flow cell for the study of gas-solid reactions via in situ powder X-ray diffraction. Rev. Sci. Instrum. 2017, 88, 105104. Doi: org/10.1063/1.4996940.

[33] He, H., Barnes, P., Munn, J., Turrillas, X., Klinowski, J. Autoclave synthesis and thermal transformations of the aluminophosphate molecular sieve VPI-5: an in situ X-ray diffraction study. Chem. Phys. Lett. 1992, 196(3), 267–273. Doi: 10.1016/0009-2614(92)85966-E.

[34] Christensen, A.N., Jensen, T.R., Norby, P., Hanson, J.C. In situ synchrotron x-ray powder diffraction studies of crystallization of microporous aluminophosphates and Me2+-substituted aluminophosphates. Chem. Mater. 1998, 10(6), 1688–1693. Doi: 10.1021/cm980049s.

[35] Piarristeguy, A.A., Cuello, G.J., Yot, P.G., Ribes, M., Pradel, A. Neutron thermodiffraction study of the crystallization of Ag–Ge–Se glasses: evidence of a new phase. J. Phys.: Condens. Matter. 2008, 20, 155106. Doi: 10.1088/0953-8984/20/15/155106.

[36] Storm, M.M., Johnsen, R.E., Norby, P. In situ X-ray powder diffraction studies of the synthesis of graphene oxide and formation of reduced graphene oxide. J. Solid State Chem. 2016, 240, 49–54. Doi: 10.1016/j.jssc.2016.05.019.

[37] Jaffe, A., Li, Y., Beavers, C.M., Voss, J., Mao, W.L., Karunadasa, H.I. High-pressure single-crystal structures of 3D lead-halide hybrid perovskites and pressure effects on their electronic and optical properties. ACS Cent. Sci. 2016, 2, 201–209. Doi: 10.1021/acscentsci.6b00055.

[38] Halder, G.J., Kepert, C.J. In Situ single-crystal x-ray diffraction studies of desorption and sorption in a flexible nanoporous molecular framework material. J. Amer. Chem. Soc. 2005, 127(219), 7891–7900. Doi: 10.1021/ja042420k.

[39] Hyun, S.M., Lee, J.H., Jung, G.Y., Kim, Y.K., Kim, T.K., Jeoung, S., Kwak, S.K., Moon, D., Moon, H.R. Exploration of gate-opening and breathing phenomena in a tailored flexible metal-organic framework. Inorg. Chem. 2016, 55(4), 1920–1925. Doi: 10.1021/acs.inorgchem.5b02874.

[40] Maity, A., Dutta, R., Penkala, B., Ceretti, M., Letrouit-Lebranchu, A., Chernyshov, D., Perichon, A., Piovano, A., Bossak, A., Meven, M., Paulus, W. Solid-state reactivity explored in situ by synchrotron radiation on single crystals: From $SrFeO_{2.5}$ to $SrFeO_3$ via electrochemical oxygen intercalation. J. Phys. D: Appl. Phys 2015, 48, 504004. Doi: 10.1088/0022-3727/48/50/504004.

[41] Sun, Y., Zhang, H., Wang, J., Guo, J., Peng, Q. Atomic-scale imaging of incipient interval-layered hydrogenation of single crystal magnesium. Scr. Mater. 174, 77–79. Doi: 10.1016/j.scriptamat.2019.08.023.

[42] Turner, J.F.C., Done, R., Dreyer, J., David, W.I.F., Catlow, C.R.A. On apparatus for studying catalysts and catalytic processes using neutron scattering. Rev. Sci. Instrum. 1999, 70(5), 2325–2330. Doi: 10.1063/1.1149758.

[43] Albers, P., Prescher, G., Seibold, K., Parker, S.F. Applications of neutron scattering for investigating heterogeneous catalysts. Chem. Eng. Technol. 1999, 22(2), 135–137.

[44] Kandemir, T., Girgsdies, F., Hansen, T.C., Liss, K.D., Kasatkin, I., Kunkes, E.L., Wowsnick, G., Jacobsen, N., Schlögl, R., Behrens, M. In situ study of catalytic processes: Neutron diffraction of a methanol synthesis catalyst at industrially relevant pressure. Angew. Chem. Int. Ed. 2013, 52, 5166–5170. Doi: 10.1002/anie.201209539.

[45] Kandemir, T., Wallacher, D., Hansen, T., Liss, K.D., Naumann d'Alnoncourt, R., Schlögl, R., Behrens, M. In situ neutron diffraction under high pressure – providing an insight into working catalysts. Nucl. Instrum. Meth. A. 2012, 673, 51–55. Doi: 10.1016/j.nima.2012.01.019.

[46] Auer, H., Weber, S., Hansen, T.C., Többens, D.M., Kohlmann, H. Reversible hydrogenation of the Zintl phases BaGe and BaSn studied by in situ diffraction. Z. Kristallogr. 2018, 233(6), 399–409. Doi: 10.1515/zkri-2017-2142.

[47] Yu, D.J., Chen, Y., Huang, L., An, K. Tracing phase transformation and lattice evolution in a TRIP sheet steel under high-temperature annealing by real-time In Situ neutron diffraction. Crystal 2018, 8(9), 12. Doi: 10.3390/cryst8090360.

[48] Boulet-Roblin, L., Sheptyakov, D., Borel, P., Tessier, C., Novák, P., Villevieille, C. Crystal structure evolution via operando neutron diffraction during long-term cycling of a customized 5 V full Li-ion cylindrical cell $LiNi_{0.5}Mn_{1.5}O_4$ vs. graphite. J. Mater. Chem. A. 2017, 5(48), 25574–25582. Doi: 10.1039/C7TA07917F.

[49] Latroche, M., Chabre, Y., Decamps, B., Percheron-Guégan, A., Noreus, D. In situ neutron diffraction study of the kinetics of metallic hydride electrodes. J. Alloys Compds. 2002, 334 (1), 267–2276. Doi: 10.1016/S0925-8388(01)01799-6.

[50] Tagliazucca, V., Schlichte, K., Schüth, F., Weidenthaler, C. Molybdenum-based catalysts for the decomposition of ammonia: In situ X-ray diffraction studies, microstructure, and catalytic properties. J. Catal. 2013, 305, 277–289. Doi: 10.1016/j.jcat.2013.05.011.

[51] Tagliazucca, V., Leoni, M., Weidenthaler, C. Crystal structure and microstructural changes of molybdenum nitrides traced during catalytic reaction by in situ X-ray diffraction studies. Phys. Chem. Chem. Phys. 2014, 16(13), 6182–6188. Doi: 10.1039/C3CP54578D.

[52] Egami, T., Bilinge, S.J.L. Underneath the Bragg peak: structural analysis of complex materials. Pergamon Materials Series. Amsterdam: Elsevier, 2003.

[53] Debye, P. Zerstreuung von Röntgenstrahlen. Ann. Phys. 1915, 46, 809–823. Doi: 10.1002/andp.19153510606.

[54] Zernicke, F., Prins, J.A. Die Beugung von Röntgenstrahlen in Flüssigkeiten als Effekt der Molekülanordnung. Z. Phys. A-Hadron Nucl. 1927, 41(2), 184–194. Doi: 10.1007/bf01391926.

[55] Egami, T. Atomic correlations in non-periodic. Matter. Mater. Trans. JIM 1990, 31(3), 163–176. Doi: 10.2320/matertrans1989.31.163.

[56] Toby, B.H., Egami, T. Accuracy of the pair distribution function-analysis applied to crystalline and noncrystalline materials. Acta Crystallogr. Sect. A. 1992, 48, 336–346. Doi: 10.1107/s0108767391011327.

[57] Billinge, S.J.L. Nanostructure studied using the atomic pair distribution function. Z. Kristallogr. Suppl. 2007;26:17–26.

[58] White, C.E., Provis, J.L., Llobet, A., Proffen, T., Van Deventer, J.S.J. Evolution of local structure in geopolymer gels: An in situ neutron pair distribution function analysis. J. Am. Ceram. Soc. 2011, 94(10), 3532–3539. Doi: 10.1111/j.1551-2916.2011.04515.x.

[59] White, C.E., Page, K., Henson, N.J., Provis, J.L. In situ synchrotron X-ray pair distribution function analysis of the early stages of gel formation in metakaolin-based geopolymers. Appl. Clay Sci. 2013, 73, 17–25. Doi: 10.1016/j.clay.2012.09.009.

[60] Saha, D., Jensen, K.M.O., Tyrsted, C., Bojesen, E.D., Mamakhel, A.H., Dippel, A.C., Christensen, M., Iversen, B.B. In situ total X-ray scattering study of WO3 nanoparticle formation under hydrothermal conditions. Angew. Chem. Int. Ed. 2014, 53, 3667–3670. Doi: 10.1002/anie.201311254.

[61] Rønning, M., Tsakoumis, N.E., Voronov, A., Johnsen, R.E., Norby, P., Van Beek, W., Borg, Ø., Rytter, E., Holmen, A. Combined XRD and XANES studies of a re-promoted $Co/\gamma-Al_2O_3$ catalyst at Fischer–Tropsch synthesis conditions. Catal. Today. 2010, 155(3), 289–295. Doi: 10.1016/j.cattod.2009.10.010.

[62] Grunwaldt, J.D., Clausen, B.S. Combining XRD and EXAFS with on-line catalytic studies for characterization of catalysts. Top. Catal. 2002, 18(1), 37–43. Doi: 10.1023/a:1013838428305.

Sebastian Bette and Robert E. Dinnebier

2 Understanding stacking disorder in layered functional materials using powder diffraction

Abstract: Stacking fault disorder is a widespread phenomenon in layered materials. It can have a significant impact on the substance properties and usually inhibits its structural characterization. Due to recent advances in instrumentation and analytical tools, even small diffraction effects, such as line broadenings, peak asymmetry, and low-intensity superstructure reflections, caused by faulting can be detected and analyzed by using X-ray powder diffraction. Improvements in computational hardware and software algorithms enable quantitative analyses on the degree of faulting by recursive routines. In this chapter, an introduction into the phenomena of stacking fault disorder and their effect on powder diffraction patterns is given by using brucite-type hydroxides and the layered honeycomb materials as examples.

Keywords: layered materials, stacking fault disorder, interstratification, turbostratic disorder, X-ray powder diffraction

2.1 Introduction

The crystal structures of many technologically important materials contain two-dimensional layered structural motifs, periodically repeated in the third dimension, giving the crystal a distinct stacking sequence. These include clays, which are essential components for ceramics, construction and composite materials, and layered double hydroxides phases (LDH phases) that are used as electrode materials and precursors for catalysts. In numerous cases, these structure models provide only an idealized picture of the atom arrangement. The real structures of these materials exhibit irregular stacking motifs involving the occurrence of defects called stacking faults. The structure and quantity of these faults can have a strong influence on the properties and performance of the materials, and must therefore be characterized.

Acknowledgment: The authors thank Dr. Maxwell W. Terban from the Max-Planck Institute for Solid State Research for proof reading the book chapter.

Sebastian Bette, Robert E. Dinnebier, Max-Planck Institute for Solid State Research, Heisenbergstrasse 1, 70569 Stuttgart, Germany

https://doi.org/10.1515/9783110674910-002

Conventional crystal structure analysis is performed using larger [1] (X-ray or neutron diffraction) or smaller (electron diffraction [2]) individual single crystals or using powder diffraction methods [3] (X-ray or neutron) on micro and nano crystalline powders. However, the characterization of stacking faults and their sequences is challenged by the loss of periodicity in the third dimension. This leads to diffuse elastic scattering generated at arbitrary positions in reciprocal space, and as a consequence, scattered intensity is no longer observed only at distinct reciprocal lattice points. In three-dimensional reciprocal space, new features such as streaks or superstructure reflections appear. The investigation of these complicated features for stacking faulted single crystals is preferably carried out using high energy synchrotron diffraction and large 2D-detectors to collect information over a wide range of reciprocal space with high signal-to-noise [4].

Often, the type of faulting can be different for individual single crystals of the same batch of a certain material. By simultaneously collecting scattering patterns averaged over a large number of crystallites, as with X-ray powder diffraction (XRPD) or neutron powder diffraction (NPD), structural details can be obtained that are more representative of the bulk properties, making these techniques more suitable. Since powder diffraction is the one-dimensional projection of three-dimensional reciprocal space, the above mentioned effects can appear as new peaks, peak shifts, peak asymmetries, and/or peak broadening all over the powder pattern.

In this chapter, an introduction into the phenomena of stacking fault disorder and their effect on powder diffraction patterns is given. Examples of stacking fault disorder in technologically important layered brucite-type and honeycomb materials are presented.

2.2 Stacking orders of close-packed spheres

The simplest type of stacking faults occurs in close-packed spheres, for which two basic types of packing are possible: cubic close packing (*ccp*) and hexagonal close packing (*hcp*) (Fig. 2.1). Both packing types can be distinguished by their stacking sequence. In a *ccp* structure that contains only one atom type, for example, copper metal, the layers are stacked perpendicular to the [111] direction in an αβγ-stacking sequence, with metal position indicated by small Greek letters (Fig. 2.2, left). An *hcp* structure, e.g. magnesium metal, exhibits layers stacked perpendicular to the [001] directions with an αβαβ-stacking sequence (Fig. 2.2, right).

A useful description of a stacking sequence can be achieved by using stacking vectors, which describe the transition from a given layer to the subsequent one (Fig. 2.2, magenta and blue arrows). For convenience, the unit cells (Fig. 2.2, black lines) with lattice parameters a, b, c, α, β, γ are transformed into pseudo-orthorhombic or pseudo-

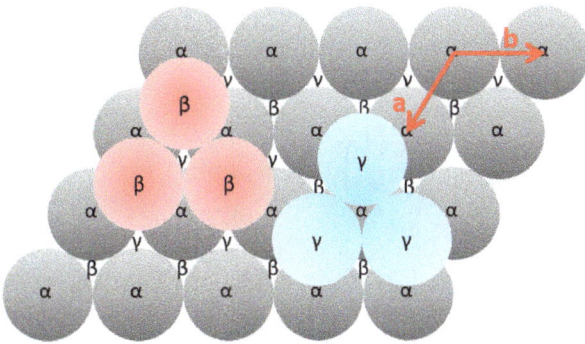

Fig. 2.1: Close-packed hexagonal layer α, with the spheres centered at the origin of the hexagonal unit cell, and additional hexagonal layers β and γ, with spheres centered above the trigonal voids located at (2/3 1/3 z) and (1/3 2/3 z) of the hexagonal unit cell.

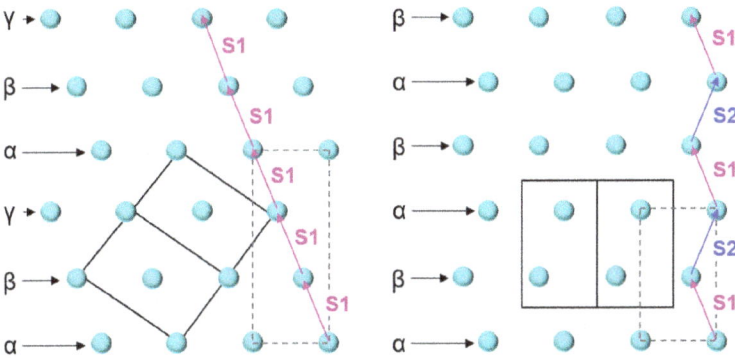

Fig. 2.2: Stacking sequence in *ccp* (left) and *hcp* packing (right). Unit cell edges are displayed in black and transformed unit cells in gray dashed lines. The stacking order is indicated by stacking vectors **S1** and **S2** describing layer to layer transitions. Positions of metal atoms are indicated by lower case Greek letters (from [5]).

trigonal unit cells with triclinic lattice parameters a', b', c', $\alpha' = \beta' = 90°$, $\gamma' = 90°$ or $120°$, in which the layers are perpendicular to the c'-axis (Fig. 2.2, dashed gray lines). The stacking vectors, **Si**, are given in fractional coordinates of the transformed unit cell. Each stacking vector consists of 3 components (eq. (2.1)): Si_x, Si_y and Si_z, with Si_x and Si_y describing the shift within the $a'b'$ plane and Si_z describing the shift parallel to c':

$$Si = \begin{pmatrix} Si_x \\ Si_y \\ Si_z \end{pmatrix}. \qquad (2.1)$$

A *ccp* structure can be described by using only one stacking vector, **S1**, whereas for an *hcp*-structure, two stacking vectors, **S1** and **S2,** are required and repeated in an alternating fashion (Fig. 2.2, magenta, blue):

$$ccp:S1 = \begin{pmatrix} 1/3 \\ -1/3 \\ 1/3 \end{pmatrix}, \quad hcp:S1 = \begin{pmatrix} 1/3 \\ -1/3 \\ 1/2 \end{pmatrix}, \quad S2 = \begin{pmatrix} -1/3 \\ 1/3 \\ 1/2 \end{pmatrix}. \tag{2.2}$$

2.3 Types of stacking faults

Many inorganic materials, for example, NaCl or MgO, exhibit close-packed structures. In these structures, layers of metal cations denoted by small Greek letters, and layers of anions denoted by capital Roman letters, are arranged in an alternating fashion (Fig. 2.3a). Layered structures like $Mg(OH)_2$ (brucite) or $CdCl_2$ can be derived from close-packed motifs by removing half of the cation layers (Fig. 2.3b). As a consequence, the stacking sequence transforms from CαBγAβCαBγAβ ... (NaCl-type) to CαB□AβC□BγA□ ... ($CdCl_2$-type) with "□" indicating a cation vacancy. Usually, the vacancies are not explicitly shown, and layers are indicated by parentheses. Therefore, the stacking order in a $CdCl_2$-type lattice can also be expressed as (CαB)(AβC)(BγA). Like close-packed structures, layered structures exhibit several types of packing. In Fig. 2.4, three examples are shown: C19-type ($CdCl_2$-type), C6-type (CdI_2/brucite-type), and CrOOH-type (sometimes denoted a 3R-type). There are many more basic stacking types in other layered compounds; for a good overview, see Hulliger [6].

Several types of stacking faults can occur in a structure based on close-packed spheres. In a *ccp* structure, the stacking order can have two different orientations: αβγ (stacking vector **S1**, Fig. 2.5a) and γβα (stacking vector **S2**). When two differently oriented crystals are intergrown, the stacking vector switches from **S1** to **S2** at the interface. This interface can be considered a *stacking fault*. The switch of the stacking vectors creates a local *hcp*-like stacking, for example, βγβ (Fig. 2.5a) in the stacking order. This type of intergrowth is also termed *twinning*.

If the stacking vector switches from **S1** to **S2** and afterwards back to **S1**, then *ccp*-like packing is interrupted by a *hcp*-like transition (Fig. 2.5b). This can be considered a *local stacking fault*. Sometimes, local faults exhibit a certain extension, for example, after several *ccp*-like layer-to-layer transitions, more than one *hcp*-like transition occurs before the stacking order switches back to *ccp* stacking (Fig. 2.5c). Thus, distinct structural motifs are *crystallographically intergrown,* and the interfaces between the intergrown sections can be considered stacking faults.

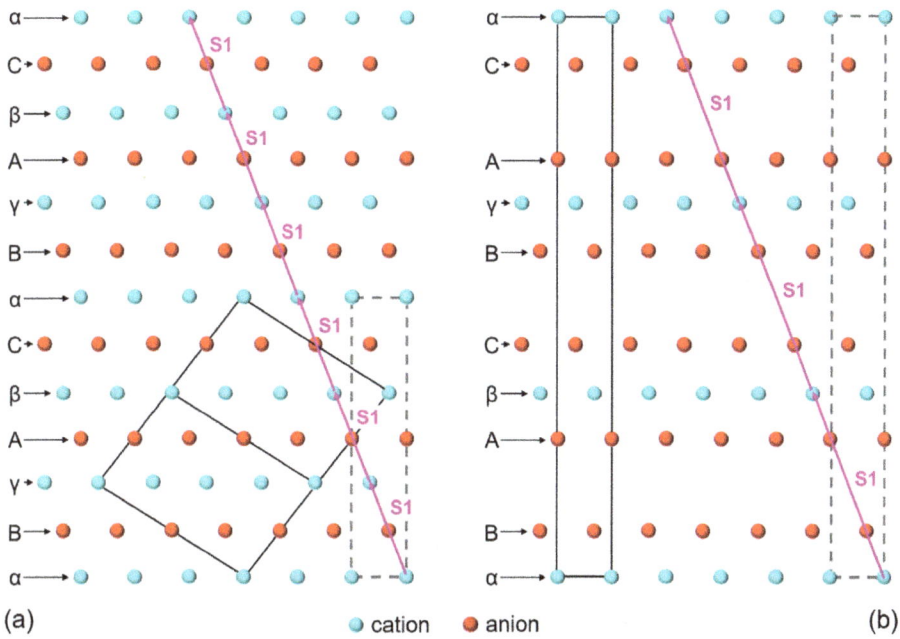

Fig. 2.3: Stacking sequences in a binary MX *ccp* structure (a) and in a related MX$_2$ structure (b). Unit cell edges are displayed as black solid lines and edges of transformed unit cells as gray dashed lines. The stacking order is indicated by stacking vector **S1** describing layer to layer transitions. Positions of metal cations are indicated by lower case Greek letters; anion positions are indicated by capital Roman letters (from [5]).

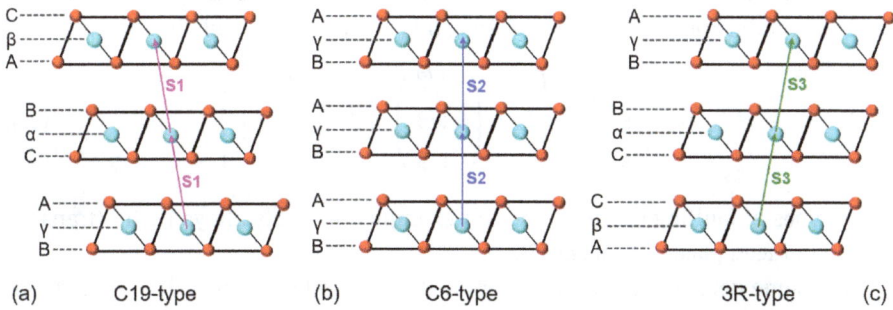

Fig. 2.4: Examples of stacking types in layered structures. MX$_{6/3}$-octahedra with M = metal cation and X = anion are indicated by black solid lines (from [5]).

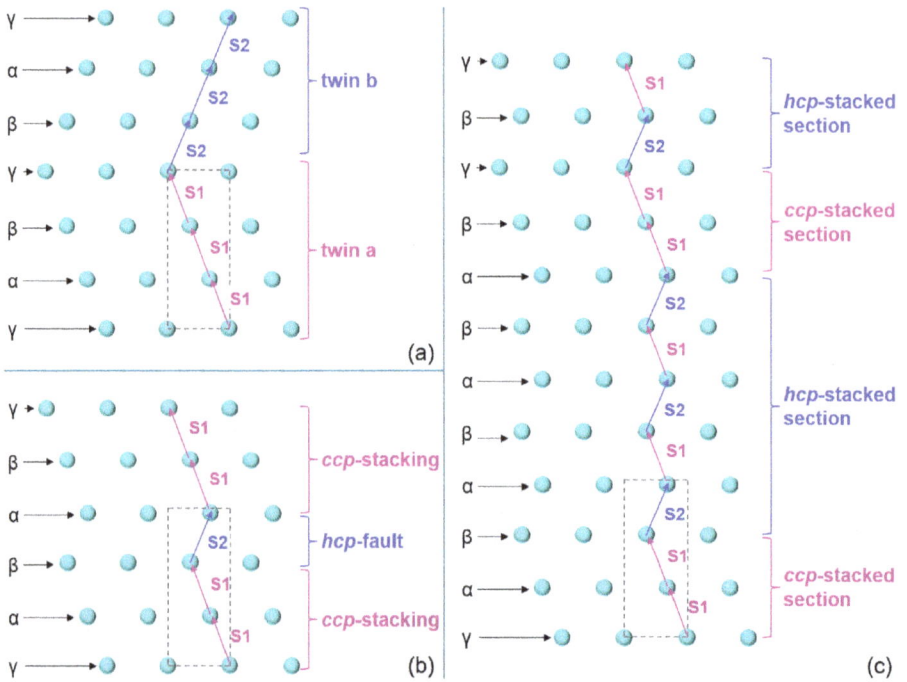

Fig. 2.5: Types of stacking faults. Transitions between different stacking vectors, leading to twinning (a), a single *hcp*-fault in a *ccp*-stacking (b), and (c) crystallographic intergrowth of *ccp*- and *hcp*-stacked sections (from [5]).

In each of these cases, the stacking vector can be described by a *finite* array of distinct vectors. For close-packed spheres, this array contains only two vectors[1]:

$$S = \begin{pmatrix} s_x \\ s_y \\ s_z \end{pmatrix} \in \left\{ S1 = \begin{pmatrix} -1/3 \\ 1/3 \\ 1/3 \end{pmatrix}, S2 = \begin{pmatrix} 1/3 \\ -1/3 \\ 1/3 \end{pmatrix} \right\}. \tag{2.3}$$

These types of faulting can be found, for example, in highly alloyed TRIP (transformation induced plasticity) steels [7].

The same considerations can be applied to layered structures. The main difference is that more than two basic packing types exist. If, for example, the packing types presented in Fig. 2.4 form a microstructure, the array that describes the stacking vector consists of three vectors:

---

1 $s_z = 1/3$ assumes c' is 3 times the layer spacing.

$$S = \begin{pmatrix} S_x \\ S_y \\ S_z \end{pmatrix} \in \left\{ S1 = \begin{pmatrix} -1/3 \\ 1/3 \\ 1/3 \end{pmatrix}, \ S2 = \begin{pmatrix} 0 \\ 0 \\ 1/3 \end{pmatrix}, \ S3 = \begin{pmatrix} 1/3 \\ -1/3 \\ 1/3 \end{pmatrix} \right\}. \qquad (2.4)$$

A larger number of possible packing types in the microstructure of a layered compound contribute a larger array of possible stacking vectors. Although the structure remains closely related to a close-packing, this array will always be finite.

In layered structures with a large interlayer spacing, weak interlayer interactions or high r_{cation}:r_{anion} ratios, the stacking order of the layers often deviates from close packing. Hence, layers can be randomly displaced from their ideal position by shifts within the layer plane or by rotation within the layer plane, leading to so-called *turbostratic* disorder (Fig. 2.6). Accordingly, a random component, Δ, defined as

$$\Delta = \begin{pmatrix} \Delta_x \\ \Delta_y \\ \Delta_z \end{pmatrix}, \qquad (2.5)$$

can be added to a given stacking vector. As turbostratic-like disorder does not affect the interlayer spacing, $\Delta_z = 0$:

$$S = Si + \Delta = \begin{pmatrix} S_x \\ S_y \\ S_z \end{pmatrix} + \begin{pmatrix} \Delta_x \\ \Delta_y \\ 0 \end{pmatrix} \text{ with } 0 < \Delta x, \ \Delta y < 1 \text{ and } \Delta x, \ \Delta y \in \{\mathbb{R}\}. \qquad (2.6)$$

Since the x and y components can adopt all real values between 0 and 1, the stacking is described by an *infinite* array of distinct vectors.

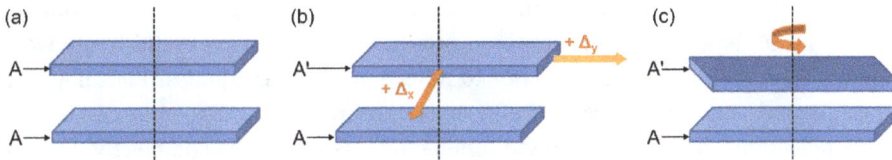

Fig. 2.6: Ideally stacked layers (a); top layer randomly dislocated by a shift in the *ab*-plane (b); (c) top layer randomly dislocated by a rotation in the *ab*-plane (from [5]).

Some layered compounds can take up atoms, ions, or molecules between the layers in a process called intercalation. Intercalation can be inhomogeneous, that is, some layers are intercalated and others are not, or homogeneous, and is usually accompanied by an increase of the interlayer distance (Fig. 2.7). When the distribution of intercalated particles is random or inhomogeneous, intercalation appears as a form of stacking faulting called *interstratification*. Interstratification affects the z-component of the stacking vector and often goes along with turbostratic-like disorder in materials, such as brucite-type magnesium and nickel hydroxides [8–10].

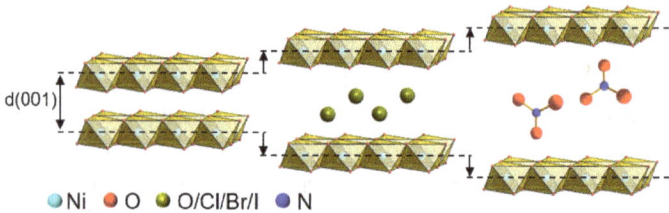

d(001)

● Ni ● O ● O/Cl/Br/I ● N

Fig. 2.7: Disorder in α/β-Ni(OH)$_2$ by intercalation of ions and molecules ≡ interstratification (from [5]).

2.4 Modeling the influence of stacking faults on diffraction

2.4.1 Turbostratic disorder

Several approaches for modeling the intensity of a stacking faulted crystal exist. One of the most fundamental and significant theoretical works on the effects of stacking disorder on XRPD patterns was carried out by B.E. Warren [11]. He considered completely random stacking (turbostratic disorder) in a structure consisting of equidistant layers. As rotations of layers (Fig. 2.8a) do not cause any additional reflections in an XRPD pattern, only random translations are considered.

In his work, Warren showed that only sharp 00l and diffuse hk0 reflections remain. The 2θ dependent reflection intensities (i.e. the broadened reflections) were derived from the Debye formula. For hk0 reflections, the randomness does not play any role and the intensity is that of a crystalline reflection. The phase factors of these reflections, however, are in the summation, over all layers completely random (Fig. 2.8a), resulting in incoherent scattering from the individual layers. This turns the reciprocal lattice points into a series of parallel lines perpendicular to the layer, called truncation rods [11] (Fig. 2.8b). For stacking along **c* (perpendicular to the ab-plane)**, these rods intersect the plane at the hk0 points. This leads to anisotropic broadening of the hk0 reflections. Warren demonstrated that these reflections are always broadened toward higher diffraction angles leading to characteristic Warren-type triangular or saw-tooth peak shapes (Fig. 2.8c). The unaffected 00l reflections remain sharp.

In the following, the general calculation of structure factors for layered structures is derived by closely following the work of Treacy, Newsam, and Deem [14]: a layered crystal structure shall consist of N stacked layers of M distinct types stacked in **c***

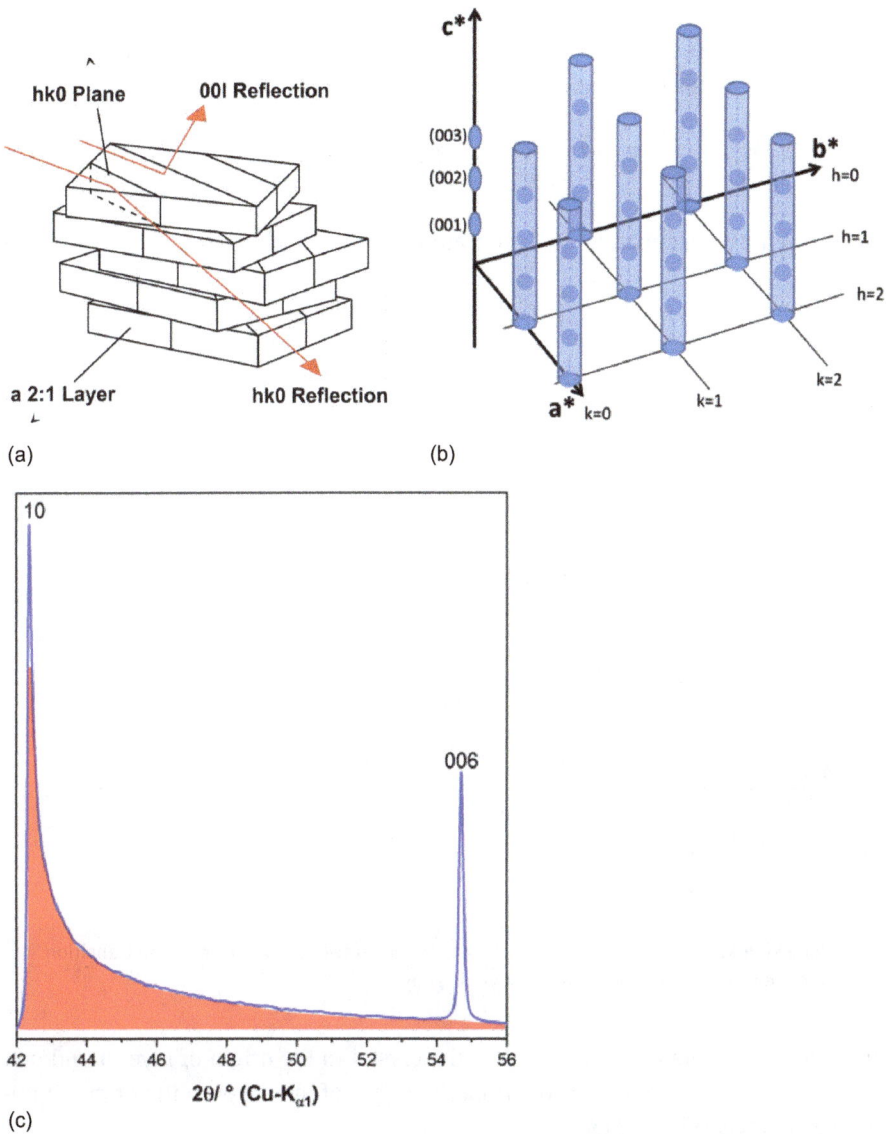

(a)

(b)

(c)

Fig. 2.8: Real (a) and reciprocal (b) space of a turbostratically disordered layer structure [12, 13] (from [5]), and (c) a small region of the simulated diffraction pattern of turbostratically disordered graphite are shown. The characteristic triangular peak shape of the two-dimensional 10 peak forbidden by lattice symmetry (≡ "Warren-peak") is highlighted by red color.

direction. The basis vectors within the layer (**a, b**) are identical for all layers, but translation in **c*** direction is disturbed. In most cases, unit cells need to be transformed to meet the conditions above.

The structure factor for a scattering vector $\mathbf{s} = (hkl)$ of a single layer of type m can be defined as the sum of the (complex) atomic scattering factors f_j of all atoms in the unit cell of this layer weighted by the complex phase factor,

$$\mathbf{F}_m(\mathbf{s}) = \sum_{j=1}^{atoms_m} O_{mjs} T_{mjs} f_{mjs} e^{2\pi i(\mathbf{s} \cdot \mathbf{r}_{mj})}, \qquad (2.7)$$

with the relative occupancy O_{mjs}, the atomic displacement factor T_{mjs}, and the atomic coordinates $\mathbf{r}_{mj} = \begin{pmatrix} x_{mj} \\ y_{mj} \\ z_{mj} \end{pmatrix}$ relative to the length of the unit cell axes.

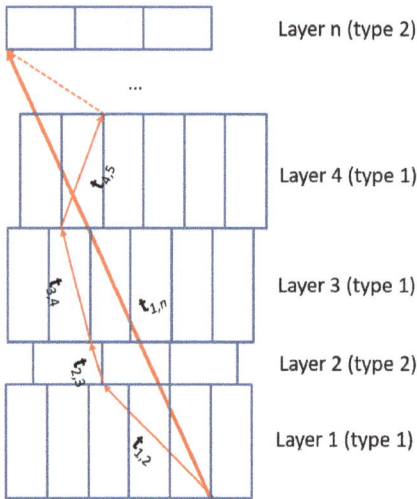

Fig. 2.9: Stacking sequence of a layered crystal consisting of two types of layers. The transition vector from the origin of the first to the nth layer is shown.

The vector $\mathbf{t}_{1,n}$ points from the origin of the crystal to the origin of layer n and can be written as the sum of all vectors from the origin of one layer to the origin of the consecutive layer (Fig. 2.9) as

$$\mathbf{t}_{1,n} = \sum_{j=1}^{n-1} \mathbf{t}_{j,j+1}. \qquad (2.8)$$

For the hypothetical case that the stacking sequence $(i,j,k,l \ldots)$ of such a layered crystal with N layers is known (Fig. 2.9), the scattered wave function of the layered crystal can be calculated as

$$\Phi^{(N)}_{ijk,\ldots}(\mathbf{s}) = \frac{1}{N} \left\{ \begin{array}{c} \mathbf{F}_i(\mathbf{s}) + \\ \mathbf{F}_j(\mathbf{s})e^{2\pi i(\mathbf{s}\cdot\mathbf{t}_{ij})} + \\ \mathbf{F}_k(\mathbf{s})e^{2\pi i(\mathbf{s}\cdot(\mathbf{t}_{ij}+\mathbf{t}_{jk}))} + \\ \mathbf{F}_l(\mathbf{s})e^{2\pi i(\mathbf{s}\cdot(\mathbf{t}_{ij}+\mathbf{t}_{jk}+\mathbf{t}_{kl}))} + \\ \cdots \end{array} \right\}, \tag{2.9}$$

and the diffracted intensity distribution for this stacking sequence is

$$I^{(N)}_{ijk\ldots}(\mathbf{s}) = \Phi^{(N)}_{ijk\ldots}(\mathbf{s})\Phi^{*(N)}_{ijk\ldots}(\mathbf{s}). \tag{2.10}$$

In the following, the diffraction effects caused by stacking fault disorder are demonstrated using the examples of turbostratically, disordered graphite and stacking faulted copper metal.

2.5 Example: turbostratic disorder in graphite

In Fig. 2.10, the diffraction effects of increasing turbostratic disorder in 3R-graphite are visualized. Starting from a hexagonal basis layer, six stacking vectors with components $Sx = 1/3 + \Delta$, $Sy = 2/3 + \Delta$, and $Sz = 1/3$ are defined. For each component, a random number Δ is used that can vary between 0 and Δ_{max}. During the simulations, Δ_{max} was incrementally increased from 0 ($\equiv$ no turbostratic disorder) to 0.5 ($\equiv$ completely disordered). Since turbostratic disorder does not modulate the z-component of the stacking vector, no $00l$ reflections are affected. At a low degree of turbostratic disorder, all non-$00l$ reflections exhibit peak broadening and splitting. With growing disorder, these reflections are finally smeared out. As the stacking faults break the hexagonal $R\bar{3}m$ symmetry, the 010 reflection, also called Warren-peak that is forbidden by the space group symmetry, appears. This reflection remains sharp, even at a growing degree of disorder. At its basis, it shows diffuse scattering appearing as peak tailing in the direction of higher diffraction angles. The resulting triangular peak-shape, indicative of stacking fault disorder, hence the 010 reflection, is also commonly denoted as "Warren-type peak shape."

2.5.1 Distinct stacking faults

In most cases, the shifts between layers are not random. Rather, they can be described by a limited set of stacking vectors. Since it is impossible to know the exact stacking sequence of a crystal, transition probabilities are defined, which are used to describe the degree of faulting, and are expressed through a square

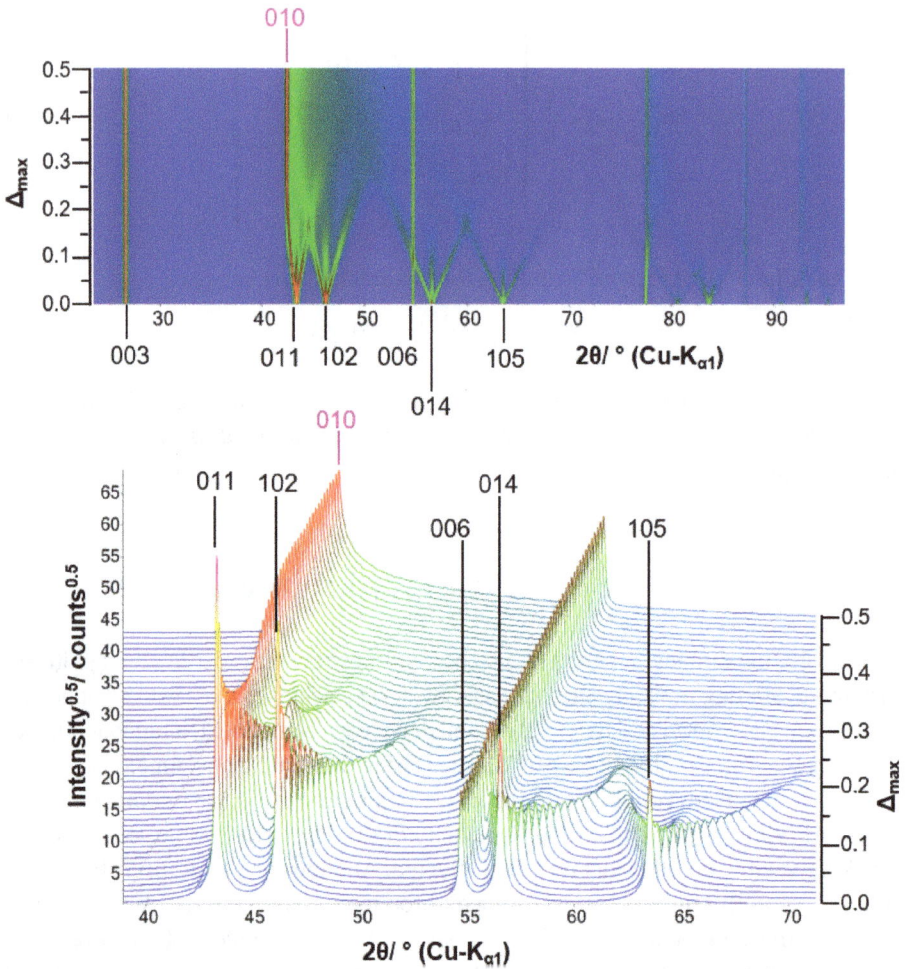

Fig. 2.10: 2D and waterfall plots of a series of diffraction patterns of graphite, for which the turbostratic disorder, Δ_{max}, was incrementally increased starting from regular 3R-graphite. The increasing amount of stacking faults breaks the translational symmetry, which eventually leads to the appearance of the 010, although it is forbidden by symmetry in the regular 3R-graphite structure.

probability matrix. In this matrix, the elements P_{ij} describe the probability of layer i being followed by layer j. The description is often set up so that the diagonal terms P_{ii} correspond to the probability of a faultless stacking and the off-diagonal terms describe a stacking fault, though this does not have to be the case. An example of a transition probability matrix including three stacking vectors is given in Tab. 2.1:

Tab. 2.1: Example of a 3×3 transition probability matrix for a 3-layer system $(P_{n1} + P_{n2} + P_{n3} = 1)$.

Transition from↓/ to→	Layer 1	Layer 2	Layer 3
layer 1	P_{11}	P_{12}	P_{13}
layer 2	P_{21}	P_{22}	P_{23}
layer 3	P_{31}	P_{32}	P_{33}

When generating stacking sequences, the position/type of the preceding layer is always taken into account. Thus, the transition probability matrix can be non-symmetric. The whole approach can be considered *recursive*. As a statistical constraint, all transition probabilities within each row of the matrix must sum up to unity,

$$\sum_{j=1}^{n} P_{ij} = 1. \tag{2.11}$$

Since the stacking of layers can start with different layer types, G_i is the probability that the i-type layer exists (which is equivalent to the concentration of layer type i). This leads to the two normalization equations:

$$G_i = \sum_{j=1}^{n} G_j P_{ij}, \quad \sum_{j=1}^{n} G_j = 1. \tag{2.12}$$

The scattering intensity for a statistical ensemble is then the weighted incoherent sum over *all M^N stacking permutations*:

$$I^{(N)}(\mathbf{s}) = \sum_{i,j,k,l,\,\ldots} G_i P_{ij} P_{jk} P_{kl} \ldots \Phi^{(N)}_{ijk,\,\ldots}(\mathbf{s}) \Phi^{*(N)}_{ijk,\,\ldots}(\mathbf{s}) \tag{2.13}$$

The scattered wave function, which represents the average of all crystals containing exactly N layers, beginning on i-type, is

$$\Phi^{(N)}_i(\mathbf{s}) = \left\{ \begin{array}{l} \mathbf{F}_i(\mathbf{s}) + \\ \sum_j P_{ij}(\mathbf{s}) e^{2\pi i(\mathbf{s} \cdot \mathbf{t}_{i,j})} [\mathbf{F}_j(\mathbf{s}) + \\ \sum_k P_{jk}(\mathbf{s}) e^{2\pi i(\mathbf{s} \cdot \mathbf{t}_{j,k})} [\mathbf{F}_k(\mathbf{s}) + \\ \sum_l P_{kl}(\mathbf{s}) e^{2\pi i(\mathbf{s} \cdot \mathbf{t}_{k,l})} [\mathbf{F}_l(\mathbf{s}) + \\ \cdots \end{array} \right\}. \tag{2.14}$$

from which the recursive relation follows,

$$\Phi_i^{(N)}(\mathbf{s}) = F_i(\mathbf{s}) + \sum_j P_{ij}(\mathbf{s})e^{2\pi i(\mathbf{s}\cdot\mathbf{r}_{i,j})}\Phi_i^{(N-1)}(\mathbf{s}), \quad \Phi_i^{(0)}(\mathbf{s}) = 0 \tag{2.15}$$

The mean scattered wave function from a stacked crystal can thus be divided into the contribution to the wave function from the layer at the origin and the mean scattered wave function from a crystal that is displaced by one layer (Fig. 2.11). Treacy et al. [14] made use of the recursive nature of stacking faults in their program DIF-FaX, which is based on the self-similarity of stacking sequences. For details on the mathematical formalism, see [14] or [15].

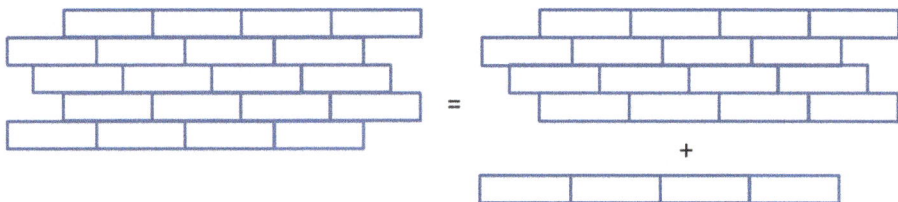

Fig. 2.11: A layered crystal can be viewed as consisting of a basis layer and the original crystal minus the base layer. Repeating this procedure shows the recursive nature of a layered crystal.

The so-called supercell approach is alternative formalism used, for example, in the program TOPAS v6 and onward [16, 17]. Whilst DIFFaX provides a straightforward simulation tool for single-phase systems, it does not offer the power of full Rietveld refinement. There have been a number of Rietveld-compatible methods published, each with its own advantages and disadvantages (e.g. [15, 18–21]). Other approaches to deriving the layer constitution from vastly stacking faulted samples exist, for example, by using two-dimensional projections of the crystal structure and geometric considerations [22]. In the supercell approach, an extended unit cell is defined that contains a large (several 100 to several 1,000) number, N_v, of layers that are stacked in the c'-direction. Individual faults can also be built into the supercell. With a sufficiently large c^* axis, the large number of hkl reflections of the supercell add up to give an excellent approximation of the broadened, smooth peak-shapes observed experimentally.

In contrast to the notation used above, all layers of type m are displaced by a set of V_m stacking vectors $\mathbf{t}_{m,v} = \begin{pmatrix} t_{x,m,v} \\ t_{y,m,v} \\ t_{z,m,v} \end{pmatrix}$, eq. (2.13) can now be rewritten as

$$\Phi_{\mathbf{s}}^{(N)} = \sum_{m=1}^{M} \sum_{v=1}^{V_m} \sum_{j=1}^{atoms_m} O_{m,j,s} T_{m,j,s} f_{m,j,s} e^{2\pi i \left(\mathbf{s} \cdot (\mathbf{r}_{m,j} + \mathbf{t}_{m,v})\right)}$$

$$= \sum_{m=1}^{M} \sum_{v=1}^{V_m} \sum_{j=1}^{atoms_m} O_{m,j,s} T_{m,j,s} f_{m,j,s} e^{2\pi i \left(\mathbf{s} \cdot \mathbf{r}_{m,j}\right)} e^{2\pi i (\mathbf{s} \cdot \mathbf{t}_{m,v})}$$

$$= \sum_{m=1}^{M} \left[\sum_{v=1}^{V_m} e^{2\pi i (\mathbf{s} \cdot \mathbf{t}_{m,v})} \mathbf{F}_m(\mathbf{s}) \right], \tag{2.16}$$

where, $\mathbf{r}_{m,j}$ are the relative atomic coordinates of layer m, and the "unit cell" of a layer type is described by two vectors within the layer and a perpendicular vector in the stacking direction with length given by the sum of the heights of all layers in the structure. The summation over the phase shifts caused by displacement can be re-written as

$$\sum_{v=1}^{V_m} e^{2\pi i (\mathbf{s} \cdot \mathbf{t}_{m,v})} = \sum_{v=1}^{V_m} e^{2\pi i \left(h t_{x,m,v} + k t_{y,m,v} + l t_{z,m,v} \right)} = \sum_{v=1}^{V_m} e^{2\pi i \left(h t_{x,m,v} + k t_{y,m,v} \right)} e^{2\pi i (l t_{z,m,v})}. \tag{2.17}$$

If the displacement vectors $\mathbf{t}_{m,v}$ are not randomly offset, but discrete, there is only a limited V_{mxy} set of unique x and y coordinates $t_{x,m,v}$ and $t_{y,m,v}$, which occur V_{mz} times. In such a case, eq. (2.17) can be rewritten as:

$$\sum_{v=1}^{V_m} e^{2\pi i (\mathbf{s} \cdot \mathbf{t}_{m,v})} = \sum_{vxy=1}^{V_{mxy}} e^{2\pi i \left(h t_{x,vxy} + k t_{y,vxy} \right)} \sum_{vz=1}^{V_{mz}} e^{2\pi i (l t_{z,vz})} \tag{2.18}$$

which considerably reduces the required number of calculations.

2.6 Example: Stacking faults in copper: recursive routine versus supercell approach

As an example, diffraction profiles of polycrystalline copper with faultless *ccp*-stacking and hexagonal intergrowths have been simulated using both the supercell and the recursive approach in Fig. 2.12. To achieve this, the space group was reduced to $P1$, and the unit cell of copper was transformed as described in Tab. 2.2. As copper is situated at the origin in both the original and the transformed unit cell, the fractional coordinates do not need to be transformed. Diffraction profiles of polycrystalline copper with faultless *ccp*-stacking (black line) and with a hexagonal intergrowth (colored lines) were simulated by applying a 2×2 transition probability matrix with $P_{12} = P_{22} = 0.02$ (implying $P_{11} = P_{21} = 0.98$) and stacking vectors as given in eq. (2.3). For the simulations, the *ccp* unit cell of copper was transformed to a hexagonal one with $\mathbf{c}' = [111]$ from the ccp structure. It can be seen that a supercell approach becomes more

Tab. 2.2: Transformation of the *ccp* structure of copper to a pseudo-hexagonal supercell.

Copper	Original unit cell	Transformed unit cell
Space group	$Fm\bar{3}m$	$P1$
Unit cell	$a = b = c = 3.62$ Å; $\alpha = \beta = \gamma = 90°$	$a' = b' = 1/\sqrt{2}a = 2.56$ Å; $c' = 3/\sqrt{3}\,a = 2.09\,N_v$ Å; $\alpha = \beta = 90; \gamma = 120$
Volume	47.6 Å [3]	$\tfrac{1}{3} \cdot N_v \cdot 47.6$ Å [3]
Stacking direction	[111]	[001]
Number of layers	3	N_v

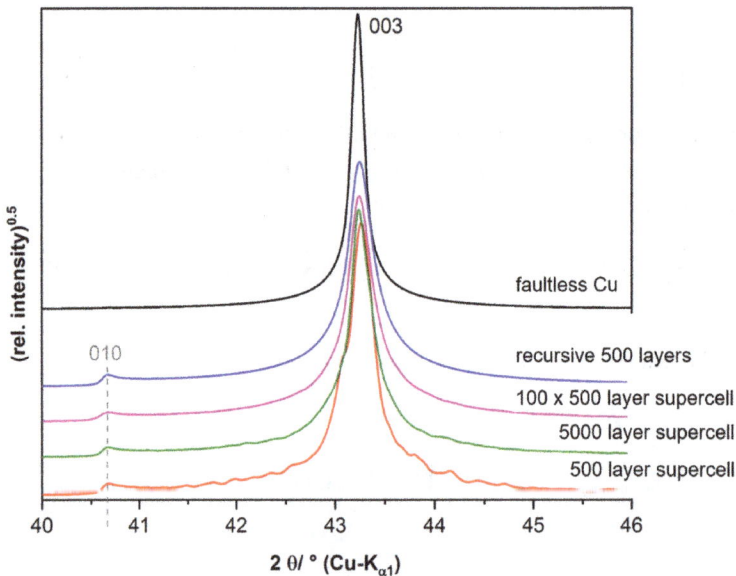

Fig. 2.12: Simulations of diffraction profiles for *ccp*-Cu. Top to bottom: faultless *ccp*-packing (black), 2% hexagonal intergrowth 500 layer recursive calculation (blue), averaging 100 stacks of 500 layers by explicit calculation (magenta), 5,000 layers explicit (green), and 500 layers explicit (red). Corresponding hexagonal intergrowths cause the appearance of a peak close to the *ccp*-forbidden 010 reflection. All indices are given in the notation of the transformed (Tab. 2.2) unit cell. (from [5]).

similar to the recursive method if either the number of layers, that is, the size of the supercell, is increased to a large number (thousands of layers), or if many smaller supercells are averaged (i.e. pseudo-recursive approach). The latter approach is computationally less expensive.

A final proof of the equivalence of both methods is given in Fig. 2.13. Here, the diffraction pattern (blue line) of stacking faulted copper was simulated using a recursive approach with the microstructural parameters given above. The pattern was then refined using a supercell approach, in which 100 stacks of 500 layers were averaged (calculated pattern: red line). The fit is almost perfect (difference curve: gray line). Hence, the supercell approach provides the opportunity for local and global optimization of the microstructure of stacking faulted materials.

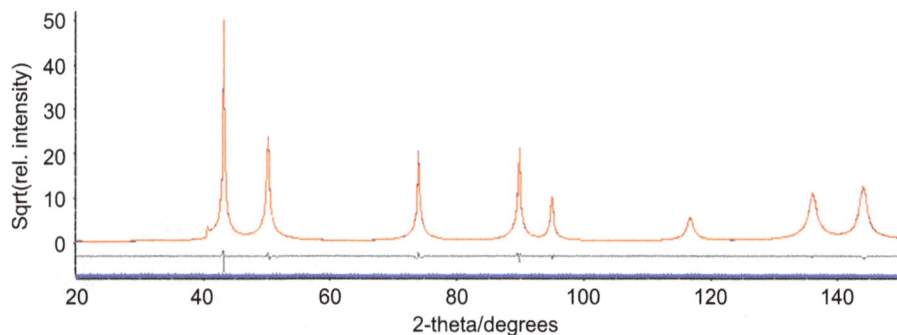

Fig. 2.13: Full Rietveld fit of DIFFaX-simulated data for 2% faulted Cu. Note the effectively continuous blue bar of *hkl* tick marks (from [5]).

2.7 Stacking faults in technologically important layered materials

In the following sections, two technologically important, notoriously stacking faulted materials are discussed in detail: layered double hydroxides (LDH) with brucite-type layers and materials with honeycomb motifs.

2.7.1 Stacking faults in layered double hydroxides with brucite-type layers

Stacking fault disorder is a common phenomenon in LDH-phases. The material properties are strongly affected by disorder. Well-ordered β-Ni(OH)$_2$ shows a higher electrochemical capacity [23] and a lower solubility [24] than strongly faulted β-Ni(OH)$_2$, whereas stacking faulted β-Ni(OH)$_2$ exhibits a higher electrochemical activity [25, 26]. As a consequence, it is very important to determine and therefore understand the different types of stacking faults in these materials and to properly use the tools available to quantify the amount of faulting.

2.7.1.1 The brucite-type structure

In a brucite-type crystal structure, the metal cation exhibits a six-fold coordination. The anions, usually oxygen containing oxide or hydroxide build up an octahedral coordination sphere. Edge sharing $MO_{6/3}$ octahedra form layers that are usually situated within the ab-plane (Fig. 2.14). In brucite-type structures, the layers are stacked in an eclipsed fashion. This means that the cations in the cation substructure are situated on identical layer positions (Fig. 2.14, γ). The anions are stacked in an ABAB fashion leading to an overall AγB□AγB□ stacking order. Examples for brucite-type compounds are: $Mg(OH)_2$ (brucite) [27], $Ca(OH)_2$ (portlandite) [28], $Mn(OH)_2$ [29], $Fe(OH)_2$ [30], $Co(OH)_2$ [31], and β-$Ni(OH)_2$ [32].

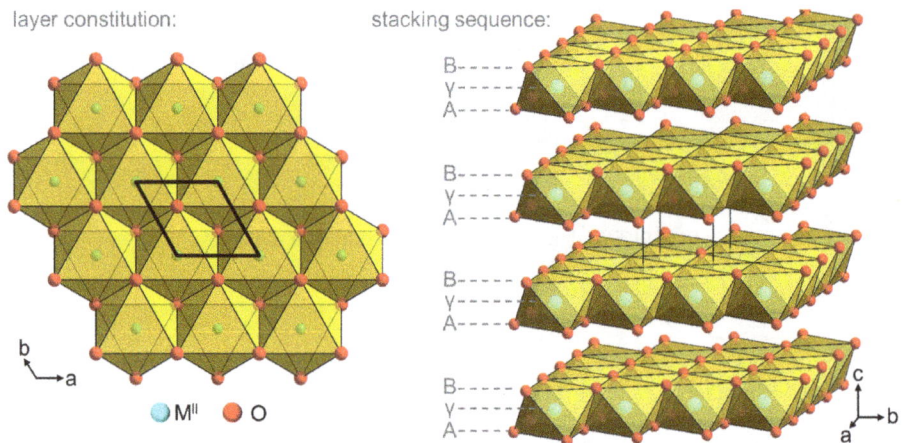

Fig. 2.14: Layer constitution and stacking order in brucite type structures.

2.7.1.2 Possible stacking orders

The stacking order of brucite-type (also known as CdI_2-type) structures can also be denoted as C6-type stacking. The eclipsed (AγB)□(AγB)□ stacking order leads to a maximization of the distance between anions of neighboring layers (Fig. 2.15). In the case of divalent hydroxides, that is, $M(OH)_2$ phases, the hydroxide related protons are situated in tetrahedral voids. This coordination environment of the protons can also be achieved by a staggered (AγB)□(CβA)□(BαC)□ stacking order, which is also known as $CdCl_2$- or C19-type stacking or by a staggered (AγB)□(AβC)□(BαC)□(BγA)□(CβA)□(CαB)□ stacking order that is denoted as 2 H_2-type. Some LDH-phases also contain metal ions with a higher valence than two, like $Cr^{III}OOH$ or $Co^{III}OOH$ [33]. If divalent metal cations in a brucite-type lattice are replaced by tri- or tetravalent ions, then hydroxide ions must be replaced by oxide ions in order to maintain charge balance. Due to the partial replacement of hydroxide, other stacking orders that provide short distances

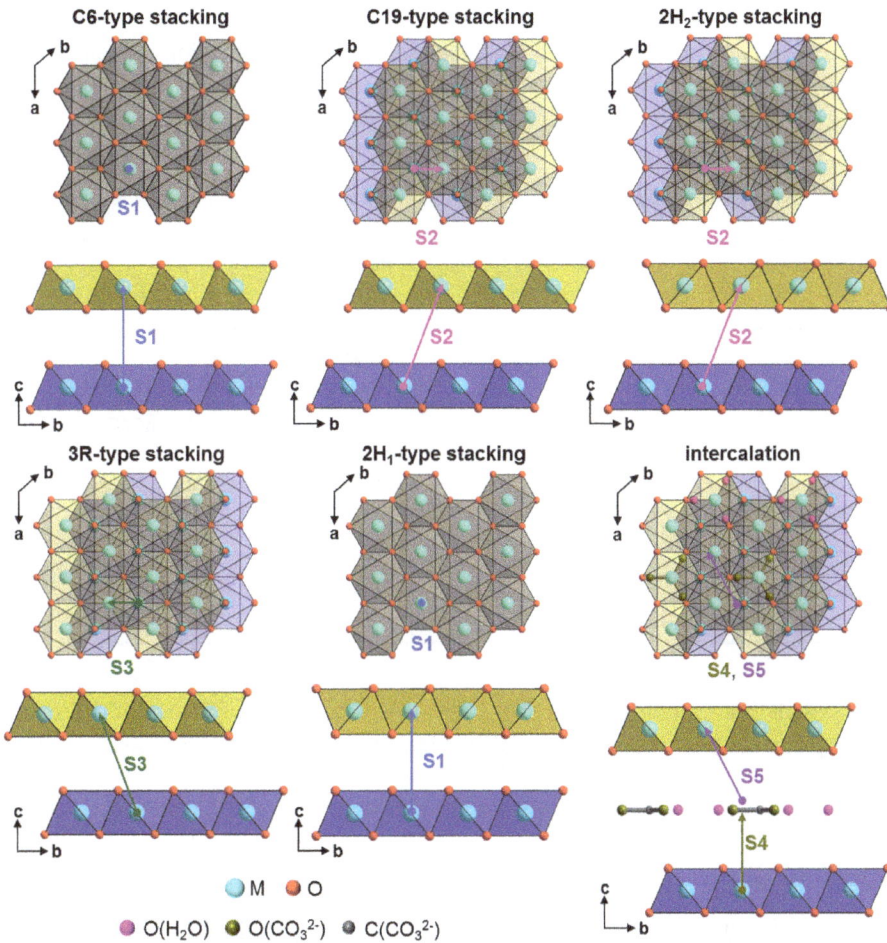

Fig. 2.15: Examples of possible stacking orders in LDH-materials. Blue polyhedra represent bottom layers and yellow polyhedra subsequent layers; the stacking order is indicated by stacking vectors (arrows). The stacking vectors shown in this image are presented in detail in Tab. 2.3.

between anions of neighboring layers, enabling the formation of strong interlayer H-bonds, become more favorable. This can be achieved by an $(A\gamma B)\square(B\alpha C)\square(C\beta A)\square$ stacking order (3R- or CrOOH-type stacking), or by an $(A\gamma B)\square(B\gamma A)\square$ stacking order (2H$_1$-type) that is realized in CoO$_{1.7}$(OH)$_{0.3}$ [34]. In between brucite-type layers, molecules like water or ions like carbonate can be intercalated (Fig. 2.16) as seen in both minerals, for example, Takovite (Ni$_{1.869}$ Al$_{1.131}$ (OH)$_6$ (CO$_3$)$_{1.020}$) [35], Iowaite ((Mg$_4$Fe(OH)$_{10}$Cl(H$_2$O)$_3$)$_{0.6}$) [36], Hydrotalcite ((Mg$_4$Al$_2$(OH)$_{12}$CO$_3$(H$_2$O)$_3$)$_{0.5}$) [37], or in α-Ni(OH)$_2$ [38, 39], and artificially-synthesized, mixed Co-Fe-, Co-Al-, Ni-Fe-, and Ni-Al-LDH phases [40]. This leads to an $(A\gamma B)a(B\alpha C)b(C\beta A)c$ stacking order, where small

Latin letters indicate the positions of the intercalated layers. Usually, stacking vectors S1, S2, S3, S4, and S5 (Fig. 2.15) are used to describe a certain stacking order. An overview on the stacking vectors that can be used to create the discussed stacking-types can be found in Tab. 2.3.

Tab. 2.3: Stacking vector components for different stacking sequences in a brucite-type unit cell.

Stacking vector	Components			Stacking sequence
	x-	y-	z-	
S1	0	0	1	**C6-type:** (AγB)□(AγB)□ **2 H$_1$-type:** (AγB)□(BγA)□
S2	1/3	2/3	1	**C19-type:** (AγB)□(CβA)□(BαC)□ **2H$_1$-type:** (AγB)□(AβC)□(BαC)□(BγA)□(CβA) □(CαB)□
S3	−1/3	−2/3	1	**3R-type:** (AγB)□(BαC)□(CβA)□
S4	0	0	≈ 0.82	**Intercalation-type:** (AγB)a(BαC)b(CβA)c
S5	−1/3	−2/3	≈ 0.82	

The possible stacking orders of the brucite-type layers lead to fundamentally different space group symmetries and unit cell metrics, and therefore to different diffraction patterns (Fig. 2.16).

2.7.1.3 Diffraction effects of different faulting scenarios

In real structures of many brucite-type LDH phases, the stacking order is not homogeneous. Different types of layer stackings occur in the material, in other words, brucite-type LDH phases like β-Ni(OH)$_2$ [41] are extremely stacking faulted. Well crystalline, faultless brucite-type hydroxides can only be obtained by precipitation from very diluted salt solutions, and the synthesis procedure includes several aging and washings steps [24]. Common synthesis routes of LDH phases or mixed metal LDH phases by fast precipitation or co-precipitation [42] usually lead to highly faulted materials.

As already mentioned, XRPD patterns often indicate stacking faults by anisotropic peak broadening [43]. As an example, the measured diffraction pattern of a mixed nickel-cobalt-aluminum hydroxide (NCA precursor material) is presented in Fig. 2.17 (blue circles). In this pattern, some reflections like the 010 reflection are very sharp, while other reflections like 001 or 011 are broadened, or like the 012 reflection, are extremely broadened. The pattern can be indexed with a trigonal brucite-type unit cell (space group $P\bar{3}m1$), but any attempt to perform a Rietveld [44] refinement using the brucite-type structure leads to large misfits (Fig. 2.17 red

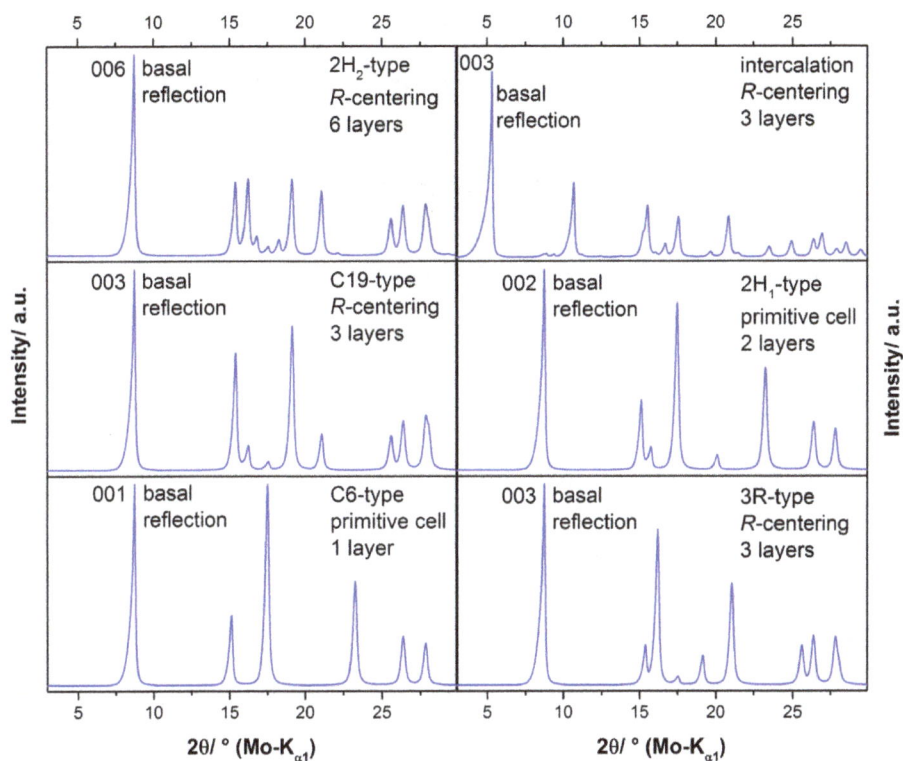

Fig. 2.16: Simulated diffraction patterns of structures composed of layers of edge sharing NiO$_{6/3}$ octahedra that are stacked in different ways. For the simulations, the patterns were convoluted with a typical instrumental function of a diffractometer using Debye-Scherrer geometry and Mo-K$_{\alpha1}$ radiation.

and gray lines). Broadening of the 011 and 012 reflections indicates that the layer orientation, that is. the stacking order is faulted, and broadening of the 001 reflection points to faulting that affects the interlayer distance. The complex diffraction line shapes of stacking faulted samples can sometimes be modeled by anisotropic microstrain broadening models [45, 46] that are used to refine strongly idealized and averaged structure models of stacking faulted substances [47]. Information on the degree of faulting, however, cannot be directly extracted from these approaches.

In the following, the diffraction effects of different kinds of stacking faults on the XRPD pattern of brucite-type β-Ni(OH)$_2$ are demonstrated (Fig. 2.18). C19-type faults within the brucite-type stacking order lead to strong broadening of the 012 reflection and to broadening of the 011 reflection that is also shifted toward higher 2θ angles (Fig. 2.18a). The 011 and 012 reflections are also broadened by 2H$_2$-type faults, but the broadening occurs symmetrically (Fig. 2.18b). The 001 and 010 reflections are not affected in either faulting scenario. Randomly occurring 3R-type faults cause the same type of peak broadening, but the shifting of the 011 reflection is

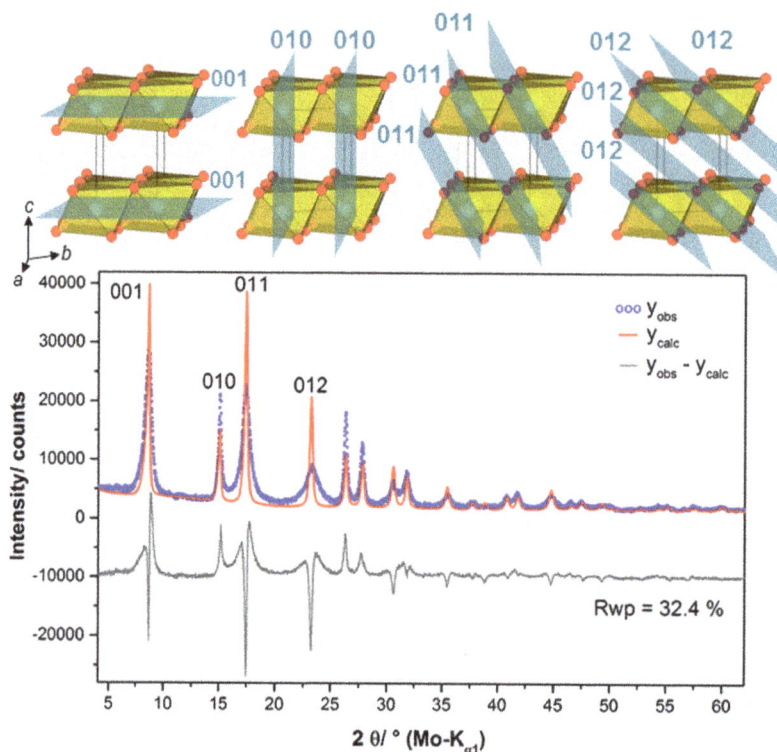

Fig. 2.17: Rietveld refinement of a NCA-precursor material (from [48]) using a brucite-type average structure. The material exhibits different kinds of stacking faults leading to broadening of most reflections. For selected reflections, the indices are given and the corresponding lattice planes are shown in the upper part.

inverted; this peak is shifted toward lower 2θ angles (Fig. 2.18c). Due to the identical stacking vectors for $2H_1$-type faults and the C6-type basic stacking order, this kind of faulting leads to less pronounced diffraction effects. It mainly leads to a decrease in the intensity of the 011 reflection. When the entire stacking order is shifted more toward the 2H-type, new reflections appear at approx. 15.8° 2θ and 20.0° 2θ for Mo-$K_{\alpha1}$ radiation (Fig. 2.18d). It should be noted that a smaller scattering contrast between anions and cations, as in $Mg(OH)_2$, leads to more pronounced diffraction effects for this kind of faulting. Interstratification between brucite-type layers strongly affects the 001 reflection (Fig. 2.18e). It leads to significant anisotropic peak broadening with tailing toward lower 2θ angles. As a result, the peak maximum is seemingly shifted toward higher 2θ angles. Additionally, the intensities of the 010, 011, and 012 reflections are affected and the last peak is also broadened. Stacking faults can also appear as a crystallographic intergrowth of differently stacked domains. In this case, peak broadening due to the domain boundaries appears to a lesser degree. However, new peaks also appear (Fig. 2.18f). Depending on the extension of the intergrown

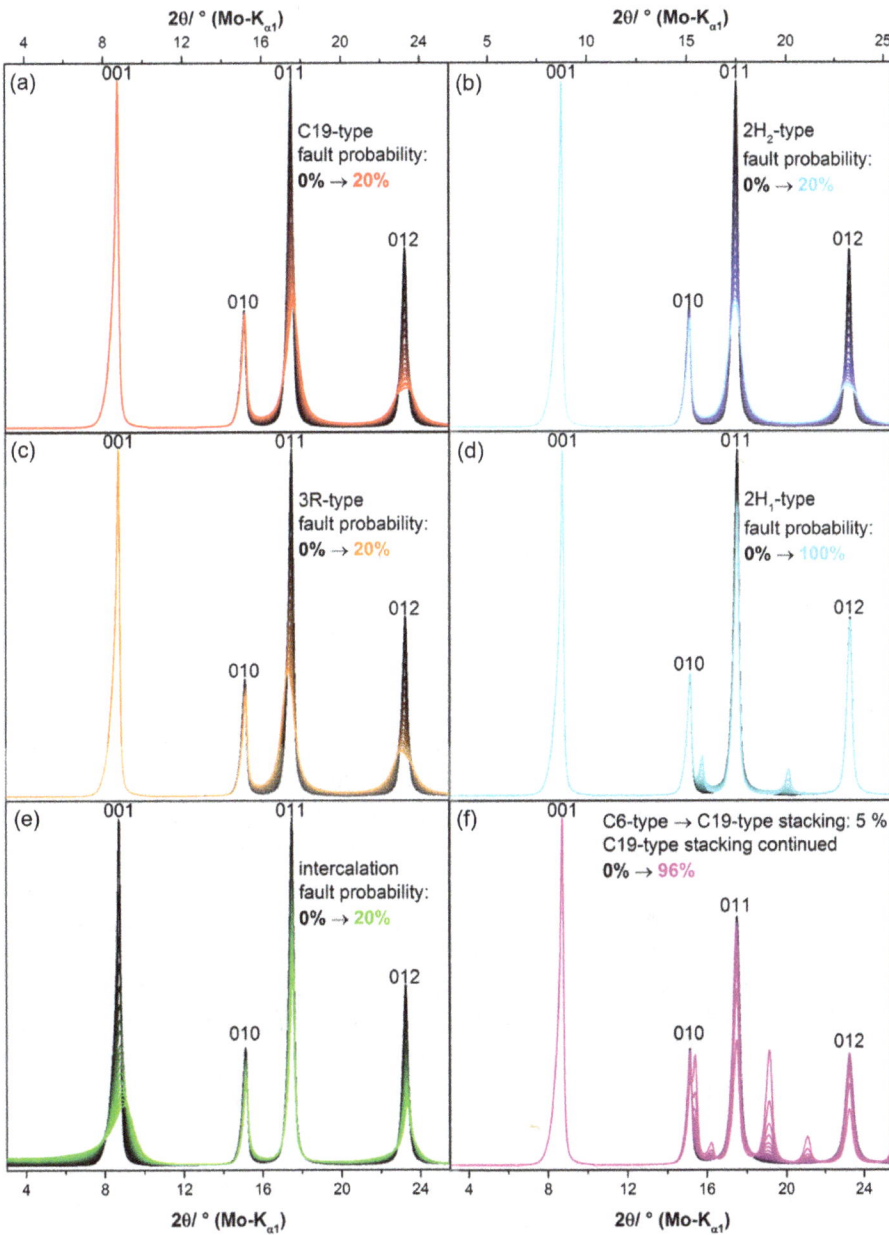

Fig. 2.18: Simulated XRPD patterns of stacking faulted β-Ni(OH)$_2$ with H-atoms omitted using different faulting scenarios. For the simulations, the patterns were convoluted with a typical instrumental function of a diffractometer using Debye-Scherrer geometry and Mo-K$_{\alpha1}$ radiation.

domains, the additional reflections can exhibit a different width. In reality, multiple types of stacking faults often occur simultaneously in different ratios [48, 49].

2.7.1.4 Implications on the determination of the crystallite size

The determination of crystallite sizes from XRPD data, for example, by using the Scherrer equation, is a common practice in materials research. Due to the pronounced plate-like morphology (Fig. 2.19a) of the LDH phases, two crystallite sizes, one referring to the height of the crystals (Fig. 2.20a (h)) and the other referring to the lateral dimensions (Fig. 2.20a (d)) are often determined, with the *aspect ratio* commonly reported as a morphological property. All these approaches lead to the determination of the extension of the coherently scattering domain. As stacking faults break the coherence in the crystal structure, the degree of faulting has to be taken into account for all crystallite size determinations.

As seen in Fig. 2.18, C19-, 3R-, 2H-, and intercalation-type faults usually do not, or only slightly, affect the width of the 010 reflection. Accordingly, this reflection may be used for an approximate estimation of the lateral dimension of the coherently scattering domain. The width of the 001 reflection, as an indicator for the vertical extension of the coherently scattering domain, is affected by the total heights of the crystallites, and also by the amount of intercalation faults, as they break the coherence in c-direction. Due to the statistical distribution of faults in an ensemble of crystallites, that is, in a powder, it is not trivial to derive a distinct value for the extension of the coherently scattering domain in the [001] direction. Fault probabilities are usually obtained as results of microstructural refinements. For given intercalation fault probabilities (Pcar), it is possible to calculate the probabilities for a certain number of layers that are stacked homogeneously, that is, to calculate the probability for certain heights of the coherently scattering domains (Fig. 2.19b). The mean size of the coherently scattering domain in the stacking direction is indicated by the number of layers that can be stacked without intercalation at a given Pcar with a probability of >50% (Fig. 2.19b, b, gray line) multiplied by the z-component of the stacking vector.

2.7.1.5 Examples

Some examples of faulting types that were observed in different brucite-type materials are given in Tab. 2.4. Most investigations were performed on β-Ni(OH)$_2$ due to its technological relevance. As seen in Fig. 2.18, different types of faulting can cause similar diffraction effects, and due to the sensitivity of these materials toward a high vacuum environment and the electron beam, usually the stacking faults cannot be directly visualized by HR-STEM. Accordingly, a microstructural analysis based on XRPD data can

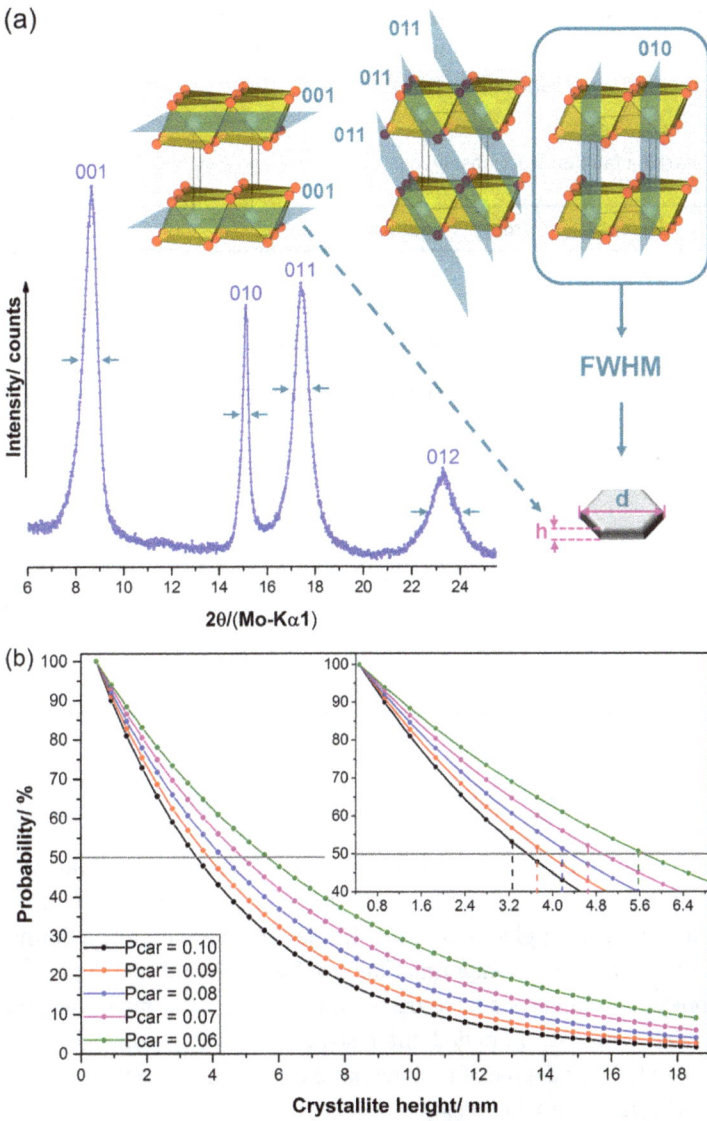

Fig. 2.19: (a) Excerpt of the measured diffraction pattern (Mo-K$_{\alpha1}$ radiation, Debye-Scherrer geometry) of a brucite-type, stacking-faulted NCA-precursor material (see Fig. 2.17) including selected reflection indices and illustration of the corresponding lattice planes. (b) Height (crystallographic c-direction) of a coherently stacked domain depending on the intercalation fault probability Pcar.

only lead to unambiguous results if all possible faulting scenarios are considered. If only a limited number of faulting types are included into the microstructural analysis, the result might be incomplete as shown for β-Ni(OH)$_2$ (Tab. 2.4).

Tab. 2.4: Examples of stacking fault disorder observed in some brucite-type materials.

Compound	Base stacking order	Faulting type	References
Mg(OH)$_2$ (brucite)	C6-type	turbostratic disorder, intercalation faults	[8]
β-Ni(OH)$_2$	C6-type	C19-type, 2H$_2$-type faults	[10, 25, 50]
		turbostratic disorder, intercalation faults	[9]
Ni$_{1-x-y}$Co$_x$Al$_y$(OH)$_{2-y}$O$_y$ (NCA-precursor)	C6-type	intercalation faults	[51]
		3R-type, C19-type, intercalation faults	[48]
NiCl(OH)	3R-type	turbostratic disorder, C19-type faults, C6-type domains	[49]
CoO(OH)	3R-type	C6-type and C19-type faults	[52]

2.7.2 Stacking faults in layered honeycomb materials

Layered honeycomb compounds can be used for various applications, for example, Li$_2$PtO$_3$ as a potential cathode material for lithium ion batteries [53] or H$_3$LiI$_2$O$_6$ as a potential material for pH-sensing electrodes [54]. The latter honeycomb compound has also attracted interest as the first experimental proof of the quantum spin liquid state [55]. Many honeycomb materials have been shown to be notoriously stacking faulted [56–59], which seriously impeded their structural characterization [47]. Moreover, it was shown that the degree of faulting has a direct impact on the material properties, like the pH-sensing capability [54].

2.7.2.1 The honeycomb lattice

The constitution of the layers in honeycomb materials is comparable to the brucite-type layer, as they consist of edge sharing M^{a}O$_6$ octahedra that form layers usually within the ab-plane. It should be noticed that the anion does not necessarily need to be oxide. There are also layered honeycomb compounds known, based on tellurides like ScSiTe$_3$ or BiSiTe$_3$. In contrast to brucite-type layers, only $2/3$ of the intra-layer cation positions are occupied by M^a-type metal cations, which yield the characteristic honeycomb motif (Fig. 2.21, a, yellow octahedra). The remaining $\frac{1}{3}$ of the intra-layer cation positions can be occupied by different metal cations, for example, M^b as in α-Na_2IrO_3 or vacant as in $SnTiO_3$ or sodium deficient $H_{3+x}Na_{1-x}Ir_2O_6$ (Fig. 2.20a, blue octahedra). Additional M^b-type metal cations can be situated in octahedral voids between the layers, leading to an (AγB)α'(CβA)γ'(BαC)β' stacking order (Fig 2.20b) that is similar to the C19-type stacking in brucite-type compounds. The honeycomb layers can be also arranged in a 3R-type stacking with an (AγB)□(BαC)□(CβA)□ stacking order that provides trigonal prismatic interlayer voids, which are usually vacant (Fig. 2.20c).

The symmetry of the honeycomb lattice can be either monoclinic (Fig. 20, a, magenta cell edges) or trigonal (Fig. 2.20, black cell edges). In the following, we will limit the discussion to trigonal lattice symmetries.

2.7.2.2 Possible stacking orders

Due to the extension of the honeycomb lattice that is considerably larger within the *ab*-plane than in the brucite-type lattice, more complex stacking orders can occur. The intrinsic three-fold symmetry of the honeycomb lattice means that the 3R-type stacking, which leads to a staggered arrangement of the honeycombs, can be realized in three different ways (Fig. 2.21) using three different stacking vectors (Tab. 2.5, S1-1, S1-2, and S1-3). In a non-distorted lattice, all three types of the 3R-type stacking order are symmetrically equivalent. This means that they lead to the same diffraction pattern (Fig. 2.23). A circular sequence of S1-1, S1-2, and S1-3 leads to the same diffraction pattern as a homogeneous stacking using the S1-1 vector. These considerations are also applicable for the C19-type stacking, which can be described by three symmetrically equivalent stacking vectors (Fig. 2.21, Tab. 2.5, S2-1, S2-2, and S2-3). As this stacking order can usually be observed when metal cations are intercalated between the layers, the interlayer distance of a honeycomb lattice that is stacked in a C19-type order is larger than the interlayer spacing of 3R-type stacked layers with empty voids between the layers. Accordingly, the basal reflection of the C19-type stacked material is shifted toward lower 2θ angles (Fig. 2.23). The C6-type stacking represents a completely eclipsed conformation of the honeycombs (Fig. 2.21), and only one stacking vector (Tab. 2.5, S3) provides this kind of stacking order. As the unit cell of a layered honeycomb material contains only one layer, the diffraction pattern exhibits considerably fewer reflections than an identical material stacked in the 3R- or C19-type (Fig. 2.22). Another possible stacking type is a completely staggered conformation of the honeycomb layers (Fig. 2.21). This can be achieved by two stacking vectors (Tab. 2.5, S4-1, and S4-2) that are symmetrically equivalent. An alternating stacking order using these stacking vectors, however, exhibits large differences in the diffraction pattern from a homogeneous, completely-staggered stacking order (Fig. 2.22).

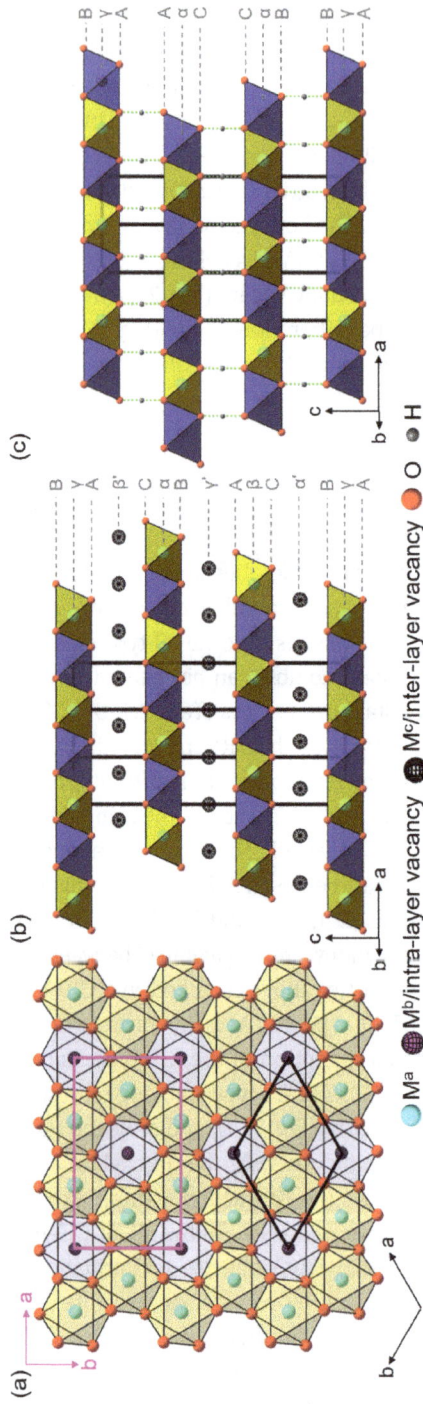

Fig. 2.20: (a) Setup of a honeycomb layer that is composed of $M^aO_{6/3}$ octahedra (yellow) and $M^bO_{6/3}$ octahedra (blue). The monoclinic layer setup is indicated by magenta unit cell edges and the trigonal layer setup by black unit cell edges. (b) C19-type stacking of the honeycomb layers is shown; additional M^b-type cations can be intercalated in the octahedral interlayer voids; (c) 3R-type stacking of the honeycomb layers that provides strong interlayer H-bonds.

○ M^a ● M^b/intra-layer vacancy ⊕ M^c/inter-layer vacancy ○ O ● H

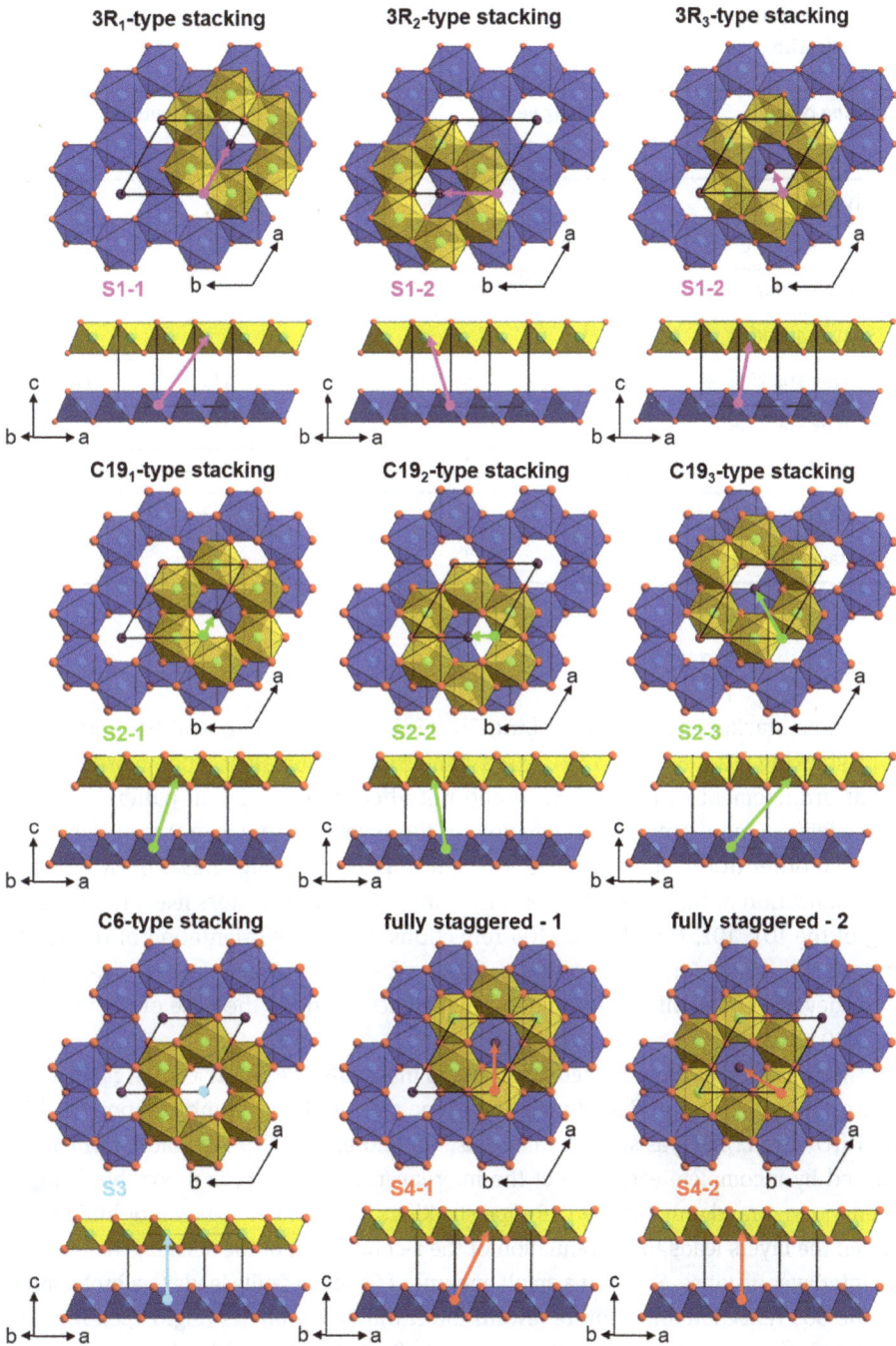

Fig. 2.21: Examples of possible stacking orders in layered honeycomb-materials. Blue polyhedra represent the bottom layer and yellow polyhedra subsequent layers. The stacking order is indicated by stacking vectors (arrows). The stacking vectors shown in this image are given in Tab. 2.5.

Tab. 2.5: Stacking vector components for different stacking sequences in layered honeycomb materials with trigonal unit cells.

Stacking type	Stacking vector	Components		
		x-	y-	z-
$3R_1$-type stacking	S1-1	2/3	0	1/3
$3R_2$-type stacking	S1-2	0	2/3	1/3
$3R_3$-type stacking	S1-3	1/3	1/3	1/3
$C19_1$-type stacking	S2-1	1/3	0	$1/3 \cdot 1.14$
$C19_2$-type stacking	S2-2	0	1/3	$1/3 \cdot 1.14$
$C19_3$-type stacking	S2-3	2/3	2/3	$1/3 \cdot 1.14$
2H-type stacking	S3	0	0	1/3
fully staggered – 1	S4-1	2/3	1/3	$1/3 \cdot 1.14$
fully staggered – 2	S4-2	1/3	2/3	$1/3 \cdot 1.14$

2.7.2.3 Diffraction effects of different types of faulting

The three stacking vectors describing 3R- (Tab. 2.5, S1-1, S1-2, and S1-3) and C19-type stacking (S2-1, S2-2, and S2-3) are symmetrically equivalent, and an ordered circular arrangement of the vectors would not affect the diffraction pattern. Despite these facts, a more random arrangement of these vectors, that is, a loss in the orientational order of the stacking vectors has a huge impact (Fig. 2.23a, b). A growing randomization of the orientation of the S1 and S2 stacking vectors leads to broadening of the $\bar{1}01$, 102, 003 and the $\bar{1}04$ reflections. When the orientation of the stacking vectors is more and more random, these reflections are completely smeared out, and they appear as diffuse, triangular-shaped scattering at the base of the 003 and the 010 reflections.

Many of the 3R-type stacked honeycomb materials like $H_3LiIr_2O_6$[54], $Ag_3LiIr_2O_6$[47], or $Ag_3LiRu_2O_6$[56] are produced using C19-type stacked honeycomb compounds like α-Li_2IrO_3 and Li_2RuO_3 as starting materials. Therefore, local C19-type faults, which are caused by incomplete exchange of the inter-layer lithium ions, is a possible faulting scenario in a randomly oriented 3R-type stacking. Random intercalation of lithium between the layers leads to a modulation of the z-component of the stacking vector, that is, interlayer distance. So, even a small amount of C19-type faults leads to a broadening of the 003 reflection and shifting toward lower diffraction angles (Fig. 2.23c). In addition, reflections like 111, $2\bar{1}2$, $2\bar{1}5$ that were not affected by the randomization of the direction of the S1-stacking vector, now become much broadened.

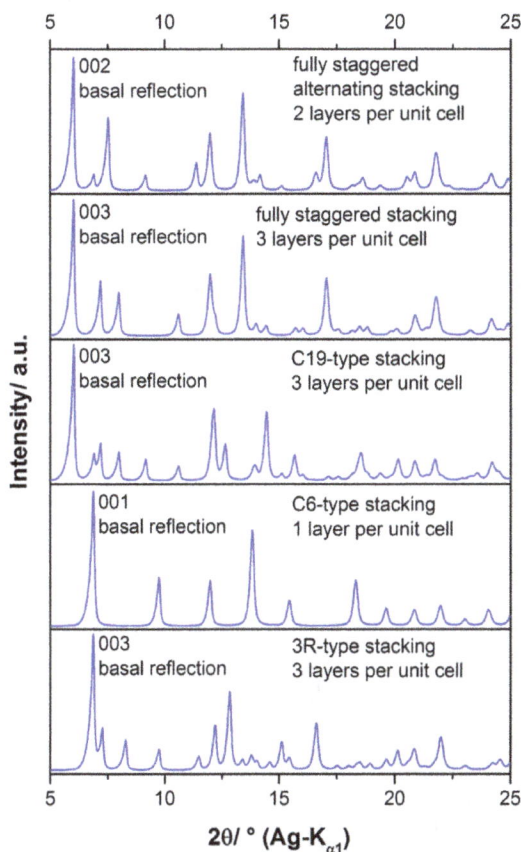

Fig. 2.22: Simulated diffraction patterns of structures composed of honeycomb-layers of edge sharing $IrO_{6/3}$ and $LiO_{6/3}$ octahedra that are stacked in different ways. For the simulations, the patterns were convoluted with a typical instrumental function of a diffractometer using Debye-Scherrer geometry and Ag-$K_{\alpha 1}$ radiation.

A complete exchange of the intra- and interlayer M^b cations of α-Li_2IrO_3, α-Na_2IrO_3, and Li_2RuO_3 most likely leads to a ruthenium or iridium oxide hydroxide ($IrO(OH)_2$ or $RuO(OH)_2$) with a C6-type stacking order. Due to an incomplete protonation or a statistical distribution of the hydroxide ions within the anion substructure, the C6-type stacking order can exhibit layers stacked in a randomly oriented 3R-type fashion. This leads to a strong symmetrical broadening of the 101, 111, 012, and 112 reflections, even at a small degree of faulting (Fig. 2.23d), whereas other, non-00 l reflections like 110 are practically unaffected.

Randomized orientations of the stacking vector, in a fully staggered-stacked honeycomb compound also leads to strong peak broadening (Fig. 2.23e). Some diffraction lines like the 101, 012, or 104 reflections are completely smeared out, whereas other reflections like 110 and 113 are not affected. New, broad peaks additionally appear,

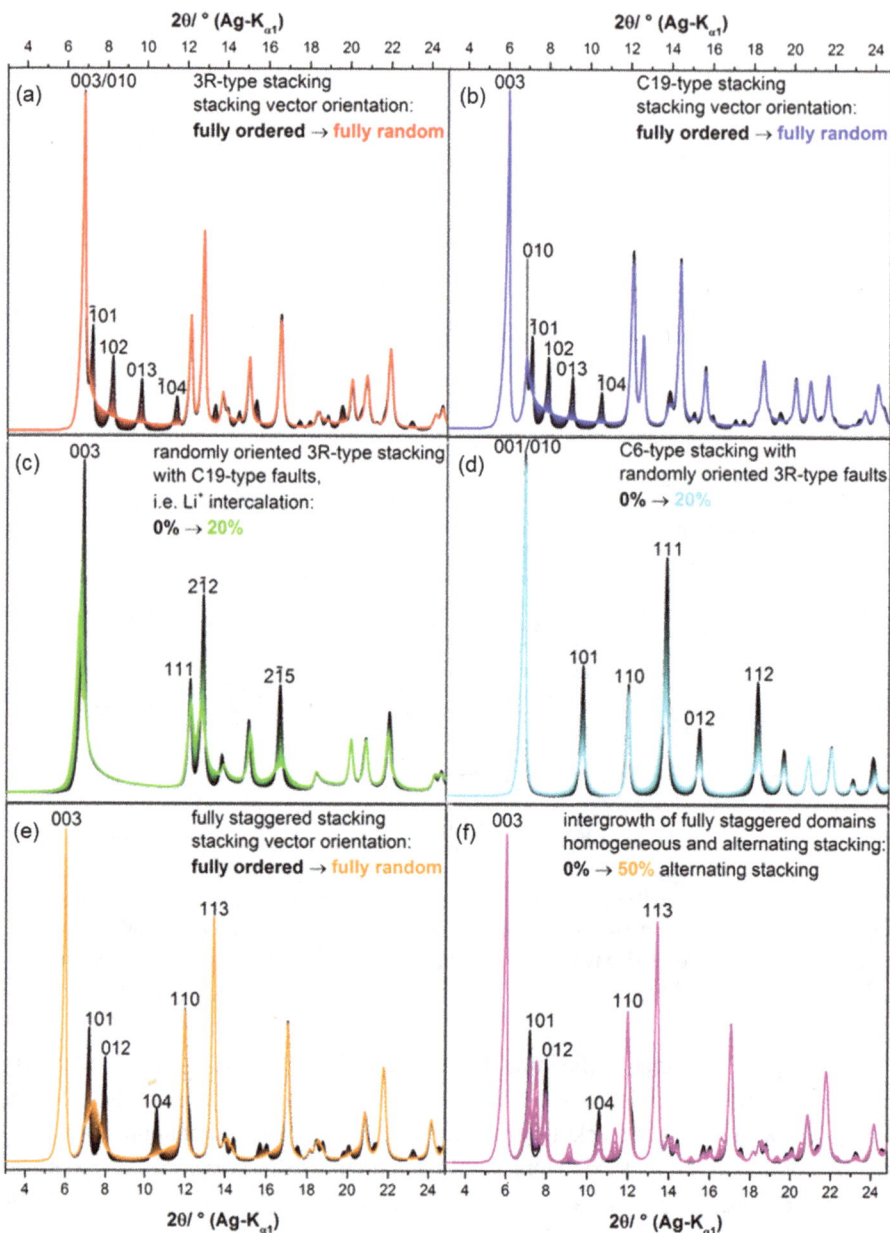

Fig. 2.23: Simulated XRPD patterns of stacking faulted layered honeycomb compounds composed of edge sharing $IrO_{6/3}$ and $LiO_{6/3}$ octahedra using different faulting scenarios. For the simulations, the patterns were convoluted with a typical instrumental function of a diffractometer using Debye-Scherrer geometry and Ag-$K_{\alpha1}$ radiation.

for example, between the original positions of the 101 and 012 reflections. This is attributed to the fact that a complete random distribution of the two stacking vectors
S4-1 and S4-2 (Tab. 2.5) leads to local domains of a honeycomb lattice stacked in an
alternating fashion. Due to the small extension of the domains, these additional diffraction lines are very broad.

A faulting scenario describing a crystallographic intergrowth of domains with homogeneously, fully-staggered stacked honeycomb layers versus finite domains with
staggered-stacked honeycomb layers that are arranged in an alternating fashion,
leads to a diffraction pattern (Fig. 2.23f) that appears to be a superposition of the
patterns of the single stacking types (Fig. 2.22). With growing extension of the domain
of the alternatingly stacked layers, the additional reflections become sharper.

2.7.2.4 Examples

Many reports on stacking-faulted layered honeycomb compounds can be found in the
literature. Some examples are given in Tab. 2.6. The most common type of faulting that
has been observed is the loss of the orientation of the stacking vector within a C19- or
3R-type stacking. As this kind of disorder leads to a significant broadening of certain

Tab. 2.6: Examples of stacking fault disorder observed in some layered honeycomb materials.

Compound	Base stacking order	Faulting type	References
Li_2MnO_3	C19-type stacking	random orientation of the stacking vector	[61]
α-Li_2IrO_3	C19-type stacking		[62]
α-Na_2IrO_3	C19-type stacking		[63]
$Na_{3-\delta}(Na_{1-x}Mg_x)Ir_2O_6$	C19-type stacking		[64]
Li_2PtO_3	C19-type stacking		[65]
$H_3LiIr_2O_6$	3R-type stacking		[60]
$Ag_3LiIr_2O_6$	3R-type stacking		[47, 59]
$Ag_3LiRh_2O_6$	3R-type stacking		[47]
$Ag_3LiRu_2O_6$	3R-type stacking		[59]
$SnTiO_3$	fully staggered	crystallographic domain-like intergrowth of fully staggered 1 and 2	[58]
$NaIrO_3$	C6-type stacking	C19-type faults	[57]
$H_{3+x}Na_{1-x}Ir_2O_6$	C6-type stacking	3R- and C19-type faults	[63]

reflections (Fig. 2.23a, b), which sometimes merge with the background, some of the honeycomb compounds were described with artificially high symmetries and small unit cells, for example, by using a complete occupational disorder of the intra-layer cations [54, 60]. Also, the domain-like crystallographic intergrowth of different kinds of a fully-staggered stacking was observed in $SnTiO_3$ [58]. The additional reflections make an *ab initio* indexing of the powder pattern impossible. Some compounds like $NaIrO_3$ or $H_{3+x}Na_{1-x}Ir_2O_6$ (Tab. 2.6) that show an overall eclipsed stacking of the honeycombs were found with 3R- and C19-type faults identified in their microstructures.

2.7.3 Synopsis

Stacking faults occur in layered compounds more often than one may think. The advances in equipment for laboratory powder diffraction provide the possibility for reliable detection of even small diffraction effects caused by faulting (i.e. line broadenings, peak asymmetry, and low-intensity superstructure reflections). In previous works, stacking faults have often been ignored or compensated by meaningless corrections during structural analyses. Simulations of diffraction patterns and local and global optimization of microstructural parameters by using either recursive or supercell approaches are still computationally expensive. Over the last decade, substantial increases in computational power of the hardware and speed of software algorithms have been achieved. Hence, quantitative microstructural analysis on simple layered compounds like the presented brucite-type hydroxides, or the layered honeycomb materials, can now be considered state-of-the-art. With advancing cloud computing approaches, the microstructural analyses of more complex materials of technological interest like zeolites, metal organic frameworks (MOFs), or active pharmaceutical ingredients (APIs) will likely become possible in the near future.

References

[1]   Ilyushin, A.S., Kovalchuk, M.V. The 100th anniversary of the discovery of X-ray diffraction. Crystallogr. Rep. 2012, 57, 617–627.
[2]   Kolb, U., Mugnaioli, E., Gorelik, T.E. Automated electron diffraction tomography – a new tool for nano crystal structure analysis. Cryst. Res. Technol. 2011, 46, 542–554.
[3]   Etter, M., Dinnebier, R.E. A century of powder diffraction: A brief history. Z. Anorg. Allg. Chem. 640, 3015–3028.
[4]   Ramsteiner, I.B., Schöps, A., Reichert, H., Dosch, H., Honkimäki, V., Zhong, Z., Hastings, J.B. High-energy X-ray diffuse scattering. J. Appl. Cryst. 2009, 42, 392–400.
[5]   Dinnebier, R.E., Leineweber, A., Evans, J.S.O. Rietveld Refinement: Practical Pattern Analysis using Topas 6.0 DeGruyter. Berlin, 2018.
[6]   Hulliger, F. Structural Chemistry of Layer-Type Phases. Dordrecht: Springer Netherlands, 1976.

[7]   Martin, S., Ullrich, C., Šimek, D., Martin, U., Rafaja, D. Stacking fault model of ∈-martensite and its DIFFaX implementation. J. Appl. Crystallogr. 2011, 44, 779–787.

[8]   Radha, A.V., Vishnu Kamath, P., Subbanna, G.N. Disorder in layered hydroxides: Synthesis and DIFFaX simulation studies of $Mg(OH)_2$. Mater. Res. Bull. 2003, 38, 731–740.

[9]   Ramesh, T.N., Jayashree, R.S., Kamath, P.V. Disorder in layered hydroxides: Diffax simulation of the X-ray powder diffraction patterns of nickel hydroxide. Clays Clay Miner. 2003, 51, 570–576.

[10] Ramesh, T.N., Kamath, P.V. Planar defects in layered hydroxides: Simulation and structure refinement of β-nickel hydroxide. Mater. Res. Bull. 2008, 43, 3227–3233.

[11] Warren, B.E. X-Ray Diffraction in Random Layer Lattices. Phys. Rev. 1941, 59, 693–698.

[12] Moore, D.M., Reynolds, R.C. X-ray Diffraction and the Identification and Analysis of Clay Minerals. Oxford; New York: Oxford University Press, 1997.

[13] Ufer, K., Kleeberg, R., Bergmann, J., Dohrmann, R. Rietveld refinement of disordered illite-smectite mixed-layer structures by a recursive Algorithm. II: Powder-pattern refinement and quantitative phase analysis. Clays Clay Miner. 2012, 60, 535–552.

[14] Treacy, M.M.J., Newsam, J.M., Deem, M.W. A general recursion method for calculating diffracted intensities from crystals containing planar faults. Proc. R. Soc. London, Ser. A. 1991, 433, 499–520.

[15] Casas-Cabanas, M., Reynaud, M., Rikarte, J., Horbach, P., Rodríguez-Carvajal, J. Faults: A program for refinement of structures with extended defects. J. Appl. Crystallogr. 2016, 49, 2259–2269.

[16] Coelho, A.A., Evans, J.S.O., Lewis, J.W. Averaging the intensity of many-layered structures for accurate stacking-fault analysis using Rietveld refinement. J. Appl. Crystallogr. 2016, 49, 1740–1749.

[17] Ainsworth, C.M., Lewis, J.W., Wang, C.-H., Coelho, A.A., Johnston, H.E., Brand, H.E.A., Evans, J. S.O. 3D transition metal ordering and rietveld stacking fault quantification in the new oxychalcogenides $La_2O_2Cu_{2-4x}Cd_{2x}Se_2$. Chem. Mater. 2016, 28, 3184–3195.

[18] Leoni, M., Gualtieri, A.F., Roveri, N. Simultaneous refinement of structure and microstructure of layered materials. J. Appl. Crystallogr. 2004, 37, 166–173.

[19] NEWMOD, a Computer Program for the Calculation of One-Dimensional Diffraction Patterns of Mixed-Layered Clays. 1985.

[20] Ufer, K., Roth, G., Kleeberg, R., Stanjek, H., Dohrmann, R., Bergmann, J. Description of X-ray powder pattern of turbostratically disordered layer structures with a Rietveld compatible approach. Z. Kristallogr. – Cryst. Mater. 2004, 219.

[21] Wang, X., Hart, R.D., Li, J., McDonald, R.G., Van Riessen, A. Quantitative analysis of turbostratically disordered nontronite with a supercell model calibrated by the PONKCS method. J. Appl. Crystallogr. 2012, 45, 1295–1302.

[22] Leineweber, A., Kreiner, G., Grüner, D., Dinnebier, R., Stein, F. Crystal structure, layer defects, and the origin of plastic deformability of Nb_2Co_7. Intermetallics. 2012, 25, 34–41.

[23] Gourrier, L., Deabate, S., Michel, T., Paillet, M., Hermet, P., Bantignies, J.-L., Henn, F. Characterization of unusually large "pseudo-single crystal" of β-nickel hydroxide. J. Phys. Chem. C. 2011, 115, 15067–15074.

[24] Palmer, D.A., Gamsjäger, H. Solubility measurements of crystalline β-$Ni(OH)_2$ in aqueous solution as a function of temperature and pH. J. Coord. Chem. 2010, 63, 2888–2908.

[25] Delmas, C., Tessier, C. Stacking faults in the structure of nickel hydroxide: A rationale of its high electrochemical activity. J. Mater. Chem. 1997, 7, 1439–1443.

[26] Ramesh, T.N. Crystallite size effects in stacking faulted nickel hydroxide and its electrochemical behaviour. Mater. Chem. Phys. 2009, 114, 618–623.

[27] Zigan, F., Rothbauer, R. Neutron diffraction measurements on brucite. Neues Jahrb. Mineral., Monatsh. 1967, 137–143.

[28] Petch, H.E. The hydrogen positions in portlandite, $Ca(OH)_2$, as indicated by the electron distribution. Acta Cryst. 1961, 14, 950–957.

[29] Christensen, A.N., Ollivier, G. Hydrothermal preparation and low temperature magnetic properties of $Mn(OH)_2$. Solid State Commun. 1972, 10, 609–614.

[30] Lutz, H.D., Möller, H., Schmidt, M. Lattice vibration spectra. Part LXXXII. Brucite-type hydroxides $M(OH)_2$ (M = Ca, Mn, Co, Fe, Cd) – IR and Raman spectra, neutron diffraction of Fe $(OH)_2$. J. Mol. Struct. 1994, 328, 121–132.

[31] Pertlik, F. The distortion of the hexagonal close packing of oxygen atoms in $Co(OH)_2$ compared to isotypic brucite-type structures. Monatsh. Chem. 1999, 130, 1083–1088.

[32] Kazimirov, V.Y., Smirnov, M.B., Bourgeois, L., Guerlou-Demourgues, L., Servant, L., Balagurov, A.M., Natkaniec, I., Khasanova, N.R., Antipov, E.V. Atomic structure and lattice dynamics of Ni and Mg hydroxides. Solid State Ionics. 2010, 181, 1764–1770.

[33] Delaplane, R.G., Ibers, J.A., Ferraro, J.R., Rush, J.J. Diffraction and spectroscopic studies of the cobaltic acid system $HCoC_2$–$DCoO_2$. J. Chem. Phys. 1969, 50, 1920–1927.

[34] Poltavets, V.V., Croft, M., Greenblatt, M. Charge transfer, hybridization and local inhomogeneity effects in $Na_xCoO_2 \bullet yH_2O$: An x-ray absorption spectroscopy study. Phys. Rev. B. 2006, 74, 075107–1–075108–8.

[35] Mills, S.J., Whitfield, P.S., Kampf, A.R., Wilson, S.A., Dipple, G.M., Raudsepp, M., Favreau, F. Contribution to the crystallography of hydrotalcites: The crystal structures of woodallite and takovite. J. Geosci. 2012, 57, 273–279.

[36] Allmann, R., Donnay, J.D.H. Structure of iowaite. Amer. Mineral. 1969, 54, 296–299.

[37] Allmann, R., Jepsen, H.P. Structure of hydrotalcite. Neues Jahrb. Mineral., Monatsh. 1969, 544–551.

[38] Taibi, M., Ammar, S., Jouini, N., Fiévet, F., Molinié, P., Drillon, M. Layered nickel hydroxide salts: Synthesis, characterization and magnetic behaviour in relation to the basal spacing. J. Mater. Chem. 2002, 12, 3238–3244.

[39] Taibi, M., Jouini, N., Rabu, P., Ammar, S., Fiévet, F. Lamellar nickel hydroxy-halides: Anionic exchange synthesis, structural characterization and magnetic behavior. J. Mater. Chem. C. 2014, 2, 4449–4460.

[40] Radha, A.V., Shivakumara, C., Kamath, P.V. DIFFaX simulations of stacking faults in layered double hydroxides (LDHs). Clays Clay Miner. 2005, 53, 520–527.

[41] Hall, D.S., Lockwood, D.J., Bock, C., MacDougall, B.R. Nickel hydroxides and related materials: A review of their structures, synthesis and properties. Proc. R. Soc. London, Ser. A. 2014, 471, 20140792–20140792.

[42] Kim, Y., Kim, D. Synthesis of high-density nickel cobalt aluminum hydroxide by continuous coprecipitation method. ACS Appl. Mater. Interfaces. 2012, 4, 586–589.

[43] Welberry, T.R., Butler, B.D. Interpretation of diffuse X-ray scatteringviamodels of disorder. J. Appl. Crystallogr. 1994, 27, 205–231.

[44] Rietveld, H.M. A profile refinement method for nuclear and magnetic structures. J. Appl. Crystallogr. 1969, 2, 65–71.

[45] Stephens, P.W. Phenomenological model of anisotropic peak broadening in powder diffraction. J. Appl. Crystallogr. 1999, 32, 281–289.

[46] Leineweber, A. Anisotropic diffraction-line broadening due to microstrain distribution: Parametrization opportunities. J. Appl. Crystallogr. 2006, 39, 509–518.

[47] Todorova, V., Leineweber, A., Kienle, L., Duppel, V., Jansen, M. On $AgRhO_2$, and the new quaternary delafossites $AgLi_{1/3}M_{2/3}O_2$, syntheses and analyses of real structures. J. Solid State Chem. 2011, 184, 1112–1119.

[48] Bette, S., Hinrichsen, B., Pfister, D., Dinnebier, R.E. A routine for the determination of the microstructure of stacking-faulted nickel cobalt aluminium hydroxide precursors for lithium nickel cobalt aluminium oxide battery materials. J. Appl. Crystallogr. 2020, 53, 76–87.

[49] Bette, S., Dinnebier, R.E., Freyer, D. Structure solution and refinement of stacking-faulted NiCl(OH). J. Appl. Crystallogr. 2015, 48, 1706–1718.

[50] Casas-Cabanas, M., Palacín, M.R., Rodríguez-Carvajal, J. Microstructural analysis of nickel hydroxide: Anisotropic size versus stacking faults. Powder Diffr. 2005, 20, 334–344.

[51] Ramesh, T.N. X-ray diffraction studies on the thermal decomposition mechanism of nickel hydroxide. J. Phys. Chem. B. 2009, 113, 13014–13017.

[52] Kudielka, A., Bette, S., Dinnebier, R.E., Abeykoon, M., Pietzonka, C., Harbrecht, B. Variability of composition and structural disorder of nanocrystalline CoOOH materials. J. Mater. Chem. C. 2017, 5, 2899–2909.

[53] Asakura, K., Okada, S., Arai, H., Tobishima, S.-I., Sakurai, Y. Cathode properties of layered structure Li$_2$PtO$_3$. J. Power Sources. 1999, 81–82, 388–392.

[54] O'Malley, M.J., Woodward, P.M., Verweij, H. Production and isolation of pH sensing materials by carbonate melt oxidation of iridium and platinum. J. Mater. Chem. 2012, 22, 7782.

[55] Kitagawa, K., Takayama, T., Matsumoto, Y., Kato, A., Takano, R., Kishimoto, Y., Bette, S., Dinnebier, R., Jackeli, G., Takagi, H. A spin–orbital-entangled quantum liquid on a honeycomb lattice. Nature. 2018, 554, 341–345.

[56] Kimber, S.A.J., Ling, C.D., Morris, D.J.P., Chemseddine, A., Henry, P.F., Argyriou, D.N. Interlayer tuning of electronic and magnetic properties in honeycomb ordered Ag$_3$LiRu$_2$O$_6$. J. Mater. Chem. 2010, 20, 8021.

[57] Wallace, D.C., McQueen, T.M. New honeycomb iridium(v) oxides: NaIrO3 and Sr$_3$CaIr$_2$O$_9$. Dalton Trans. 2015, 44, 20344–20351.

[58] Diehl, L., Bette, S., Pielnhofer, F., Betzler, S., Moudrakovski, I., Ozin, G.A., Dinnebier, R., Lotsch, B.V. Structure-directing lone Pairs: Synthesis and structural characterization of SnTiO$_3$. Chem. Mater. 2018, 30, 8932–8938.

[59] Bette, S., Takayama, T., Duppel, V., Poulain, A., Takagi, H., Dinnebier, R.E. Crystal structure and stacking faults in the layered honeycomb, delafossite-type materials Ag$_3$LiIr$_2$O$_6$ and Ag$_3$LiRu$_2$O$_6$. Dalton Trans. 2019, 48, 9250–9259.

[60] Bette, S., Takayama, T., Kitagawa, K., Takano, R., Takagi, H., Dinnebier, R.E. Solution of the heavily stacking faulted crystal structure of the honeycomb iridate H$_3$LiIr$_2$O$_6$. Dalton Trans. 2017, 46, 15216–15227.

[61] Bréger, J., Jiang, M., Dupré, N., Meng, Y.S., Shao-Horn, Y., Ceder, G., Grey, C.P. High-resolution X-ray diffraction, DIFFaX, NMR and first principles study of disorder in the Li$_2$MnO$_3$–Li[Ni$_{1/2}$Mn$_{1/2}$]O$_2$ solid solution. J. Solid State Chem. 2005, 178, 2575–2585.

[62] O'Malley, M.J., Verweij, H., Woodward, P.M. Structure and properties of ordered Li$_2$IrO$_3$ and Li$_2$PtO$_3$. J. Solid State Chem. 2008, 181, 1803–1809.

[63] Laha, S., Bette, T.T., Plass, M., Falling, L., Weber, D., Moudrarkovski, I., Schützendübe, P., Duppel, V., Takagi, H., Dinnebier, R.E., Vargas-Barbosa, N.M., Lotsch, B.V. 2020. Layered Iridates for the Oxygen Evolution Reaction: A Structure-Activity Study. In preparation.

[64] Wallace, D.C., Brown, C.M., McQueen, T.M. Evolution of magnetism in theNa$_{3-\delta}$(Na$_{1-x}$Mg$_x$)Ir$_2$O$_6$ series of honeycomb iridates. J. Solid State Chem. 2015, 224, 28–35.

[65] Casas-Cabanas, M., Rodríguez-Carvajal, J., Canales-Vázquez, J., Laligant, Y., Lacorre, P., Palacín, M.R. Microstructural characterisation of battery materials using powder diffraction data: DIFFaX, FAULTS and SH-FullProf approaches. J. Power Sources. 2007, 174, 414–420.

Alain Lafond, Catherine Guillot-Deudon, Michaël Paris,
Maria Teresa Caldes, and Stéphane Jobic

3 Crystal chemistry investigations on photovoltaic chalcogenides

Abstract: Chalcogenides have strong potential applications for thin-film photovoltaics due to their high light absorption coefficient and because their band gap can be easily tuned as desired via appropriate chemical substitutions. The thin-film fabrication processes lead unavoidably to local slight variations of the compositions. The ability of photovoltaic materials to accommodate such disparities in composition of the absorber is a critical point to achieve high photovoltaic conversion efficiency. This chapter sums up the results of fundamental chemical crystallographic investigations on compounds derived from Cu_2ZnSnS_4 (CZTS) and from $CuInS_2$ (CIGS). From this study, the pseudo-ternary diagrams ($Cu_2(S,Se)–Zn(S,Se)–Sn(S,Se)_2$, and $Cu_2S–In_2S_3–Ga_2S_3$) can be proposed and could serve as support to better understand the lower photovoltaic performances of sulphides compared to selenides in both systems.

Keywords: CIGS, CZTS, copper chalcogenides, ternary diagrams, X-ray diffraction, solid-state NMR

3.1 Introduction

Photovoltaics (abbreviated PV hereafter) is a generic term to describe the direct conversion of light (very often from the sun) into electricity. Basically, a PV solar cell is a *pn* junction (semiconducting materials) in which photons are absorbed and give rise to electron-hole pairs. These carriers are collected on both sides of the cell leading to a photovoltage and a photocurrent in the outside circuit, thus available to operate

Acknowledgments: The results presented in this chapter are parts of the PhD thesis of Léo Choubrac, Pierre Bais, and Angélica Thomere. The authors acknowledge them for their contributions. Many thanks also to Nicolas Barreau for fruitful discussions about the preparation of thin film CIGS solar cells.

This work was supported by different research programs, namely NovACEZ ANR-10-HABISOL-008 and ANRIEED-002-01 from the French Government. The French National Centre for Scientific Research (CNRS), the University of Nantes, the Région Pays de la Loire, the Synchrotron Soleil, the Institut Photovoltaïque d'Ile-de-France (IPVF) and the French Electricity Company (EDF) are also thanked for their financial supports.

Alain Lafond, Catherine Guillot-Deudon, Michaël Paris, Maria Teresa Caldes, and
Stéphane Jobic, Université de Nantes, CNRS, Institut des Matériaux Jean Rouxel, IMN, F-44000 Nantes, France

https://doi.org/10.1515/9783110674910-003

devices. The market is, by far, still dominated by the silicon-based technology [1] although emerging materials are of interest. Among them, copper-based chalcogenides have strong potential applications specifically for thin film solar cells (i.e., low amount of active materials) due to their high light absorption coefficient [2] and their band gap tunability [3, 4].

To date, solar cells based on $Cu(In,Ga)Se_2$ (CIGSe) and CdTe have high energy conversion efficiency [5, 6]. However, some of the chemical elements involved in these materials suffer from either scarcity or toxicity issues. Thus, compounds derived from Cu_2ZnSnS_4 (CZTS) have been proposed as alternative absorbers [7]. On the other hand, as the efficiency of a solar cell based on a single *pn* junction is intrinsically limited to about 30% (known as Shockley–Queisser limit [8]), the concept of tandem solar cell has been proposed [9] to overpass this limit. In such devices, more of the solar spectrum can be absorbed; the photons of lowest and highest energy are absorbed in the bottom cell (band gap of about 1.1 eV) and in the top cell (E_g around 1.7 eV), respectively. In this scenario, $Cu(In,Ga)S_2$ (CIGS) compounds receive a strong incentive to be used as absorber in the top cell, the bottom one being based on Si technology.

Anyway, before launching of a new product on the market, fundamental knowledge on the chemical and physical properties of materials have to be amassed. In particular, to optimize optoelectronic properties, the relationship between crystal structure (long and short ranges) and material properties have to be clarified. In this context, we have thoroughly studied CZTS and CIGS compounds through a crystal chemistry approach. Indeed, the crystal structure is the key venue to rationalize and move forward. In this chapter, we have selected two specific topics: the study of the cationic disorder in CZTS materials and the investigation of the phase equilibria in the $Cu_2S-In_2S_3-Ga_2S_3$ pseudo-ternary system. Our aim is clearly to highlight the capital importance of X-ray diffraction techniques (especially on single crystals) and nuclear magnetic resonance (NMR) spectroscopy in the in-depth characterization of solid materials.

3.2 Experimental

In the two thorough crystal chemistry investigations presented here, the use of highly pure and very well characterized samples is of high importance. Thus the experimental procedures (synthesis and chemical analyses) are key steps and are briefly described hereafter. In addition, few words are given to contextualize the inputs of the single-crystal resonant X-ray diffraction and of the solid-state NMR spectroscopy.

3.2.1 Synthesis

The studied compounds in the $Cu_2S–ZnS–SnS$ and $Cu_2S–In_2S_3–Ga_2S_3$ systems were synthesized by ceramic route, which consists of a high-temperature reaction between the precursors in evacuated quartz ampoules, namely, pure elements or binary sulfides are weighted in an appropriate ratio and pressed into pellets (when possible), and introduced in a silica tube. Once sealed under vacuum, the mixture is heated up to 750–900 °C for periods ranging from 2 to 3 days to 1 to 2 weeks. Afterward, the samples may be homogenized by grinding, pressed into pellets again and annealed at the same temperature. At this step, samples can be slow cooled or quenched in water. This synthesis route was adapted from the one proposed by Bernardini [10] and Schorr [11]. In a number of cases, iodine is added to favor crystal growth and get crystals large enough (typically 10–20 μm in each dimension) for single-crystal X-ray studies.

3.2.2 Elemental chemical analyses

The energy-dispersive X-ray (EDX) analyses were done using a scanning electron microscope (JEOL 5800 LV) with an accelerating voltage of 20 kV. As the elemental composition is a crucial point as part of this study particularly, careful analyses were systematically carried out on polished sections of the powders imbedded in epoxy. The elemental percentages were calculated using internal calibration in order to achieve both accurate and precise results. The standards were Cu metal (Cu–Kα), InAs (In–Lα), GaP (Ga–Kα), ZnS (Zn–Kα), SnO_2 (Sn–Lα), FeS_2 (S–Kα). Collected EDX data were systematically validated by checking the charge balance equilibrium with Cu^+, In^{3+}, Ga^{3+}, Zn^{2+}, Sn^{4+}, and S^{2-} ions. In the case of the CIGS derivatives, the stoichiometric compound ($Cu_{1.005(5)}In_{0.483(4)}Ga_{0.512(3)}S_{2.000(9)}$), well characterized by powder X-ray diffraction (PXRD) and electron microprobe analysis, was used as a secondary standard.

3.2.3 Conventional X-ray diffraction in laboratory

PXRD measurements were carried out in a Bruker D8 Advance diffractometer in a Bragg-Brentano geometry using monochromatic radiation Cu(K-L3, $\lambda = 1.540598$ Å) and a LynxEye detector. Powder patterns were recorded in the 8–100° 2θ range, 2θ step = 0.008°. The full pattern matching and Rietveld analyses were performed thanks to Jana2006 program [12].

For the single-crystal structure determinations, data collections were done on a Brüker Kappa CCD instrument, using graphite monochromated Mo $KL_{2,3}$ radiation (0.71073 Å) with the ρ- and ω-scan techniques. The recorded images were processed with the set of programs of Nonius. The structures were also refined using Jana2006 program.

3.2.4 Single-crystal X-ray resonant diffraction

In X-ray diffraction experiments, scattering comes from the interaction of the imping-ing light with electrons of matter. The power of such scattering by a given atom de-pends on its atomic scattering factor f which is a complex term, $f = f_o + f' + if''$. The nonresonant factor f_o is related to the atomic number of the corresponding atom and is energy independent but decreases when the scattering angle $(\sin(\theta)/\lambda)$ increases. On the contrary, the f' and f'' terms, known as the resonant factors, are quite independent on the scattering angle but hugely dependent on the X-ray photon energy near the edge of the scattered atom (see Fig. 3.1). In conventional X-ray diffraction experiment, very often the contributions of f' and f'' remain very low compared to that of f_o. In those conditions, neighbor elements in the periodic table have very close atomic scat-tering factors and appear to be roughly identical in X-ray diffraction analyses. This situation occurs in CZTS compounds for copper and zinc (identical electronic configu-rations for Zn^{2+} and Cu^+ cations). Thus, Cu/Zn distribution in the crystal structure of CZTS compounds is hard to address by conventional X-ray diffraction. In contrast, X-ray resonant diffraction (i.e., close to absorption edge of one of the constituent element) is known to enhance the contrast between elements which are close in the periodic table [13]. This method requires finely tuned photon energy which is readily available at synchrotron facilities. Let us notice that neutron diffraction could also be employed as reported by Schorr [14].

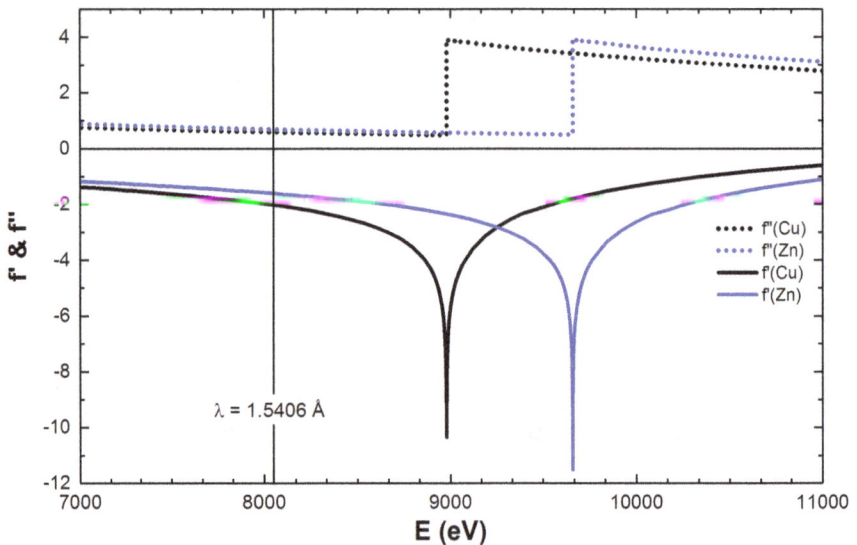

Fig. 3.1: Calculated f' and f'' scattering factors for copper and zinc close to their K absorption edge (Dispano software with Sasaki data).

In CZTS compounds, the single-crystal X-ray resonant diffraction technique has been used to investigate the copper–zinc distribution (see below). Practically, single-crystals were picked up from powders and sorted out via preliminary diffraction data collection at the laboratory to check their quality. The best ones were selected for resonant diffraction investigations on the CRISTAL beamline at the Soleil French synchrotron. In order to increase the contrast between Cu and Zn data, sets were collected just below both the Cu-K and Zn-K absorption edges, at $\lambda = 1.3825$ (2) Å and $\lambda = 1.2844$ (2) Å, respectively. An extra data set was collected far from the absorption edges at $\lambda = 0.66842$ (4) Å. All crystal structure refinements were performed using JANA2006 [12], using the anomalous scattering factor values from Sasaki data given by the Dispano program [15]. More details about these experiments can be found in the original paper [16].

3.2.5 Solid-state NMR spectroscopy

The ^{119}Sn spectra were acquired on Bruker Avance III 300 MHz (7 T) spectrometer using a 4 mm CP-MAS probe. This setup appears as a good compromise since the use of low magnetic field avoids the too large spreading of spectra due to chemical shift anisotropy. Moreover, the T_1 relaxation time, then the recycle time, which can be quite long for ½ spin nucleus in crystalline compound is lowered. Finally, the large volume of 4 mm rotor allows to reduce the number of scan while keeping a high MAS speed.

Spectra were acquired by using a full shifted echo acquisition ($\pi/2$–1.9 ms–π– acq) to obtain a pure absorption mode line shape (then, an absolute criterion for adjusting phases of the spectra). The radio frequency strength was 80 kHz. The MAS frequency was set to 14 kHz and the recycle delay to 120 s. Spectra were referenced to Me_4Sn using Ph_4Sn as a secondary reference (–121.15 ppm). Note that when line broadening occurs, the use of CPMG (Carr–Purcell–Meiboom–Gill) approach [17, 18] followed by spectral reconstruction can be beneficial to save spectrometer time [19].

3.3 Cationic disorder in CZTS materials

As aforementioned, Cu_2ZnSnS_4 (CZTS) material has been proposed as suitable absorber for thin film solar cell successively by Ito [2], Nakayama [20], and Katagiri [7]. During the last two decades, the efficiency of CZTSSe-based solar cells dramatically increased but the world record of 12.6% [21] reached a ceiling and is far below the best performance of its CIGSe counterpart [5].

Formally, several intrinsic limitations of CZTSSe materials have been found such as cationic disorder [22–24], shift from the nominal stoichiometry due to the volatility of SnS [25], and the formation of secondary phases [26] to account the far from expected PV performances.

The crystal structure of the CZTS compounds derives from that of natural kesterite, $Cu_2(Zn,Fe)SnS_4$, in the $I\bar{4}$ space group [27] for low Fe/Zn ratios. For Fe-rich compounds, an alternative stannite crystal structure can occur, corresponding to the more symmetric $I\bar{4}2\,m$ space group [27]. These two crystal structures are very similar; the main difference originates from the ordering of the metal atoms. In kesterite, the $z = 0$ and $z = 1/2$ layers are comprised of well-ordered Cu and Sn, whereas mixed (Zn,Fe) sites are found on the $z = 1/4$ and $z = 3/4$ layers (2c and 2d Wyckoff positions of the $I\bar{4}$ space group). In the alternate stannite structure, mixed (Fe, Zn) sites are found in the $z = 0$ and $z = 1/2$ layers, while copper is located on a unique crystallographic position (4d of the $I\bar{4}2m$ space group) at the $z = 1/4$ and $z = 3/4$ (see Fig. 3.2). The occurrence of both kesterite and stannite crystal structures in the case of the synthetic CZTSSe compounds has been mentioned in the literature [14, 27, 28]. In the kesterite structure, the 2c and 2d sites are very similar. Thus, a specific cationic disorder can occur where Cu and Zn atoms occupy both sites. The first experimental evidence of such Cu/Zn disorder in kesterites was obtained by S. Schorr *et al.* through neutron diffraction experiments [14].

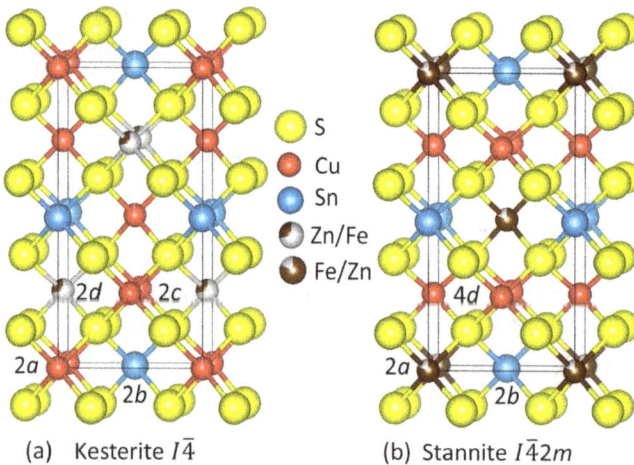

Fig. 3.2: Unit cell content for the kesterite (a) and stannite (b) minerals [27] (Wyckoff positions of metal atoms are given).

The CZTSSe materials have been thoroughly investigated through theoretical calculations specifically to determine the nature of defects (and their concentration), which could explain their semiconducting properties [29, 30]. These studies demonstrated that at the thermodynamic equilibrium for the stoichiometric Cu_2ZnSnS_4 composition, many defects in high concentration may exist in this material. Nevertheless, its p-type behavior is correlated to Cu'_{Zn} antisites, the concentration of isolated $Zn^{\bullet}_{Cu}$ donor defects being strongly limited thanks to the concomitant and competing formation of neutral [V'$_{Cu}$ + Zn$^{\bullet}$$_{Cu}$] complexes (Kröger–Vink notation [31]).

On the other hand, best performances are observed so far, for Cu-poor/Zn-rich composition, for which, on a chemical point of view, two main substitutions can occur:
- the A-type where two copper atoms are replaced by one zinc atom as aforementioned (2Cu^x$_{Cu}$ → V'$_{Cu}$ + Zn$^{\bullet}$$_{Cu}$),
- the B-type where two copper atoms and one tin atom are simultaneously replaced by three zinc atoms (2Cu^x$_{Cu}$ + Sn^x$_{Sn}$ → 2Zn$^{\bullet}$$_{Cu}$ + Zn''$_{Sn}$).

The two corresponding substitution lines are drawn in the Cu$_2$S–ZnS–SnS$_2$ pseudo-ternary diagram displayed in Fig. 3.3.

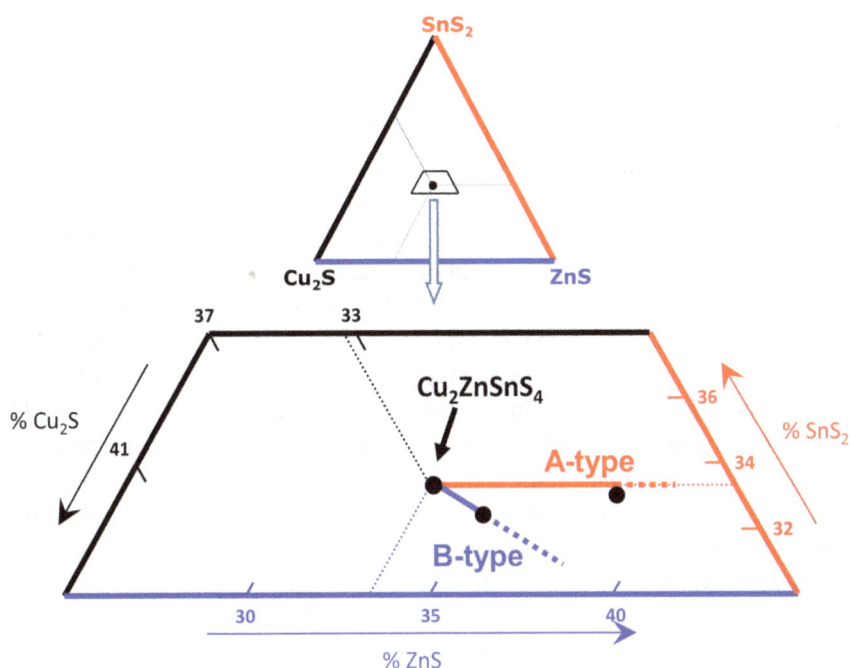

Fig. 3.3: Substitution lines, A-type and B-type, in the Cu$_2$S–ZnS–SnS$_2$ pseudo-ternary diagram. The black solid circles indicate the compositions of three samples (from EDX analyses). The two copper-poor samples are quite pure of A-type and B-type, respectively, and correspond to the experimental limits of these substitution processes.

In the following sections, we focus our attention on the experimental investigation of the relationship between the synthesis conditions and the Cu/Zn disorder in both stoichiometric and copper poor CZTS materials. For that goal, two powerful techniques, the X-ray resonant single-crystal diffraction and the solid-state NMR spectroscopy were used.

3.3.1 Single-crystal X-ray resonant diffraction

As introduced in the experimental section, conventional X-ray diffraction techniques cannot address the issue of the metal atom distribution in the crystal structure of CZTSSe compounds. By the use of neutron diffraction, Schorr et al. [14] succeeded in verifying that the correct crystal description is indeed kesterite (i.e., with two distinct crystallographic sites for copper, see Fig. 3.2). However, both experimental [32–34] and theoretical [29] studies suggested that copper and zinc can occupy both $2c$ and $2d$ crystallographic sites. Cu and Zn atoms can be fully randomly distributed leading to the so-called disordered-kesterite [35, 36].

A single crystal, picked out of the powder from a quenched sample of Cu_2ZnSnS_4, was selected for X-ray resonant diffraction. This crystal appeared to be slightly twinned, and this feature was taken into account both in the data processing and structure refinements. The diffraction data sets collected near the Cu-K edge ($\lambda = 1.3825$ Å) and Zn-K edge ($\lambda = 1.2844$ Å) were first evaluated separately. At these energies, the overall atomic scattering factors of copper and zinc are distinct enough to clearly compare the ordered cationic distribution (100% of Cu on $2c$ and 100% of Zn on $2d$) and the random distribution (50% of Cu and 50% of Zn both on $2c$ and $2d$). The two data sets were then used in combined refinements showing that the best residual factors are obtained for the fully disordered distribution, R/wR(all) = 0.048/0.112 vs R/wR(all) = 0.068/0.172 (for 605 hkl values and 21 parameters). Due to this indisputable random distribution of Cu and Zn cations, atoms located at $z = 1/4$ and $z = 3/4$ are equivalent (an average of 4 Cu/Zn atoms) leading to a higher symmetry structure. The correct space group is then definitely $I\bar{4}2m$ (instead of $I\bar{4}$) in which the four-fold site corresponds to the $4d$ Wyckoff position in the so-called disordered-kesterite structure (see Fig. 3.4). In that context, the eight sulfur atoms occupy the special position x,x,z ($8i$) instead of the general position x,y,z of the $I\bar{4}$ space group. The final refinement of this structural model with the combined data sets converged to R/R_w (all) = 0.043/0.101 values (377 unique reflections and 16 parameters).

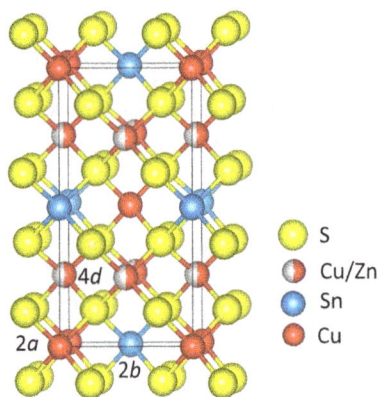

Fig. 3.4: Unit cell content for Cu_2ZnSnS_4 with disordered-kesterite structure, in space group $I\bar{4}2m$.

This study has shown that the resonant effect at Cu and Zn absorption edges is large enough to enhance the contrast between Cu and Zn and to give a direct evidence of the disordered-kesterite structure for Cu_2ZnSnS_4 in the case of a quenched sample. The same investigation was done on a sample which was very slowly cooled from 750 °C down to the room temperature, with the same results. At the long range scale (i.e., average structure obtained from diffraction experiments), Cu_2ZnSnS_4 adopts consequently the disordered-kesterite structure, regardless the thermal history of the sample. Nevertheless, there are experimental evidences showing that slow cooled samples could exhibit a lower Cu/Zn disorder [37, 38]. This prompted us to investigate the ordering at short range scale (local order) via solid-state NMR spectroscopy, indeed a very complementary technique to diffraction for that purpose.

3.3.2 Solid-state NMR

Solid-state NMR was successfully applied to structural characterization of a large variety of CZTSSe compounds [19, 36, 39], as well as related materials like Li_2ZnSnS_4 or Cu_2SnS_3 [40, 41]. The ^{65}Cu, ^{67}Zn, ^{119}Sn, and ^{77}Se nuclei have been used to probe the local structure of CZTSSe compounds. ^{65}Cu static NMR was shown to be a quick and very efficient way to identify the kesterite structure (versus the stannite one) from the spectral signature of the two crystallographic sites, $2a$ and $2c$ for Cu atoms [36, 42]. Although more demanding, ^{67}Zn MAS NMR appeared as a valuable tool to prove the presence of B-type substitutions in Cu-poor Zn-rich CZTS materials [43]. ^{77}Se MAS NMR has suggested the random distribution of S and Se in mixed S/Se compounds [39]. Finally, the ^{119}Sn MAS NMR signature of the defect complexes associated with the A-type substitution was used to show their propensity to segregate [43].

For all of the four nuclei, the modification of the local environment (geometry and chemistry) by Cu/Zn disorder leads to line broadening reflecting the distribution of chemical shift. For the two quadrupolar nuclei ^{65}Cu and ^{67}Zn, an additional broadening comes from the distribution of electric field gradient (EFG). As a non-quadrupolar nucleus, ^{119}Sn is not subject to EFG. Moreover, under MAS condition, the broadening of the ^{119}Sn NMR lines is only governed by the distribution of isotropic chemical shift. In addition, as a heavy nucleus, the chemical shift of ^{119}Sn is very sensitive to its local environment, that is, up to its second coordination sphere. Thus, ^{119}Sn appears as the nucleus of choice to probe the Cu/Zn disorder in kesterite structure.

The ^{119}Sn MAS NMR spectrum of the quenched stoichiometric Cu_2ZnSnS_4 sample (Fig. 3.5) exhibits a single line showing an unusual asymmetric broadening, which is the signature of the occurrence of Cu/Zn disorder. On the contrary, for its slowly cooled variant, the narrower line width indicates a strong reduction of the level of Cu/Zn disorder. However, even carefully slow-cooled, stoichiometric Cu_2ZnSnS_4 samples cannot reach the full Cu/Zn ordering expected at thermodynamic equilibrium as pointed out by the existence of a low intensity tail at low frequency (from −130 to −160 ppm). This situation dramatically changes for Cu-poor Zn-rich CZTS materials where A-type substitution ($2Cu^x_{Cu} \rightarrow V'_{Cu} + Zn^{\bullet}_{Cu}$) was shown to be able to decrease the level of Cu/Zn disorder. This is illustrated in Fig. 3.5, where the ^{119}Sn MAS NMR spectrum of the slow cooled Cu-poor Zn-rich CZTS sample no longer shows the low frequency tail, and where the ^{119}Sn NMR line of the quenched Cu-poor Zn-rich CZTS sample shows only moderate broadening. Using ^{119}Sn NMR characterization, the segregation of $[V'_{Cu} + Zn^{\bullet}_{Cu}]$ defect complexes and their role on Cu/Zn disorder have also been proven in pure selenide [42] or mixed S/Se CZTSSe compounds synthesized from colloidal or ceramic routes [19, 39].

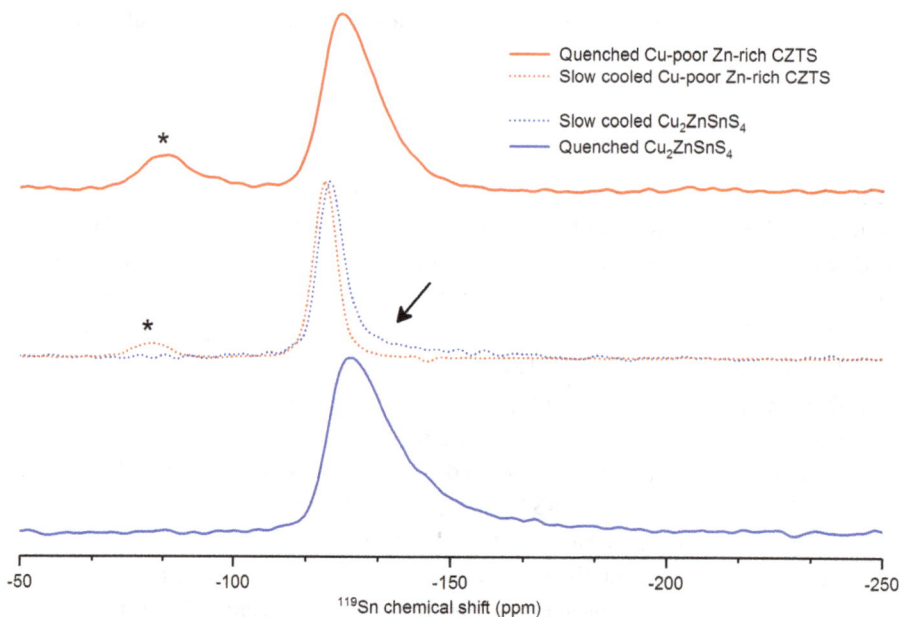

Fig. 3.5: ^{119}Sn NMR MAS (14 kHz) spectra of quenched or slow cooled samples of Cu_2ZnSnS_4 and Cu-poor Zn-rich CZTS. Asterisks indicate the signal from Sn nuclei experiencing the presence of Cu vacancies at the 2a site. The arrow points out the tail demonstrating that slow cooled Cu_2ZnSnS_4 compounds are not fully ordered.

3.3.3 Concluding remarks on CZTS

The aim of the above section was to illustrate the use of two very powerful techniques to investigate the Cu/Zn disorder in CZTS phases. For the studied samples, whatever the thermal treatment, the CZTS samples remain highly disordered at long range, that is, at the X-ray diffraction scale, while there are evidences of local cationic Cu/Zn ordering at short range, that is, at the NMR scale, this last observation being more pronounced in the slow cooled sample than the quenched one. It is worth noticing that the use of the NMR spectroscopy requires pure and well-characterized samples in a significant amount (ca. 100 mg). Anyway, NMR can help a lot in the material characterization even for thin film, for which area and thickness have to be large enough. Investigations devoted to the Cu/Zn disorder in CZTS materials are today still in progress to refine the image of disorder in these phases and their impact on properties [44–46]. In parallel, other materials for thin film solar cells are studied in Nantes (France) and other laboratories, especially materials with different requisites for application in tandem solar cells. In this context, let us consider now the $Cu_2S–In_2S_3–Ga_2S_3$ system.

3.4 Phase equilibria in $Cu_2S–In_2S_3–Ga_2S_3$ pseudo-ternary system

A promising way to improve conversion efficiencies beyond the limit for a single junction is by combining a silicon cell with a wide-bandgap absorber material in a tandem solar cell [47]. In that framework, copper indium gallium sulfide (CIGS) compounds, with the chalcopyrite crystal structure, appear to be good candidates since their band gap can be easily tuned between 1.5 and 2.4 eV by adjusting the Ga/(In + Ga) ratio. Up to now, the CIGS-based solar cells have efficiencies far below those of their selenide counterparts (CIGSe). One important feature of the selenides compared to sulfides originates from their ability to accommodate copper deficiencies without major crystal structure changes [48]. Although the CIGS compounds have been already studied through a solid-state chemistry approach [49–51] and through thermal analyses [52, 53], no systematic investigation on the copper-poor limits of the chalcopyrite phase domain is reported in the literature. Here, we present an overall study of the whole $Cu_2S–In_2S_3–Ga_2S_3$ pseudo-ternary system through a chemical crystallography approach on bulk materials.

In these compounds, the charge balance is quite well fulfilled with normal valence for the cations, that is, Cu(I), In(III), Ga(III), and S(–II). In copper-deficient compounds, the charge balance is achieved when 3 copper atoms are replaced by one trivalent atom leading to the formation of two vacancies, which is illustrated by the general formulation $Cu_{1-z}(In_{1-x}Ga_x)_{1+z/3}S_2$, where x is the Ga/(In+Ga) ratio ($0 \le x \le 1$) and z is the copper deficiency from the stoichiometric composition ($0 \le z \le 0.75$), see Fig. 3.6. For examples, the value $z = 0$ corresponds to the so-called stoichiometric $Cu(In,Ga)S_2$ compounds and the thiospinel $CuIn_5S_8$ can be represented by $Cu_{0.25}In_{1.25}S_2$, that is, $z = 0.75$.

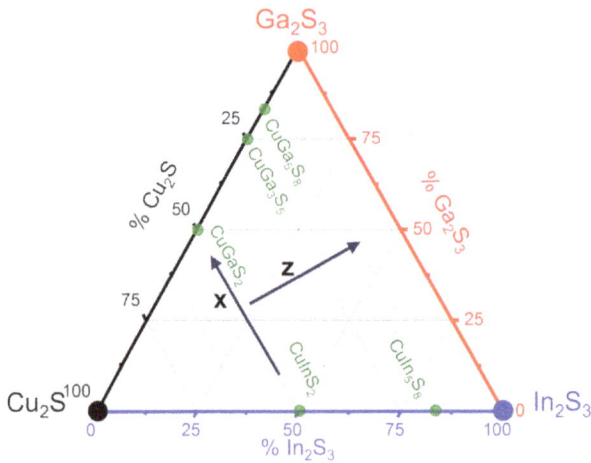

Fig. 3.6: $Cu_2S-In_2S_3-Ga_2S_3$ pseudo-ternary diagram. The general composition of the studied samples is $Cu_{1-z}(In_{1-x}Ga_x)_{1+z/3}S_2$, where x is the Ga/(In + Ga) ratio and z the copper deficiency compared to Cu(In,Ga)S_2.

3.4.1 Stoichiometric compounds Cu(In,Ga)S_2

Several papers report on the crystal structure of the stoichiometric CIGS series from PXRD analyses [50, 54, 55]. This series is claimed to be a solid solution from $CuInS_2$ ($x = 0$) to $CuGaS_2$ ($x = 1$) with the chalcopyrite structure (see Fig. 3.7). Copper and indium atoms are located on two distinct crystallographic sites and gallium progressively replaces indium with a random distribution on the $4b$ site. For the pure gallium compound, the very low difference between X-ray scattering factors of Cu and Ga atoms could prevent definitive conclusions about possible cation disorder especially from X-ray powder diffraction technique.

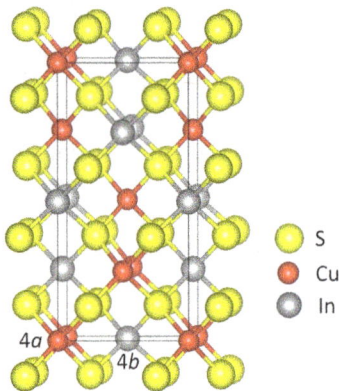

Fig. 3.7: Representation of the unit cell content of the chalcopyrite crystal structure of $CuInS_2$ (space group $I\bar{4}2d$).

We have investigated the $Cu(In_{1-x}Ga_x)S_2$ series through single crystal X-ray diffraction that is known as a powerful technique to evaluate very tiny structural modifications that can occur especially for the gallium pure derivative. For that purpose, crystals were picked from powders with nominal compositions of $x = 0$, 0.25, 0.50, and 1. Table 3.1 gives some of the refinement results for these four crystals.

From this study, it is clear that the indium-rich compounds crystallize in the chalcopyrite structure with the $I\bar{4}2d$ space group in which Cu and (In,Ga) are respectively located on $4a$ (0,0,0) and $4b$ (0,0,1/2) Wyckoff positions while sulfur atoms are on the $8d$ position ($x \approx 0.25,1/4,1/8$). In this structure, the atomic radii of copper and the trivalent metals (In,Ga) are different enough to lead to a clear segregation of both atoms on the two distinct crystallographic sites ($4a$ and $4b$). On the other hand, for the indium-free compound ($CuGaS_2$), a mixing of copper and gallium is likely due to their very similar atomic radii. If such atomic mixing would occur, the crystal structure would be modified. In the $I\bar{4}2d$ space group, the hhl reflections with $2h + l \neq 4n$ are forbidden (intensities are exactly equal to 0) due to the presence of the d glide plane (further details about systematic absences can be found, for example, in the *International Tables of Crystallography*, 5th edition, Vol. A, p. 832). Thus, the systematic absence of such reflections is the signature of the chalocopyrite structure. Special attention was paid to these reflections in the case of $CuGaS_2$. They are all found to be below the standard $I/\sigma(I) = 3$ threshold, meaning they are so weak and so noisy that they can be considered as absent reflections, confirming the $I\bar{4}2d$ space group for $CuGaS_2$ and, thus, the chalcopyrite structure. It is worth noting that such analysis cannot be done from powder diffraction pattern because the corresponding hhl reflections are still of very low intensities even if there was a structural modification from the chalcopyrite structure (see Section 3.4.3).

Tab. 3.1: Unit cell volume (V), x atomic coordinate of sulfur ($x(S)$), equivalent atomic displacement parameters (U_{eq}) for the stoichiometric $Cu(In_{1-x}Ga_x)S_2$ series.

Ga/ (Ga + In) (XRD)	V (Å³)	$x(S)$	U_{eq}(Cu) (Å²)	U_{eq}(In/Ga) (Å²)	U_{eq}(S) (Å²)	R_{obs}/ wR_{obs}	Diff. Fourier Map $\Delta\rho_{max}$, $\Delta\rho_{min}$ (e⁻/Å³)
0	339.3	0.22984(15)	0.02020(12)	0.01203(7)	0.0113(2)	1.58/4.41	0.44/−0.34
0.305(1)	326.7	0.2372(3)	0.0209(2)	0.01239(11)	0.0150(6)	3.10/5.67	0.91/−0.99
0.495(13)	319.8	0.2429(3)	0.0207(2)	0.01274(14)	0.0157(15)	3.27/ 7.14	2.12/−1.97
1	301.3	0.2556(3)	0.0162(2)	0.00964(14)	0.009(2)	1.85/4.82	1.01/−1.03

The structural refinements have been done from single-crystal XRD data in the chalcopyrite structure ($I\bar{4}2d$ space group).

When Ga is substituted for In, the unit cell volume decreases (see Fig. 3.8a), consistent with the fact that the Ga–S bonds are shorter than the In–S ones. In addition, the position of the sulfur atom is slightly shifted toward $x \approx 0.25$. Consequently, the Cu–S bond length is roughly constant (Fig. 3.8b) meaning that the bond valence of copper is also constant, very close to 1, along the series. The chalcopyrite structure is flexible enough to accommodate quite large metal(III)–S bond distance variation from 2.46 Å in CuInS$_2$ to 2.29 Å in CuGaS$_2$. It is worth noting that in these sulfides, the metal–S bond distances cannot be well evaluated from ionic radii, the concept of bond valence [56, 57] is much more suited.

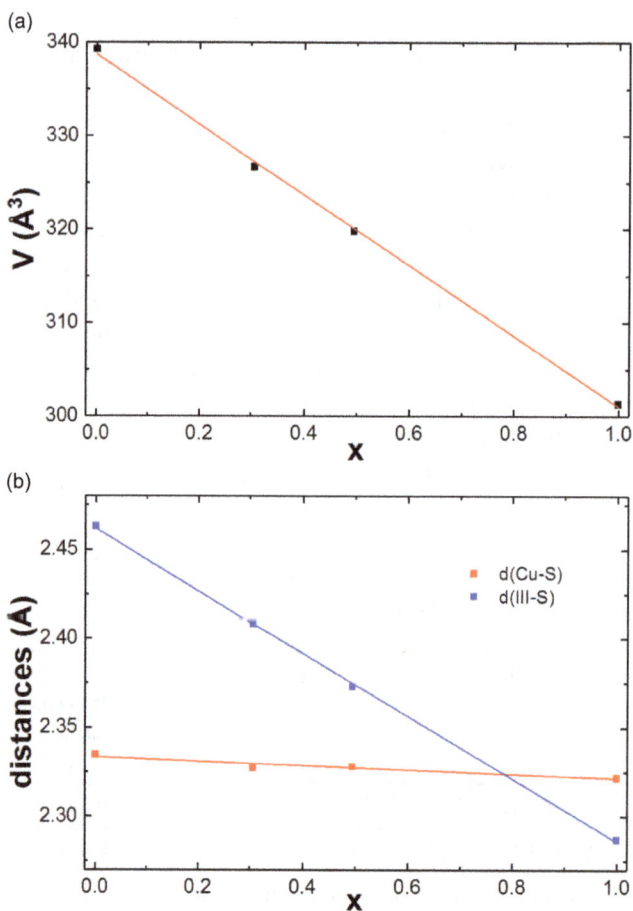

Fig. 3.8: Evolution of the unit cell volume (a) and the Cu–S bond distance (b) in the Cu(In$_{1-x}$Ga$_x$)S$_2$ series from single-crystal X-ray diffraction analyses. Note that III stands for the trivalent atoms Ga and In.

3.4.2 $CuInS_2$–In_2S_3 system

In the literature, the Cu_2S–In_2S_3 pseudo-binary system ($Cu_{1-z}In_{1+z/3}S_2$) has been well investigated. Here again, charge balance is maintained in the formula through the decrease of Cu(I) (minus z value) and the corresponding increase of In(III) (plus $z/3$ value). The first published results indicate a large two-phase region between $CuInS_2$ and $CuIn_5S_8$ [58]. A tentative temperature-composition phase diagram in the Cu_2S–In_2S_3 system, published by Binsma [49], also shows the absence of a single phase in a large composition range between $CuInS_2$ (chalcopyrite) and $CuIn_5S_8$ (thiospinel) at room temperature. This large two-phase feature has recently been confirmed and the limit of the chalcopyrite phase domain has been found in the range of $0 \leq z \leq 0.09$ (see Fig. 3.9) [59]. Additionally, slightly copper-rich (relative to $CuInS_2$) samples were shown to contain stoichiometric $CuInS_2$ and binary copper sulfides.

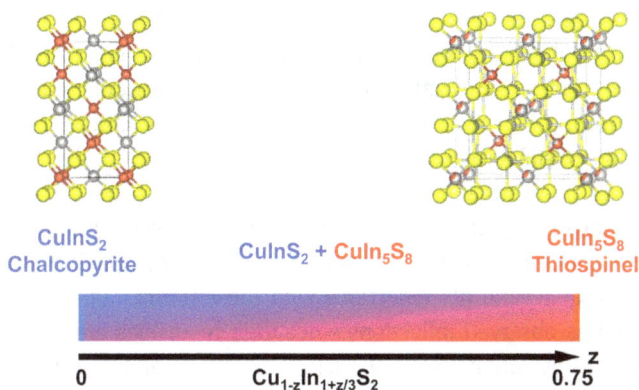

$CuInS_2$
Chalcopyrite

$CuInS_2$ + $CuIn_5S_8$

$CuIn_5S_8$
Thiospinel

0        $Cu_{1-z}In_{1+z/3}S_2$ 0.75   z

Fig. 3.9: Phase diagram on the Cu_2S–$CuIn_5S_8$ line showing the large two-phase domain between $Cu_{0.91}In_{1.03}S_2$ with the chalcopyrite structure and $CuIn_5S_8$ (equivalent to $Cu_{0.25}In_{1.25}S_2$) with the thiospinel structure.

It is worth noting that, unlike for copper-indium selenides (CISe) compounds, the specific composition of $CuIn_3X_5$ definitely does not exist for sulfides (X = S). Thus, if CIS thin films are prepared in copper-poor conditions, to avoid the formation of copper sulfide, it is very likely that the film will contain the cubic thiospinel $CuIn_5S_8$ compound. In the crystal structure of this compound, copper atoms share two distinct tetrahedral sites with a part of indium atoms ($4a$ and $4c$ sites) while the remaining indium atoms are located on the $16e$ octahedral site in the $F\bar{4}3m$ cubic space group [60]. The crystal structure difference between chalcopyrite and thiospinel gives rise to very different electronic properties. For instance, the $CuIn_5S_8$ compound is found to be an n-type semiconductor and to be detrimental to the PV application [61].

From a sample with target composition of $Cu_{0.70}In_{1.10}S_2$ (i.e., $z = 0.30$) a single crystal of the thiospinel phase was used for X-ray diffraction experiment. The obtained crystal structure is very close to that given in the literature but with a slight difference on the distribution of copper and indium atoms on the $4a$ and $4c$ tetrahedral sites ($F\bar{4}3m$ space group). The copper/indium site occupancy factors (s.o.f.) are $4a$ 0.30/0.70 and $4c$ 0.70/0.30, instead of $4a$ 0.52/0.48 and $4c$ 0.48/0.52 as published by Gastaldi [60]. This discrepancy may originate from distinguishable synthesis conditions.

3.4.3 CuGaS$_2$–Ga$_2$S$_3$ system

From the literature, it is clear that the copper-gallium sulfides (CGS) behave very differently than the copper indium ones [62–64]. First, the compounds of interest ($Cu_{1-z}Ga_{1+z/3}S_2$ with $0 \le z \le 0.75$), all adopt a tetragonal crystal structure in which copper and gallium are in tetrahedral environments (while In may accept Td and Oh chemical environments). Copper and gallium are hard to be distinguished by conventional X-ray diffraction, this feature could explain some discrepancies into the literature about the actual crystal structure of these compounds. Then, we decided to reinvestigate the CuGaS$_2$–CuGa$_5$S$_8$ system especially through the single-crystal X-ray diffraction technique. The tetragonal symmetry for all the studied crystals is confirmed. On the other hand, for a slight copper-poor CGS compound (target composition of $Cu_{0.90}Ga_{1.03}S_2$) it was found that several hhl reflections with the condition $2h + l \ne 4n$ are to be definitely classified as observed reflections since their intensities are larger than three times their standard deviations (see Tab. 3.2). Thus, the $I\bar{4}2d$ space group has to be discarded for this crystal. The structure was then refined with the stannite model, space group $I\bar{4}2m$, in which copper and gallium share the $4d$ crystallographic position. In order to ensure the charge equilibrium, the copper deficiency (mainly located on the $2a$ position) is compensated by a gallium excess on $4d$.

Tab. 3.2: Some of the strongest forbidden reflections in the $I\bar{4}2d$ space group for a crystal with composition close to $Cu_{0.90}Ga_{1.03}S_2$.

H	k	l	I	$\sigma(I)$	$I/\sigma\,(I)$
1	1	4	266.8	6.5	41.0
0	0	6	326.8	12.6	25.8
1	1	4	346.1	15.4	22.4
0	0	6	355.6	16.8	21.2
−2	2	6	223.8	11.5	19.5

These reflections cannot definitely be considered as absent reflections.

The crystal structure parameters are gathered in Tab. 3.3. The refined composition is $Cu_{0.874}Ga_{1.042}S_2$.

Tab. 3.3: Main structural parameters for a crystal with refined composition of $Cu_{0.874}Ga_{1.042}S_2$ in the $I\bar{4}2m$ space group (#121).

Atom	Wyckoff site	s.o.f.	x	y	z	$U_{eq}(Å^2)$
Cu	2a	0.839(3)	0	0	0	0.0182(2)
Ga	2b	1	0.5	0.5	0	0.00900(14)
Ga	4d	0.5419(14)	0.5	0	0.25	0.01183(15)
Cu	4d	0.455(–)	0.5	0	0.25	0.01183(–)
S	8i	1	0.74772(7)	0.25228(–)	0.12432(4)	0.00986(18)

Residual factors $R(obs)/R(all) = 2.07/3.92$, goodness of fit $= 1.15$, Fourier-difference largest peaks: $0.50/-0.69$ e$^-$/Å^3, reflections obs/all 241/368, 16 refined parameters. In order to fulfil the charge balance equilibrium, the normalized occupancies (ai) of copper and gallium are restrained according to the equation $ai[Cu(4d)] = 1-ai[Cu(2a)]-3*ai[Ga(2b)]-3*ai[(Ga(4d)]$, where $ai = $ sof × site multiplicity/general multiplicity. In the $I\bar{4}2m$ space group, the general multiplicity is 16.

For very copper-poor compositions, that is, $z = 0.60$ ($CuGa_3S_5$) and $z = 0.75$ ($CuGa_5S_8$) the powder diffraction patterns clearly show the presence of 002 and 110 reflections with their intensities increasing with copper deficiency (Fig. 3.10). The structure of these compounds has been refined with the stannite model already proposed by Maeda [63]. Additional samples have been prepared with intermediate compositions z in the range of 0.60–0.75. They are found to be of pure phases meaning that a solid solution occurs in the composition range $z = 0.60-0.75$. Because copper and gallium have very close crystallographic behavior, the Cu/Ga distribution on the cationic sites of the stannite structure can vary in a large range according the composition as soon as the charge equilibrium is achieved: $3Cu^+ \Leftrightarrow Ga^{3+} + 2V_{Cu}$.

In the case of intermediate copper-poor target compositions ($0.10 < z < 0.60$), the corresponding powder patterns can be indexed with two unit cells, close to that of $Cu_{0.90}Ga_{1.03}S_2$ ($z = 0.10$) and $Cu_{0.40}Ga_{1.20}S_2$ ($z = 0.60$), see Fig. 3.11). It is worth nothing that this two-phase feature disappears if the sample is quenched from 800 °C to room temperature at the end of the synthesis procedure.

3.4.4 Copper-poor CIGS compounds

Let us consider now the $Cu_2S-In_2S_3-Ga_2S_3$ ternary system, where CIGS samples are copper-poor and contain both indium and gallium as trivalent element, corresponding to the general formulation: $Cu_{1-z}(In_{1-x}Ga_x)_{1+z/3}S_2$, with $0 < x < 1$ and $0 < z \le 0.60$. So far, no information about these compositions is reported in the literature. From

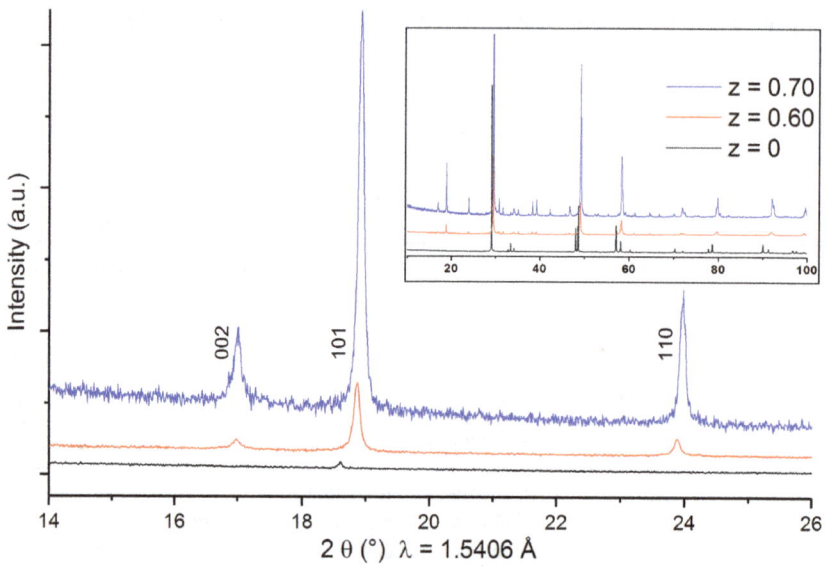

Fig. 3.10: Powder X-ray diffraction pattern of CuGaS$_2$ (with chalcopyrite structure) and two very copper-poor CGS compounds with $z = 0.60$ and 0.70 (with stannite structure). The labeled reflections are forbidden in the chalcopyrite structure.

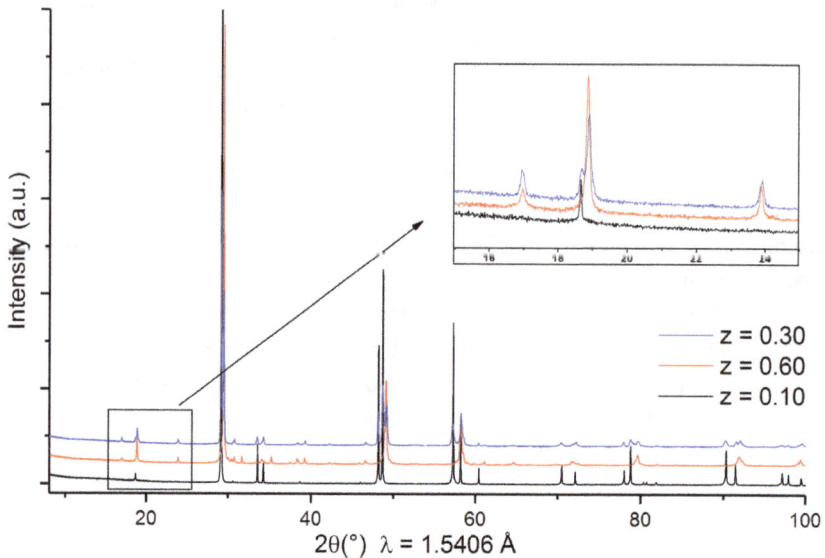

Fig. 3.11: Powder X-ray diffraction pattern of slight Cu-poor ($z = 0.10$) Cu-poor ($z = 0.30$) and very Cu-poor ($z = 0.60$) CGS samples. The two-phase nature of the sample with intermediate composition is emphasized in the low-2θ zoom (see inset).

the observation of powder-ray diffraction patterns, one can conclude that all, these prepared materials were no pure phase. In other words, as soon as the target composition is copper-poor, the obtained powdered sample is, at least, formed of two phases. Hence, for the targeted $Cu_{0.70}(In_{0.70}Ga_{0.30})_{1.10}S_2$ ($x = 0.30$, $z = 0.30$) composition, two compounds can be detected on the backscattered electron SEM image (Fig. 3.12). From the EDX analyses, the actual compositions of the two phases in equilibrium are $Cu_{0.96(4)}In_{0.61(3)}Ga_{0.39(1)}S_{2.00(2)}$ ($x = 0.39$, $z = 0.04$) and $Cu_{0.41(2)}In_{0.93(3)}Ga_{0.26(1)}S_{2.00(1)}$ ($x = 0.26$, $z = 0.59$). It is worth noticing that the almost stoichiometric $Cu(In,Ga)S2$ phase contains more (less) gallium (indium) than the copper-poor one. The two-phase nature of $Cu_{0.3}(In_{0.7}Ga_{0.3})_{1+z/3}S_2$ samples ($z = 0$, 0.21, 0.30, 0.60) was confirmed by powder XRD analyses. Figure 3.13 gives the X-ray diffraction patterns of samples with target compositions of $Cu_{0.79}(In_{0.70}Ga_{0.30})_{1.07}S_2$ ($x = 0.30$, $z = 0.21$) and $Cu_{0.70}(In_{0.70}Ga_{0.30})_{1.10}S_2$ ($x = 0.30$, $z = 0.30$) compared to those of the single phase samples with compositions close to $Cu(In_{0.70}Ga_{0.30})S_2$ (chalcopyrite crystal structure) and $Cu_{0.40}(In_{0.70}Ga_{0.30})_{1.20}S_2$ (trigonal structure, see next section). One can definitely conclude that there is a large miscibility gap for $0.04 < z < 0.60$ for mixed In/Ga CIGS compounds.

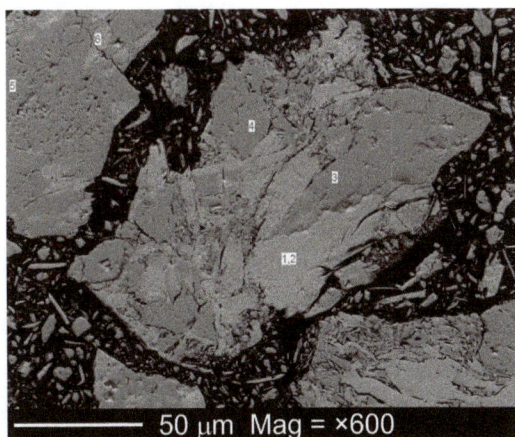

Fig. 3.12: SEM backscattered electron micrograph of sample with target composition of $x = 0.30$ and $z = 0.30$ showing the equilibrium between an almost stoichiometric phase ($Cu_{0.96}In_{0.61}Ga_{0.39}S_2$ in dark gray) and a very copper-poor one ($Cu_{0.41}In_{0.93}Ga_{0.26}S_2$ in light gray).

3.4.5 Very copper-poor In/Ga phases

As already mentioned, attempts to synthesize $Cu_{1-z}(In_{1-x}Ga_x)_{1+z/3}S_2$ compounds systematically lead to multi-phase samples for $z > 0.1$ with the formation of very copper-poor new CIGS compounds. To solve the structure of these new quaternary compounds, four suitable crystals were picked up in powdered samples with the

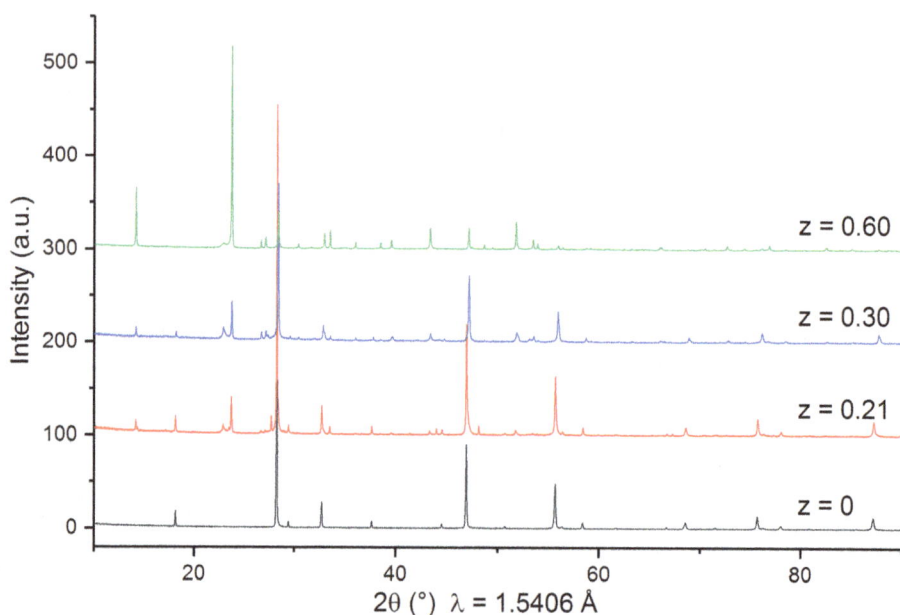

Fig. 3.13: Powder X-ray diffraction pattern of CIGS samples with target compositions of $z = 0$, 0.21, 0.30, and 0.60. The $z = 0$ pattern can be indexed in the chalcopyrite crystal structure while the $z = 0.60$ pattern corresponds to a trigonal phase. The samples with intermediate compositions are clearly a mixing of these two phases.

targeted $Cu_{0.7}(In_{0.7}Ga_{03})_{1.1}S_2$ ($z = 0.3$, $x = 0.3$), $Cu_{0.4}(In_{0.7}Ga_{0.3})_{1.2}S_2$ ($z = 0.6$, $x = 0.3$) and $Cu_{0.4}(In_{0.5}Ga_{0.5})_{1.2}S_2$ ($z = 0.6$, $x = 0.5$) compositions [65]. Four-layered structure types, depicted in Fig. 3.14, were then identified and named $CIGS_4$, $CIGS_5$, $CIGS_6$, and $CIGS_7$. In the $CIGS_n$ notations, n corresponds to the number of anionic layers of the slab. Two consecutive slabs are separated by a van der Waals gap. The EDX analyses concluded to $Cu_{0.32}In_{1.74}Ga_{0.84}S_4$ ($CIGS_4$ type), $Cu_{0.65}In_{1.75}Ga_{1.4}S_5$ ($CIGS_5$ type), $Cu_{1.44}In_{2.77}Ga_{0.76}S_6$ ($CIGS_6$ type), and $Cu_{1.1}In_{2.49}Ga_{1.8}S_7$ ($CIGS_7$ type) chemical compositions. All the structures exhibit a 2D character with layers built upon (InS_6) octahedra sharing edges on which condense on both sides one, two, or three layers made of (MS_4) tetrahedra (M = Cu, In, Ga) sharing corner. This generic formulae can then be written $(M_{(Td)})_{n-2}(In_{(Oh)})S_n$. Cell parameters for all compounds are given in Tab. 3.4.

For all structures, the octahedral sites (Oh) are exclusively and fully occupied by indium atoms. Contrarily, tetrahedral sites (Td) can be partially vacant and a mixed occupancy is often observed. Interatomic distances match with those reported in the literature for phases with similar chemical environments [49, 63, 66]. As expected, the In–S bond distances in octahedral site (~2.6 Å) are larger than the In–S bonds in tetrahedral environment (~2.4 Å) that are larger than Cu(Ga)–S ones (~2.3 Å). The presence of cationic vacancies leads systematically to a slight shortening of the mean cation-sulfur bond distances compared to the expected ones. The vacancy rate of Td

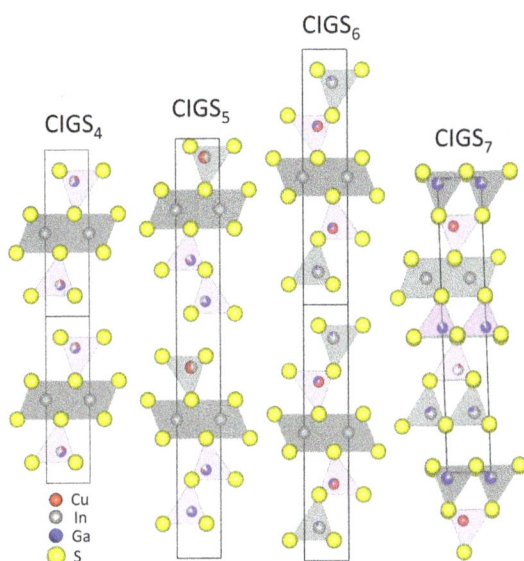

Fig. 3.14: Polyhedra representation of the structure for $Cu_{0.32}In_{1.74}Ga_{0.84}S_4$ (CIGS4) $Cu_{0.65}In_{1.75}Ga_{1.4}S_5$ (CIGS5), $Cu_{1.44}In_{2.77}Ga_{0.76}S_6$ (CIGS6), and $Cu_{1.1}In_{2.49}Ga_{1.8}S_7$ (CIGS7) compounds.

Tab. 3.4: Cell parameters for CIGS$_n$ phases.

	CIGS$_4$	CIGS$_5$	CIGS$_6$	CIGS$_7$
a (Å)	3.8203(6)	3.7898(2)	3.8506(2)	3.7933(3)
c (Å)	12.182(2)	30.664(3)	18.704(2)	21.540(14)
S G	$P\bar{3}m1$	$P6_3mc$	$P\bar{3}m1$	$P3m1$

sites never exceeds 20% except for CIGS$_7$ in which a mixed site (In–Cu) exhibits a vacancy rate of 50%. It is also worth noticing that indium atoms in tetrahedral sites are preferentially located at the van de Waals gap frontier, probably for steric reasons.

CIGS$_4$, CIGS$_5$, and CIGS$_6$ structure types with ($Td_{(Cu,In,Ga,\square)}$-$Oh_{(In)}$-$Td_{(Cu,In,Ga,\square)}$), ($Td_{(In,Ga)}$-$Td_{(Cu,Ga,\square)}$-$Oh_{(In)}$-$Td_{(Cu,In)}$) and ($Td_{(Cu,Ga)}$-$Td_{(Ga,In,\square)}$-$Oh_{(In)}$-$Td_{(Cu,Ga)}$-$Td_{(Ga,In,\square)}$) sequences are isostructural to ZnIn$_2$S$_4$-type I, Zn$_2$In$_2$S$_5$-type IIa, and Zn$_3$In$_2$S$_6$, respectively [67]. The CIGS$_5$ stacking type has been also already observed for Ag$_{1.25}$Ga$_{2.5}$In$_{3.75}$S$_{10}$ [68]. Conversely CIGS$_7$ ($Td_{(In,Ga)}$-$Td_{(Cu,In,\square)}$-$Td_{(Ga,\square)}$-$Oh_{(In)}$-$Td_{(In,Ga)}$-$Td_{(Cu)}$) exhibits an unprecedented layered stacking.

Due to the strong structural similarity in between $(M_{(Td)})_{n-2}In_{(Oh)}$ S$_n$ blocks and, on the other hand, to the 2D character of these building entities, intergrowths and stacking faults could exist. In fact, the quality of the single crystals was often poor and the proposed structural models (CIGS$_4$, CIGS$_5$, CIGS$_6$, and CIGS$_7$) should probably be considered as average structures. Due to the multiple possibilities to distribute cations over tetrahedral sites, these materials are subject to chemical composition deviations (In/Ga), which open up an avenue for the stabilization of new materials. To illustrate this purpose, in Fig. 3.15, the HAADF-STEM image of a faulted crystal and its corresponding [100] electron diffraction pattern are shown. The streaking of diffraction spots observed along c*-axis testify of a strong structural disorder. In addition, the Z-contrast image (HAADF-STEM) figures out intergrowths. The In-based octahedral layers appear as brighter spot rows and tetrahedral (MS_4) ones like gray rows. The van der Waals gap is imaged as a dark area. The number of tetrahedral layers on each side of the In-based layers can be then easily counted and the different members of the series $(M_{(Td)})_{n-2}(In_{(Oh)})S_n$ can thus be identified (see annotations in Fig. 3.15).

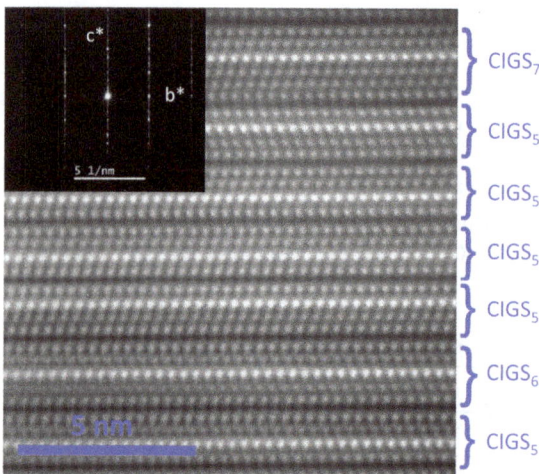

Fig. 3.15: HAADF-STEM image of a faulted crystal highlighting the propensity of $(M_{(Td)})_{n-2}(In_{(Oh)})S_n$ materials to contain intergrowth defects (the structure type, according to the $CIGS_n$ notation is given for each slab). The van der Waals gaps are imaged as dark contrast lines. The inset gives the corresponding [100] electron diffraction pattern.

3.4.6 CIGS phase diagram

Taking into account all the above presented results (combined with data from literature), we can now propose a ternary phase diagram for the $Cu_2S-In_2S_3-Ga_2S_3$ system [59] limited to the copper-poor side of the stoichiometric line $CuInS_2-CuGaS_2$ (see Fig. 3.16). As detailed above, except for the indium-free compounds (i.e., $Cu_{1-z}Ga_{1+z/3}S_2$) the tetragonal structure (either chalcopyrite or stannite) with only tetrahedral cationic environments definitely does not exist for copper-poor compositions. For such target compositions, the obtained samples are made of at least two phases: a phase close to stoichiometry $Cu(In,Ga)S_2$ is systematically in equilibrium with a very copper-poor one. The main structural feature of these very-copper poor phases is the distribution of indium both on tetrahedral and octahedral sites while Cu and Ga remain with tetrahedral coordination [65].

As already indicated, the aim of this investigation on CIGS sulfide materials was their potential application as top cell in a silicon based tandem solar cell. Although this concept is quite simple, the gain in energy conversion efficiency compared to the silicon based single junction needs to have a high efficient top cell with efficiencies close to 20% [69]. This target is very challenging taking into account that for CIGSSe material, the performance rapidly drops when the bandgap increases [70]. Even though cell efficiencies up to 10% have been achieved with pure sulfide CIGS based solar cells for quite a long time [71], the record is, nowadays, just above 15% [72], which is far below the efficiencies for the selenide CIGSe counterpart materials (>23%). We have investigated crystal chemistry issues, which can be correlated to this large difference between sulfide and selenide. The proposed ternary phase diagram definitely shows that mixed In/Ga copper-poor CIGS materials do not exist with a tetragonal symmetry (in contrast with selenide derivatives). In other words, as soon as the prepared composition is Cu-poor, the resulting sample corresponds to the equilibrium of an almost stoichiometric phase and very copper-poor phases with other crystal structures. Thus, in the case of thin films preparation, the control of the synthesis conditions has to be very precise to obtain single phase materials.

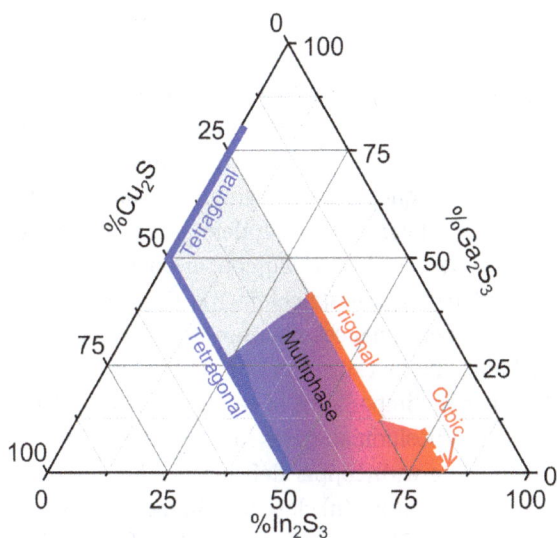

Fig. 3.16: CIGS ternary phase diagram. The single phase domains (according EDX and PXRD analyses) are indicated in blue. The hatched area indicates the polyphase domain. The high gallium content domain has not been studied, except the indium-free CGS line.

3.5 Summary

Due to the very strong dependence of properties of a material with its composition and its structural arrangement, much attention has to be devoted to its characterization. Indeed, for solid-state chemists, crystal chemistry is the essence of their work. Once the long range ordering of atoms is determined, physical characteristics can be anticipated, and change in composition may be envisioned if needed to achieve desired properties. For sure, materials contain naturally intrinsic defects, which may drastically impact their properties. These defects are in so low concentrations that they request specific techniques to be identified and quantified (out of the scope of a crystal chemistry investigation). Nevertheless, crystal X-ray analyses can be appropriate to account defect in larger concentration, specifically when crystal sites are randomly occupied. If scattering factors are quite different, regular single-crystal analyses are sufficient. On the contrary, resonant X-ray (or neutron) diffraction may be requested to distinguish ions with identical (or almost identical) electronic configuration but this requires facility access to a synchrotron radiation (or neutron) source. Nevertheless, let us mention that much information can be extracted from a careful crystal analysis with a conventional X-ray diffraction source in laboratory if characteristics of chemical species are taken into account, for example, their oxidation states, their preferential chemical environment, or their ionic radius. NMR is a powerful complementary technique to X-ray diffraction to get local information. Whatever, when both techniques are coupled, a better characterization of the materials is achievable that can be extrapolated to thin films and PV devices.

As illustrations of the above general description of the crystal chemistry approach, this chapter is dealing with the investigation of two compounds, which can be used as absorber in thin film solar cells. In the case of CZTS material, we have shown how advanced characterization techniques (resonant single-crystal X-ray diffraction and solid-state NMR spectroscopy) can help to understand the complex Cu/Zn distribution. Additionally, this study has shown that copper-poor CZTS compounds are less prone to Cu/Zn disorder than the stoichiometric Cu_2ZnSnS_4. This point can be correlated to the better PV performances obtained with such copper-poor CZTS absorbers. On the other hand, the investigation of the $Cu_2S–In_2S_3–Ga_2S_3$ system, pseudo-ternary system illustrates the importance of careful structural determinations coupled to precise chemical composition analyses in order to evaluate the ability of such compounds to accommodate with copper deficiencies. For compositions of interest for tandem solar cells (i.e., Ga/(Ga + In) close to 0.3), the single phase stability domain is shown to be very narrow. The consequence is that, for thin film preparations, the control of the synthesis conditions has to be very precise to obtain single phase materials.

According to the literature, both for CZTS and CIGS compounds, selenides lead always to better PV efficiencies than the sulfide counterparts. This fact is likely correlated to the higher structural flexibility of selenides, which can keep the tetrahedral metal coordination in a wide copper-composition range.

References

[1] Global Market Outlook 2018-2022, www.solarpowereurope.org/global-market-outlook-2018-2022/ (accessed March 16 2020).

[2] Ito, K., Nakazawa, T. Electrical and optical properties of stannite-type quaternary semiconductor thin films. Jpn. J. Appl. Phys. 1988, 27, 2094–2097.

[3] Woods-Robinson, R., Han, Y., Zhang, H., Ablekim, T., Khan, I., Persson, K., Zakutayev, A. Review of wide band gap chalcogenide semiconductors. arXiv:1910.08153 [cond-mat] 2020.

[4] Bodnar, I.V., Lukomskii, A.I. The concentration dependence of the band gap for $CuGa_xIn_{1-x}S_2$ and $AgGa_xIn_{1-x}S_2$ solid solutions. Phys. Status Solidi A. 1986, 98(2), K165–K169. Doi: https://doi.org/10.1002/pssa.2210980255.

[5] Nakamura, M., Yamaguchi, K., Kimoto, Y., Yasaki, Y., Kato, T., Sugimoto, H. Cd-Free $Cu(In,Ga)(Se,S)_2$ thin-film solar cell with record efficiency of 23.35%. IEEE J. Photovoltaics. 2019, 9(6), 1863–1867. Doi: https://doi.org/10.1109/JPHOTOV.2019.2937218.

[6] First Solar Achieves Yet Another Cell Conversion Efficiency World Record https://investor.firstsolar.com/news/press-release-details/2016/First-Solar-Achieves-Yet-Another-Cell-Conversion-Efficiency-World-Record/default.aspx (accessed Jan 31, 2020).

[7] Katagiri, H., Sasaguchi, N., Hando, S., Hoshino, S., Ohashi, J., Yokota, T. Preparation and evaluation of Cu_2ZnSnS_4 thin films by sulfurization of E-B evaporated precursors. Sol. Energy Mater. Sol. Cells. 1997, 49(1–4), 407–414. Doi: https://doi.org//10.1016/S0927-0248(97)00119-0.

[8] Shockley, W., Queisser, H.J. Detailed balance limit of efficiency of P-n junction solar cells. J. Appl. Phys. 1961, 32(3), 510–519. Doi: https://doi.org//10.1063/1.1736034.

[9] Vos, A.D. Detailed balance limit of the efficiency of tandem solar cells. J. Phys. D: Appl. Phys. 1980, 13(5), 839–846. Doi: https://doi.org//10.1088/0022-3727/13/5/018.

[10] Bernardini, G.P., Borrini, D., Caneschi, A., Di Benedetto, F., Gatteschi, D., Ristori, S., Romanelli, M. EPR and SQUID magnetometry study of Cu_2FeSnS_4 (Stannite) and Cu_2ZnSnS_4 (Kesterite). Phys. Chem. Miner. 2000, 27(7), 453–461. Doi: https://doi.org//10.1007/s002690000086.

[11] Schorr, S., Höhne, R., Wagner, G., Riede, V., Kockelmann, W. Investigation of the solid solution series $2(MnX)–CuInX_2$ (X=S, Se). J. Phys. Chem. Solids. 1966–1969, 66(11). Doi: https://doi.org//10.1016/j.jpcs.2005.09.066.

[12] Petříček, V., Dušek, M., Palatinus, L. Crystallographic computing system JANA2006: General features. Z. Kristallogr. 2014, 229(5), 345–352. Doi: https://doi.org//10.1515/zkri-2014-1737.

[13] Hodeau, J.-L., Favre-Nicolin, V., Bos, S., Renevier, H., Lorenzo, E., Berar, J.-F. Resonant diffraction. Chem. Rev. 2001, 101(6), 1843–1868. Doi: https://doi.org//10.1021/cr0000269.

[14] Schorr, S., Hoebler, H.J., Tovar, M. A neutron diffraction study of the stannite-kesterite solid solution series. Eur. J. Mineral. 2007, 19(1), 65.

[15] Laugier, J.; Bochu, B. Dispano program, part of LMGP-Suite. Suite of Programs for the Interpretation of X-Ray Experiments; ENSP/Laboratoire des Matériaux et du Génie Physique:

BP 46. 38042 Saint Martin d'Hères, France, 1999. http://mill2.chem.ucl.ac.uk/ccp/web-mir rors/lmgp-laugier-bochu/.

[16] Lafond, A., Choubrac, L., Guillot-Deudon, C., Fertey, P., Evain, M., Jobic, S. X-Ray resonant single-crystal diffraction technique, a powerful tool to investigate the kesterite structure of the photovoltaic Cu_2ZnSnS_4 compound. Acta Crystallographica Section B Structural Science, Crystal Engineering and Materials. 2014, 70(2), 390–394. Doi: https://doi.org//10.1107/S2052520614003138.

[17] Larsen, F.H., Jakobsen, H.J., Ellis, P.D., Nielsen, N.C. QCPMG-MAS NMR of half-integer quadrupolar nuclei. J. Magn. Reson. 1998, 131(1), 144–147. Doi: https://doi.org//10.1006/jmre.1997.1341.

[18] Hung, I., Gan, Z. On the practical aspects of recording wideline QCPMG NMR spectra. J. Magn. Reson. 2010, 204(2), 256–265. Doi: https://doi.org//10.1016/j.jmr.2010.03.001.

[19] Paris, M., Larramona, G., Bais, P., Bourdais, S., Lafond, A., Choné, C., Guillot-Deudon, C., Delatouche, B., Moisan, C., Dennler, G. ^{119}Sn MAS NMR to assess the cationic disorder and the anionic distribution in sulfoselenide $Cu_2ZnSn(S_xSe_{1-x})_4$ compounds prepared from colloidal and ceramic routes. J. Phys. Chem. C. 2015, 119(48), 26849–26857. Doi: https://doi.org//10.1021/acs.jpcc.5b08938.

[20] Nakayama, N., Ito, K. Sprayed films of stannite Cu_2ZnSnS_4. Appl. Surf. Sci. 1996, 92, 171–175. Doi: https://doi.org//10.1016/0169-4332(95)00225-1.

[21] Wang, W., Winkler, M.T., Gunawan, O., Gokmen, T., Todorov, T.K., Zhu, Y., Mitzi, D.B. Device characteristics of CZTSSe thin-film solar cells with 12.6% efficiency. Adv. Energy Mater. 2014, 4(7), 1301465. Doi: https://doi.org//10.1002/aenm.201301465.

[22] Rau, U., Werner, J.H. Radiative efficiency limits of solar cells with lateral band-gap fluctuations. Appl. Phys. Lett. 2004, 84(19), 3735–3737. Doi: https://doi.org//doi:10.1063/1.1737071.

[23] Scragg, J.J.S., Larsen, J.K., Kumar, M., Persson, C., Sendler, J., Siebentritt, S., Björkman, C.P. Cu–Zn disorder and band gap fluctuations in $Cu_2ZnSn(S,Se)_4$: Theoretical and experimental investigations. Phys. Status Solidi B. 2016, 253(2), 247–254. Doi: https://doi.org//10.1002/pssb.201552530.

[24] Quennet, M., Ritscher, A., Lerch, M., Paulus, B. The order-disorder transition in Cu_2ZnSnS_4: A theoretical and experimental study. J. Solid State Chem. 2017, 250, 140–144. Doi: https://doi.org//10.1016/j.jssc.2017.03.018.

[25] Scragg, J.J., Ericson, T., Kubart, T., Edoff, M., Platzer-Björkman, C. Chemical insights into the instability of Cu_2ZnSnS_4 films during annealing. Chem. Mater. 2011, 23(20), 4625–4633. Doi: https://doi.org//10.1021/cm202379s.

[26] Buffière, M., Brammertz, G., Batuk, M., Verbist, C., Mangin, D., Koble, C., Hadermann, J., Meuris, M., Poortmans, J. Microstructural analysis of 9.7% Efficient $Cu_2ZnSnSe_4$ thin film solar cells. Appl. Phys. Lett. 2014, 105(18), 183903. Doi: https://doi.org//10.1063/1.4901401.

[27] Hall, S.R., Szymanski, J.T., Stewart, J.M. Kesterite, $Cu_2(Zn,Fe)SnS_4$, and stannite, $Cu_2(Fe,Zn)SnS_4$ structurally similar but distinct minerals. Can. Mineral. 1978, 16(2), 131.

[28] Schorr, S. Crystallographic Aspects of Cu_2ZnSnS_4 (CZTS). In: Copper Zinc Tin Sulfide-Based Thin-Film Solar Cells, John Wiley & Sons, Ltd., 2015, 53–74, Doi: https://doi.org//10.1002/9781118437865.ch3.

[29] Chen, S., Yang, J.H., Gong, X.G., Walsh, A., Wei, S.H. Intrinsic point defects and complexes in the quaternary kesterite semiconductor Cu_2ZnSnS_4. Phys. Rev. B. 2010, 81(24), 245204.

[30] Kosyak, V., Postnikov, A.V., Scragg, J., Scarpulla, M.A., Platzer-Björkman, C. Calculation of point defect concentration in Cu_2ZnSnS_4: Insights into the high-temperature equilibrium and quenching. J. Appl. Phys. 2017, 122(3), 035707. Doi: https://doi.org//10.1063/1.4994689.

[31] Cox, P.A. Transition Metal Oxides: An Introduction to Their Electronic Structure and Properties. Oxford University Press, 2010.

[32] Fontané, X., Izquierdo-Roca, V., Saucedo, E., Schorr, S., Yukhymchuk, V.O., Valakh, M.Y., Pérez-Rodríguez, A., Morante, J.R. Vibrational properties of stannite and kesterite type compounds: Raman scattering analysis of $Cu_2(Fe,Zn)SnS_4$. J Alloys Compd. 2012, 539, 190–194. Doi: https://doi.org//10.1016/j.jallcom.2012.06.042.

[33] Levcenko, S., Tezlevan, V.E., Arushanov, E., Schorr, S., Unold, T. Free-to-bound recombination in near stoichiometric Cu_2ZnSnS_4 Single Crystals. Phys. Rev. B. 2012, 86(4), 045206. Doi: https://doi.org//10.1103/PhysRevB.86.045206.

[34] Valakh, M.Y., Kolomys, O.F., Ponomaryov, S.S., Yukhymchuk, V.O., Babichuk, I.S., Izquierdo-Roca, V., Saucedo, E., Perez-Rodriguez, A., Morante, J.R., Schorr, S., Bodnar, I.V. Raman scattering and disorder effect in Cu_2ZnSnS_4. Phys. Status Solidi RRL. 2013, 7(4), 258–261. Doi: https://doi.org//10.1002/pssr.201307073.

[35] Schorr, S. The crystal structure of kesterite type compounds: A neutron and X-Ray diffraction study. Sol. Energy Mater. Sol. Cells. 2011, 95(6), 1482–1488. Doi: https://doi.org//10.1016/j.solmat.2011.01.002.

[36] Choubrac, L., Paris, M., Lafond, A., Guillot-Deudon, C., Rocquefelte, X., Jobic, S. Multinuclear (^{67}Zn, ^{119}Sn and ^{65}Cu) NMR spectroscopy – an ideal technique to probe the cationic ordering in Cu_2ZnSnS_4 photovoltaic materials. Phys. Chem. Chem. Phys. 2013, 15, 10722. Doi: https://doi.org//10.1039/C3CP51320C.

[37] Scragg, J.J.S., Choubrac, L., Lafond, A., Ericson, T., Platzer-Björkman, C. A low-temperature order-disorder transition in Cu_2ZnSnS_4 thin films. Appl. Phys. Lett. 2014, 104(4), 041911. Doi: https://doi.org//10.1063/1.4863685.

[38] Timmo, K., Kauk-Kuusik, M., Pilvet, M., Raadik, T., Altosaar, M., Danilson, M., Grossberg, M., Raudoja, J., Ernits, K. Influence of order-disorder in Cu_2ZnSnS_4 powders on the performance of monograin layer solar cells. Thin Solid Films 2017, 633, 122–126. Doi: https://doi.org//10.1016/j.tsf.2016.10.017.

[39] Bais, P., Caldes, M.T., Paris, M., Guillot-Deudon, C., Fertey, P., Domengès, B., Lafond, A. Cationic and anionic disorder in CZTSSe kesterite compounds: A chemical crystallography study. Inorg. Chem. 2017, 56(19), 11779–11786. Doi: https://doi.org//10.1021/acs.inorgchem.7b01791.

[40] Lafond, A., Guillot-Deudon, C., Vidal, J., Paris, M., La, C., Jobic, S. Substitution of Li for Cu in Cu_2ZnSnS_4: Toward wide band gap absorbers with low cation disorder for thin film solar cells. Inorg. Chem. 2017, 56(5), 2712–2721. Doi: https://doi.org//10.1021/acs.inorgchem.6b02865.

[41] Pogue, E.A., Paris, M., Sutrisno, A., Lafond, A., Johnson, N., Shoemaker, D.P., Rockett, A.A. Identifying short-range disorder in crystalline bulk Cu_2SnS_3 phases: A solid-state nuclear magnetic resonance spectroscopic investigation. Chem. Mater. 2018. https://doi.org//10.1021/acs.chemmater.8b01182.

[42] Choubrac, L., Lafond, A., Paris, M., Guillot-Deudon, C., Jobic, S. The stability domain of the selenide kesterite photovoltaic materials and NMR investigation of the Cu/Zn disorder in $Cu_2ZnSnSe_4$ (CZTSe). Phys. Chem. Chem. Phys. 2015. https://doi.org//10.1039/C5CP01709B.

[43] Paris, M., Choubrac, L., Lafond, A., Guillot-Deudon, C., Jobic, S. Solid-state NMR and raman spectroscopy to address the local structure of defects and the tricky issue of the Cu/Zn disorder in Cu-Poor, Zn-Rich CZTS materials. Inorg. Chem. 2014, 53(16), 8646–8653. Doi: https://doi.org//10.1021/ic5012346.

[44] Ritscher, A., Hoelzel, M., Lerch, M. The order-disorder transition in Cu_2ZnSnS_4 – A neutron scattering investigation. J. Solid State Chem. 2016, 238, 68–73. Doi: https://doi.org//10.1016/j.jssc.2016.03.013.

[45] Rudisch, K., Ren, Y., Platzer-Björkman, C., Scragg, J. Order-disorder transition in B-Type Cu₂ZnSnS₄ and limitations of ordering through thermal treatments. Appl. Phys. Lett. 2016, 108(23), 231902. Doi: https://doi.org//10.1063/1.4953349.

[46] Kandare, S.P., Dahiwale, S.S., Dhole, S.D., Rao, M.N., Rao, R. Order-disorder transition in nano-Cu₂ZnSnS₄: A Raman spectroscopic study. Mater. Sci. Semicond. Process. 2019, 102, 104594. Doi: https://doi.org//10.1016/j.mssp.2019.104594.

[47] Essig, S., Steiner, M.A., Allebé, C., Geisz, J.F., Paviet-Salomon, B., Ward, S., Descoeudres, A., LaSalvia, V., Barraud, L., Badel, N., Faes, A., Levrat, J., Despeisse, M., Ballif, C., Stradins, P., Young, D.L. Realization of GaInP/Si dual-junction solar cells with 29.8% 1-sun efficiency. IEEE J. Photovoltaics. 2016, 6(4), 1012–1019. Doi: https://doi.org//10.1109/JPHOTOV.2016. 2549746.

[48] Souilah, M., Lafond, A., Guillot-Deudon, C., Harel, S., Evain, M. Structural investigation of the Cu₂Se–In₂Se₃–Ga₂Se₃ phase diagram, X-Ray photoemission and optical properties of the Cu₁₋ᵤ(In₀.₅Ga₀.₅)₁₊ᵤ/₃Se₂ compounds. J. Solid State Chem. 2010, 183(10), 2274–2280. Doi: https://doi.org//10.1016/j.jssc.2010.08.014.

[49] Binsma, J.J.M., Giling, L.J., Bloem, J. Phase relations in the system Cu₂S-In₂S₃. J. Cryst. Growth. 1980, 50(2), 429–436. Doi: https://doi.org//10.1016/0022-0248(80)90090-1.

[50] Yamamoto, N., Miyauchi, T. Growth of single crystals of CuGaS₂ and CuGa₁₋ₓInₓS₂ in In Solution. Jpn. J. Appl. Phys. 1972, 11(9), 1383. Doi: https://doi.org//10.1143/JJAP.11.1383.

[51] Stephan, C. Structural Trends in off Stoichiometric Chalcopyrite Type Compound Semiconductors. Germany: Freie Universität Berlin, Freie Universität Berlin, 2011.

[52] Bodnar, I.V., Bologa, A.P., Korzun, B.V., Makovetskaya, L.A. Melting temperatures of the AIBIIICVI₂-Type (AI-Cu, Ag; BIII-Al, Ga, In; CVI-S, Se) Compounds and phase diagrams of their solid solutions. Thermochim Acta. 1985, 93, 685–688. Doi: https://doi.org//10.1016/0040-6031(85)85172-8.

[53] Marushko, L.P., Romanyuk, Y.E., Piskach, L.V., Parasyuk, O.V., Olekseyuk, I.D., Volkov, S.V., Pekhnyo, V.I. The reciprocal system CuGaS₂ + CuInSe₂ ⇌ CuGaSe₂ + CuInS₂. Chem. Metals Alloys 2010, 3, 18–23.

[54] Kato, T., Hayashi, S., Kiuchi, T., Ishihara, Y., Nabetani, Y., Matsumoto, T. Structural properties of Cu(Ga₁₋ₓInₓ)ᵧSᵤ bulk alloys. J. Cryst. Growth. 2002, 237–239, 2005–2008. Doi: https://doi. org//10.1016/S0022-0248(01)02304-1.

[55] Oishi, K., Yoneda, K., Yoshida, O., Yamazaki, M., Jimbo, K., Katagiri, H., Araki, H., Kobayashi, S., Tsuboi, N. Characterization of Cu(In,Ga)S₂ crystals grown from the melt. Thin Solid Films. 2007, 515(15), 6265–6268. Doi: https://doi.org//10.1016/j.tsf.2006.12.145.

[56] Brown, I.D. Predicting bond lengths in inorganic crystals. Acta Cryst B. 1977, 33(5), 1305–1310. Doi: https://doi.org//10.1107/S0567740877005998.

[57] Brese, N.E., O'Keeffe, M. Bond-valence parameters for solids. Acta Cryst B. 1991, 47(2), 192–197. Doi: https://doi.org//10.1107/S0108768190011041.

[58] Verheijen, A.W., Giling, L.J., Bloem, J. The region of existence of CuInS₂. Mater. Res. Bull. 1979, 14(2), 237–240. Doi: https://doi.org//10.1016/0025-5408(79)90124-7.

[59] Thomere, A., Guillot-Deudon, C., Caldes, M.T., Bodeux, R., Barreau, N., Jobic, S., Lafond, A. Chemical crystallographic investigation on Cu₂S-In₂S₃-Ga₂S₃ ternary system. Thin Solid Films. 2018, 665, 46–50. Doi: https://doi.org//10.1016/j.tsf.2018.09.003.

[60] Gastaldi, L., Scaramuzza, L. Single-crystal structure analysis of the spinel copper pentaindium octasulphide. Acta Cryst B. 1980, 36(11), 2751–2753. Doi: https://doi.org//10. 1107/S0567740880009880.

[61] Orlova, N.S., Bodnar, I.V., Kudritskaya, E.A. Crystal growth and properties of the CuIn₅S₈ and AgIn₅S₈ compounds. Cryst. Res. Technol. 1998, 33(1), 37–42. Doi: https://doi.org/10.1002/ (SICI)1521-4079(1998)33:1<37::AID-CRAT37>3.0.CO;2-M.

[62] Kokta, M., Carruthers, J.R., Grasso, M., Kasper, H.M., Tell, B. Ternary phase relations in the vicinity of chalcopyrite copper gallium sulfide. J. Electron. Mater. 1976, 5(1), 69–89. Doi: https://doi.org//10.1007/BF02652887.

[63] Maeda, T., Yu, Y., Chen, Q., Ueda, K., Wada, T. Crystallographic and optical properties and band diagrams of $CuGaS_2$ and $CuGa_5S_8$ Phases in Cu-Poor $Cu_2S–Ga_2S_3$ pseudo-binary system. Jpn. J. Appl. Phys. 2017, 56(4S), 04CS12. Doi: https://doi.org//10.7567/JJAP.56.04CS12.

[64] Ueda, K., Maeda, T., Wada, T. Crystallographic and optical properties of $CuGa_3S_5$, $CuGa_3Se_5$ and $CuIn_3(S,Se)_5$ and $CuGa_3(S,Se)_5$ systems. Thin Solid Films. 2017, 633, 23–30. Doi: https://doi.org//10.1016/j.tsf.2017.01.036.

[65] Caldes, M.T., Guillot-Deudon, C., Thomere, A., Penicaud, M., Gautron, E., Boullay, P., Bujoli-Doeuff, M., Barreau, N., Jobic, S. Lafond, A. Layered Quaternary Compounds in the $Cu_2S–n_2S_3–a_2S_3$ System. Inorg. Chem. 2020, 59(7), 4546–4553. https://doi.org/10.1021/acs.inorgchem.9b03686.

[66] Paorici, C., Zanotti, L., Gastaldi, L. Preparation and structure of the $CuIn_5S_8$ single-crystalline phase. Mater. Res. Bull. 1979, 14(4), 469–472. Doi: https://doi.org//10.1016/0025-5408(79)90187-9.

[67] Haeuseler, H., Srivastava, S.K. Phase equilibria and layered phases in the Systems A_2X_3-M_2X_3-M'X (A = Ga, In; M = Trivalent Metal; M' = Divalent Metal; X = S, Se). Zeitschrift für Kristallographie – Crystalline Materials. 2000, 215(4), 205–221. Doi: https://doi.org//10.1524/zkri.2000.215.4.205.

[68] Ivashchenko, I.A., Danyliuk, I.V., Olekseyuk, I.D., Pankevych, V.Z., Halyan, V.V. Phase equilibria in the quasiternary system $Ag_2S–Ga_2S_3–In_2S_3$ and optical properties of $(Ga_{55}In_{45})_2S_{300}$, $(Ga_{54.59}In_{44.66}Er_{0.75})_2S_{300}$ single crystals. J. Solid State Chem. 2015, 227, 255–264. Doi: https://doi.org//10.1016/j.jssc.2015.04.006.

[69] Lal, N.N., White, T.P., Catchpole, K.R. Optics and light trapping for tandem solar cells on silicon. IEEE J. Photovoltaics. 2014, 4(6), 1380–1386. Doi: https://doi.org//10.1109/JPHOTOV.2014.2342491.

[70] Gloeckler, M., Sites, J.R. Efficiency limitations for wide-band-gap chalcopyrite solar cells. Thin Solid Films. 2005, 480–481, 241–245. Doi: https://doi.org//10.1016/j.tsf.2004.11.018.

[71] Kaigawa, R., Neisser, A., Klenk, R., Lux-Steiner, M.-C. Improved performance of thin film solar cells based on Cu(In,Ga)S_2. Thin Solid Films. 2002, 415(1), 266–271. Doi: https://doi.org//10.1016/S0040-6090(02)00554-0.

[72] Hiroi, H., Iwata, Y., Adachi, S., Sugimoto, H., Yamada, A. New world-record efficiency for pure-sulfide Cu(In,Ga)S_2 thin-film solar cell with cd-free buffer layer via KCN-free process. IEEE J. Photovoltaics. 2016, 6(3), 760–763. Doi: https://doi.org//10.1109/JPHOTOV.2016.2537540.

Susan Schorr and Galina Gurieva

4 Energy band gap variations in chalcogenide compound semiconductors: influence of crystal structure, structural disorder, and compositional variations

Abstract: The high efficient thin-film solar cell technologies are based on chalcogenide compound semiconductor absorber layers. The aim of this chapter is to give an overview of the crystal structure, structural disorder, and intrinsic point defects of some important ternary and quaternary chalcogenide semiconductors applied successfully in thin-film photovoltaic (PV) technologies. The experimentally determined band gap energy, an important parameter of a PV device, is correlated to structural features in these semiconductor materials demonstrating the importance of the crystal structure for material properties.

Keywords: chalcopyrites, kesterites, compound semiconductors, crystal structure, structural disorder, band gap energy

4.1 Introduction

Since the industrial revolution began in the eighteenth century, fossil fuels in the form of coal, oil, and natural gas have powered the technology and transportation networks that drive society. The world's demand for energy is projected to double by 2050, but the supply of fossil fuels is limited. Thus reliable, cost-effective, and environmentally friendly supply of energy has emerged as one of the most pressing challenges for global society.

Photovoltaics (PV), the direct conversion from sunlight into electricity, has developed into a mature technology during the recent past, and it is predicted that PV-based energy supply will develop into the multi-terawatts (TW) range. During 2018, more than 100 GW of PV energy was installed (compared to a cumulative total of 512 GW by the end of 2018), leading to a remarkable 2.9% of PV contribution to electricity demand [1]. Second-generation solar cells are based on thin films of materials, as compound semiconductors forming the absorber layer. Thin-film photovoltaic technologies based on CdTe and chalcopyrite-type $Cu(In,Ga)Se_2$ (CIGS), contributed around 3% to the total

Susan Schorr, Galina Gurieva, Helmholtz-Zentrum Berlin für Materialien u. Energie GmbH,
Department Structure and Dynamics of Energy Materials, Hahn-Meitner-Platz 1, 14109 Berlin,
Germany, susan.schorr@helmholtz-berlin.de

https://doi.org/10.1515/9783110674910-004

PV market in 2018 (corresponding to 3.5 GW) [1]. But both technologies use elements that are considered critical raw materials (CRM) by the European Commission (In, Ga, Te) [2] and, in addition, Cd is a very toxic element. Quaternary compound semiconductors, such as kesterite-type $Cu_2ZnSn(S,Se)_4$ (CZTSSe), were developed for thin-film PV solutions based on earth abundant and non-toxic elements. The materials due to their crystal structure generally named as kesterites ($Cu_2ZnSnSe_4$, Cu_2ZnSnS_4), and their solid solution ($Cu_2ZnSn(S,Se)_4$) have achieved the highest PV conversion efficiencies so far among the emerging CRM-free technologies, with values in the 11–13% range [3–5]. This CRM-free technology has a high potential, but research is needed to raise the device efficiency to such levels that it can compete with the PV market.

Thin film solar cells based on chalcopyrite-type or kesterite-type chalcogenides are made of a stack of metal and semiconductor thin films. On a substrate (sola lime glass, flexible polyimide foils, or flexible steel foils) follows a molybdenum back contact (metal), a polycrystalline absorber layer (p-type semiconductor), a buffer layer (n-type semiconductor), and ZnO front contact (n-type semiconductor).

One of the reasons for the success of CIGS-based thin-film solar cells is the remarkable flexibility of its chalcopyrite crystal structure, accepting large stoichiometry deviations. CIGS absorbers for high efficiency solar cells (record efficiency 23.35% [6]) are fabricated with an off-stoichiometric Cu-poor chemical composition. The resulting material is a p-type semiconductor material which is highly compensated. Thus acceptor- and donor-type point defects (such as vacancies, anti-sites, or interstitials) are present within the material, and p-type conductivity is established due to the exceptionally low formation energy of Cu vacancies. The structural flexibility is a key for the quaternary kesterite-type compounds too. CZTSSe-based thin-film solar cells reach the highest efficiency (record efficiency of 12.6%) [4] with a Cu-poor and Zn-rich composition of the absorber layer.

4.1.1 Crystal structure of ternary and quaternary chalcogenide semiconductors

Ternary $A^I B^{III} X_2^{VI}$ and quaternary $A_2^I B^{II} C^{IV} X_4^{VI}$ chalcogenide semiconductors, which will be considered here, belong to the adamantine family of compounds [7]. The parent structures of this family are the diamond-type and the lonsdaleite-type structure, respectively. Any compound with a crystal structure derived from one or both of these tetrahedrally bonded parents is called adamantine [8].

Pamplin [8] has derived empirical rules which are valid for adamantine phases. These state that the crystal structure can be described by two interpenetrating Bravais lattices populated with cations and anions. The metals are ordered on defined cation sites, comprising one or more unit cells of the parent structures. This led to typical superstructures with lowered symmetry.

In ternary compounds of the adamantine family, the cation sites in the Bravais lattice are occupied by two different cation species A and B resulting in the general formulae $A^{N-1}B^{N+1}X_2^{8-N}$. The both most important representatives are $A^I B^{III} X_2^{VI}$ ($N = 2$) and $A^{II} B^{IV} X_2^V$ ($N = 3$) compounds. Depending on the ionicity of the chemical bonds between A-X and B-X, they crystallize in the chalcopyrite-type crystal structure (space group $I\bar{4}2d$; see Fig. 4.1, left) or in the β-NaFeO$_2$-type structure (space group $Pna2_1$; see Fig. 4.1, right) [9]. In the chalcopyrite-type crystal structure, the A cation occupies the Wyckoff position $4a$ at $(0,0,0)$ and the B cation occupies the $4b$ position at $(0,0,\frac{1}{2})$. These both are special positions without the freedom to shift (besides thermal movement around the average position). The anions occupying the Wyckoff position $8d$ at $(x,\frac{1}{4},\frac{1}{8})$. The anion x-coordinate varies in dependence of the anion type (S, Se, Te) reflecting a shift of the anion position from its ideal position in the center of the cation tetrahedron (which would be at $(\frac{1}{4},\frac{1}{4},\frac{1}{8})$). The tetragonal chalcopyrite-type structure is closely related to the zinc-blende (or sphalerite)-type structure [10] but with a non-ideal tetragonal strain parameter $\eta = c/2a \neq 1$ (called tetragonal deformation) and the anion displacement parameter $u = 1/4 + \alpha/a^2$ (called tetragonal distortion) reflecting the unequal cation–anion bond lengths R_{AX} and R_{BX}. Here, a and c are the lattice parameters and $\alpha = R_{AX}^2 - R_{BX}^2$.

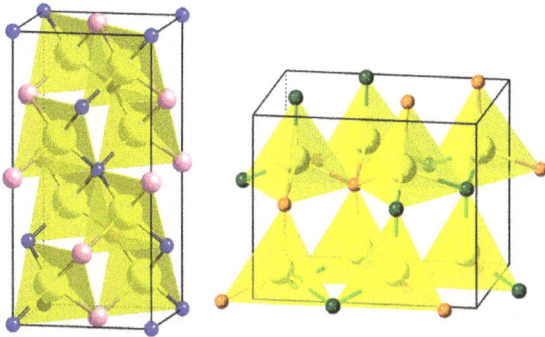

Fig. 4.1: Representation of the crystal structure of ternary $A^I B^{III} X_2^{VI}$ (left) and $A^{II} B^{IV} X_2^V$ (right) compounds. Left: chalcopyrite-type crystal structure (A^I – blue, B^{III} – pink, X^{VI} – yellow). Right: β-NaFeO$_3$-type structure (A^{II} – orange, B^{IV} – green, X^V – yellow). In both structures the anion tetrahedra are shown (XA$_2$B$_2$ tetrahedra). The black lines show the unit cell.

Quaternary compounds of the adamantine family can be represented by the general formula $A_2^{N-1}B^{2N-2}C^{N+2}X_4^{8-N}$. Most important are $A_2^I B^{II} C^{IV} X_4^{VI}$ compounds ($N = 2$), and their crystal structures are based on the same building block: the A$_2$BCX anion tetrahedron (see Fig. 4.2). They crystallize in the stannite-type (space group $I\bar{4}2m$; see Fig. 4.3, left) or in the kesterite-type (space group $I\bar{4}$; see Fig. 4.3, right) crystal structure, as well as in the Wurtz-stannite (space group $Pmn2_1$; see Fig. 4.4, left) or Wurtz-kesterite-type crystal structure (space group Pn; see Fig. 4.4, right). The kesterite-type and stannite-type crystal structures can be seen as tetragonal super-structures of the zinc-blende structure, whereas the orthorhombic Wurtz-stannite and the monoclinic Wurtz-kesterite structure can be considered as superstructures of the Wurtzite-type structure. More details on these crystal structures can be found here [11].

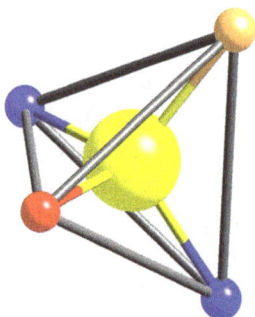

Fig. 4.2: Presentation of the A_2BCX anion tetrahedron which is the building block in quaternary A_2BCX_4 compounds (A – blue, B – orange, C – red, anion X – yellow). The gray lines represent the edges of a tetrahedron.

The stannite-type and the kesterite-type structures are closely related but show a different cation distribution which leads to different symmetries and thus different space groups. In the stannite-type structure, the monovalent A cation occupies the Wyckoff position $4d$ at $(0,\frac{1}{2},\frac{1}{2})$, the divalent B cation occupies the Wyckoff position $2a$ at $(0,0,0)$ and the four valent C cation occupies the $2b$ position at $(\frac{1}{2},\frac{1}{2},0)$. This cation arrangement leads to alternating cation layers perpendicular to the crystallographic **c**-axis in the form of Zn–Sn, Cu–Cu, Zn–Sn, Cu–Cu, and Zn–Sn at $z = 0$, $\frac{1}{4}$, $\frac{1}{2}$, $\frac{3}{4}$ and 1, respectively. The anion is placed on the $8g$ position at (x,x,z), thus

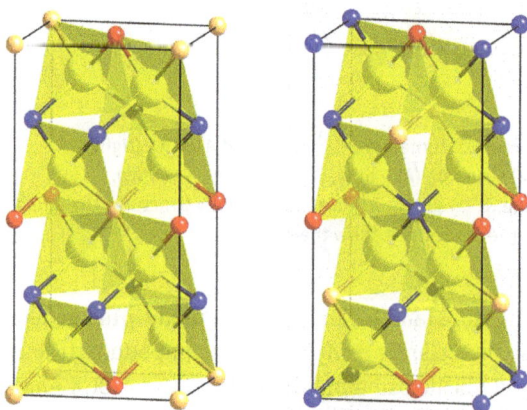

Fig. 4.3: Representation of the crystal structure of quaternary $A_2^I B^{II} C^{IV} X_4^{VI}$ compounds. Left: stannite-type crystal structure; right: kesterite-type crystal structure (A^I – blue, B^{II} – orange, C^{IV} – red, X^{VI} – yellow). The black lines show the unit cell.

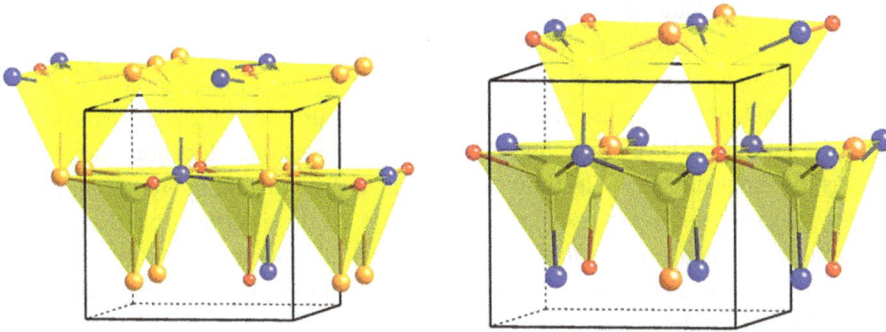

Fig. 4.4: Representation of the crystal structure of quaternary $A_2^I B^{II} C^{IV} X_4^{VI}$ compounds. Left: Wurtz-stannite-type crystal structure; right: Wurtz-kesterite-type crystal structure (A^I – blue, B^{II} – orange, C^{IV} – red, X^{VI} – yellow). The black lines show the unit cell.

it is significantly stronger and is shifted from the center of the cation tetrahedron as the anion in the chalcoprite-type crystal structure. In the kesterite-type crystal structure, the monovalent A cation occupies the Wyckoff position $2a$ at $(0,\frac{1}{2},\frac{1}{2})$ as well as the $2c$ position at $(0,\frac{1}{2},\frac{1}{4})$, and the divalent B cation occupies the Wyckoff position $2d$ at $(0,\frac{1}{2},\frac{3}{4})$ and the four valent C cation occupies the $2b$ position at $(\frac{1}{2},\frac{1}{2},0)$. This cation arrangement leads to alternating cation layers perpendicular to the crystallographic **c**-axis in the form of Cu–Sn, Cu–Zn, Cu–Sn, Cu–Zn, and Cu–Sn at $z = 0$, $\frac{1}{4}$, $\frac{1}{2}$, $\frac{3}{4}$, and 1. The anion is placed on the $8i$ position at (x,y,z). This is a general position and shows the significant shift of the anion from the center of the cation tetrahedron.

4.1.2 Crystal structure and band gap energy

The photoactive band gap energy E_g of a solar cell device is determined by the band gap of the absorber layer material which has to be optimized to reach high conversion efficiencies. The solar cell energy conversion efficiency is defined as the percentage of power converted from sunlight to electrical energy under certain conditions. The theoretical efficiency limit was first introduced by Shockley and Queisser [12], which is one of the most fundamental parameters to solar cell technology. It took the solar radiation by a 6,000 K black body into consideration. Not all photons of the solar radiation can create an electron-hole pair in the absorber layer. Photons with energy less than the band gap energy E_g are transmitted without being absorbed in the semiconductor or they will be converted into heat. On the other hand, photons with an energy higher than E_g of the semiconductor absorber material can excite electrons at the highly laying conduction band (CB) state, but such electrons will quickly relax toward the bottom of the CB states via the electron–photon interaction or carrier-

carrier scattering. Thus, the most important semiconductor properties for solar energy conversion devices are the band gap energy E_g and the optical absorption.

Every material has its own characteristic electronic energy band structure, which is strongly determined by the material's crystal structure. The electronic band structure of a solid material defines the range of energy levels that electrons can have as well as the range of energy electrons may not have. A band gap, or energy gap, is an energy range where no electronic states can exist. In the electronic band structure, the band gap generally refers to the energy difference between the top of the valence band (VB) and the bottom of the CB in semiconductors. It corresponds to the energy required to lift a valence electron to become a conduction electron which can freely move within the crystal structure and serve as a charge carrier to conduct electrical current. The energy bands and band gaps are derived by the band theory [13, 14] by examining the allowed quantum mechanical wave functions for an electron in the periodic arrangement of atoms or molecules in the crystal structure.

Ternary $A^I B^{III} X_2^{VI}$ semiconductors show remarkable anomalies in the energy band gaps relative to their binary zinc-blende analogs, $A^{II} X^{VI}$, in terms of a chemical factor and a structural factor. It was shown [15] that the chemical factor is controlled by a p-d hybridization effect and by a cation electronegativity effect, whereas the structural contribution to the anomaly is controlled by the existence of bond alteration $(R_{AX} \neq R_{BX})$ manifested by nonideal anion displacements $u - \frac{1}{4} \neq 0$ (here u is the anion x-coordinate).

Cation and anion alloying (forming solid solutions or mixed crystals) is widely used as a strategy for band gap engineering. It is also applied in thin-film solar cell technology. While the lattice parameters in $AB_{1-x}C_x X_2$ solid solution series depend linearly on the chemical composition of the mixed crystals fulfilling Vegard's law [16], the band gap energy shows a nonlinear behavior. This band gap bowing which can be expressed by

$$E_g(x) = E_g^{x=0}(1-x) + E_g^{x=1}x + bx(1-x) \tag{4.1}$$

($E_g^{x=0}$ and $E_g^{x=1}$ are the band gap energies of the end members of the alloy series and b is the bowing parameter) is a common effect in semiconductor alloy systems.

Cu(In,Ga)Se$_2$ (CIGS) is a solid solution between the chalcopyrite-type semiconductors, CuInSe$_2$ and CuGaSe$_2$, used to optimize optoelectronic properties achieving highly efficient solar cell devices. Best efficiencies are obtained for Cu-poor material and an average Ga content of about 20–35 mol% [17]. The lattice parameter as well as the structural deformation (c/2a) in the solid solution series CuIn$_{1-x}$Ga$_x$Se$_2$ fulfills Vegard's law (see Fig. 4.5). The bang gap energy of these mixed crystals shows a bowing behavior with a bowing parameter b in the range 0.15–0.24 [18].

$Cu_2 B^{II} C^{IV} X_4^{VI}$ semiconductors, such as Cu$_2$ZnSnS$_4$ or Cu$_2$ZnSnSe$_4$, are direct semiconductors with the corresponding optical transition occurring at the G point of the Brillouin zone. In these materials, the conduction band (CB) minimum consists of IV ns states (e.g. Sn 5s) and VI np states (e.g. S 3p). Considering quaternary compounds

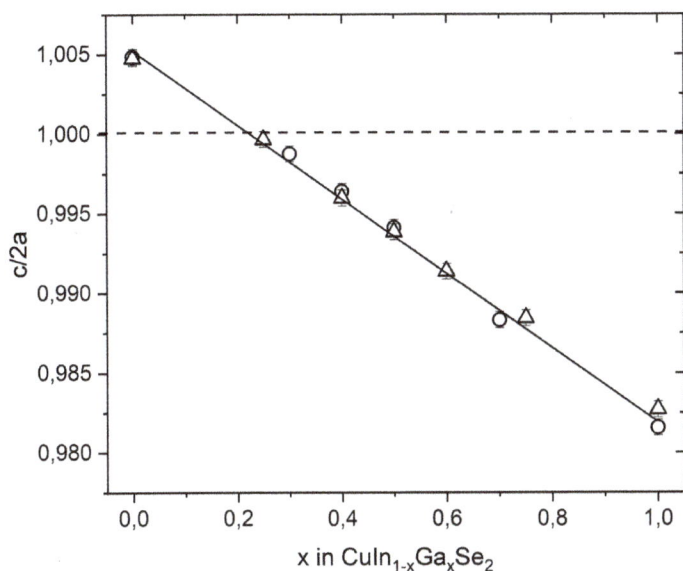

Fig. 4.5: Tetragonal deformation $c/2a$ in Cu(In,Ga)Se$_2$ mixed crystals from [19] (circles) and [20] (triangles), as functions of the Ga concentration (x). The solid line represents a linear fir, and the dashed line shows the non-deformed case $c/2a = 1$.

with the same divalent cation and the same anion, the structure of the CB, especially the CB minimum, depends critical on the four-valent cation which is Si, Ge, or Sn. The valence band maximum consists of Cu 3d states and VI np states. Therefore, the valence band structure and especially the valence band maximum depends strongly on the anion, which is S, Se, or Te. Tab. 4.1 shows an overview of the crystal structure and the band gap energy of quaternary $A_2^I B^{II} C^{IV} X_4^{VI}$ compound semiconductors.

Within a homologous series the S-based semiconductor has usually a larger band gap energy than the Se- or Te-based semiconductor (see Fig. 4.6). Splitting of the top of the valence band is due to spin-orbit interaction but also by the crystal-field interaction. It was obtained theoretically [21] that the strength of the crystal field interaction is stronger in the stannite-type structure than in the kesterite-type structure.

Tab. 4.1: Overview of crystal structure and band gap energy E_g of quaternary $A_2^I B^{II} C^{IV} X_4^{VI}$ compound semiconductors. K – kesterite structure, S – stannite structure, w-S – Wurtz-stannite structure and w-K – Wurtz-kesterite structure.

Material	Crystal structure	Reference	E_g [eV]	Reference
Cu_2ZnSnS_4	K	[22]	1.43 (5)	this work
$Cu_2ZnSnSe_4$	K	[23]	0.94 (5)	this work
Ag_2ZnSnS_4	S	[24]	2.01 (5)	[25]
$Ag_2ZnSnSe_4$	K	[26]	1.31 (5)	this work
Cu_2MnSnS_4	S	[27]	1.62 (K)*	[28]
			1.52 (S)*	[29]
			1.59 (5)	[29]
$Cu_2MnSnSe_4$	K	this work	0.88 (5)	this work
Cu_2CdSnS_4	S	this work	1.29 (5)	this work
$Cu_2CdSnSe_4$	S	[27]	0.94 (5)	[29]
Cu_2ZnSiS_4	w-S	[30]	2.97 ($\perp \vec{c}$)	[30]
			3.07 ($\|\vec{c}$)	
$Cu_2ZnSiSe_4$	w-K	[31]	2,22 (5)	this work
Cu_2ZnGeS_4	w-S	this work	2.25 (5)	this work
	K	this work	1.85 (5)	this work
$Cu_2ZnGeSe_4$	K	[32]	1.31 (5)	this work

*calculated

4.1.3 Cu/Zn disorder in kesterite-type Cu_2ZnSnS_4, $Cu_2ZnSnSe_4$, $Cu_2ZnSn(S,Se)_4$, and $Cu_2ZnGeSe_4$

The quaternary semiconductors Cu_2ZnSnS_4 (CZTS), $Cu_2ZnSnSe_4$ (CZTSe), Cu_2ZnSn $(S,Se)_4$ (CZTSSe), and $Cu_2ZnGeSe_4$ (CZGeSe) crystallize in the kesterite-type structure which has been proven experimentally by neutron diffraction [22, 23, 32, 33]. Additionally, the existence of Cu_{Zn} and Zn_{Cu} antisites involving the Wyckoff positions $2c$ and $2d$ has been observed. Thus, a Cu/Zn disorder exists in cation layers at $z = \frac{1}{4}$ and $z = \frac{3}{4}$, which are perpendicular to the crystallographic **c**-axis. The Cu/Zn disorder has been found experimentally for the first time in a neutron diffraction-based study of CZTS [22]. Due to this disorder phenomenon, the term "disordered kesterite" was established [34]. The degree of Cu/Zn disorder can be expressed by the order parameter Q with [35]

$$Q = \frac{[Cu_{2c} + Zn_{2d}] - [Zn_{2c} + Cu_{2d}]}{[Cu_{2c} + Zn_{2d}] + [Zn_{2c} + Cu_{2d}]} \qquad (4.2)$$

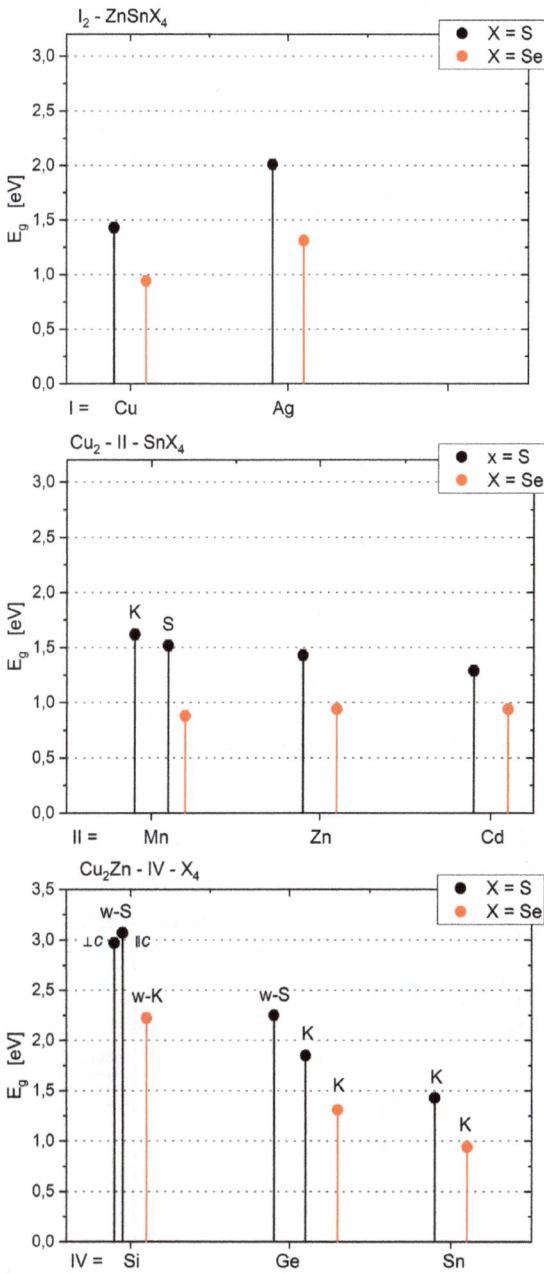

Fig. 4.6: Band gap energy in $A_2^I B^{II} C^{IV} X_4^{VI}$ semiconductors.

The first hint of a temperature dependent change in the cation arrangement within the kesterite-type structure was observed in CZTS by in situ synchrotron X-ray diffraction (synXRD) [36]. Later on, an order–disorder phase transition was proved at 259 ± 10 °C in CZTS [37] by Raman spectroscopy and at 203 ± 6 °C in CZTSe by anomalous synXRD [35]. The latter method allows not only a qualitative but also a quantitative investigation of the Cu/Zn disorder determining the order parameter Q.

The Cu/Zn disorder influences the optoelectronic properties such as band gap energy and band gap fluctuations [38].

4.1.4 Off-stoichiometry and point defects in kesterite-type semiconductors

As the chalcopyrite-type crystal structure, the kesterite-type structure also shows a remarkable flexibility which originates from the propensity of the crystal structure to stabilize intrinsic point defects as vacancies, anti-sites, and interstitials. Deviations from the stoichiometric composition lead to the formation of such intrinsic point defects in the compound, which significantly influence the electrical and optical properties of the material (see Fig. 4.7). The formation of point defects is driven thermodynamically by minimizing the Gibbs free energy of the crystal. Moreover, the formation of secondary phases is highly probable. The correlation between off-stoichiometry and intrinsic point defects is of great importance for the understanding and rational design of solar cell devices.

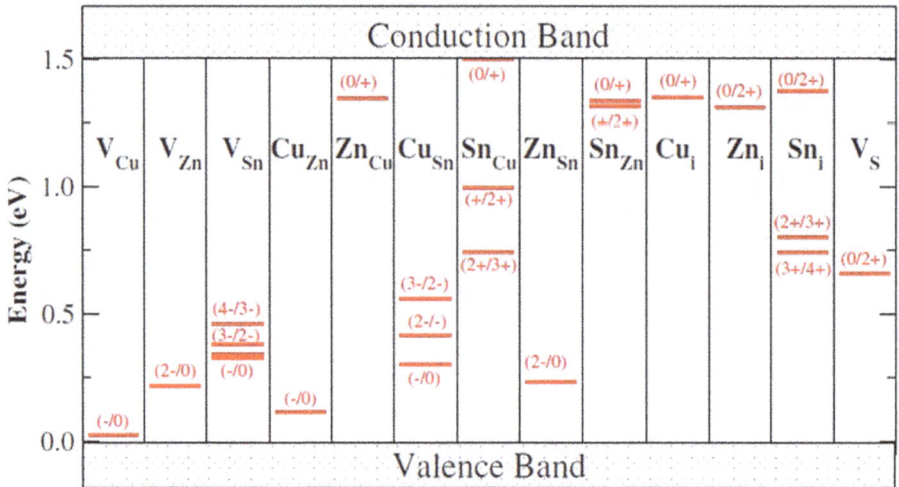

Fig. 4.7: Transition-energy levels of intrinsic point defects in the band gap of Cu_2ZnSnS_4 [39].

Forming off-stoichiometric kesterite-type compounds, the charge balance has to be insured. Thus, only certain substitutions can be envisioned to account for the charge balance in the off-stoichiometric material assuming the oxidation states of cations and anions are retained. According to this model, the off-stoichiometric composition corresponds to certain point defects [40]. This correlation is the basis for the definition of off-stoichiometry types. A summary of the off-stoichiometry types, named A to L [41], is presented in Tab. 4.2.

Tab. 4.2: Overview of off-stoichiometry types in kesterite type Cu_2ZnIVX_4, where IV = Sn or Ge and X = S or Se.

Type	Composition	Cation substitution	Corresponding intrinsic point defects	General chemical formulae
A	Cu-poor, Zn-rich, IV-const.	$2\,Cu^+ \rightarrow Zn^{2+}$	$V_{Cu} + Zn^{2+}_{Cu}$	$Cu_{2-2a}Zn_{1+a}IVX_4$
B	Cu-poor, Zn-rich, IV-poor	$2\,Cu^+ + IV^{4+} \rightarrow 3Zn^{2+}$	$2\,Zn^{2+}_{Cu} + Zn^{2+}_{IV}$	$Cu_{2-2b}Zn_{1+3b}IV_{1-b}X_4$
C	Cu-rich, Zn-poor, IV-rich	$3\,Zn^{2+} \rightarrow 2Cu^+ + IV^{4+}$	$2\,Cu^+_{Zn} + IV^{4+}_{Zn}$	$Cu_{2+2c}Zn_{1-3c}IV_{1+c}X_4$
D	Cu-rich, Zn-poor, IV-const.	$Zn^{2+} \rightarrow 2Cu^+$	$Cu^+_{Zn} + Cu_i$	$Cu_{2+2d}Zn_{1-d}IVX_4$
E	Cu-poor, Zn-poor, IV-rich	$2\,Cu^+ + Zn^{2+} \rightarrow IV^{4+}$	$2\,V_{Cu} + Sn^{4+}_{IV}$ or $IV^{4+}_{Cu} + V_{Cu} + V_{Zn}$	$Cu_{2+2e}Zn_{1-e}IV_{1+e}X_4$
F	Cu-rich, Zn-rich, IV-poor	$IV^{4+} \rightarrow Zn^{2+} + 2Cu^+$	$Zn^{2+}_{IV} + 2\,Cu^+_i$ or $Cu^+_{IV} + Cu^+_i + Zn^{2+}_i$	$Cu_{2+2f}Zn_{1+f}IV_{1-f}X_4$
G	Cu-const., Zn-rich, IV-poor	$IV^{4+} \rightarrow 2Zn^{2+}$	$Zn^{2+}_{IV} + Zn^{2+}_i$	$Cu_2Zn_{1+2g}IV_{1-g}X_4$
H	Cu-const., Zn-poor, IV-rich	$2\,Zn^{2+} \rightarrow IV^{4+}$	$IV^{4+}_{Zn} + V_{Zn}$	$Cu_2Zn_{1-h}IV_{1+1/2h}X_4$
I	Cu-rich, Zn-const., IV-poor	$IV^{4+} \rightarrow 4Cu^+$	$Cu^+_{IV} + 3Cu^+_i$	$Cu_{2(1+2i)}ZnIV_{1-i}X_4$
J	Cu-poor, Zn-const., IV-rich	$4\,Cu^+ \rightarrow IV^{4+}$	$IV^{4+}_{Cu} + 3V_{Cu}$	$Cu_{2-2j}Zn_{1-j}IV_{1+1/2j}X_4$
K	Cu-rich, Zn/IV = 1 = const.	$Zn^{2+} + IV^{4+} \rightarrow 6Cu^+$	$Cu^+_{Zn} + Cu^+_{IV} + 4Cu^+_i$	$Cu_{2+6k}Zn_{1-k}IV_{1-k}X_4$
L	Cu-poor, Zn/IV = 1 = const.	$6\,Cu^+ \rightarrow Zn^{2+} + IV^{4+}$	$Zn^{2+}_{Cu} + IV^{4+}_{Cu}$	$Cu_{2-2l}Zn_{1+1/3l}IV_{1+1/3l}X_4$

The existence of off-stoichiometric kesterite-type CZTS, CZTSe, CZTSSe, and CZGeSe was shown in extended and systematic studies of powder series [23, 32, 33, 42]. These investigations based on neutron diffraction experiments proved the ability of kesterite-type compounds to tolerate deviations from stoichiometric composition keeping the kesterite type structure but showing cation ratios Cu/(Zn + Sn) and Zn/Sn lower or higher than 1. Thus four possible pairs of cation ratios for defining off-stoichiometry in kesterite-type compounds in general exist: (i) Cu/(Zn + IV) < 1 and Zn/Sn > 1, (ii) Cu/(Zn + IV) < 1 and Zn/Sn < 1, (iii) Cu/(Zn + IV) > 1 and Zn/Sn > 1, (iv) Cu/(Zn + IV) > 1 and Zn/Sn < 1, Cu/(Zn + IV) < 1 and Zn/Sn > 1. These pairs form the four quadrants of the cation-ratio plot (see Fig. 4.8).

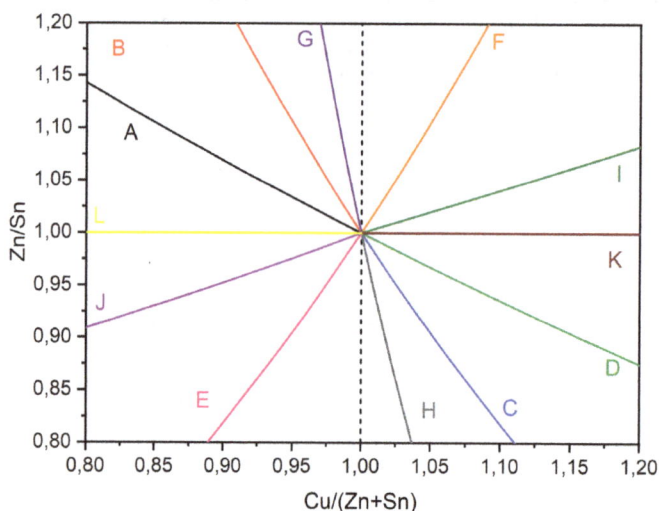

Fig. 4.8: Overview of off-stoichiometry types A–L in the cation-ratio plot.

4.1.5 Experimental determination of the band gap energy by diffuse reflectance spectroscopy

The most developed optical tool applicable to the study of powders is diffuse reflectance spectroscopy (DRS) [43]. Diffuse reflectance refers to radiation, which is reflected in all directions, as opposed to specular reflectance which has a defined angle of reflection to the incident radiation. The former occurs in samples with particles oriented in different directions and the latter occurs at an optically smooth surface. The generally accepted theory of DRS was developed by Kubelka and Munk [44, 45]. Based on the assumption that diffuse reflectance (R) arises from absorption and scattering of light by a surface, these authors have developed a simple expression which translates the diffuse reflectance spectrum to the absorption spectrum, so called the Kubelka–Munk pseudo-absorption function $F(R)$:

$$F(R) = \frac{(1-R)^2}{2R} \tag{4.3}$$

The validity of $F(R)$ relies upon the condition that the sample is thick, densely packed and constituted by randomly shaped particles whose sizes are comparable or bigger than 10 μm. It should be noted that DRS has been reported to systematically underestimate the band gap, with an error which increases with decreasing crystallite size [46].

DRS measurements were carried out in air at room temperature in a spectrophotometer equipped with an integrating sphere – PerkinElmer UV/Vis-spectrometer Lambda 750S. The measurement range was adjusted for each set of samples (e.g. 700–1,400 nm with a step size of 1 nm for CZTS).

Tauc plots were obtained by plotting $(F(R)^*h\nu)^2$ versus the photon energy hv [47], the linear part of the curve was extrapolated to the baseline, and the optical band gap was extracted from the value of intersection, an example for Cu_2ZnSnS_4 is given in Fig. 4.9.

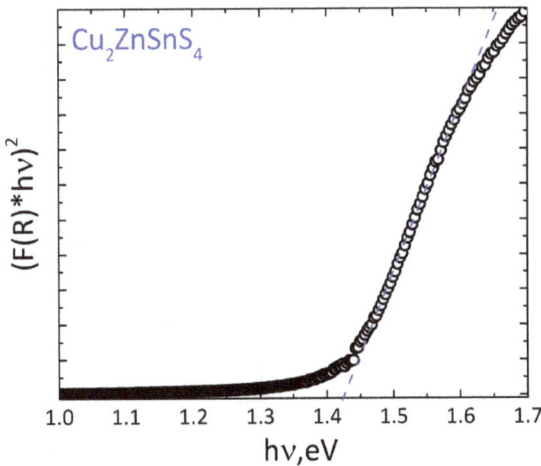

Fig. 4.9: Exemplarily Tauc plot for a direct transition (dots) with a linear fit (blue line) of a $Cu_2ZnSnSe_4$ powder, resulting in $E_g = 1.42$ eV.

4.2 Influence of the crystal structure on the band gap energy in quaternary chalcogenide semiconductors

According to the phase diagram [48], the quaternary compound semiconductor Cu_2ZnGeS_4 (CZGeS) can crystallize into two different polymorphs with a tetragonal or an orthorhombic crystal structure, respectively. The latter is a metastable phase that can only be prepared in the bulk form by extensive heating at high temperatures (≥790 °C) in a conventional solid-state synthesis route [49].

For our investigations, we have synthesized CZGeS by solid-state reaction of the elements in a closed silica tube applying temperatures up to 950 °C. With an excess of sulfur in comparison to the initial weight, various sulfur pressure in the tube could be realized.

The differentiation of the electronic similar elements Cu, Zn and Ge is not possible by X-ray diffraction, but the diffraction pattern of tetragonal kesterite-type CZGeS (space group $I\bar{4}$) and orthorhombic Wurtz-stannite-type CZGeS (space group $Pmn2_1$) are very different (see Fig. 4.10). Thus, it became already obvious from the experimental X-ray diffraction pattern, that in dependence on the sulfur excess the two polymorphs of CZGeS, crystallizes in two different structures, can be synthesized.

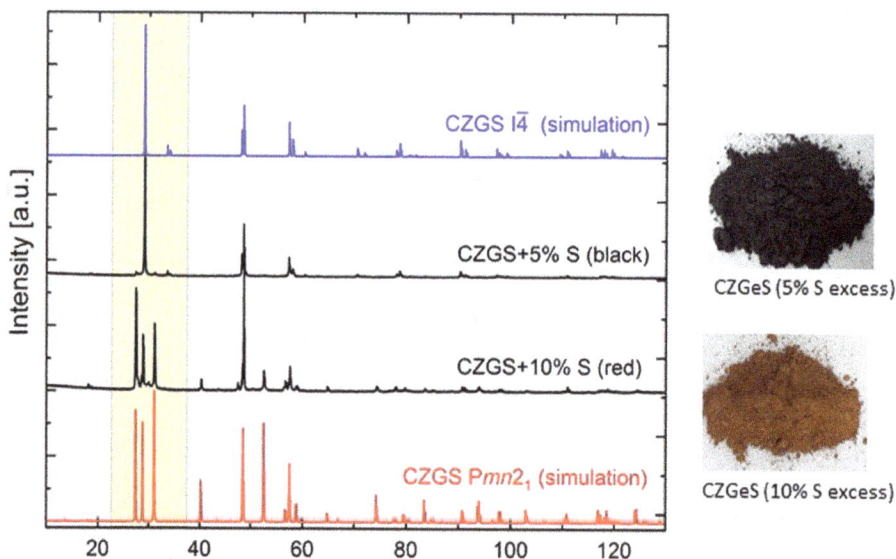

Fig. 4.10: Measured X-ray diffraction pattern of Cu_2ZnGeS_4 synthesized with sulfur excess (5% and 10%, respectively) in comparison with X-ray diffraction pattern simulated using the tetragonal kesterite-type structure (space group $I\bar{4}$) and the orthorhombic Wurtz-stannite structure (space group $Pmn2_1$). The powder samples are shown on the right.

The band gap energy E_g of the CZGeS polymorphs, which are chemically the same but structurally different, has been determined by UV–ViS spectroscopy. Fig. 4.11 shows the Tauc plot established to evaluate the band gap energy. As it is already visible from the different color of the powder samples, the band gap energy is different for CZGeS synthesized with different sulfur excess: CZGeS synthesized with 5% sulfur excess and crystallizing in the tetragonal kesterite-type structure shows a band gap of

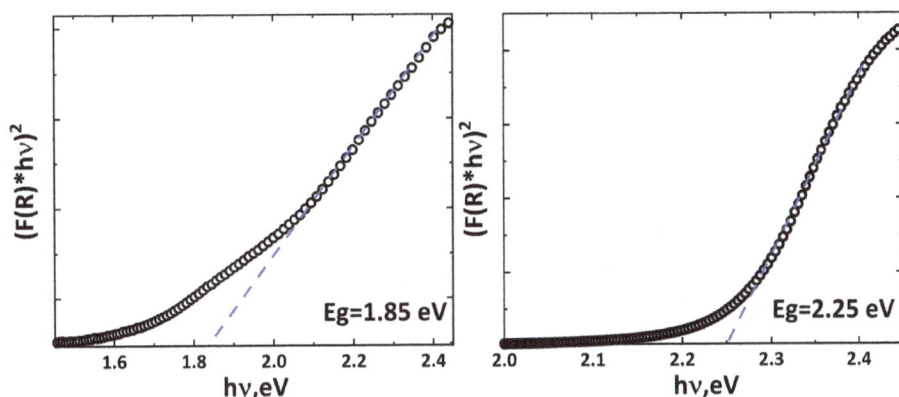

Fig. 4.11: Tauc plot of the UV–VIS spectroscopy data of Cu_2ZnGeS_4 synthesized with 5% (left) and 10% (right) sulfur excess.

1.85 eV whereas CZGeS synthesized with 10% sulfur excess and crystallizing in the orthorhombic Wurtz-stannite-type structure shows a band gap energy of 2.25 eV.

4.3 Evaluation of the band gap energy in solid solution series

As introduced in Section 4.1.2, the band gap energy E_g variation in solid solution series shows a nonlinear dependency on the chemical composition of the mixed crystals which is the band gap energy bowing.

The band gap energy bowing in the $CuIn_{1-x}Ga_xS_2$ (CIGS) and $CuIn_{1-x}Ga_xSe_2$ (CIGSe) solid solution was studied experimentally using powder samples synthesized by solid-state reaction of the elements. The band gap energy E_g was determined by UV-ViS spectroscopy applying the Kubelka–Munk method and the Tauc plot data analysis.

A band gap energy bowing (see Fig. 4.12) has been observed for both solid solution series. The bowing parameter b was determined with −0.26 (4) for CIGS and −0.30 (5) for CIGSe, respectively.

A similar investigation of kesterite-type solid solution series, such as $Cu_2ZnSn(S_{1-x}Se_x)_4$ (CZTSSe), is very complex due to the fact, that these kesterite-type compounds show a very high flexibility concerning the chemical composition which is expressed by large stoichiometry deviations (off-stoichiometry) observed in these materials (see part 1.4 of this chapter). As it will be discussed in the following (Section 4.4 of this chapter), the stoichiometry deviations influence the band gap energy in a crucial way. To discuss the bowing feature in the CZTSSe solid solution series properly it would be necessary to compare CZTSSe mixed crystals with the same off-stoichiometry.

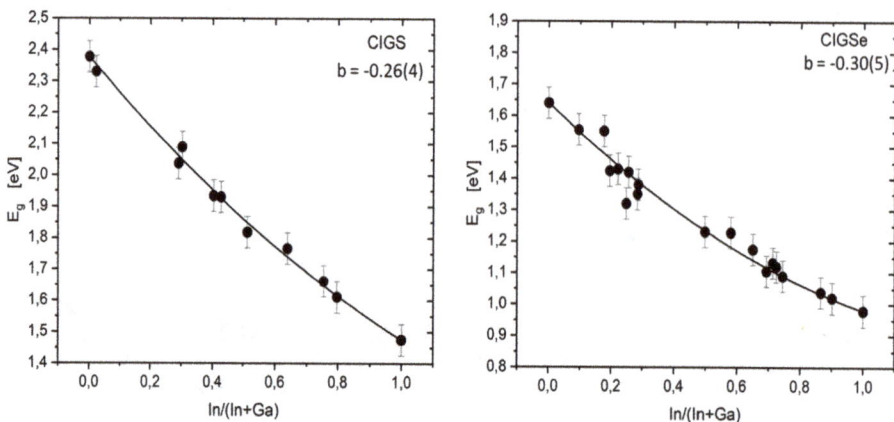

Fig. 4.12: Band gap energy E_g in dependence on the In/(In + Ga) ration in chalcopyrite-type CIGS and CIGSe solid solution series. The line represents a fit according to the bowing formulae (eq. (4.1)).

4.4 Influence of stoichiometry deviations on the band gap energy in kesterite-type chalcogenide compound semiconductors

For our detailed systematic study, off-stoichiometric Cu_2ZnSnS_4, $Cu_2ZnSnSe_4$, and $Cu_2ZnGeSe_4$ compound semiconductors have been synthesized by solid-state reaction of the elements in evacuated silica tubes at temperatures between 750 and 900 °C (details of the sample synthesis can be found in [29, 37, 38]). The chemical composition of the quaternary main phase has been determined by wavelength dispersive X-ray spectroscopy (WDX) using an electron microprobe system (JEOL–JXA 8200). About 10–15 measured points per grain were averaged if the deviation of the composition was smaller than the error of the electron microprobe analyzer. Afterwards, the chemical compositions of grains with the same chemical composition were averaged if the deviation was smaller than 1%. It has to be noted, that each powder sample consists of one homogeneous quaternary phase characterized by the cation ratios Cu/(Zn + IV) and Zn/IV (where IV stands for the group-IV element Sn and Ge, respectively). WDX spectroscopy was also applied to determine the composition of secondary phases which coexist to the off-stoichiometric homogeneous main phase. Structural parameters of the quaternary phase have been determined by X-ray diffraction and Rietveld analysis of the diffraction data. The cation site occupancy factors in the kesterite-type phase were determined by neutron diffraction and Rietveld analysis of the diffraction data. The cation distribution was deduced by the average neutron scattering length analysis method [50]. The type and concentration of intrinsic point defects has been obtained on the basis of the cation distribution (for details see [51]).

The band gap energy of off-stoichiometric kesterite-type Cu_2ZnSnS_4, $Cu_2ZnSnSe_4$, and $Cu_2ZnGeSe_4$ has been determined by UV–VIS spectroscopy applying the Kubelka–Munk method as described above.

4.4.1 Band gap energy variations in Cu-poor/Zn-rich $Cu_2ZnSnSe_4$

In sum 7 off-stoichiometric $Cu_2ZnSnSe_4$ compounds with stoichiometry deviations between $0.8 \leq \frac{Cu}{Zn+Sn} \leq 1$ and $1 \leq \frac{Zn}{Sn} \leq 1.15$, thus belonging to the A-B off-stoichiometry type, have been investigated (Fig. 4.13). According to the off-stoichiometry type

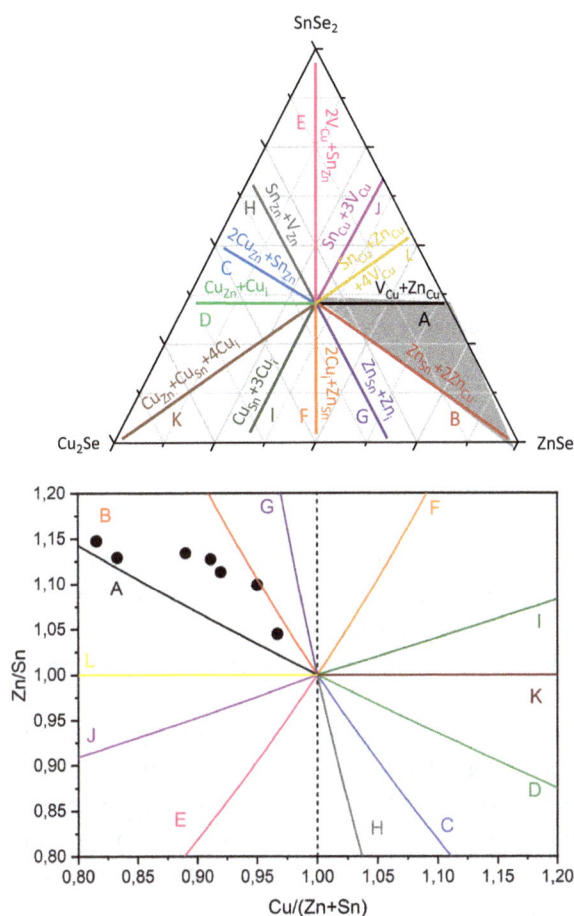

Fig. 4.13: Ternary diagram Cu_2Se-$SnSe_2$-$ZnSe$ (upper) and the cation-ratio plot (lower) both with the off-stoichiometry types A-L (lines). The data points in the cation-ratio plot correspond to the 7 A-B type off-stoichiometric CZTSe compounds investigated. The gray area in the ternary diagram marks the existence region of A-B type off-stoichiometric kesterite-type compounds.

model, in this kesterite-type compounds, the main intrinsic point defects are V_{Cu} and the anti-sites Zn_{Cu} and Zn_{Sn}.

In the studied A-B type CZTSe the band gap energy E_g increases from 0.88 eV to 1.0 eV in dependence on decreasing $Cu/(Zn + Sn)$ and increasing Zn/Sn ratio (see Fig. 4.14). The band gap variations are in the range of 0.1 eV.

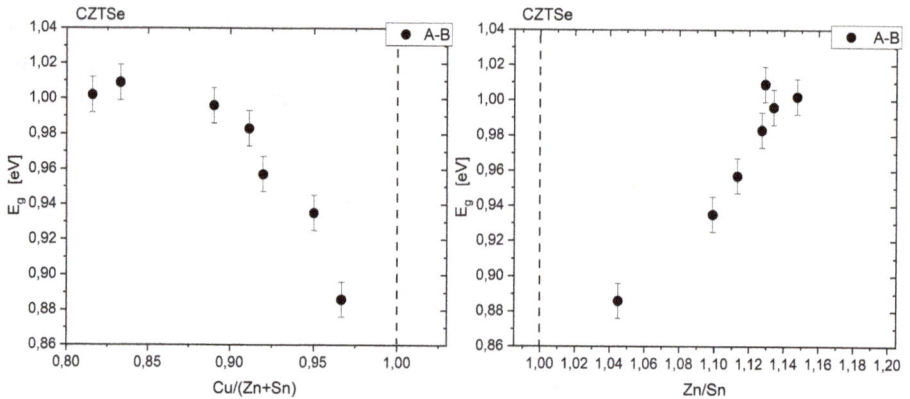

Fig. 4.14: Band gap energy E_g in Cu-poor/Zn-rich $Cu_2ZnSnSe_4$ semiconductors (according to the off-stoichiometry type model A-B type) in dependence on the cation ratios $Cu/(Zn + Sn)$ and Zn/Sn.

4.4.2 Band gap energy variations in Cu-poor/Zn-rich and Cu-rich/Zn-rich Cu_2ZnSnS_4

The quaternary phase is assigned to an off-stoichiometry type according to the experimentally determined cation ratios $Cu/(Zn-Sn)$ and Zn/Sn. In the most cases, the cation ratios lie in between two type lines in the cation-ratio plot. In this case the occurrence of intrinsic point defects according to two off-stoichiometry types are expected and there is a mix of types.

We studied the band gap energy variation in a number of Cu-poor/Zn-rich Cu_2ZnSnS_4 (CZTS) compounds. According to their cation ratios they represent the mix of two off-stoichiometry types, A-B, B-G, and G-F, respectively (see Fig. 4.15 (right)). Thus, the main intrinsic point defects which can be expected (and have been found experimentally [51]) are V_{Cu}, Zn_{Cu}, and Zn_{Sn} (A-B type), Zn_{Cu}, Zn_{Sn}, and Zn_i (B-G type) as well as Zn_{Sn}, Zn_i, and Cu_i (G-F type).

As it becomes obvious from Fig. 4.16, the band gap energy E_g is different for different off-stoichiometry types in CZTS. The main difference which can be reached is about 0.15 eV.

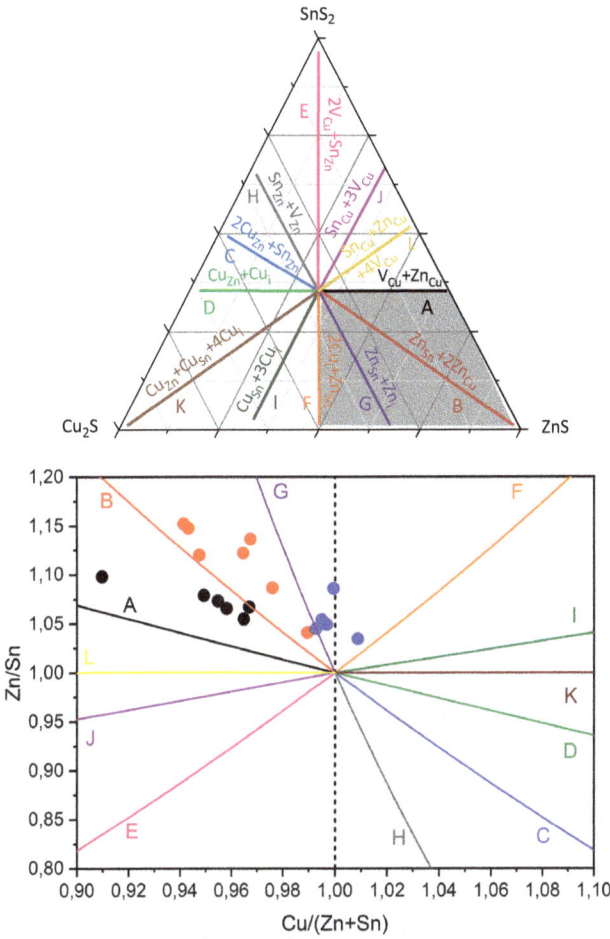

Fig. 4.15: Ternary diagram (upper) and the cation-ratio plot (lower) both with the off-stoichiometry types A–L (lines). The data points in the cation-ratio plot correspond to the A-B, B-G, and G-F type off-stoichiometric CZTS compounds investigated. The gray area in the ternary diagram marks the existence region of A-B, B-G and G-F type off-stoichiometric kesterite-type compounds.

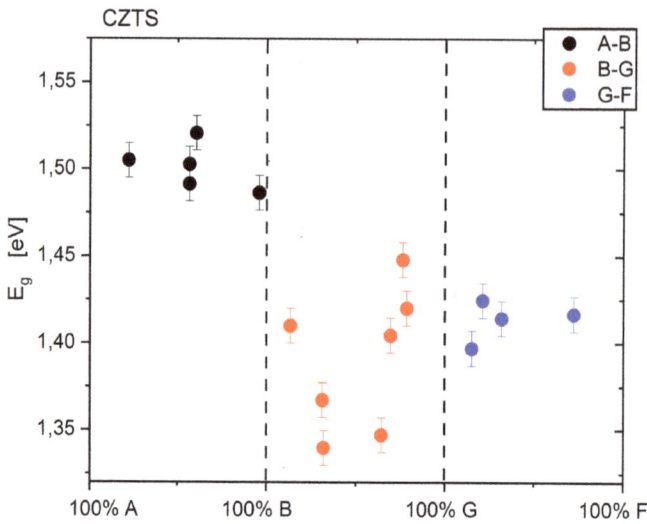

Fig. 4.16: Band gap energy E_g in Cu-poor/Zn-rich and Cu-rich/Zn-rich Cu_2ZnSnS_4 semiconductors (according to the off-stoichiometry type model A-B, B-G, and G-F type) in dependence on the according type fraction.

Variations in the band gap energy can also be obtained in Cu-rich/Zn-rich CZTS. Here, three compounds have been studied, two of them are a D-C type mixture and one presents a C-H type mixture according to the cation ratios of the quaternary phase (Fig. 4.17). Thus, the main intrinsic point defects which can be expected (and have been found experimentally [51]) are Cu_{Zn}, Sn_{Zn}, and Cu_i (D-C type) as well as Cu_{Zn}, Sn_{Zn}, and V_{Zn} (C-H type).

In the studied D-C and C-H type CZTS, the band gap energy E_g increases from 1.35 eV to 1.45 eV in dependence on increasing off-stoichiometry expressed by increasing Cu/(Zn + Sn) and decreasing Zn/Sn ratios (see Fig. 4.18).

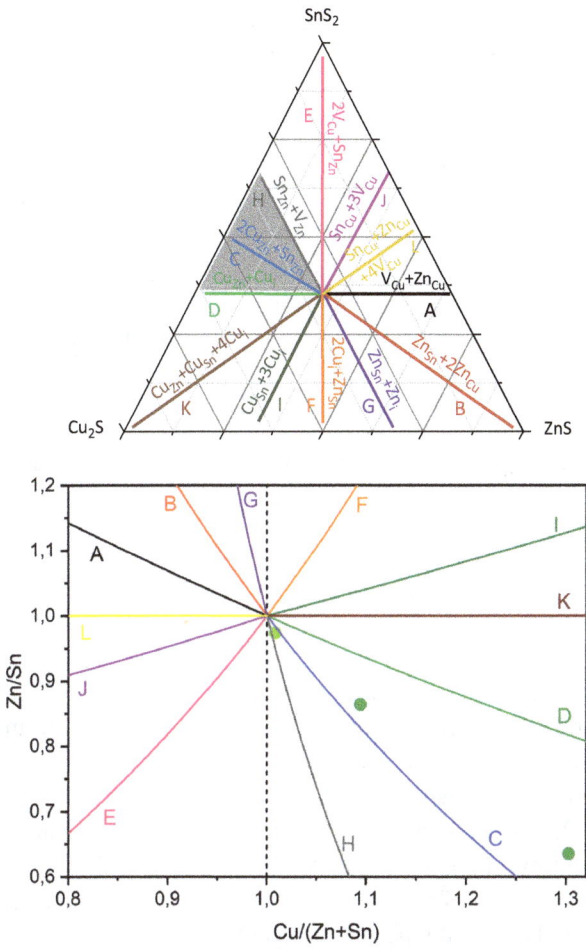

Fig. 4.17: Ternary diagram (upper) and the cation-ratio plot (lower) both with the off-stoichiometry types A-L (lines). The data points in the cation-ratio plot correspond to the D-C and C-H type off-stoichiometric CZTS compounds investigated. The gray area in the ternary diagram marks the existence region of C-D type off-stoichiometric kesterite-type compounds.

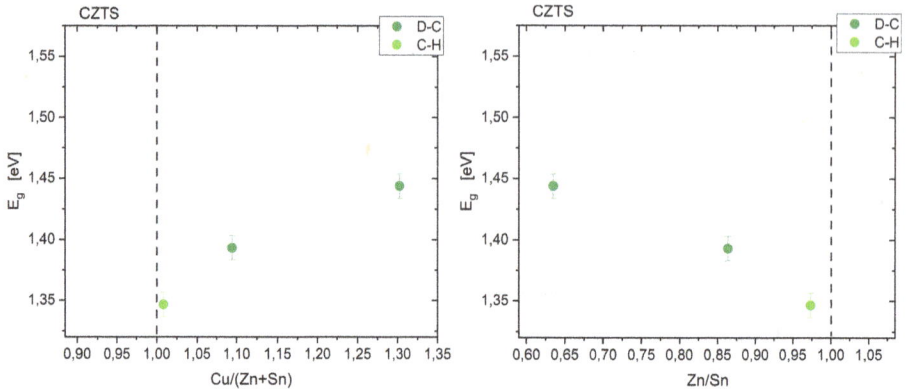

Fig. 4.18: Band gap energy E_g in Cu-rich/Zn-rich $Cu_2ZnSnSe_4$ semiconductors (according to the off-stoichiometry type model D-C as well as C-H type mixtures) in dependence on the cation ratios Cu/(Zn + Sn) and Zn/Sn.

4.4.3 Band gap energy variations in Cu-poor/Zn-rich and Cu-rich/Zn-rich $Cu_2ZnGeSe_4$

We studied the band gap energy variation in a number of Cu-poor/Zn-rich Cu-$_2$ZnGeSe$_4$ (CZGeSe) compounds which span a wide off-stoichiometry region from Cu-poor/Zn-rich to Cu-rich/Zn-rich compositions. According to their cation ratios, they represent the mix of two off-stoichiometry types, A-B, B-G, G-F, F-I, I-K, and K-D, respectively (see Fig. 4.19 (right)). Thus, the main intrinsic point defects which can be expected (and have been found experimentally [51]) are V_{Cu}, Zn_{Cu}, and Zn_{Ge} (A-B type), Zn_{Cu}, Zn_{Ge}, and Zn_i (B-G type), Zn_{Ge}, Zn_i, and Cu_i (G-F type), Zn_{Ge}, Cu_i, and Cu_{Ge} (F-I type), as well as Cu_i, Cu_{Ge}, and Cu_{Zn} (I-K type and K-D type).

Within the G-F and F-I type CZGeSe, the value of the band gap E_g somehow scatters, but a general trend can be obtained (see Fig. 4.20). Considering the band gap energy variation over the whole range from $0.8 < Cu(Zn + Ge) < 1.26$ and $1 \leq Zn/Ge < 1.4$, the band gap energy E_g decreases from Cu-poor to Cu-rich compositions as well as E_g increases with increasing Zn/Ge ratio.

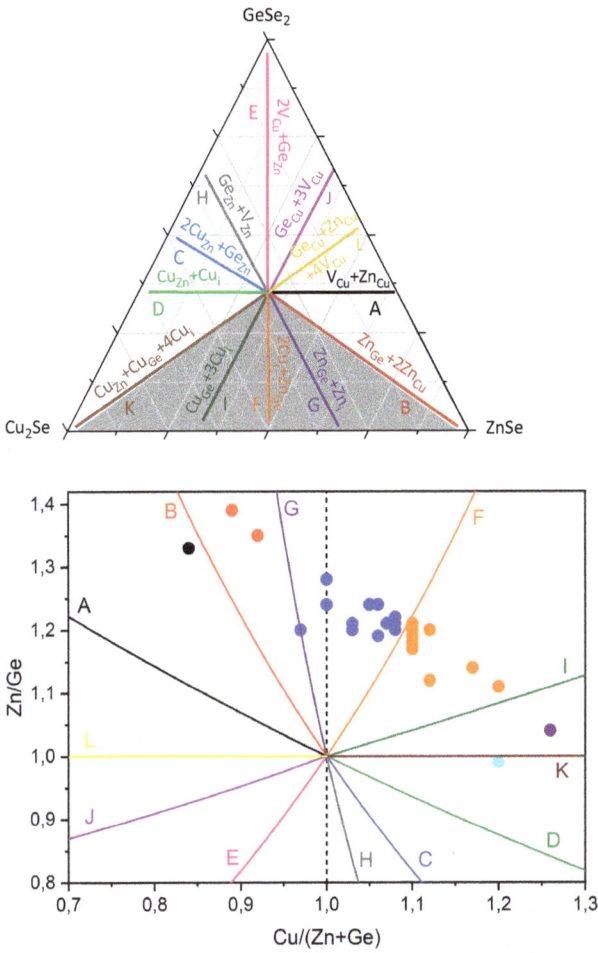

Fig. 4.19: Ternary diagram (upper) and the cation-ratio plot (lower) both with the off-stoichiometry types A–L (lines). The data points in the cation-ratio plot correspond to the A-B, B-G, G-F, F-I, I-K, and K-D type off-stoichiometric CZGeSe compounds investigated. The grey area in the ternary diagram marks the existence region of the off-stoichiometric kesterite-type compounds discussed here.

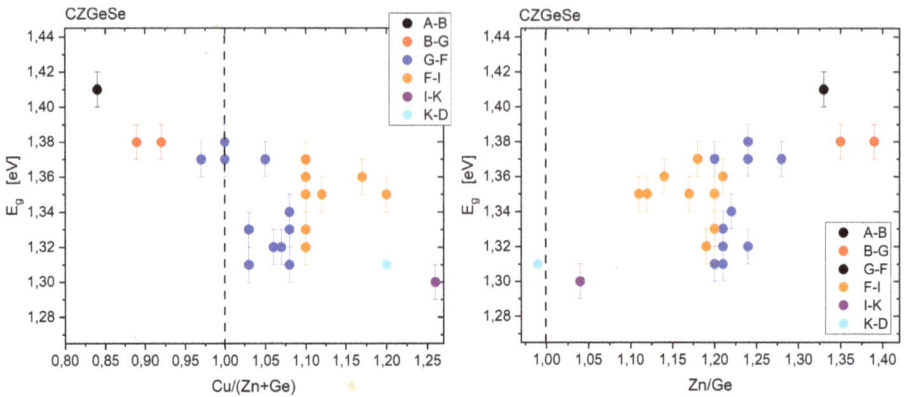

Fig. 4.20: Band gap energy E_g in Cu-poor/Zn-rich as well as Cu-rich/Zn-rich $Cu_2ZnGeSe_4$ semiconductors in dependence on the cation ratios Cu/(Zn + Ge) and Zn/Ge.

4.4.4 Comparison of Cu-poor multinary compounds with chalcopyrite- as well as kesterite-type crystal structure

The remarkable flexibility of the chalcopyrite but also the kesterite crystal structure is a key point for multinary compounds for their application as absorber layer in thin-film solar cells. The growth of thin films is in essence a non-equilibrium process that goes along with local inhomogeneities in chemical composition to be balanced by the structure type.

The highest power conversion efficiencies of CIGSe-based thin-film solar cells are achieved with a Cu-poor composition resulting in a p-type absorber layer [52]. For example, systematic investigations of $CuInSe_2$ (CISe) have shown, that Cu-poor chalcopyrite-type CISe exists even up to a Cu/In ratio of 0.75 [53]. Thin-film solar cells based on kesterite-type CZTSSe reach their highest power conversion efficiency with a Cu-poor/Zn-rich composition of the absorber layer [4].

In the following section, the influence of the compositional flexibility on the band gap energy E_g in Cu-poor chalcopyrites ($CuInS_2$-CIS, $CuInSe_2$-CISe, $CuGaS_2$-CGS) as well as in Cu-poor/Zn-rich kesterite (CZTSe) will be compared using materials with the same off-stoichiometry range. For CIS, CISe, and CGS powder samples with a Cu/III ratio between 0.98 and 0.75 are studied, for CZTSe a series of A-B type materials which are Cu-poor and Zn-rich with a Cu/(Zn + Sn) ratio between 0.97 and 0.83 (see Section 4.4.1) are considered. Because the absolute band gap energy values of these materials are different, the spread of the band gap energy around an average (mean) value was considered for the comparison. For this the band gap energies determined for the different off-stoichiometric compounds were averaged, and the difference between this average and the band gap energy of a specific off-stoichiometric compound was calculated as $E_g - E_g(average)$. This difference can have positive and negative values because the specific band gap energy value can be larger or smaller than the average (mean value). A comparison of the spread of the band gap energy around a mean value in CIS, CISe, CGS, and CZTSe, (all with a Cu-poor composition) shows only small band gap energy variations in the Cu-poor chalcopyrites, but large variations in the Cu-poor kesterites (see Fig. 4.21). This remarkable difference in the behavior of the band gap energy of these materials has crucial consequences for the absorber layer in a thin-film solar cell. As it was shown experimentally [54] by energy dispersive X-ray spectroscopy (EDX) and synchrotron X-ray fluorescence spectroscopy (XRF), the local composition of the CZTSSe phase within the absorber layer shows strong variations at the nanoscale. This causes the existence of multiple off-stoichiometry types of the CZTSSe phase which give rise to strong fluctuations in the band gap energy (as it is shown in Sections 4.4.1 to 4.4.4).

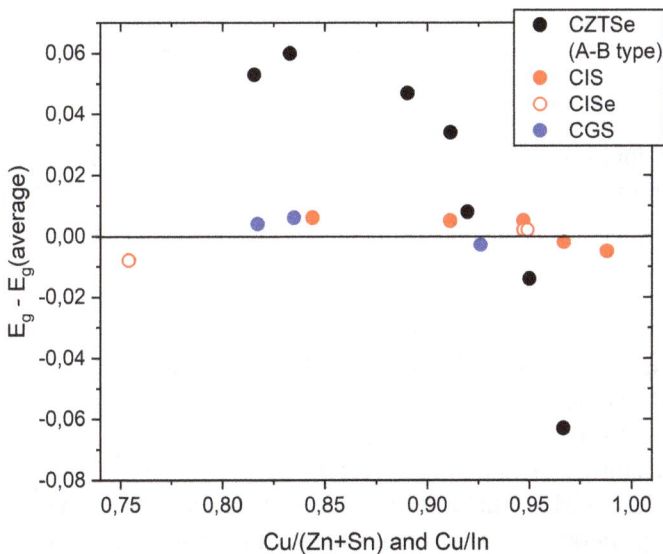

Fig. 4.21: Variation of the band gap energy E_g in off-stoichiometric A-B type CZTSe as well as Cu-poor CIS, CISe, and CGS.

4.5 Summary

The photoactive band gap energy E_g of a solar cell device depends on the band gap of the absorber layer semiconductor and is a crucial property to reach high power conversion efficiencies. The band gap energy is determined by the electronic structure of the material which is in turn determined by the material's crystal structure. Thus, an understanding of the crystal structure of a semiconductor as well as special features such as structural disorder or point defects is of high importance for the optimization of device properties.

Thin-film solar cells based on the CIGS absorber layer exhibits high conversion efficiencies, both in the laboratory and in production. This high efficiency together with long-term stability enables this technology to play a central role in the global renewable energy sector. There is also an interest in the development of CRM-free thin-film PV technologies. Solar cells based on the kesterite-type $Cu_2ZnSn(S,Se)_4$ (CZTSSe) absorber layer have achieved the highest conversion efficiencies amongst the emerging CRM-free technologies.

Cation and anion alloying (forming solid solutions or mixed crystals) is a common strategy to change the band gap energy of a material to an optimum value. This approach is applied in successful thin-film solar cell technologies as CIGS is a solid solution of $CuInSe_2$ and $CuGaSe_2$, and CZTSSe is a solid solution of Cu_2ZnSnS_4 and $Cu_2ZnSnSe_4$. Whereas the structural parameters in these solid solution series follow Vegard's law by showing a linear dependency on the chemical composition, the band gap energy obey a nonlinear behavior. This band gap energy bowing is a common effect in semiconductor alloy systems.

The remarkable flexibility of the chalcopyrite but also the kesterite crystal structure is a key point for these multinary compounds for their application as absorber layer in thin-film solar cells. The thin-film growth is a non-equilibrium process and thus indicated by local inhomogeneities concerning the chemical composition of the absorber phase. On the other hand, band gap fluctuations which can be caused by structural disorder and electrostatic potential fluctuations which can be caused by charged point defects [55] are reasons for significant losses in thin-film solar cells. We have demonstrated that strong variations of the band gap energy exist in off-stoichiometric kesterite-type chalcogenide semiconductors. Considering the band gap energy in dependence on the cation ratios which express the off-stoichiometry of the material (and can be described by the off-stoichiometry type model), significant changes of the band gap energy values can be obtained. Thus, local compositional variations in a kesterite-type absorber layer imply the formation of different off-stoichiometry types associated with different intrinsic point defects which strongly affect the local electronic properties as the band gap energy. On the other hand, these band gap energy variations are very small in off-stoichiometric chalcopyrite-type chalcogenide semiconductors, which may be a reason for the success of chalcopyrite-based thin-film solar cells.

References

[1]   Trends in PV applications. 2019. REPORT IEA PVPS T1-36: 2019.

[2]   COM. 2017 490 final. Brussels, 13.9.2017.

[3]   Giraldo, S., Saucedo, E., Neuschitzer, M., Oliva, F., Placidi, M., Alcobe, X., Izquierdo-Roca, V., Kim, S., Tampo, H., Shibata, H., Perez-Rodriguez, A., Pistor, P. How small amounts of Ge modify the formation pathways and crystallization of kesterites. Energy Environ. Sci 2018, 11, 582.

[4]   Wang, W., Winkler, M.T., Gunawan, O., Gokmen, T., Todorov, T.K., Zhu, Y., Mitzi, D.B. Device characteristics of CZTSSe thin-film solar cells with 12.6% efficiency. Adv. Energy Mater 2014, 4, 1301465.

[5]   Yan, C., Huang, J., Sun, K., Johnston, S., Zhang, Y., Sun, H., Pu, A., He, M., Liu, F., Eder, K., Yang, L., Cairney, J.M., Ekins-Daukes, N.J., Hameiri, Z., Stride, J.A., Chen, S., Green, M.A., Hao, X. Cu_2ZnSnS_4 solar cells with Over 10% power conversion efficiency enabled by heterojunction heat treatment. Nat. Energy. 2018, 3, 764.

[6]   www.solar-frontier.com/eng/news/2019/0117_press.html

[7]   Pamplin, B. The adamantine family of compounds. Progr. Crystal Growth Charact. 1981, 3, 179–192.

[8]   Pamplin, B. Ternary chalcopyrite compounds. Progr. Crystal Growth Charact. 1, 331–387.

[9]   Kühn, G., Neumann, H. AIBIIIX2VI Halbleiter mit Chalcopyritstruktur. Zeitschrift für Chemie 27, 197–206.

[10]   Jaffe, J.E., Zunger, A. Anion displacements and band-gap anomaly in ternary ABC_2 chalcopyrite semiconductors. Phys. Rev. B. 1983, 27, 5176–5179.

[11]   Schorr, S. Crystallographic aspects of Cu_2ZnSnS_4 (CZTS). In: Ito, K., ed, Copper Zinc Tin Sulfide-based Thin Film Solar Cells, 2014, Wiley, 53–74.

[12]   Shockley, W., Queisser, H.J. Detailed balance limit of efficiency of p-n junction solar cells. J. Appl. Phys. 1961, 32, 510–519.

[13]   Bloch, F. Über die Quantenmechanik der Elektronen in Kristallgittern. Zeitschrift für Physik A 1929, 52, 555–600.

[14]   Kittel, C. Introduction to Solid State Physics. Wiley, 2014.

[15]   Jaffe, J.E., Zunger, A. Theory of band gap anomaly in ABC_2 chalcopyrite semiconductors. Phys. Rev. B 1984, 29, 1882–1906.

[16]   Vegard, L. Die Konstitution der Mischkristalle und die Raumfüllung der Atome. Zeitschrift für Physik 1921, 5, 17–26.

[17]   Contreras, M.A., Ramanathan, K., AbuShama, J., Hasoon, F., Young, D.L., Egaas, B., Noufi, R. Diode characteristics in state-of-the-art $ZnO/CdS/Cu(In_{1-x}Ga_x)Se_2$ solar cells. Prog. Photovolt. Res. Appl. 2005, 13, 209–2016.

[18]   Wei, S.-H., Zunger, A. Band offsets and optical bowings of chalcopyrites and Zn-based II-VI alloys. J. Appl. Phys. 1995, 78, 3846–3856.

[19]   Suri, D.K., Nagpal, K.C., Chada, G.K. X-ray study of $Cu(In_{1-x}Ga_x)Se_2$ solid solutions. J. Appl. Crystallogr. 1989, 22, 578–583.

[20]   Abou-Ras, D., Caballero, R., Kaufmann, C.A., Nichterwitz, M., Sakurai, K., Schorr, S., Unold, T., Schock, H.W. Impact of the Ga concentration on the microstructure of $CuIn_{1-x}Ga_xSe_2$. Phys. Stat. Sol. (RRL) 2008, 2, 135–137.

[21]   Persson, C. Electronic and optical properties of Cu_2ZnSnS_4 and $Cu_2ZnSnSe_4$. J. Appl. Phys. 107, 053710-1-8.

[22]   Schorr, S., Höbler, H.-J., Tovar, M. A neutron diffraction study of the stannite-kesterite solid solution series. Europ. J. Mineral. 2007, 19, 65–73.

[23]   Schorr, S. The crystal structure of kesterite type compounds: A neutron and X-ray diffraction study. Sol. Energy Mater. Sol. Cells 95, 1482–1488.

[24] Pietak, K., Jastrzbeski, C., Zberecki, K., Jastzebski, D.J., Paszkowicz, W., Podsiadlo, S. Synthesis and structural characterization of Ag_2ZnSnS_4 crystals. J. Sol. State Chem. 2020, 290, 121467.

[25] Gong, W., Tabata, T., Takei, K., Morihama, M., Maeda, T., Wada, T. Crystallographic and optical properties of $(Cu,Ag)_2ZnSnS_4$ and $(Cu,Ag)_2ZnSnSe_4$ solid solutions. Phys. Status Solidi C. 2015, 12, 700–703.

[26] Gurieva, G., Marquez, J.A., Franz, A., Hages, C., Levcenco, S., Unold, T., Schorr, S. Effect of Ag incorporation on structure and optoelectronic properties of $(Ag_{1-x}Cu_x)_2ZnSnSe_4$ solid solutions. Phys. Rev. Mat. 2020, 4, 054602.

[27] Adachi, S. Earth abundant materials for solar cells.2015. (table 2.6, page 34). Wiley.

[28] Rudisch, K., Espinosa-Garcia, W., Osorio-Guillen, J.M., Araujo, C.M., Platzer-Björkmann, C., Scragg, J.S. Structural and electronic properties of Cu_2MnSnS_4 from experiment and first-principle calculations. Phys. Stat. Sol. B. 2019, 256, 1800743.

[29] Adachi, S., Earth abundant materials for solar cells.2015. (table 5.4, page 193). Wiley Adachi,

[30] Levcenco, S., Dumcenco, D., Huang, Y.S., Arushanov, E., Tezlevan, V., Tion, K.K., Du, C.H. Near band edge anisotropic optical transitions in wide gap semiconductor Cu_2ZnSiS_4. J. Appl. Phys. 2010, 108, 073508-1-5.

[31] Többens, D., Gurieva, G., Niedenzu, S., Schuck, G., Zizak, I., Schorr, S. Cation distribution in $Cu_2ZnSnSe_4$, Cu_2FeSnS_4 and $Cu_2ZnSiSe_4$ by multiple-edge anomalous diffraction. Acta Cryst. B. 2020, 76, inprint.

[32] Gunder, R., Marquez-Prietro, J.A., Gurieva, G., Unold, T., Schorr, S. Structural characterization of off-stoichiometric kesterite-type $Cu_2ZnGeSe_4$ compound semiconductors: From cation distribution to intrinsic point defect density. Cryst. Eng. Comm. 2018, 20, 1491–1498.

[33] Gurieva, G., Dimitrievska, M., Zander, S., Pérez-Rodríguez, A., Izquierdo-Roca, V., Schorr, S. Structural characterisation of $Cu_{2.04}Zn_{0.91}Sn_{1.05}S_{2.08}Se_{1.92}$. Phys. Stat. Sol. C. 2015, 12, 588–591.

[34] Valakh, M.Y., Kolomys, O.F., Ponomaryov, S.S., Yukhymchuk, V.O., Babichuk, I.S., Izquierdo-Roca, V., Saucedo, E., Perez-Rodriguez, A., Morante, J.R., Schorr, S., Bodnar, I.V. Raman scattering and disorder effect in Cu_2ZnSnS_4. Phys. Stat. Sol. RRL 2013, 7, 258–261.

[35] Többens, D.M., Gurieva, G., Levcenko, S., Unold, T., Schorr, S. Thermal dependency of Cu/Zn ordering in CZTSe kesterites determined by anomalous diffraction. Phys. Stat. Sol. B. 2016, 253, 1890–1897.

[36] Schorr, S., Gonzalez-Aviles, G. In-situ investigation of the structural phase transition in kesterite. Phys. Stat. Sol. A 2009, 206, 1054–1058.

[37] Scragg, J.J.S., Choubrac, L., Lafond, A., Ericson, T., Platzer-Björkman, C. A low-temperature order-disorder transition in Cu_2ZnSnS_4 thin films. Appl. Phys. Lett. 2014, 104, 041911.

[38] Rey, G., Redinger, A., Sendler, J., Weiss, T.P., Thevenin, M., Guennou, M., El Adib, B., Siebentritt, S. The band gap of $Cu_2ZnSnSe_4$: Effect of order-disorder, Appl. Phys. Lett. 2014, 105, 112106.

[39] Chen, S., Yang, J.-H., Gong, X.G., Wals, A., Wei, S.H. Intrinsic point defects and complexes in the quaternary kesterite semiconductor Cu_2ZnSnS_4. Phys. Rev. B. 2010, 81, 245204.

[40] Lafond, A., Choubrac, L., Guillot-Deudon, C., Deniard, P., Jobic, S. Crystal structures of photovoltaic chalcogenides, an intricate puzzle to solve: The cases of CIGSe and CZTS materials. Z. Anorg. Allg. Chem. 2012, 638, 2571–2577.

[41] Gurieva, G., Rios, L.E.V., Franz, A., Whitfield, P., Schorr, S. Intrinsic point defects in off-stoichiometric $Cu_2ZnSnSe_4$: A neutron diffraction study. J. Appl. Phys. 2018, 213, 161519.

[42] Valle, R.L.E., Neldner, K., Gurieva, G., Schorr, S. Existence of off-stoichiometric single phase kesterite. J. Alloys Comp. 2016, 657, 408–413.

[43] Wendlandt, W.W., Hecht, H.G. 1966. In Refectance spectroscopy Chapter 3. Interscience Pub., New York.

[44] Kubelka, P., Munk, F. Ein Beitrag zur Optik der Farbanstriche. Z. Techn. Physik. 1931, 12, 593–601.
[45] Kubelka, P. New contributions to the optics of intensely light-scattering materials – Part I. J. Opt. Soc. Am. 38(5), 448–457.
[46] Delgass, W.N., Haller, G.L., Kellerman, R., Lunsford, J.H. Spectroscopy in Heterogeneous Catalysis, Academic. New York, 1979.
[47] Tauc, J., Grigorovici, R., Vancu, A. Optical properties and electronic structure of amorphous germanium. Phys. Stat. Sol. 1966, 15, 627–637.
[48] Parasyuk, O., Piskach, L., Romanyuk, Y.E., Olekseyuk, I., Zaremba, V., Pekhnyo, V. Phase relations in the quasi-binary Cu_2GeS_3-ZnS and quasi-ternary Cu_2S-Zn(Cd)S-GeS_2 systems and crystal structure of Cu_2ZnGeS_4. J. All. Com. 2005, 397, 85–94.
[49] Fan, C.-M., Regulacio, M.D., Ye, C., Lim, S.H., Lua, S.K., Xu, Q.-H., Dong, Z., Xu, A.-W., Han, M.-Y. Colloidal nanocrystals of orthorhombic Cu_2ZnGeS_4: Phase-controlled synthesis, formation mechanism and photovoltaic behavior. Nanoscale 2015, 7, 3247–3253.
[50] Schorr, S., Stephan, C., Kaufmann, C.A. Chalcopyrite thin film solar cell devices. In: Kearley, G. J., Peterson, V. K., ed., Neutron Scattering Applications and Techniques, 2015, Springer.
[51] Schorr, S., Gurieva, G., Guc, M., Dimitievska, M., Perez-Rodriguez, A., Izquierdo-Roca, V., Schnohr, C., Kim, J., Jo, W., Merino, J.M. Point defects, compositional fluctuations, secondary phases in non-stoichiometric kesterites. J. Phys.: Energy. 2, 012002.
[52] Noufi, R., Axton, R., Herrington, C., Deb, S.K. Electronic properties versus composition of thin films of $CuInSe_2$. Appl. Phys. Lett. 1984, 45, 668–670.
[53] Stephan, C., Schorr, S., Tovar, M., Schock, H.-W. Comprehensive insights into point defects and defect cluster formation in $CuInSe_2$. Appl. Phys. Lett. 2011, 98, 091906.
[54] Ritzer, M., Schönherr, S., Schöppe, P., Larramona, G., Chone, C., Gurieva, G., Johannes, A., Ritter, K., Martinez-Criado, G., Schorr, S., Ronning, C., Schnohr, C.S. Interplay of performance-limiting nanoscale features in $Cu_2ZnSn(S,Se)_4$ solar cells. Phys. Stat. Sol. A. 2020, 2017, 2000456.
[55] Gokmen, T., Gunawan, O., Todorov, T.K., Mitzi, D.B. Band tailing and efficiency limitation in kesterite solar cells. Appl. Phys. Lett. 2013, 103, 103506.

Joachim Breternitz

5 Halide semiconductors: symmetry relations in the perovskite type and beyond

Abstract: Halide perovskites are considered as one of the most promising material systems for photovoltaic absorbers. With the steep rise in interest, a vast number of compounds that are only vaguely related are being subsumed as perovskite semiconductors. The aim of this chapter is to give an overview of the different classes of materials that belong to the halide semiconductors and to give a structural-systematic view of the halide perovskite family and its structural variations. The latter is realized in the form of a Bärnighausen tree alongside and seconded by considerations as to why this classification is important for halide perovskite materials.

Keywords: halide perovskites, group–subgroup relationships, materials for energy conversion

5.1 Halide semiconductors versus halide perovskites

A novel class of solar absorbers for solar cells has emerged over the last decade, peaking at efficiencies of over 25% (and hence close to the theoretical maximum) within this period of time [1]. This development started with methylammonium lead iodide ($CH_3NH_3PbI_3$ or $MAPbI_3$) [2], which still is the signature compound of this class of materials. Since $CH_3NH_3PbI_3$ crystallizes in a perovskite-type structure (Fig. 5.1) [3], the class of materials was accordingly called halide perovskites to specify their common chemical factor: halide anions. Since an important part of those absorber materials also contain organic cations on the A-site, they are also called hybrid halide perovskites [4]. The structural features of these compounds and the relationships between the different phases of this class of materials will be at the core of this chapter.

The structural diversity of halide materials being considered as semiconductors has, nonetheless, far outreached those materials with a clear structural link to the perovskite type [5]. From a structural crystallographic perspective, the term "halide semiconductors" is probably the better one. It includes perovskite-type materials, but is not

Acknowledgments: The author would like to express his sincere gratitude to Dr. Dennis Wiedemann for discussion and advice on the manuscript.

Joachim Breternitz, Helmholtz-Zentrum Berlin für Materialien und Energie, Hahn-Meitner-Platz 1, 14109 Berlin, Germany

https://doi.org/10.1515/9783110674910-005

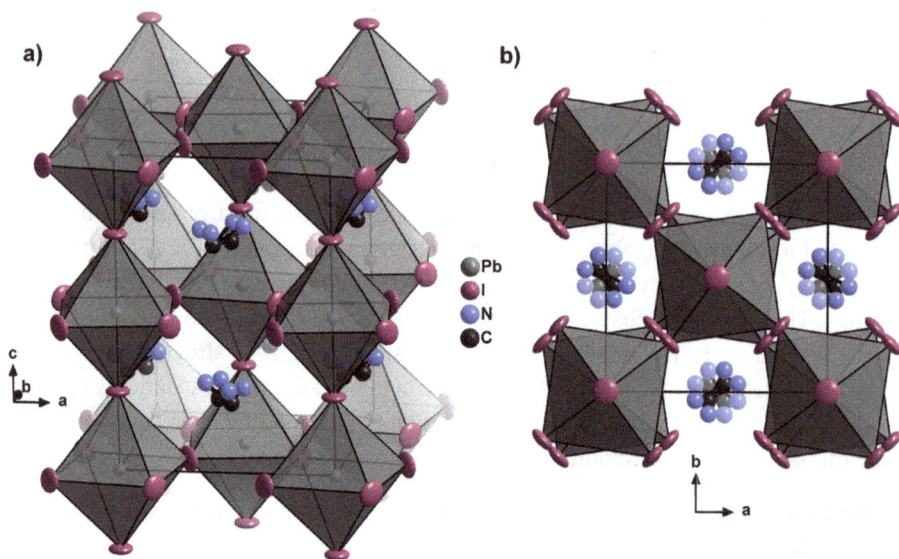

Fig. 5.1: Unit cell representation of MAPbI$_3$ at room temperature (space group *I4cm*) in a general view (a) and along the crystallographic *c*-axis (b) [3].

restricted to those materials and still highlights the unique properties of this class of materials, that is, their halide anions. It is important to make that distinction since the features of the crystal structure do translate into its properties, and especially into its electronic properties that, naturally, play a predominant role as semiconductor material. Therefore, it is beneficial to commence with a short definition of perovskite and further classes of materials that are important in the halide semiconductors. Herein, the terminology for symmetry relations may be briefly mentioned, but the reader is pointed to the next section for a more conclusive explanation of the underlying group theory in crystallographic space group types.

5.1.1 A perovskite: as far as structure is concerned

The name perovskite stems from the mineral perovskite, which was found in Achmatovsk in the Ural region, and sent to the German mineralogist Gustav Rose. He first characterized the mineral to contain Ca and Ti, probably as an oxide, and named it after the Russian mineralogist Lev Perovski in 1839 [6]. The chemical composition of the compound was later determined to be CaTiO$_3$ and it should be discovered in the following years that this material is the archetype of one of the largest (and most important) classes of structure types. Most of the compounds in this class were oxide materials, although the existence of halide perovskites was known for quite a while. Some of the seminal work, mainly focused on oxide perovskites [7–11], should be highlighted as it was vital for the understanding of the different crystal structures and their relationships within this structural type, especially for oxides. The general composition of all perovskites is ABX_3.

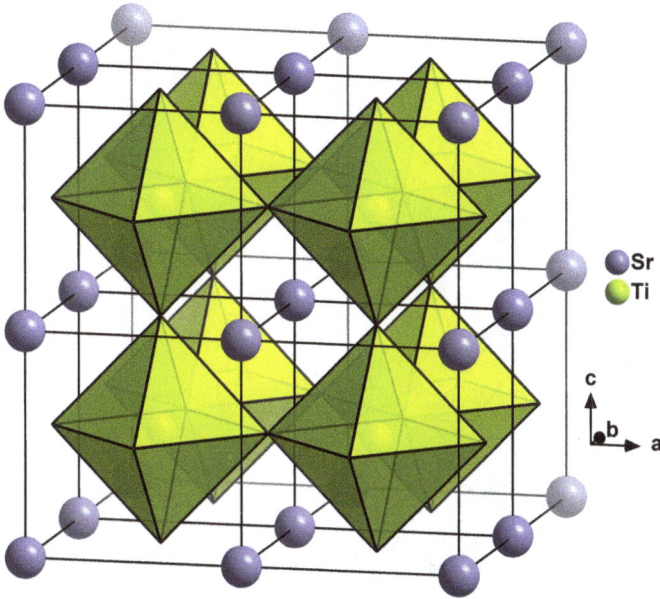

Fig. 5.2: Structural view of $SrTiO_3$ [12] in the cubic aristotype in a 2 × 2 × 2 supercell. Atoms are represented as arbitrary spheres, and oxide ions at the octahedron edges are omitted for clarity.

But from a structural point of view, what is it that determines a perovskite-type structure? The simplest way to look at a perovskite is to regard its *aristotype*, that is, the crystal structure with the highest possible symmetry (Fig. 5.2). However, the mineral perovskite itself ironically does not crystallize in the cubic aristotype, but a lower symmetry subgroup at ambient conditions [13]. The two different cations occupy two distinct positions in the crystal structure: the A-cation (Sr^{2+} in the case of $SrTiO_3$, Fig. 5.2) has 12 nearest anion neighbors forming a cuboctahedral coordination, while the B-cation (Ti^{4+} in $SrTiO_3$) is sixfold coordinated in an octahedron. These octahedra are corner-connected and form a three-dimensional corner-sharing network. This coordination in halide perovskites is vital for the electronic properties of the materials and, therefore, is the most important structural feature of a material to be considered as perovskite [14]. It should be emphasized that this does not mean that other materials cannot work as semiconductors or are intrinsically worse, but it means that they are working differently, which should reflect in the nomenclature.

5.1.2 Halide semiconductors beyond pure perovskites

The range of further halide semiconductors is very rich and spans from octahedrally coordinated cations organized in different motifs to materials that do not feature octahedrally coordinated cations whatsoever. Therefore, only examples of other structural types can be highlighted herein. It should be emphasized that for some of these classes, a clear chemical and structural relationship can be derived that is

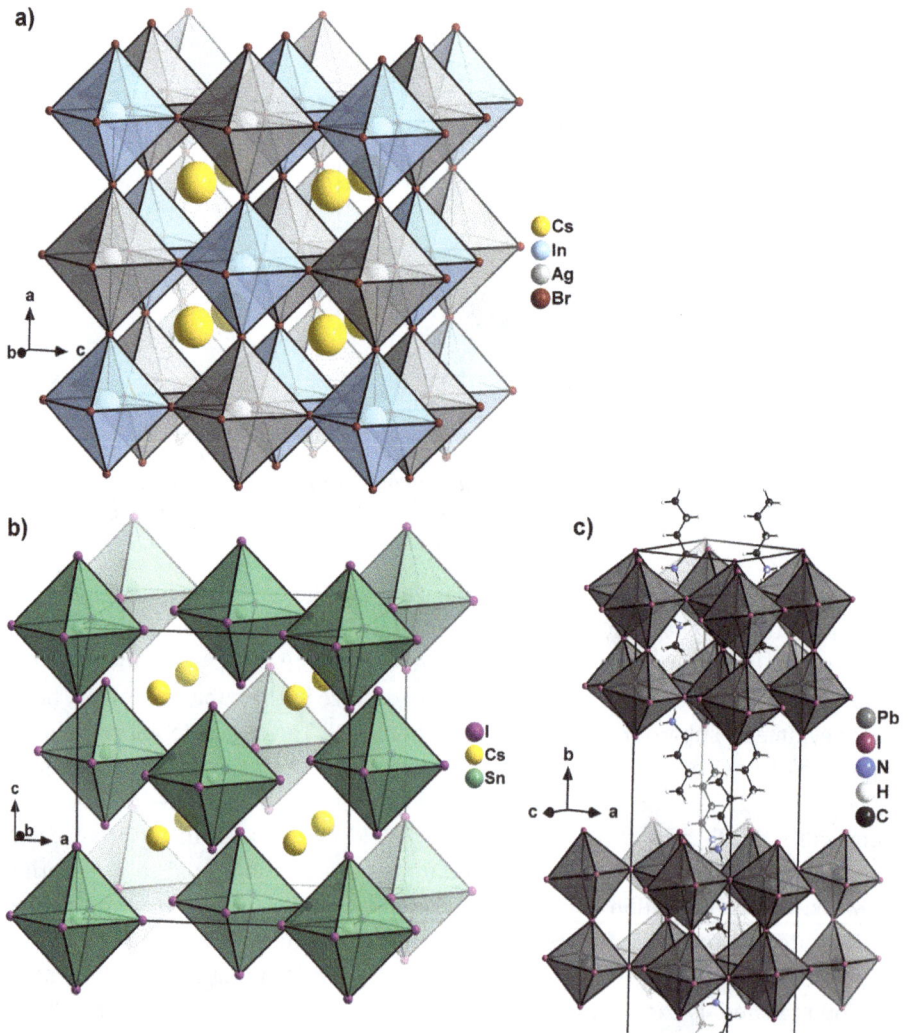

Fig. 5.3: Crystal structure representations of the double perovskite $Cs_2[AgIn]Br_6$ (a) [15], the vacancy-ordered perovskite Cs_2SnI_6 (b) [16], and the layered perovskite $(CH_3(CH_2)_3NH_3)_2(CH_3NH_3)$ Pb_2I_7 (c) [17]. Atoms are represented as generic spheres.

beyond a group–subgroup relationship. These material classes are commonly referred to by attaching an additional explanation to the term perovskite, such as "double perovskites," "layered perovskites," or "defect-ordered perovskites." From a purely structural point of view, this terminology is not correct; however, it can often be found in the less structurally focused literature and, therefore, should be mentioned for completeness. It should, nonetheless, be noted that these terms have to be used with extreme caution and, if at all, are to be used *as is*. In particular, this means as a composed term to emphasize that these materials may be linked to perovskite materials but do not belong to the perovskite structure type. The same applies, for instance, to the so-called hexagonal perovskites, in which the BX_6 octahedra are face-sharing rather than corner-sharing. This prudence is, again, not recommended out of sheer structural reasoning, but because the structural changes between these classes of materials cause clear electronic changes that need to be addressed properly.

Some of these materials beyond classical perovskites have gained quite some attention and should nonetheless be briefly mentioned with the perovskite-derived names:

- *Double perovskites* [18] probably still bear the clearest structural link to the perovskite type and still completely fulfill the prerequisite of corner-sharing BX_6 octahedra. But instead of a single type of B-cations, they possess two distinct types (usually denoted B and B′) that order in a specific way, and therefore form a quaternary, rather than a ternary compound (Fig. 5.3a). This was identified as one way to replace toxic divalent lead ions with a mixture of less toxic monovalent and trivalent cations. It should further be noted that this is different from mixed-cation perovskites, where different B-cations are alloyed on the same crystallographic site, that is, they are occupationally disordered. The aristotype of the double perovskites is a direct subgroup of the perovskite aristotype and is named after the mineral elpasolite (K_2NaAlF_6) [19]. Structural relationships between these compounds will briefly be outlined later.

- *Vacancy-ordered (double) perovskites* [20] can best be understood from the perspective of a double perovskite, but with a notable difference: Instead of two different cations, B and B′, these compounds only contain one sort, while the other sites of B′ remain unoccupied (Fig. 5.3b). This arrangement leads to the reduction of the dimensionality from a three-dimensional BX_6 octahedral network to isolated (zero-dimensional) octahedra, therefore, dramatically impacting the electronic properties. From a chemical point of view, this structure goes along with a doubling of the charge of the B-cations to account for the decreased number of B-cations per formula unit. The aristotype of these vacancy-ordered perovskites is of K_2PtCl_6 type [21]. A prominent example of such compounds is Cs_2SnI_6, which forms via rapid oxidation of the perovskite $CsSnI_3$ in air [22].

- *Layered perovskites* no longer bear a direct group–subgroup relationship to the perovskite type [23]. Rather, they can be rationalized from a more chemical and visual point of view. They still contain the motif of edge-sharing octahedra, but they no longer form a three-dimensional network. Instead, layers are formed that stack along one dimension, separated from each other by bulkier cations (Fig. 5.3c). Saparov and Mitzi rationalized the different layered perovskite variants according to the cleavage direction, which results in different orientations of the perovskite layers relative to each other [24]. According to the relative stacking of the layers, the most prominent structures are divided into Ruddlesden–Popper phases, Dion–Jacobson phases, and "alternating cation interlayer" phases. (The latter was introduced by Soe et al. [25].) However, this classification has been questioned recently and the reader is pointed to those publications for further reading [26].

– A number of further structures that have very little structural relationship to the perovskite type can be subsumed as *lower dimensional structures with octahedral coordination*. One prominent example of these compounds is $Cs_3Bi_2I_9$ (Fig. 5.4a), which crystallizes in the $Cs_3Cr_2Cl_9$ type [27]. Herein, the Bi^{3+} cations are still coordinated octahedrally, but the octahedra are no longer forming a corner-sharing network. Instead, two octahedra share faces to constitute isolated double-octahedra $[Bi_2I_9]^{3-}$ entities, which are surrounded by Cs^+ cations. These compounds are neither structurally nor chemically linked to the perovskite type and, therefore, using the term perovskite for these structures is strongly discouraged.

– Unfortunately, there are a small number of compounds that are being called "perovskites" although they do not even contain octahedrally coordinated cations, whatsoever. $(H_3N\{CH_2\}_4NH_3)[CoCl_4]$, as shown in Fig. 5.4b, for instance, is being intituled as "2D hybrid perovskite," although it is clearly not two-dimensional but contains isolated, tetrahedral $[CoCl_4]^{2-}$ anions [28]. These compounds should, by no means, be called perovskites.

a)

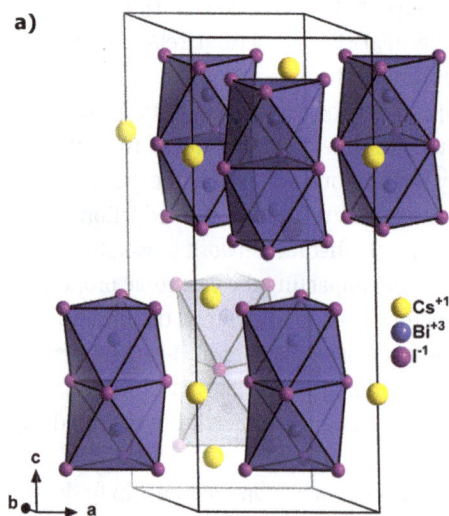

Fig. 5.4: Representations of the crystal structures of $Cs_3Bi_2I_9$ (a) [27] and $(H_3N\{CH_2\}_4NH_3)[CoCl_4]$ (b) [28]. Atoms are represented as generic spheres and atom charges are shown if given in the cif file.

b)

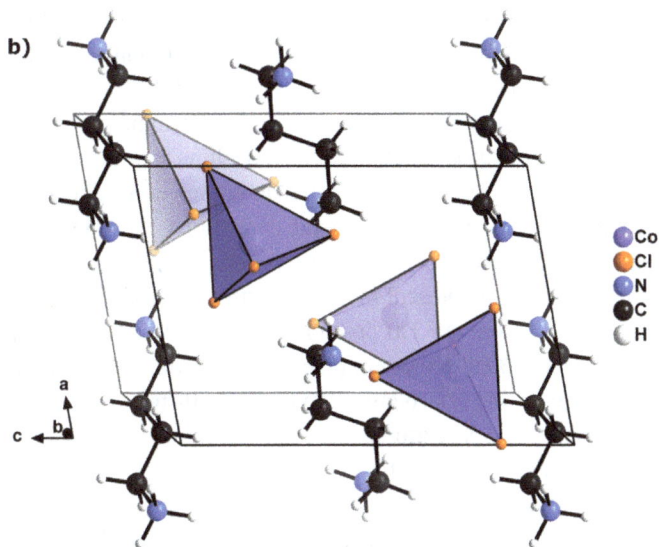

Fig. 5.4 (continued)

5.2 Crystallographic group theory in a nutshell

It is, of course, impossible to give a comprehensive overview over the whole field of crystallographic group theory within this document, and the reader is pointed to some excellent literature on the matter [7, 29, 30], especially the *International Tables for Crystallography* are most useful in establishing group–subgroup relationships [31, 32]. Very briefly, Volume A contains information on the symmetry in the space group types, while Volume A1 establishes their relationships. The tools of the Bilbao crystallographic server are also noteworthy, as they supply routines for the computer-aided group–subgroup analysis [33–35] and the seminal work of Campbell and Stokes aids in identifying distortion modes in group–subgroup relationships [36, 37]. Nevertheless, the bare minimum necessary for the proper understanding of symmetry relations between crystal structures will be briefly mentioned in this section. The unique property of crystals is the periodic structure governed by symmetry relations between the atoms. Therefore, the structure of a solid can be described using unit cells, that is, small fractions of the overall volume that contain the full information of the atomic structure and that are repeated periodically in three dimensions.[1] In addition to this *translational symmetry* (of which unit cell centerings are another type), further symmetry elements occur in crystallography that link positions

---

1 For the sake of completeness, it should be mentioned that this does not apply for aperiodic crystals, although they are crystalline, too.

in the unit cell to each other: pure *point symmetry* elements are centers of inversion, rotation axes, and mirror planes. Finally, some symmetry elements combine translational and point symmetries. These are screw axes and glide planes. It is important to point out the difference between point symmetry and translational symmetry, as this plays an important role in the nature of group–subgroup relationships and impacts the expected properties. While translational symmetry elements determine unit cell centerings, the positions of the diffraction maxima and systematic extinctions, the point symmetry elements determine, to which of the 32 crystal classes the structure belongs.

The entirety of all symmetry elements present in a crystal structure constitutes the space group type of that structure. There are 230 possible combinations of symmetry elements in 3 dimensions and hence 230 space-group types. It is important to point out the difference between space group and space-group types, as both terms are commonly interchanged and confused [38]. Space groups are a specific set of symmetry elements and their concrete distribution in space, that is, the lattice parameters are an integral part of the space group. Space-group types, however, only contain the set of symmetry elements and their relative ratio. In a way, one can think that the space-group type is scaled by the lattice parameters to the space group. Some space-group types bear the same point symmetry but differ in translational symmetry elements. These space-group types belong to the same crystal class. A comprehensive overview of space-group types with their corresponding crystal classes is given in Tab. 5.1. The knowledge of which is vital for the understanding of the different possible symmetry transitions. Space-group types can further be understood as a group in the mathematical (i.e., group-theoretical) sense that is defined by the symmetry operations constituting it [29]. It is possible to define subsets, which constitute space-group types themselves. These *subgroups* signify a lower symmetry than the original space-group type as they contain less symmetry elements, but it is important to note that they do not contain any elements that are not elements of the higher symmetry group (also called *group*). Therefore, a structure represented in a certain group can always be described in one of its subgroups. The reduction of the number of symmetry elements between the group and the subgroup has one important consequence; however, the structure described gains more degrees of freedom, either by reducing the symmetry of special positions (e.g., by no longer restricting the associated coordinates) or by splitting a position into multiple ones, thereby allowing their occupation by atoms that are crystallographically distinct.

Tab. 5.1: Overview of the crystal classes with the corresponding space-group types.

Crystal system	Crystal class	Space-group types
Triclinic	1	$P1$
	$\bar{1}$	$P\bar{1}$
Monoclinic	2	$P2, P2_1, C2$
	m	Pm, Pc, Cm, Cc
	$2/m$	$P2/m, P2_1/m, C2/m, P2/c, P2_1/c, C2/c$
Orthorhombic	222	$P222, P222_1, P2_12_12, P2_12_12_1, C222_1, C222, F222, I222, I2_12_12_1$
	$mm2$	$Pmm2, Pmc2_1, Pcc2, Pma2, Pca2_1, Pnc2, Pmn2_1, Pba2, Pna2_1, Pnn2, Cmm2, Cmc2_1, Ccc2, Amm2, Abm2, Ama2, Aba2, Fmm2, Fdd2, Imm2, Iba2, Ima2$
	mmm	Pmmm, Pnnn, Pccm, Pban, Pmma, Pnna, Pmna, Pcca, Pbam, Pccn, Pbcm, Pnnm, Pmmn, Pbcn, Pbca, pnma, Cmcm, Cmca, Cmmm, Cccm, Cmma, Ccca, Fmmm, Fddd, Immm, Ibam, Ibca, Imma
Tetragonal	4	$P4, P4_1, P4_2, P4_3, I4, I4_1$
	$\bar{4}$	$P\bar{4}, I\bar{4}$
	$4/m$	$P4/m, P4_2/m, P4/n, P4_2/n, I4/m, I4_1/a$
	422	$P422, P42_12, P4_122, P4_12_12, P4_222, P4_22_12, P4_322, P4_32_12, I422, I4_122$
	$4mm$	$P4mm, P4bm, P4_2cm, P4_2nm, P4cc, P4nc, P4_2mc, P4_2bc, I4mm, I4cm, I4_1md, I4_1cd$
	$\bar{4}2m$	$P\bar{4}2m, P\bar{4}2c, P\bar{4}2_1m, P\bar{4}2_1c, P\bar{4}m2, P\bar{4}c2, P\bar{4}b2, I\bar{4}m2, I\bar{4}c2, I\bar{4}2m, I\bar{4}2d$
	$4/mmm$	$P4/mmm, P4/mcc, P4/nbm, P4/nnc, P4/mbm, P4/mnc, P4/nmm, P4/ncc, P4_2/mmc, P4_2/mcm, P4_2/nbc, P4_2/nnm, P4_2/mbc, P4_2/mnm, P4_2/nmc, P4_2/ncn, I4/mmm, I4/mcm, I4_1/amd, I4_1/acd$
Trigonal	3	$P3, P3_1, P3_2, R3$
	$\bar{3}$	$P\bar{3}, R\bar{3}$
	32	$P312, P321, P3_112, P3_121, P3_212, P3_221, R32$
	$3m$	$P3m1, P31m, P3c1, P31c, R3m, R3c$
	$\bar{3}m$	$P\bar{3}1m, P\bar{3}1c, P\bar{3}m1, P\bar{3}c1, R\bar{3}m, R\bar{3}c$
Hexagonal	6	$P6, P6_1, P6_5, P6_2, P6_4, P6_3$
	$\bar{6}$	$P\bar{6}$
	$6/m$	$P6/m, P6_3/m$
	622	$P622, P6_122, P6_522, P6_222, P6_422, P6_322$
	$6mm$	$P6mm, P6cc, P6_3cm, P6_3mc$
	$\bar{6}m2$	$P\bar{6}m2, P\bar{6}c2, P\bar{6}2m, P\bar{6}2c$
	$6/mmm$	$P6/mmm, P6/mcc, P6_3/mcm, P6_3/mmc$
Cubic	23	$P23, F23, I23, P2_13, I2_13$
	$m\bar{3}$	$Pm\bar{3}, Pn\bar{3}, Fm\bar{3}, Fd\bar{3}, Im\bar{3}, Fa\bar{3}, Ia\bar{3}$
	432	$P432, P4_332, F432, F4_132, I432, P4_332, P4_132, I4_132$
	$\bar{4}3m$	$P\bar{4}3m, F\bar{4}3m, I\bar{4}3m, P\bar{4}3\,n, F\bar{4}3c, I\bar{4}3d$
	$m\bar{3}m$	$Pm\bar{3}m, Pn\bar{3}n, Pm\bar{3}n, Pn\bar{3}m, Fm\bar{3}m, Fm\bar{3}c, Fd\bar{3}m, Fd\bar{3}c, Im\bar{3}m, Ia\bar{3}d$

Centrosymmetric crystal classes are typeset in bold.

5.2.1 Representation of group–subgroup relationships

To represent group–subgroup relationships in an efficient way, the illustration in Bärnighausen trees has proven an effective way with information condensed into a single figure [7]. Interpreting this information may be confusing at first; therefore, a short exemplary guide shall be given. The different phases of $CsSnI_3$ at high temperatures, for instance, are depicted in Fig. 5.5 with their group–subgroup relationship. It should be noted that not every possible element of a Bärnighausen tree is always shown, and especially atomic positions are normally omitted in more comprehensive Bärnighausen trees, displaying numerous subgroups.

The structure with the highest symmetry may be found at the top of the table. In this case, at high temperature, $CsSnI_3$ crystallizes in the perovskite *aristotype*, that is, with the highest possible symmetry achievable in the system, in the space-group-type $Pm\bar{3}m$ [39]. The lower symmetry subgroups of the aristotype are *hettotypes*. While space-group types are normally given using their short Hermann–Mauguin symbol (such as $Pm\bar{3}m$), it is more convenient to use the complete Hermann–Mauguin symbol (such as $P\ 4/m\ \bar{3}\ 2/m$) when describing group–subgroup relationships, as this gives a more comprehensive view of the symmetry elements. It should also be noted that not all the subgroups in the table need to represent observed structures: The space-group types $P4/mmm$ and $Pbam$, for instance, link the space-group types of the observed structures, but are themselves not observed.

Group–subgroup transitions cannot happen in any random way. Instead, two main classes of symmetry reduced subgroups may be defined: *translationengleiche* and *klassengleiche* subgroups. In a *translationengleiche* (= same translations) subgroup, only point symmetry elements are lost, while the translational symmetry elements are preserved. Therefore, the crystal class necessarily changes between a group and its *translationengleiche* subgroup. A *klassengleiche* (= same class, meaning crystal class) subgroup, on the other hand, preserves the crystal class, and therefore, the point symmetry remains the same. Instead, the unit cell size increases or cell centering is lost. While the number of symmetry elements per unit cell remains the same, this may be understood as a reduction of the symmetry element density, as the same number of symmetry elements applies to a larger unit cell. With these two different group–subgroup types, it is useful to name the respective symmetry descents as *translationengleiche* and *klassengleiche* symmetry descents, respectively. The two variants of possible symmetry descents are denoted with small Latin letters (t for *translationengleich* and k for *klassengleich*) together with the index of the respective symmetry descent, which describes the reduction factor of the symmetry descent. Take t3 as an example, which describes a *translationengleiche* symmetry descent of index 3, that is, in this case, one-third of the symmetry elements of a group are preserved in its subgroup. In the case of *klassengleiche* symmetry descents in the same centering, an index 2 would imply a doubling of the unit cell volume. Finally, *isomorphic* symmetry descents (abbreviated with an i) are a special case of *klassengleiche* symmetry descents,

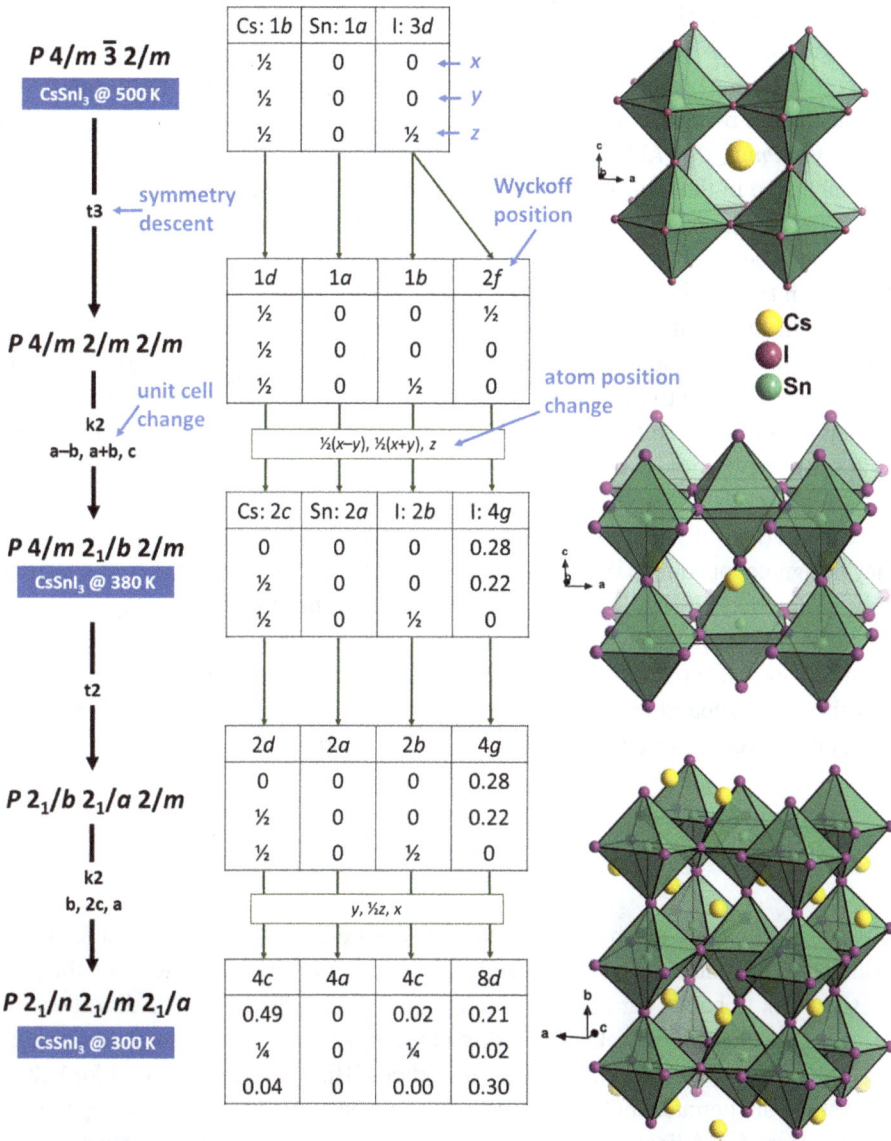

	Cs: 1b	Sn: 1a	I: 3d
P 4/m $\bar{3}$ 2/m	½	0	0 ← x
CsSnI₃ @ 500 K	½	0	0 ← y
	½	0	½ ← z

t3 ← symmetry descent

Wyckoff position

	1d	1a	1b	2f
P 4/m 2/m 2/m	½	0	0	½
	½	0	0	0
	½	0	½	0

atom position change

k2 unit cell change
a−b, a+b, c

½(x−y), ½(x+y), z

	Cs: 2c	Sn: 2a	I: 2b	I: 4g
P 4/m 2₁/b 2/m	0	0	0	0.28
CsSnI₃ @ 380 K	½	0	0	0.22
	½	0	½	0

t2

	2d	2a	2b	4g
P 2₁/b 2₁/a 2/m	0	0	0	0.28
	½	0	0	0.22
	½	0	½	0

k2
b, 2c, a

y, ½z, x

	4c	4a	4c	8d
P 2₁/n 2₁/m 2₁/a	0.49	0	0.02	0.21
CsSnI₃ @ 300 K	¼	0	¼	0.02
	0.04	0	0.00	0.30

Cs
I
Sn

Fig. 5.5: Group–subgroup relationship between the perovskite phases of CsSnI₃ at different temperatures, with splitting of the atomic positions and representations of the respective unit cells. Explanatory notes to the different elements are given in blue. Further, the high-, intermediate- [39], and low-temperature [40] structures are depicted with the atoms represented as generic spheres.

where group and subgroup belong to the same space-group type but with increased volumes in the subgroup.

It is important to factor the Wyckoff positions of the atoms in the considerations of group–subgroup relationships as they need to be consistent with the relationship in question. The fact that one can determine a group–subgroup path between two space-group types does not mean that they are linked in the respective structures; rather, the atomic positions must correspond to each other. There is, for instance, an additional non-perovskite low-temperature structure of $CsSnI_3$ described in the space-group type *Pnma* [40]. Although *Pnma* is a subgroup of $Pm\bar{3}m$, this does not constitute a membership in the family of perovskite structures. Also, it may happen that crystal structures can be expressed in different equivalent settings: The aristotype, for instance, may be described with the B-cation at the origin (Wyckoff position 1*a*) and the A-cation in the unit cell center (Wyckoff position 1*b*) as shown in Fig. 5.5, or with the A-cation on 1*a* and the B-cation on 1*b* as shown in Fig. 5.2. The anions would still arrange octahedrally around the B-cation and be on Wyckoff position 3*d* $(0,0,\frac{1}{2})$ in the first case or 3*c* $(\frac{1}{2},\frac{1}{2},0)$ in the latter case. In fact, this relationship may simply be described as an origin shift by $[\frac{1}{2},\frac{1}{2},\frac{1}{2}]$. To be consistent, it is necessary to use one unique setting. Hence, the setting with the B-cation at the origin was chosen for the following considerations (Fig. 5.5). Finally, one may find fractions (like "½") as well as decimals (like "0.5" or "~0.5") in the positional table of a Bärnighausen tree. While they designate, of course, the same numerical value, the difference lies in the restrictions of the Wyckoff site. Using a fraction means that the atom lies on a special position, which is restricted by symmetry, that is, the symmetry elements require that an atom lies on that Wyckoff site exactly at $x = \frac{1}{2}$. A decimal, on the other hand, means that the atom is found at the designated position, but there is no symmetry requirement for it; it could as well shift to, say, $x = 0.52$ or another value.

5.3 Structural distortions in the perovskite family

Distortions in perovskites have been extensively studied in the past, with a focus on oxide perovskite materials. This work builds upon these studies and extends the pioneering work to halide perovskites, which partly behave differently. The distortions in oxide perovskites are typically expressing in four ways: octahedral tilting, atom shifting, atom substitution, and Jahn–Teller distortions [11]. As the cations used for halide perovskites are normally not Jahn–Teller active, the latter point is of minor importance for this class of materials. The substitution of atoms, most importantly B-cations, in an ordered manner leads to the class of double perovskites and is treated in the next section. Therefore, two main distortion modes are similar to those in oxide perovskites:
1) shifting of individual atoms (mostly the B-cations) from their high-symmetry sites or
2) tilting of the octahedra relative to each other, which decreases the size of the coordination cuboctahedral cavity of the A-cations.

In addition to those two mechanisms of distortion, another fundamental one can occur in hybrid halide perovskites, where the A-site is occupied by a molecule rather than a single atom. This is because these molecular cations can no longer be treated as spherical and do interact with their coordination environment in a non-isotropic way. Therefore, a third distortion needs to be considered:

3) the orientation of the molecular A-cation.

It should be mentioned that these distortion modes are not necessarily isolated from each other but may be connected in a rather complex manner. The three main distortion modes outlined above will be briefly discussed below.

5.3.1 Atom shifting from high-symmetry positions

If, for instance, the B-cation shifts from the center of its octahedral coordination (Fig. 5.6), some of the symmetry elements of the high symmetry structure will be broken while others remain. In consequence, the structure will be described in a space-group type, that is, a subgroup of the pristine one. In the example depicted in Fig. 5.6, where the lead atom is shifted toward one of the iodines, this subgroup would be $P4mm$ if starting from the $Pm\bar{3}m$ aristotype. Further symmetry reductions would consequently lead to additional positional degrees of freedom and a cascade of group–subgroup descents can be established in this case. This type of distortion is, for instance, described by Spanopoulos et al. for "hollow" perovskites [41], like $MA_{0.71}en_{0.29}Pb_{0.797}I_{2.884}$ (refined as $MAPb_{0.847}I_3$; en: ethylenediamine) where lead is shifted by 11 pm from the iodine square plane. A similar effect, but through a shift of the B-cation along one of the threefold axes towards a face of the coordination octahedron, leads to the crystal structure of $CsGeX_3$ (X = Cl, Br, I) at room temperature [42], which crystallizes in the trigonal space-group type $R3m$. These shifts break the centrosymmetry, since they all occur in the same direction. The existence of a non-centrosymmetric space group is a necessary prerequisite for nonlinear optical effects. Such non-centrosymmetric space group is for instance, observed for $MAPbI_3$ at ambient conditions [43, 44].

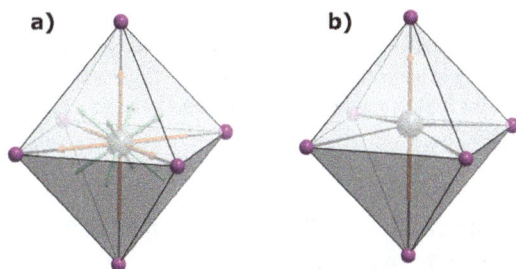

Fig. 5.6: The fourfold (orange) and threefold (green) rotation axes in the PbI_6 octahedra in (a) the cubic high-temperature structure of $CsPbI_3$ [45] and (b) the same with a lead distortion along a Pb–I bonding vector. Atoms are represented as generic spheres.

Not only the B-cation may shift but also the halide anions can move away from their high-symmetry positions to cause a symmetry reduction. This effect was, for instance, reported for the "hollow" perovskite $FA_{0.73}en_{0.27}Pb_{0.811}I_{2.892}$ (refined as $FAPb_{0.96}I_3$, space group type $Amm2$) [41]. While all halogen atoms lie on fourfold axes in the aristotype (Wyckoff position $3d$) and are crystallographically equivalent, this degeneracy is lifted in the tetragonal subgroup $P4/mmm$ (compare Fig. 5.4). In fact, the anion site splits into two inequivalent ones, $1b$ and $2f$, where the latter is occupied by four anions forming a square (and no longer lie on fourfold axes), while the $1b$ position is occupied by two anions expanding the square plane to a distorted octahedron or more exactly a square-planar bipyramid. As $1b$ anions remain on a four-fold axis, they are symmetry-restricted to a position that fixes all $X–B–X$ angles to 90° or, in other words, positions perpendicularly above the BX_4 square plane. Further symmetry breaking decreases the symmetry to the orthorhombic crystal system and also permits that the $B–X$ distances within the BX_4 square plane become inequivalent.

5.3.2 Octahedral tilting

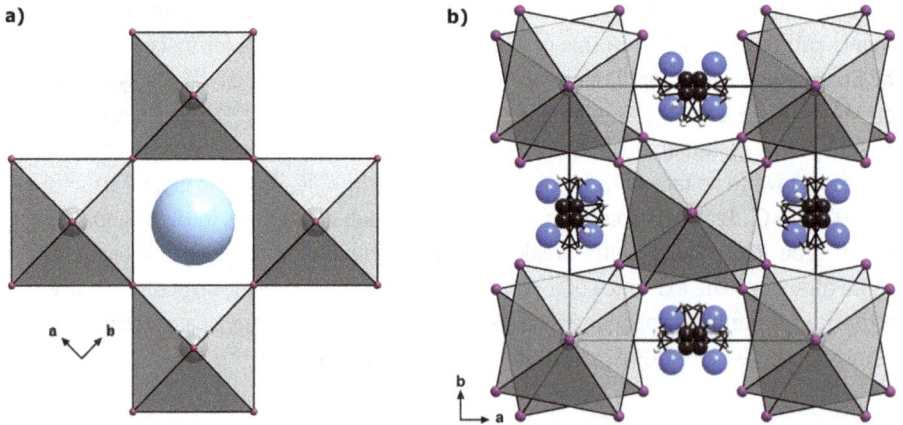

Fig. 5.7: Structural representations of $MAPbI_3$ in (a) the undistorted aristotype ($Pm\bar{3}m$) [46] and (b) the tilted structure ($I4/mcm$) [47] viewed along the crystallographic c-axis. The atoms are given as generic spheres, with methylammonium in the high temperature represented as one sphere.

One of the distinguishing features of perovskite structures is the comparably large size of the A-cation coordination polyhedra, which allows the accommodation of larger cations while the B-cations are smaller in comparison. Based on geometrical considerations, Goldschmidt defined an existence region, in which a perovskite-type structure (or a hettotype thereof) is likely to be formed [48]. For an undistorted perovskite, the ideal A-cation size should be

$$r_A = t \cdot \left[\sqrt{2} \cdot r_B + \left(\sqrt{2} - 1 \right) \cdot r_X \right]$$

where t is the tolerance factor. While t would be 1 for oxides and fluorides, a factor of 0.8–0.9 has proven more appropriate for chalcogenides and chlorides [11], and probably also for the heavier halides. For a perovskite containing Pb^{2+} ($r_{Shannon}^{[VI]}$ = 1.19 Å) and I^- ($r_{Shannon}^{[VI]}$ = 2.2 Å) [49] as B-cation and anion, respectively, the ideal size of the A-cation would hence be 2.33 Å when taking a tolerance factor of 0.9 into account. Still, the organic methylammonium cation, with which lead and iodine form the signature compound $MAPbI_3$, has an effective ionic radius that is too small (r = 2.17 Å [50]). In a case, where the A-cation is slightly smaller than ideal for the perovskite structure, this can be partially compensated through a tilting of the BX_6 octahedra relative to each other.

Glazer developed a notation scheme to describe the tilting modes in the perovskites [9]. Herein, the crystallographic directions are noted as a, b, and c. No tilting in one direction is being expressed with a superscript 0. In-phase and antiphase tiltings are being expressed with + and −, respectively. In-phase tilts mean that subsequent layers along the designated axis tilt in the same direction, while the layers twist in opposite directions in antiphase tiltings. If the tilts in different crystallographic directions are equivalent, the same letter is being used. The undistorted aristotype perovskite, for instance (Fig. 5.7a), is written in Glazer notation as $a^0a^0a^0$, as all crystallographic directions are equivalent in the cubic symmetry. The tetragonal room-temperature structure of $MAPbI_3$, on the other hand (Fig. 5.7b), is described as $a^0a^0c^-$, since the octahedra are tilted antiphase in the crystallographic c-direction. Given the octahedral network can tilt in three dimensions, 10 tilting mode combinations are possible [9]. The Glazer systematic is very useful for indexing and interpreting diffraction patterns of unknown perovskite phases in tilted hettotypes [10].

The tilting in hybrid perovskites is, however, not only influenced by the size of the A-cation but also by its shape in a rather complex manner. While $MAPbI_3$ at room temperature exhibits an $a^0a^0c^-$ tilting [47], the high-pressure phase of $MAPbI_3$ has an $a^+a^+a^+$ tilting [51]. This behavior is related to the fact that the simplification to assume a spherical A-cation as sphere is no longer valid in the case of molecular cations. This not only impacts tilting modes but also makes symmetry reductions unique to hybrid perovskites possible.

5.3.3 Molecular cation orientation

The specific feature of hybrid halide perovskites is their molecular cation. Some of the most important molecular cations that are intrinsically anisotropic are shown in Fig. 5.8. While their apparent shape in hybrid perovskites is nearly spherical at

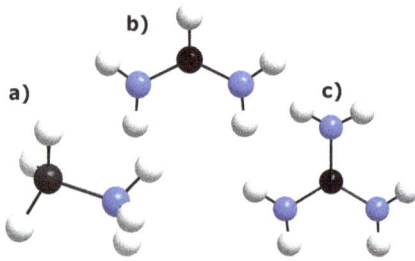

Fig. 5.8: Representations of the molecular ions (a) methylammonium [47], (b) formamidinium [41], and (c) guanidinium cations [52], respectively, as from crystal structures. Blue: nitrogen; black: carbon; white: hydrogen atoms with arbitrary radii.

higher temperatures due to fast rotation [53–55], this is no longer true in the lower temperature structures, where the thermal activation is not sufficient for fast reorientation [47]. Therefore, hybrid perovskites typically exhibit multiple phase transitions depending on temperature and pressure. Not only the intrinsic anisotropy of the cations is important but also the hydrogen interaction with the surrounding anions [54]. This is insofar important as these bonds are not isotropic either. It is, for instance, known that the hydrogen atoms bound to nitrogen tend to form stronger hydrogen bonds than the ones bound to carbon [53]. This can be explained as an effect of the difference in electronegativity between nitrogen and carbon, and hence the stronger polarization of the hydrogen bound to nitrogen. It does, in return, influence the anionic network and the orientation of the A-cation therein.

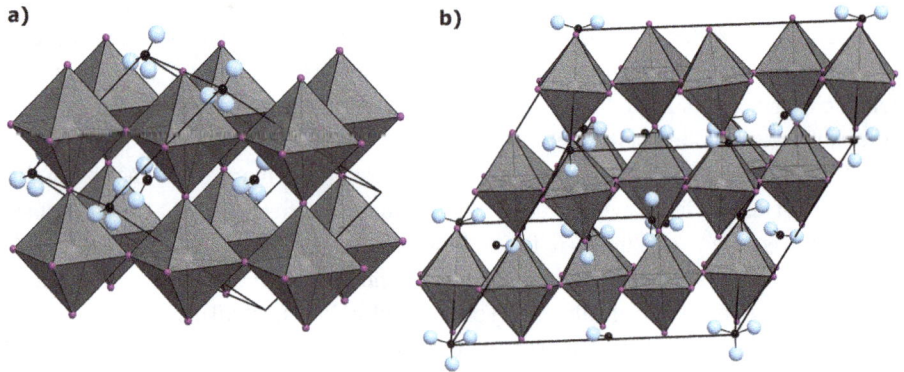

Fig. 5.9: Representation of the crystal structures of FAPbI$_3$ at room temperature (a) in $P3m$ and at 150 K (b) in $P3$ [16]. The formamidinium cations are disordered and all possible orientations are represented. Atoms are given as generic spheres.

One stunning example of the influence of cation orientation was reported by Stoumpos et al. for FAPbI$_3$ [16]. At room temperature, the formamidinium cations lie on threefold axes (Fig. 5.9a) that are seemingly contradicting their (lower) molecular symmetry (Fig. 5.8b). In fact, the molecular cation is statistically disordered over three possible orientations. At 150 K, on the other hand, half of the cationic sites are still disordered in a similar way, while the other half appears to be ordered (Fig. 5.9b).

5.4 The perovskite family

5.4.1 Bärnighausen tree for halide perovskites

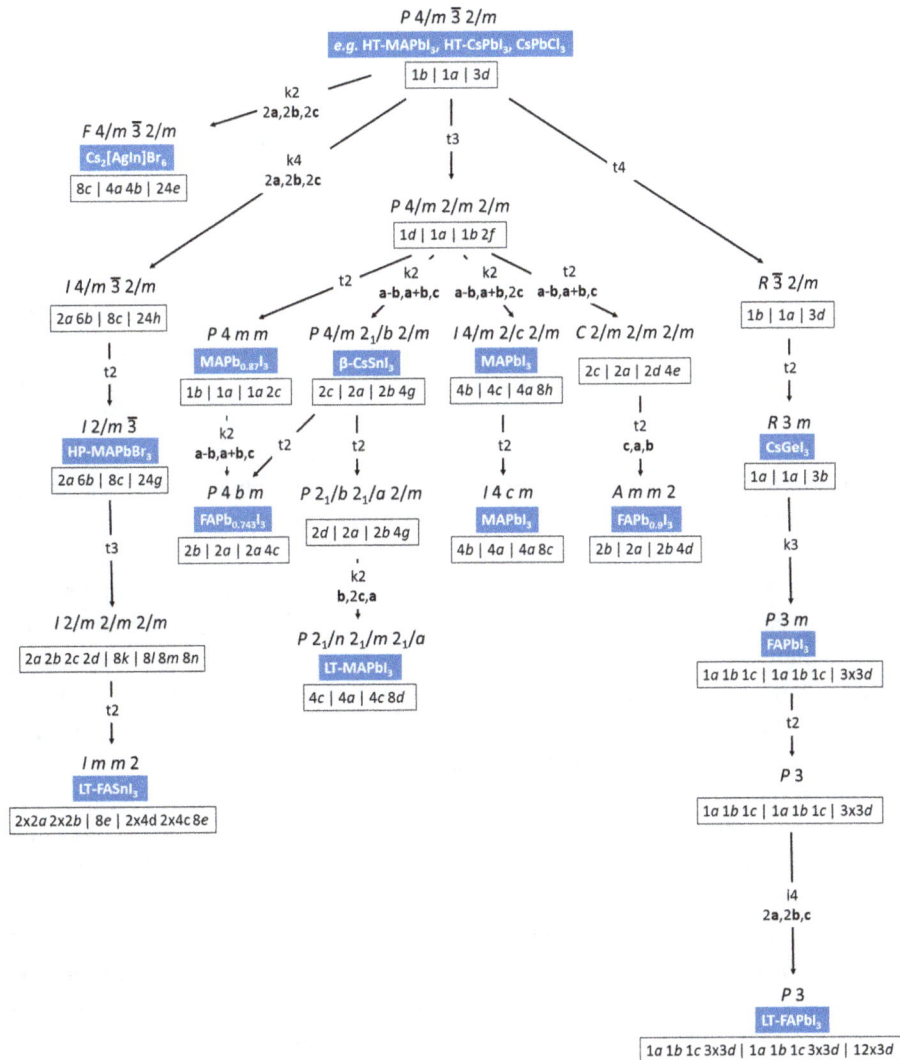

$P\,4/m\,\bar{3}\,2/m$
e.g. HT-MAPbI$_3$, HT-CsPbI$_3$, CsPbCl$_3$
$1b \mid 1a \mid 3d$

k2
2a,2b,2c

$F\,4/m\,\bar{3}\,2/m$
Cs$_2$[AgIn]Br$_6$
$8c \mid 4a\,4b \mid 24e$

k4
2a,2b,2c

t3

t4

$P\,4/m\,2/m\,2/m$
$1d \mid 1a \mid 1b\,2f$

$I\,4/m\,\bar{3}\,2/m$
$2a\,6b \mid 8c \mid 24h$

t2

k2
a-b,a+b,c

k2
a-b,a+b,2c

t2
a-b,a+b,c

$R\,\bar{3}\,2/m$
$1b \mid 1a \mid 3d$

$I\,2/m\,\bar{3}$
HP-MAPbBr$_3$
$2a\,6b \mid 8c \mid 24g$

$P\,4\,m\,m$
MAPb$_{0.87}$I$_3$
$1b \mid 1a \mid 1a\,2c$

$P\,4/m\,2_1/b\,2/m$
β-CsSnI$_3$
$2c \mid 2a \mid 2b\,4g$

$I\,4/m\,2/c\,2/m$
MAPbI$_3$
$4b \mid 4c \mid 4a\,8h$

$C\,2/m\,2/m\,2/m$
$2c \mid 2a \mid 2d\,4e$

t2

t2

k2
a-b,a+b,c

t2

t2

t2

t2
c,a,b

$R\,\bar{3}\,m$
CsGeI$_3$
$1a \mid 1a \mid 3b$

t3

$P\,4\,b\,m$
FAPb$_{0.743}$I$_3$
$2b \mid 2a \mid 2a\,4c$

$P\,2_1/b\,2_1/a\,2/m$
$2d \mid 2a \mid 2b\,4g$

$I\,4\,c\,m$
MAPbI$_3$
$4b \mid 4a \mid 4a\,8c$

$A\,m\,m\,2$
FAPb$_{0.9}$I$_3$
$2b \mid 2a \mid 2b\,4d$

$R\,3\,m$
FAPbI$_3$
$1a\,1b\,1c \mid 1a\,1b\,1c \mid 3x3d$

$I\,2/m\,2/m\,2/m$
$2a\,2b\,2c\,2d \mid 8k \mid 8l\,8m\,8n$

k2
b,2c,a

$P\,2_1/n\,2_1/m\,2_1/a$
LT-MAPbI$_3$
$4c \mid 4a \mid 4c\,8d$

k3

t2

$I\,m\,m\,2$
LT-FASnI$_3$
$2x2a\,2x2b \mid 8e \mid 2x4d\,2x4c\,8e$

t2

$P\,3$
$1a\,1b\,1c \mid 1a\,1b\,1c \mid 3x3d$

i4
2a,2b,c

$P\,3$
LT-FAPbI$_3$
$1a\,1b\,1c\,3x3d \mid 1a\,1b\,1c\,3x3d \mid 12x3d$

Fig. 5.10: Bärnighausen tree for halide perovskites. Some exemplary compounds are given in blue.

Similar to the oxides, halide perovskites show a great degree of variability that translates into many different crystal structures. Therefore, it is important to highlight the systematic work performed mainly for oxide perovskites in the past. The Bärnighausen tree for perovskites can hence be adapted from those oxide perovskites [8], bearing the differences outlined in Section 5.1.3 in mind.

Unfortunately, the lattice parameters of group and subgroup are not always straightforward to bring together, like a simple doubling of the lattice parameters. This is due to the fact that multiple settings are possible for most space-group types. The International Union of Crystallography defined one setting for each space-group type as a standard setting in the *International Tables for Crystallography* (often abbreviated as International Tables or IT) [31], and transformations into the standard setting may afford transformations of the unit cell of the subgroup. The transformation from the perovskite aristotype into the elpasolite type of the double perovskites (adapted, e.g., by $Cs_2[AgIn]Br_6$ [15]) is rather straightforward: All lattice parameters (a, b, and c) are doubled. Applying the same logic to the transformation from the aristotype to the room-temperature structure of $MAPbI_3$ via the intermediate space-group-type $P4/mmm$ would result in a structure in the space-group-type $F4/mmc$ with $a' = 2a$, $b' = 2b$, and $c' = 2c$. However, $F4/mmc$ is a nonstandard setting of the space-group-type $I4/mcm$ with a smaller unit cell. In fact, all face-centered tetragonal (nonstandard) space-group types can be described as body-centered structure, which is the standard setting for these structures. This is why the unit cell transformation is given as "$a–b$, $a + b$, $2c$" in the Bärnighausen tree (Fig. 5.10). It should be noted that these describe vector operations and correspond to a 45° rotation in the ab-plane with $|a'| = |b'| = \sqrt{2}\cdot|a|$. This is especially important when discussing space-group types in the orthorhombic crystal system: Since none of the directions are linked through symmetry (as opposed to $a = b$ in the tetragonal crystal system, for instance), their order is arbitrary, but it is often convenient to use the standard settings as defined in the IT. The group–subgroup relationship between the structures of β-$CsSnI_3$ ($P4/mbm$) and α-$MAPbI_3$ ($Pnma$) via the intermediate space-group $Pbam$ can be seen as an exemplary case: Transforming $Pbam$ into the desired subgroup would lead to the nonstandard setting $Pbnm$, which is historically often given as setting for this structure. In order to establish the standard setting, the axes must be permuted: $a' = b$, $b' = 2c$, and $c' = a$.

The splittings of the Wyckoff sites descending into a subgroup are given in the white boxes in Fig. 5.10. They are organized in the positions of A-cation, B-cation, and X-anion, respectively. The splitting can go so far that the three unique Wyckoff positions in the aristotype are split into 24 sites in the low-temperature structure of $FAPbI_3$ as reported by Stoumpos et al. [16]. For convenience, the position of the A-cation is assumed to be occupied by a single atom only. It should not be forgotten that the atoms of the molecular cations in hybrid perovskites mostly occupy general positions in the close vicinity of this special position. They, furthermore, do not necessarily obey the symmetry of the crystallographic site and therefore need to be treated with an appropriate disorder model.

Tables 5.2 and 5.3 contain some of the technologically relevant compounds with their respective space-group types corresponding to Fig. 5.10. Given the sheer amount

Tab. 5.2: Space-group types of inorganic and hybrid lead halide perovskites.

Compound	Measurement conditions	Space-group type	Reference
$MAPbI_3$	Ambient conditions[2]	$Pm\bar{3}m$	[46]
$MAPbI_3$	Ambient conditions	$I4/mcm$	[47]
		$I4cm$[3]	[59]
$MAPbI_3$	100 K	$Pnma$	[47]
$MAPbI_3$	293 K, 0.45 GPa	$Im\bar{3}$	[51]
$MA_{1-x}en_xPb_{1-0.7x}I_{3-0.4x}$	Ambient conditions	$P4mm$	[41]
$MAPbBr_3$	296 K	$Pm\bar{3}m$	[60]
$MAPbBr_3$	115 K	$Pnma$	[61]
$MAPbBr_3$	296 K, 1.7 GPa	$Im\bar{3}$	[60]
$MAPbCl_3$	Ambient conditions	$Pm\bar{3}m$	[62]
$MAPbCl_3$	100 K	$Pnma$	[63]
$FAPbI_3$	293 K	$P3m$	[16]
$FAPbI_3$	150 K	$P3$	[16]
$FAPbI_3$	300 K	$Pm\bar{3}m$	[64]
$FAPbI_3$	200 K	$P4/mbm$	[64]
$FA_{1-x}en_xPbI_3$	293 K	$P4bm$	[41]
$FA_{1-x}en_xPbI_3$	293 K	$Amm2$	[41]
$FAPbBr_3$	300 K	$Pm\bar{3}m$	[65]
$FAPbBr_3$	240 K	$P4/mbm$	[66]
$FAPbBr_3$	140 K	$Pnma$	[66]
$FAPbCl_3$	295 K	$Pm\bar{3}m$	[67]
$CsPbI_3$	634 K	$Pm\bar{3}m$	[45]

---

2 Reported as metastable and grown at 60 °C.
3 Centrosymmetric and non-centrosymmetric representations, respectively. The differences are so subtle that the higher symmetrical centrosymmetric structure representation can be used as a good approximation.

Tab. 5.2 (continued)

Compound	Measurement conditions	Space-group type	Reference
$CsPbI_3$	Ambient conditions	$Pnma$	[68]
$CsPbBr_3$	413 K	$Pm\bar{3}m$	[69]
$CsPbBr_3$	Ambient conditions	$Pnma$	[70]
$CsPbCl_3$	325 K	$Pm\bar{3}m$	[71]
$CsPbCl_3$	Ambient conditions	$Pnma$	[70]

If no pressure is given, the measurement was performed at ambient pressure
(MA^+: methylammonium, FA^+: formamidinium, en^{2+}: ethylenediammonium).

Tab. 5.3: Space group types of inorganic and hybrid tin and germanium halide perovskites.

Compound	Measurement conditions	Space-group type	Reference
$MASnI_3$	295 K	$Pm\bar{3}m$	[72]
$MASnI_3$	140 K/200 K	$I4/mcm$	[72]
		$I4cm^4$	[16]
$MASnCl_3$	478 K	$Pm\bar{3}m$	[58]
$MASnCl_3$	350 K	$R3m$	[58]
$MASnCl_3$	318 K	Pc^5	[58]
$MASnCl_3$	297 K	$P1^6$	[58]
$FASnI_3$	340 K	$Amm2$	[16]
$FASnI_3$	180 K	$Imm2$	[16]
$CsSnI_3$	500 K	$Pm\bar{3}m$	[39]
$CsSnI_3$	380 K	$P4/mbm$	[39]
$CsSnI_3$	300 K	$Pnma$	[39]
$CsSnBr_3$	300 K	$Pm\bar{3}m$	[73]

---

4 Centrosymmetric and non-centrosymmetric representations, respectively. The differences are so subtle that the higher symmetrical centrosymmetric structure representation can be used as a good approximation.
5 The octahedra are so strongly distorted that a three-dimensional network is questionable.
6 The octahedra are so strongly distorted that a three-dimensional network is questionable.

Tab. 5.3 (continued)

Compound	Measurement conditions	Space-group type	Reference
$CsSnCl_3$	Ambient conditions	$Pm\bar{3}m$	[74]
$CsGeI_3$	573 K	$Pm\bar{3}m$	[42]
$CsGeI_3$	293 K	$R3m$	[42]
$CsGeBr_3$	543 K	$Pm\bar{3}m$	[42]
$CsGeBr_3$	293 K	$R3m$	[42]
$CsGeCl_3$	443 K	$Pm\bar{3}m$	[42]
$CsGeCl_3$	293 K	$R3m$	[42]

If no pressure is given, the measurement was performed at ambient pressure (MA^+: methylammonium, FA^+: formamidinium).

of research in the field, these tables should not be seen as a complete compilation but rather as a starting point listing some of the seminal work on the structural elucidation of hybrid halide perovskites. It should also be noted that only perovskite-type structures are listed. A low-temperature phase of $CsPbI_3$, for instance, assumes a second polymorph in $Pnma$ [45], but it contains double rods of edge-sharing octahedra rather than a three-dimensional network of corner-sharing octahedra. Hence, it is not a perovskite. Furthermore, Sn^{2+} and Ge^{2+} containing halides assume perovskite-derived structures at lower temperatures. In these, the octahedra are so strongly distorted that the $B–X$ distances split into three short and three much longer ones [52, 56–58], thereby reducing the dimensionality of the BX_3 substructure.

5.5 Double perovskites or elpasolites

5.5.1 From disordered perovskites to double perovskites

One point that is unfortunately often overlooked is the difference between chemical composition and structural variation. From a chemical point of view, mixed cation perovskites like $CsSn_{0.5}Pb_{0.5}I_3$ [75] and double perovskites like $Cs_2AgInBr_6$ [76] appear rather similar, as the elemental ratio in both cases is 2:1:1:6. However, the significant difference between the two lies in their crystal structure. In $CsSn_{0.5}Pb_{0.5}I_3$, the two different B-cations are statistically occupying the same crystallographic position, and hence form an occupational disorder, but In^{3+} and Ag^+ in $Cs_2AgInBr_6$ are ordered on distinct crystallographic sites: $InBr_6$ and $AgBr_6$ octahedra alternate within the structure, thus belonging to the elpasolite type. In fact, the "vacancy-ordered perovskites" of the

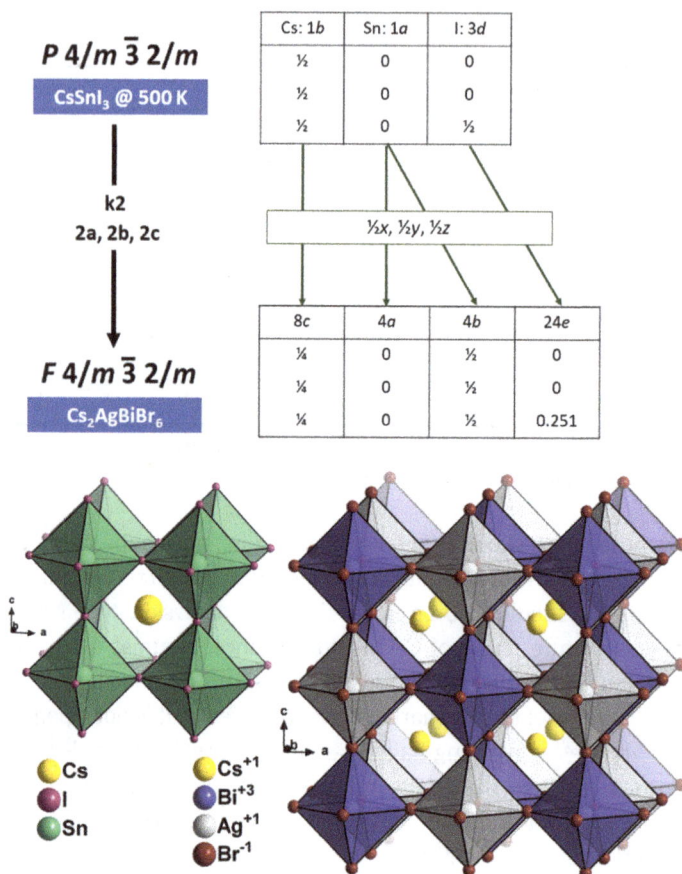

	Cs: 1b	Sn: 1a	I: 3d
	½	0	0
	½	0	0
	½	0	½

$P\,4/m\,\bar{3}\,2/m$

CsSnI₃ @ 500 K

k2
2a, 2b, 2c

½x, ½y, ½z

	8c	4a	4b	24e
	¼	0	½	0
	¼	0	½	0
	¼	0	½	0.251

$F\,4/m\,\bar{3}\,2/m$

Cs₂AgBiBr₆

- Cs
- I
- Sn
- Cs^{+1}
- Bi^{+3}
- Ag^{+1}
- Br^{-1}

Fig. 5.11: Relationship between the perovskite aristotype and the elpasolite type and structural representations of perovskite-type CsSnI₃ [39] and elpasolite-type Cs₂AgBiBr₆ [76]. Atoms are given as generic spheres.

K_2PtCl_6 type are structurally very closely linked to the elpasolite-type structure, in that they form a variant with unoccupied B′-cation positions.

The structure of $Cs_2AgInBr_6$ merits a closer look, since Ag^+ and In^{3+} are isoelectronic and hence practically indistinguishable via X-ray diffraction [15]. Although theoretically suitable for it, neutron diffraction does not allow a clear conclusion of the cation ordering as the compound can only be synthesized mechanochemically and hence has a low crystallinity. However, knowledge of the group–subgroup relationships between the perovskite and the elpasolite type (Fig. 5.11) allows to draw a conclusion: The coordinates of the anion position in the perovskite aristotype (Wyckoff position 3d) are fixed by symmetry, and hence restrict the coordination octahedra to be of a single size. The lower symmetry in the subgroup $Fm\bar{3}m$, in which the elpasolite type is described, adds a degree of positional freedom to the anion site (Wyckoff position 24e) such that it can be shifted along the vector between two next-neighboring B and B′-cations. Therefore, alternating octahedra can become different in size. Indeed, refinements of X-ray powder diffraction data show that the bromide ions are displaced towards one of the B/B′-cations making the Ag–Br and In–Br distances different. The refined values coincide with values expected for such bonds and hence corroborate the cation ordering [14].

5.5.2 Structural variations in the double perovskite family

While there is quite a few double-perovskite structures derived from the elpasolite type, their variability appears to be significantly smaller than for the perovskite family (Fig. 5.12). In fact, most double perovskites crystallize in the elpasolite type [77], which can be viewed as the aristotype for this structural class. Another reason for the scarcity of elpasolite-type structures may be lack of low-temperature and variable-pressure characterizations.

One of the examples of a lower symmetry double perovskite can be found for MA_2KBiCl_6 (Fig. 5.12b). The orientation of the methylammonium cations as well as a distortion of the octahedra around the B-cations leads to a rhombohedral structure of MA_2KBiCl_6 [78]. Herein, the fourfold axes are lost in a *translationengleiche* descent of index 4. For an appropriate description, the setting of the unit cell needs to be changed in a quite complex manner. In a simple way, one could use the rhombohedral setting of the unit cells, in which the lattice parameters of group and subgroup are equal. However, the hexagonal setting is much more common and the base-vector transformation needs to take a directional change into account. That is why the Bärnighausen tree in Fig. 5.12a shows a change from cubic to hexagonal settings in its right-hand branch. The methylammonium cations are ordered on the threefold axes within the rhombohedral subgroup structure. While the hydrogen-atom positions were not refined for the reported structure, they could be ordered in accordance with the site symmetry and hence be potentially non-disordered.

Another notable exception from the pristine elpasolite type is $Cs_2AuAuCl_6$ [79]. While its empirical formula is $CsAuCl_3$, it is not a perovskite but more complex. One glance at the structure (Fig. 5.12c) reveals the significant difference: the two crystallographically independent gold ions have different chemical environments with one sort of octahedra, around a Au^{3+} ion, elongated in one direction and the other one, around a Au^+ ion, compressed in the same direction to form square-planar bipyramids. This is because the gold ions are not divalent and uniform, as would be necessary for a simple halide perovskite, but adopt two distinct oxidation states with different crystal-chemical behavior.

The case of $Cs_2AgBiBr_6$ [76], a compound that has been studied at variable temperatures, is slightly different. While it crystallizes in the undistorted elpasolite type at ambient conditions (Fig. 5.11), it crystallizes in $I4/m$ at lower temperatures and

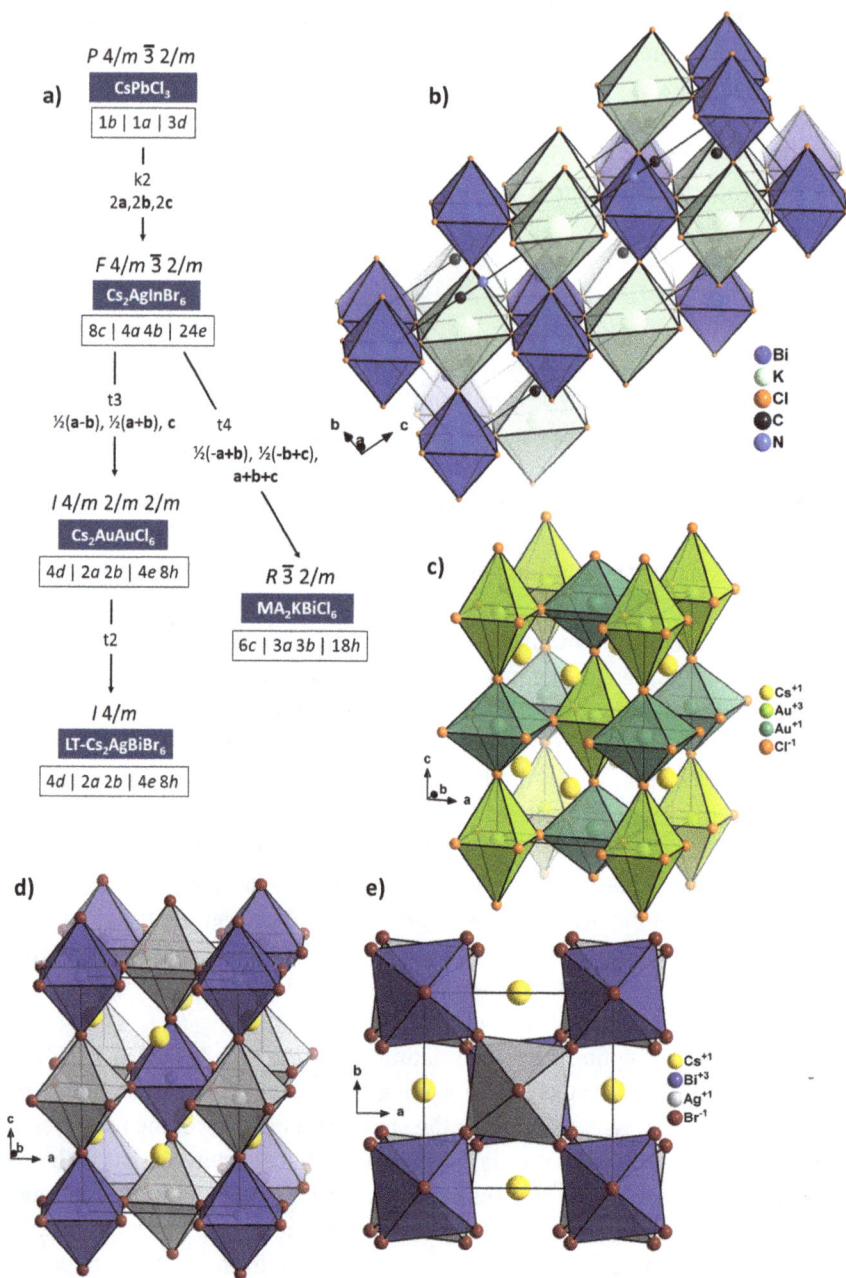

Fig. 5.12: Group–subgroup relationship for the halide double perovskites (a) and structure representations of MA_2KBiCl_6 (b) [78], $Cs_2AuAuCl_6$ (c) [79], and $Cs_2AgBiBr_6$ at 30 K [76] in general view (d) and along the crystallographic c-axis (e). Atoms are given as generic spheres and with the charges as given in the respective cif files.

shows octahedral tilting (Fig. 5.12c, d) along the c-axis in an antiphase manner, that is, $a^0a^0c^-$ in Glazer notation.

5.6 Physical implications of group–subgroup relationships

This section will attempt to tackle the probably most pressing question of this chapter: Why is this important? Indeed, group–subgroup relationships may appear as an overly formalized way of expressing structural similarities in chemically linked structures, but there are clear and very important implications that help understand material properties based on structural relationships. The real forte of this method is the understanding of phase transitions. Two examples, in which this falls into place, will be given below. Also, this formalism allows a better understanding of the materials as such, which will be elaborated on in the following section. Of course, this is not exhaustive and further applications, for instance, in computational chemistry and solid-state physics, profit from understanding the structural relationships between the different phases.

5.6.1 Structure similarities at first sight

Undoubtedly, the most efficient way of spotting a perovskite-type structure is looking at a representation and checking for a three-dimensional network of corner-sharing octahedra. However, some other details may not be quite as evident, and an inspection of the group–subgroup relations to the perovskite aristotype may be beneficial. This is especially true for hybrid halide perovskites, as the molecular cations are often statically and dynamically disordered over special positions of higher symmetry, at least at elevated temperature. The first means that molecular cations in different unit cells exhibit different orientations, thus leading to disorder in the space-averaged picture. The latter means that all molecular cations rotate within their cages. The lower the symmetry of the A-cation site is, the higher are the chances that the symmetry of the molecule reciprocates with the symmetry of the crystallographic site. The crystal symmetry of the A-cation site in a given crystal structure relates to the symmetry in other structures via the Bärnighausen tree and therefore can give a first indication on the disorder of the molecular cation. A good example for this is the metastable room-temperature structure of FAPbI$_3$ described in $P3m$ (Fig. 5.9a). Herein, the FA$^+$ ions exhibit threefold disorder, because they lie on a threefold axis. In the cubic aristotype, however, the A-cation position lies on fourfold axes, making these particular orientations impossible. Finally, such a symmetry analysis can be useful for indexing diffractograms of unknown phases, in which a deviation from the perovskite aristotype is suspected.

5.6.2 Second-order phase transitions

As mentioned above, the strong point of symmetry relationships is understanding phase transitions. For this, a short excursion into the theory of solid-state phase transitions is necessary. According to Ehrenfest, two general forms of phase transitions can be defined [80]: The first type is characterized by an abrupt change of first derivatives of the free energy and is hence called a "first-order transition." If the first derivative remains continuous through the phase transition and only higher derivatives of the free energy change abruptly, it is called a "second-order transition." According to the Landau theory, and this is where the Bärnighausen trees come in handy, a direct group–subgroup relationship is a necessary prerequisite for a second-order phase transition [30]. However, the existence of a group–subgroup relationship cannot rule out a first-order transition, but a second-order phase transition cannot exist without the group–subgroup relationship. This may be illustrated by the emblematic material $MAPbI_3$ (Fig. 5.13).

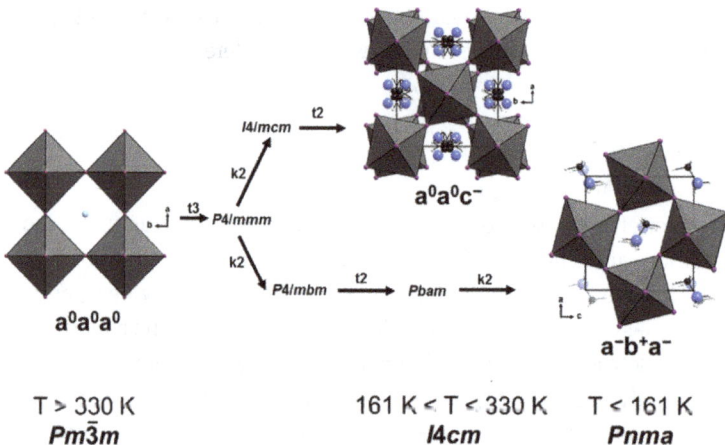

Fig. 5.13: Phase transitions of $MAPbI_3$ as a function of temperature.

While both low-temperature structures of $MAPbI_3$ belong to subgroups of the high-temperature perovskite aristotype, there is no direct group–subgroup relationship between the two low-temperature structures. Therefore, the low-temperature phase transition at 161 K can only be of first order. Commonly, first-order phase transitions are reconstructive, that is, associated with a rearrangement of the structure and breaking of bonds. One of the critical values that may change abruptly during a phase transition is the volume. This would have dramatic effects on a solar cell: An abrupt change of the absorber-layer volume in a solar cell could lead to cracks, and therefore electric failures, or even a peel off. While a transition temperature below –110 °C does not sound worrying on the Earth, this could, for instance, become important for applications in space. Furthermore, perovskites with other A-cations or cation mixtures thereof may have similar phase transitions closer to room temperature.

The phase transition between the tetragonal room-temperature structure and the cubic high-temperature structure at 330 K, on the other hand, exhibits a direct group–subgroup relationship. Plausibly, the concomitant volume change is rather smooth. This plays an important role in solar cell production: Most methods for the deposition of perovskite thin films use temperatures above the tetragonal-to-cubic phase transition and therefore cause the material to undergo it on subsequent cooling.

5.6.3 Twinning during phase transitions

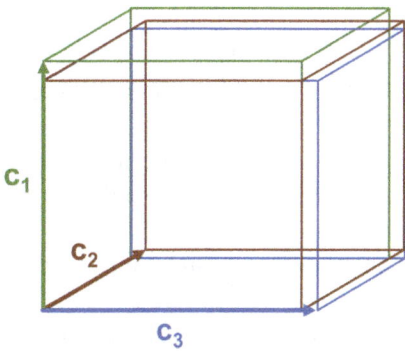

Fig. 5.14: Overlay of three possible orientations of a tetragonal unit cell derived from a cubic unit cell with a small deviation of the unique c-axis, respectively.

While the above-mentioned phase transition at 330 K is not prone to abrupt volume changes, it shows a *translationengleiche* symmetry descent of index 3. The lost point symmetry elements are linked to the four threefold axes in the cubic system and give rise to twinning along them. The transition of MAPbI$_3$ from the cubic aristotype to the tetragonal room-temperature structure commonly leads to the formation of twins. As the threefold axes, which restrict the cubic lattice parameters to be equal, are lost, one of the lattice parameters (the c-axis by convention) is free to change irrespective of the other two. Since the room-temperature phase is tetragonal, the other two axes are restricted to be equal. It is, however, arbitrary, which of the three formerly equivalent axes becomes unique during phase transition. Three possible solutions are, therefore, thinkable (Fig. 5.14), which can be transformed into one another by a threefold rotation – around the axis that was lost during the phase transition. There is no energetic reason for a preferential formation of any of the three possible orientations. They form simultaneously in the same crystal, thus leading to twinning. Unfortunately, the twin law in this case is slightly more complex, as the centering of the tetragonal unit cell affords a 45° turn in the *ab*-plane [43] to change from the non-standard all-face centering to the standard body-centered setting.

While the occurrence of twinning itself may not sound dramatic, it can make a reliable crystal structure solution and refinement more complicated and prone to errors. More importantly, twinning is accompanied by the formation of extended twin interfaces that can potentially act as recombination centers and therefore affect the electronic performance of a device based on such crystals. This is insofar important, as many film synthesis methods, like the inverse temperature method, are working at elevated temperatures. The films, therefore, undergo a phase transition to the room-temperature structure before use. Recent reports suggest that thin-film crystallization under an electric field can increase performance [81], which may be related to the suppression of twin formation.

5.7 Conclusions

The properties of crystalline solids are largely governed by their crystal structure. Understanding the group–subgroup relationships between phases of these materials allows to rationalize some of their properties and to understand temperature and pressure behavior. This is crucial for materials in which the electronic structure is essential for the performance such as in solar absorbers. That is also the reason for the emphasis of perovskite nomenclature: The research community is urged to rethink the practice of disconnecting the term "perovskite" from structural features, that is, partly observed at the moment.

The distortion modes within the perovskite family, as far as applicable to halides, have been established and their consequences have been outlined. Practical consequences of structural distortions need careful evaluation for each case; the group–subgroup relationships described herein are intended to give guidance for future systems, especially the role of the molecular cations within the structure, and (at least indirectly) for the electronic properties, merits further efforts. Variable temperature and pressure studies may help complement our understanding of non-spherical cations occupying the A-site.

Finally, I hope that this structural systematic work will help direct the efforts in tuning those materials to their peak performance. This may appear as very fundamental science, but it is these fundamental properties that aid to maximize the performance of a material. Those fundamental studies form the basis for the material systems of the future and are performed with a clear application perspective. Viable solutions for renewable energies are probably the most significant scientific question of this century.

References

[1]   NREL – National Renewable Energies Laboratory. 2020. https://www.nrel.gov/pv/cell-efficiency.html. accessed 13/04/2020.

[2]   Kojima, A., Teshima, K., Shirai, Y., Miyasaka, T. Organometal halide perovskites as visible-light sensitizers for photovoltaic cells. J. Am. Chem. Soc. 2009, 131, 6050–6051.

[3]   Breternitz, J., Lehmann, F., Barnett, S.A., Nowell, H., Schorr, S. Role of the iodide–methylammonium interaction in the ferroelectricity of $CH_3NH_3PbI_3$. Angew. Chem. Int. Ed. 2020, 59, 424–428.

[4]   Zhao, Y., Zhu, K. Organic–inorganic hybrid lead halide perovskites for optoelectronic and electronic applications. Chem. Soc. Rev. 2016, 45, 655–689.

[5]   Jodlowski, A., Rodríguez-Padrón, D., Luque, R., De Miguel, G. Alternative perovskites for photovoltaics. Adv. Energy Mater. 2018, 8, 1703120.

[6]   Rose, G. Ueber einige neue Mineralien des Urals. J. Prakt. Chem. 1840, 19, 459–468.

[7]   Bärnighausen, H. Group-subgroup relations between space groups a useful tool in crystal chemistry. MATCH 1980, 139–175.

[8]   Bock, O., Müller, U. Symmetrieverwandtschaften bei Varianten des Perowskit-Typs. Acta Crystallogr. B 2002, 58, 594–606.

[9]   Glazer, A.M. The classification of tilted octahedra in perovskites. Acta Crystallogr. B 1972, 28, 3384–3392.

[10]   Glazer, A.M. Simple ways of determining perovskite structures. Acta Crystallogr. A 1975, 31, 756–762.

[11]   Tilley, R.J.D. The Goldschmidt Tolerance factor. In: Perovskites. Chichester: John Wiley & Sons, 2016, 6–11.

[12]   Nelmes, R.J., Meyer, G.M., Hutton, J. Thermal motion in $SrTiO_3$ at room temperature: Anharmonic or disordered? Ferroelectrics 1978, 21, 461–462.

[13]   Sasaki, S., Prewitt, C.T., Bass, J.D., Schulze, W.A. Orthorhombic perovskite $CaTiO_3$ and $CdTiO_3$: structure and space group. Acta Crystallogr. C 1987, 43, 1668–1674.

[14]   Breternitz, J., Schorr, S. What Defines a Perovskite? Adv. Energy Mater. 2018, 8, 1802366.

[15]   Breternitz, J., Levcenko, S., Hempel, H., Gurieva, G., Franz, A., Hoser, A., Schorr, S. Mechanochemical synthesis of the lead-free double Perovskite $Cs_2[AgIn]Br_6$ and its optical properties. J. Phys. Energy 2019, 1, 25003.

[16]   Stoumpos, C.C., Malliakas, C.D., Kanatzidis, M.G. Semiconducting Tin and Lead Iodide Perovskites with Organic Cations: Phase Transitions, High Mobilities, and Near-Infrared Photoluminescent Properties. Inorg. Chem. 2013, 52, 9019–9038.

[17]   Stoumpos, C.C., Cao, D.H., Clark, D.J., Young, J., Rondinelli, J.M., Jang, J.I., Hupp, J.T., Kanatzidis, M.G. Ruddlesden–Popper Hybrid Lead Iodide Perovskite 2D Homologous Semiconductors. Chem. Mater. 2016, 28, 2852–2867.

[18]   Meyer, E., Mutukwa, D., Zingwe, N., Taziwa, R. Lead-free halide double perovskites: A review of the structural, optical, and stability properties as well as their viability to replace lead halide perovskites. Metals 2018, 8, 667.

[19]   Moras, L.R. Crystal structure of dipotassium sodium fluoroaluminate (elpasolite). J. Inorg. Nucl. Chem. 1974, 36, 3876–3878.

[20]   Maughan, A.E., Ganose, A.M., Scanlon, D.O., Neilson, J.R. Perspectives and design principles of vacancy-ordered double perovskite halide semiconductors. Chem. Mater. 2019, 31, 1184–1195.

[21]   Engel, G. Die Kristallstrukturen einiger Hexachlorokomplexsalze. Z. Kristallogr. Cryst. Mater. 1935, 90, 341–373.

[22]   Qiu, X., Cao, B., Yuan, S., Chen, X., Qiu, Z., Jiang, Y., Ye, Q., Wang, H., Zeng, H., Liu, J., Kanatzidis, M.G. From unstable $CsSnI_3$ to air-stable Cs_2SnI_6: A lead-free perovskite solar cell light absorber with bandgap of 1.48 eV and high absorption coefficient. Sol. Energy Mater. Sol. Cells. 2017, 159, 227–234.

[23] Zhang, L., Liang, W. How the structures and properties of two-dimensional layered perovskites $MAPbI_3$ and $CsPbI_3$ vary with the number of layers. J. Phys. Chem. Lett. 2017, 8, 1517–1523.

[24] Saparov, B., Mitzi, D.B. Organic–inorganic perovskites: structural versatility for functional materials design. Chem. Rev. 2016, 116, 4558–4596.

[25] Soe, C.M.M., Stoumpos, C.C., Kepenekian, M., Traoré, B., Tsai, H., Nie, W., Wang, B., Katan, C., Seshadri, R., Mohite, A.D., Even, J., Marks, T.J., Kanatzidis, M.G. New type of 2D perovskites with alternating cations in the interlayer space, $(C(NH_2)_3)(CH_3NH_3)_nPb_nI_{3n+1}$: Structure, properties, and photovoltaic performance. J. Am. Chem. Soc. 2017, 139, 16297–16309.

[26] Mercier, N. Hybrid halide perovskites: discussions on terminology and materials. Angew. Chem. Int. Ed. 2019, 58, 17912–17917.

[27] Chabot, B., Parthé, E. $Cs_3Sb_2I_9$ and $Cs_3Bi_2I_9$ with the hexagonal $Cs_3Cr_2Cl_9$ structure type. Acta Crystallogr. B 1978, 34, 645–648.

[28] Abdel-Aal, S.K., Abdel-Rahman, A.S. Synthesis, structure, lattice energy and enthalpy of 2D hybrid perovskite $[NH_3(CH_2)_4NH_3]CoCl_4$, compared to $[NH_3(CH_2)_nNH_3]CoCl_4$, n=3–9. J. Cryst. Growth 2017, 457, 282–288.

[29] Müller, U. Anorganische Strukturchemie. Wiesbaden: Vieweg+Teubner, 2008.

[30] Müller, U. Symmetry Relationships between Crystal Structures. Oxford: Oxford University Press, 2013.

[31] Hahn, T. ed. International Tables for Crystallography Volume A: Space-Group Symmetry. Dordrecht: Springer, 2005.

[32] Wondratschek, H., Müller, U. ed. International tables for crystallography, Volume A1. Dordrecht: Kluwer Academic Publishers, 2004.

[33] Aroyo, M.I., Perez-Mato, J.M., Capillas, C., Kroumova, E., Ivantchev, S., Madariaga, G., Kirov, A., Wondratschek, H. Bilbao crystallographic server: I. Databases and crystallographic computing programs. Z. Kristallogr. Cryst. Mater. 2006, 221, 15–27.

[34] Aroyo, M.I., Kirov, A., Capillas, C., Perez-Mato, J.M., Wondratschek, H. Bilbao crystallographic server. II. Representations of crystallographic point groups and space groups. Acta Crystallogr. A 2006, 62, 115–128.

[35] Kroumova, E., Aroyo, M.I., Perez-Mato, J.M., Kirov, A., Capillas, C., Ivantchev, S., Wondratschek, H. Bilbao crystallographic server: Useful databases and tools for phase-transition studies. Phase Trans. 2003, 76, 155–170.

[36] Campbell, B.J., Stokes, H.T., Tanner, D.E., Hatch, D.M. ISODISPLACE: a web-based tool for exploring structural distortions. J. Appl. Crystallogr. 2006, 39, 607–614.

[37] Stokes, H.T., Hatch, D.M., Campbell, B.J., accessed 2020. ISOTROPY Software Suite, https://iso.byu.edu

[38] Nespolo, M., Aroyo, M.I., Souvignier, B. Crystallographic shelves: space-group hierarchy explained. J. Appl. Crystallogr. 2018, 51, 1481–1491.

[39] Chung, I., Song, J.-H., Im, J., Androulakis, J., Malliakas, C.D., Li, H., Freeman, A.J., Kenney, J.T., Kanatzidis, M.G. $CsSnI_3$: Semiconductor or metal? high electrical conductivity and strong near-infrared photoluminescence from a single material. high hole mobility and phase transitions. J. Am. Chem. Soc. 2012, 134, 8579–8587.

[40] Yamada, K., Funabiki, S., Horimoto, H., Matsui, T., Okuda, T., Ichiba, S. Structural phase transitions of the polymorphs of $CsSnI_3$ by means of Rietveld analysis of the x-ray diffraction. Chem. Lett. 1991, 20, 801–804.

[41] Spanopoulos, I., Ke, W., Stoumpos, C.C., Schueller, E.C., Kontsevoi, O.Y., Seshadri, R., Kanatzidis, M.G. Unraveling the chemical nature of the 3D "Hollow" hybrid halide perovskites. J. Am. Chem. Soc. 2018, 140, 5728–5742.

[42] Thiele, G., Rotter, H.W., Schmidt, K.D. Kristallstrukturen und Phasentransformationen von Caesiumtrihalogenogermanaten(II) $CsGeX_3$ (X = Cl, Br, I). Z. Anorg. Allgem. Chem. 1987, 545, 148–156.

[43] Breternitz, J., Tovar, M., Schorr, S. Twinning in MAPbI$_3$ at room temperature uncovered through Laue neutron diffraction. Sci. Rep. 2020, 10, 16613.

[44] Röhm, H., Leonhard, T., Schulz, A.D., Wagner, S., Hoffmann, M.J., Colsmann, A. Ferroelectric properties of perovskite thin films and their implications for solar energy conversion. Adv. Mater. 2019, 31, 1806661.

[45] Trots, D.M., Myagkota, S.V. High-temperature structural evolution of caesium and rubidium triiodoplumbates. J. Phys. Chem. Solids. 2008, 69, 2520–2526.

[46] Luan, M., Song, J., Wei, X., Chen, F., Liu, J. Controllable growth of bulk cubic-phase CH$_3$NH$_3$PbI$_3$ single crystal with exciting room-temperature stability. CrystEngComm 2016, 18, 5257–5261.

[47] Weller, M.T., Weber, O.J., Henry, P.F., Di Pumpo, A.M., Hansen, T.C. Complete structure and cation orientation in the perovskite photovoltaic methylammonium lead iodide between 100 and 352 K. Chem. Commun. 2015, 51, 4180–4183.

[48] Goldschmidt, V.M. Die Gesetze der Krystallochemie. Die Naturwissenschaften 1926, 14, 477–485.

[49] Shannon, R.D. Revised effective ionic radii and systematic studies of interatomic distances in halides and chalcogenides. Acta Crystallogr. A 1976, 32, 751–767.

[50] Kieslich, G., Sun, S., Cheetham, A.K. Solid-state principles applied to organic–inorganic perovskites: new tricks for an old dog. Chem. Sci. 2014, 5, 4712–4715.

[51] Szafrański, M., Katrusiak, A. Mechanism of pressure-induced phase transitions, amorphization, and absorption-edge shift in photovoltaic methylammonium lead iodide. J. Phys. Chem. Lett. 2016, 7, 3458–3466.

[52] Szafrański, M., Ståhl, K. Crystal structure and phase transitions in perovskite-like C (NH$_2$)$_3$SnCl$_3$. J. Solid State Chem. 2007, 180, 2209–2215.

[53] Franz, A., Többens, D.M., Schorr, S. Interaction between cation orientation, octahedra tilting and hydrogen bonding in methylammonium lead triiodide. Cryst. Res. Technol. 2016, 51, 534–540.

[54] Schuck, G., Többens, D.M., Koch-Müller, M., Efthimiopoulos, I., Schorr, S. Infrared spectroscopic study of vibrational modes across the orthorhombic–tetragonal phase transition in methylammonium lead halide single crystals. J. Phys. Chem. C 2018, 122, 5227–5237.

[55] Whitfield, P.S., Herron, N., Guise, W.E., Page, K., Cheng, Y.Q., Milas, I., Crawford, M.K. Structures, phase transitions and tricritical behavior of the hybrid perovskite methyl ammonium lead iodide. Sci. Rep. 2016, 6, 35685.

[56] Dang, Y., Liu, Y., Sun, Y., Yuan, D., Liu, X., Lu, W., Liu, G., Xia, H., Tao, X. Bulk crystal growth of hybrid perovskite material CH$_3$NH$_3$PbI$_3$. CrystEngComm 2015, 17, 665–670.

[57] Yamada, K., Isobe, K., Okuda, T., Furukawa, Y. Successive phase transitions and high ionic conductivity of trichlorogermanate (II) salts as studied by [35]Cl NQR and powder X-Ray diffraction. Z. Naturforsch. A: Phys. Sci. 1994, 49, 258–266.

[58] Yamada, K., Kuranaga, Y., Ueda, K., Goto, S., Okuda, T., Furukawa, Y. Phase transition and electric conductivity of ASnCl$_3$ (A = Cs and CH$_3$NH$_3$). Bull. Chem. Soc. Jap. 1998, 71, 127–134.

[59] Xie, J., Liu, Y., Liu, J., Lei, L., Gao, Q., Li, J., Yang, S. Study on the correlations between the structure and photoelectric properties of CH$_3$NH$_3$PbI$_3$ perovskite light-harvesting material. J. Power Sources 2015, 285, 349–353.

[60] Jaffe, A., Lin, Y., Beavers, C.M., Voss, J., Mao, W.L., Karunadasa, H.I. High-pressure single-crystal structures of 3D lead-halide hybrid perovskites and pressure effects on their electronic and optical properties. ACS Central Sci. 2016, 2, 201–209.

[61] Mashiyama, H., Kawamura, Y., Kubota, Y. The anti-polar structure of ((CH$_3$)NH$_3$)PbBr$_3$. J. Kor. Phys. Soc. 2007, 51, 850–853.

[62] Nandi, P., Giri, C., Swain, D., Manju, U., Topwal, D. Room temperature growth of CH$_3$NH$_3$PbCl$_3$ single crystals by solvent evaporation method. CrystEngComm 2019, 21, 656–661.

[63] Chen, K., Deng, X., Goddard, R., Tüysüz, H. Pseudomorphic transformation of organometal halide perovskite using the gaseous hydrogen halide reaction. Chem. Mater. 2016, 28, 5530–5537.

[64] Weber, O.J., Ghosh, D., Gaines, S., Henry, P.F., Walker, A.B., Islam, M.S., Weller, M.T. Phase behavior and polymorphism of formamidinium lead iodide. Chem. Mater. 2018, 30, 3768–3778.

[65] Elbaz, G.A., Straus, D.B., Semonin, O.E., Hull, T.D., Paley, D.W., Kim, P., Owen, J.S., Kagan, C.R., Roy, X. Unbalanced hole and electron diffusion in lead bromide perovskites. Nano Lett. 2017, 17, 1727–1732.

[66] Franz, A., Többens, D.M., Lehmann, F., Kärgell, M., Schorr, S. The influence of deuteration on the crystal structure of hybrid halide perovskites: a temperature-dependent neutron diffraction study of FAPbBr$_3$. Acta Crystallogr. B 2020, 76, 267–274.

[67] Govinda, S., Kore, B.P., Swain, D., Hossain, A., De, C., Row, T.N.G., Sarma, D.D. Critical comparison of FAPbX$_3$ and MAPbX$_3$ (X = Br and Cl): How do they differ? J. Phys. Chem. C 2018, 122, 13758–13766.

[68] Sutton, R.J., Filip, M.R., Haghighirad, A.A., Sakai, N., Wenger, B., Giustino, F., Snaith, H.J. Cubic or orthorhombic? Revealing the crystal structure of metastable black-phase CsPbI$_3$ by theory and experiment. ACS Energy Lett. 2018, 3, 1787–1794.

[69] Rodová, M., Brožek, J., Knížek, K., Nitsch, K. Phase transitions in ternary caesium lead bromide. J. Therm. Anal. Calorim. 2003, 71, 667–673.

[70] Linaburg, M.R., McClure, E.T., Majher, J.D., Woodward, P.M. Cs$_{1-x}$Rb$_x$PbCl$_3$ and Cs$_{1-x}$Rb$_x$PbBr$_3$ solid solutions: understanding octahedral tilting in lead halide perovskites. Chem. Mater. 2017, 29, 3507–3514.

[71] Hutton, J., Nelmes, R.J., Meyer, G.M., Eiriksson, V.R. High-resolution studies of cubic perovskites by elastic neutron diffraction: CsPbCl$_3$. J. Phys. C: Solid State Phys. 1979, 12, 5393–5410.

[72] Takahashi, Y., Obara, R., Lin, -Z.-Z., Takahashi, Y., Naito, T., Inabe, T., Ishibashi, S., Terakura, K. Charge-transport in tin-iodide perovskite CH$_3$NH$_3$SnI$_3$: origin of high conductivity. Dalton Trans. 2011, 40, 5563.

[73] Fabini, D.H., Laurita, G., Bechtel, J.S., Stoumpos, C.C., Evans, H.A., Kontos, A.G., Raptis, Y.S., Falaras, P., Van Der Ven, A., Kanatzidis, M.G., Seshadri, R. Dynamic stereochemical activity of the Sn^{2+} lone pair in perovskite CsSnBr$_3$. J. Am. Chem. Soc. 2016, 138, 11820–11832.

[74] Bulanova, G.G., Podlesskaya, A.B., Soboleva, L.V., Soklakov, A.I. Study of cesium tin chloride CsSnCl$_3$ and Cs$_2$SnCl$_6$. Izv. Akad. Nauk SSSR, Neorg. Mater. 1972, 8, 1930–1932.

[75] Li, Y., Sun, W., Yan, W., Ye, S., Rao, H., Peng, H., Zhao, Z., Bian, Z., Liu, Z., Zhou, H., Huang, C. 50% Sn-based planar perovskite solar cell with power conversion efficiency up to 13.6%. Adv. Energy Mater. 2016, 6, 1601353.

[76] Schade, L., Wright, A.D., Johnson, R.D., Dollmann, M., Wenger, B., Nayak, P.K., Prabhakaran, D., Herz, L.M., Nicholas, R., Snaith, H.J., Radaelli, P.G. Structural and optical properties of Cs$_2$AgBiBr$_6$ double perovskite. ACS Energy Lett. 2018, 4, 299–305.

[77] Igbari, F., Wang, Z.-K., Liao, L.-S. Progress of lead-free halide double perovskites. Adv. Energy Mater. 2019, 9, 1803150.

[78] Wei, F., Deng, Z., Sun, S., Xie, F., Kieslich, G., Evans, D.M., Carpenter, M.A., Bristowe, P.D., Cheetham, A.K. The synthesis, structure and electronic properties of a lead-free hybrid inorganic–organic double perovskite (MA)$_2$KBiCl$_6$ (MA = methylammonium). Mater. Horiz. 2016, 3, 328–332.

[79] Eijndhoven, J.C.M.T.-V., Verschoor, G.C. Redetermination of the crystal structure of Cs$_2$AuAuCl$_6$. Mater. Res. Bull. 1974, 9, 1667–1670.

[80] Jaeger, G. The Ehrenfest classification of phase transitions: Introduction and evolution. Arch. Hist. Exact Sci. 1998, 53, 51–81.

[81] Zhang, -C.-C., Wang, Z.-K., Li, M., Liu, Z.-Y., Yang, J.-E., Yang, Y.-G., Gao, X.-Y., Ma, H. Electric-field assisted perovskite crystallization for high-performance solar cells. J. Mater. Chem. A 2018, 6, 1161–1170.

Nicole Knoblauch and Martin Schmücker

6 Structural ordering in ceria-based suboxides applied for thermochemical water splitting

Abstract: Solar-thermochemical water splitting by means of suitable redox materials is a promising way of future hydrogen supply. In the first step of the thermochemical process, the redox material is partially reduced at high temperatures (e.g., using concentrating solar facilities). In the second step which proceeds at lower temperatures, re-oxidation of the redox material takes place by water vapor, resulting in splitting of H_2O molecules and subsequent H_2 release.

To avoid energy losses, the temperature gap between reduction and oxidation step should be low as possible. For this requirement the redox entropy of the employed Redox material should be high. High redox entropies can be expected if structural disorder in the reduced condition is high in comparison to the oxidized state. Thus, suboxide forming materials such as ceria resulting in oxygen defect structures are suitable materials for thermochemical water splitting.

Reduction temperature of pure ceria, however, still is too high for technical standards. Lowering the reduction temperature can be achieved by adding suitable cations such as Zr. On the other hand, addition of Zr typically results in a drastic lowering of the oxidation temperature which is due to smaller redox entropy with respect to the one of $CeO_2 \rightarrow CeO_{2-\delta}$. This behavior can be explained intuitively by higher structural ordering of $(Ce,Zr)O_{2-\delta}$ in the reduced state. Actually Ce^{3+} and Zr^{4+} are known to form an ordered pyrochlore structure rather than a fluorite-type solid solution defect structure. Pyrochlore-forming species Ce^{3+} and Zr^{4+}, however, are highly diluted in compositions such as $Ce_{0.85} Zr_{0.15}O_{2-\delta}$ ($\delta \approx 0.03$) and hence only clusters of pyrochlore-type short-range order can be expected distributed in a ceria-based solid solution. Direct detection of the presumed clusters with higher structural order is difficult but careful dilatometric studies provide indirect evidence of fluorite-pyrochlore transitions.

In binary ceria–zirconia ceramics redox entropy and structural ordering depends significantly on processing temperature. A comparative redox study using thermogravimetry, dilatometry, and water splitting experiments reveals that samples synthesized at 1,923 K develop structural ordering during reduction in contrast to samples processed below 1,673 K. In general, a certain temperature is a precondition of pyrochlore-type ordering rather than the degree of reduction or the concentration of Ce^{3+}, respectively. This finding is of high significance for

Nicole Knoblauch, Deutsches Zentrum für Luft- und Raumfahrt (DLR), Linder Höhe, 51147 Köln
Martin Schmücker, HRW University of Applied Science, Duisburger Str. 100, 45407 Mülheim/Ruhr, Germany

https://doi.org/10.1515/9783110674910-006

processing conditions of ceria–zirconia ceramics and for the process conditions of thermochemical water splitting cycles.

Keywords: solar thermal water splitting, CeO_2–ZrO_2, pyrochlore structure, redox entropy, dilatometry

6.1 Introduction

Hydrogen produced from renewable resources is a key element of future energy technology. Electrolysis of water using electricity from photovoltaics (PV) or concentrating solar power (CSP) is already a viable technical route for producing H_2. Less established, but promising approaches of hydrogen production without intermediate energy conversion are thermolysis or thermochemical water splitting cycles [1]. For that, concentrated solar radiation provides the required heat energy to decompose water into hydrogen or carbon dioxide into carbon monoxide.

Single-step thermal dissociation of water, however, is difficult because of thermodynamic conditions requiring temperatures well above 2,500 K to achieve a reasonable degree of H_2 yield. The same is true for CO_2 decomposition. Moreover, in case of water thermolysis, an effective high-temperature separation technique is necessary to isolate H_2 from O_2 in order to prevent explosive oxyhydrogen.

To circumvent these limitations, thermochemical cycles making use of redox reactions come into play. In this concept, a multivalent metal oxide in its reduced condition is re-oxidized by water vapor at moderate temperatures, thus resulting in water splitting and hydrogen release. The water splitting redox system is regenerated in the next step at high temperatures via concentrated solar radiation and/or by low oxygen partial pressure. During the reduction step, oxygen is gradually released from the redox material, hence a lower oxidation state arises and finally the system is ready for the next water-splitting step in the redox loop process (Fig. 6.1). An overview on redox-based thermochemical cycles is given for instance by Scheffe et al. [2].

Research on appropriate thermochemical cycles for water splitting started in the 1960s driven from the idea of using waste heat from high-temperature nuclear reactors. Metal oxide-based redox materials for a two-step water-splitting with solar radiation as heat source were proposed for the first time in the late 1970s [3, 4]. Nonetheless, thermochemical cycles are still in their research phase.

From technological point of view, the reduction temperature of the redox agent should not be too high and from kinetic reasons, the water splitting temperature should not be too low. As a consequence from these conditions and to avoid energy losses in the redox cycle, the temperature gap between reduction and oxidation step should be low as possible. For this requirement, the redox entropy of the employed redox material should be high which can easily be illustrated by means of Ellingham diagrams ($\Delta G°$ vs. T) with ΔS as slope of redox equilibrium curves (Fig. 6.2).

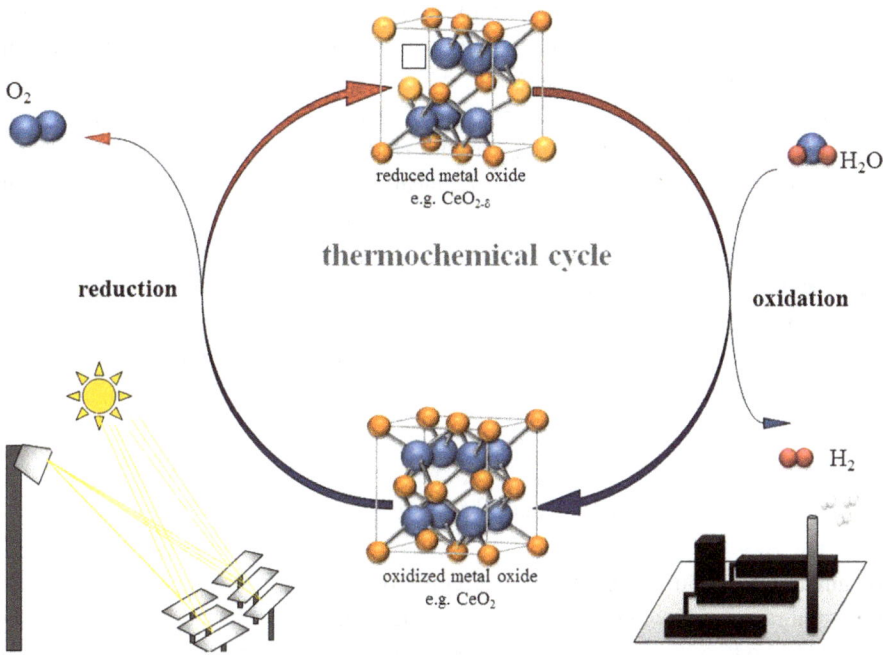

Fig. 6.1: Schematic sketch of a solar-driven thermochemical cycle using a two-step redox reaction for solar fuels production. MO_{ox} and MO_{red} denotes the oxidized and reduced metal oxide, respectively.

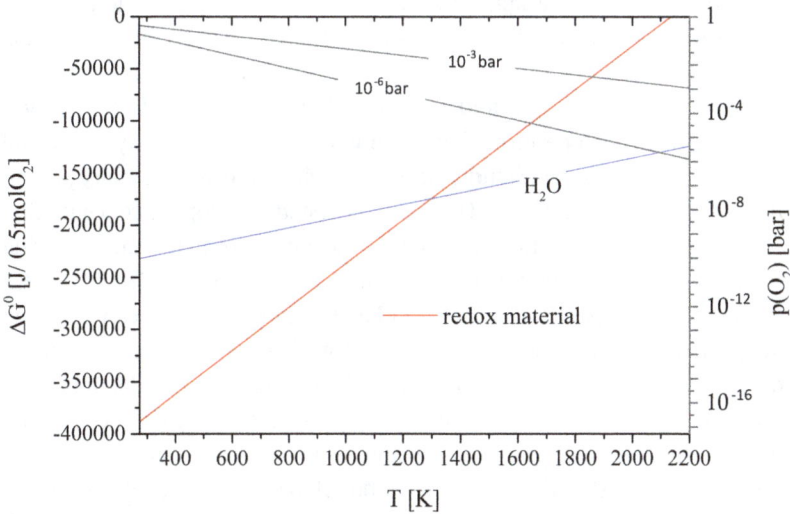

Fig. 6.2: Ellingham Diagram (ΔG^0 vs. T) for hypothetical redox system and $H_2 + 0.5O_2 \Leftrightarrow H_2O$ equilibrium line. Left from the curves intersect the water splitting reaction is thermodynamically favored.

Most solid redox systems display redox entropies of ≈100 J/K per 0.5 mole O_2 resulting from oxygen release or incorporation from or into the solid, respectively. Significant higher redox entropies can only be expected if structural disorder in the reduced condition is high in comparison to the oxidized state. Thus, suboxide forming materials such as perovskites $ABO_{3-\delta}$ or ceria $(CeO_{2-\delta})$-based materials resulting in oxygen defect structures are suitable candidates for thermochemical water splitting. Further advantages of suboxides are inherent high oxygen diffusion rates which facilitate the redox kinetics; moreover, close structural relationship between the stoichiometric phase in its oxidized condition and the non-stoichiometric phase in the reduced state circumvent nucleation phenomena and related incubation times in the course of redox reactions.

With ceria as redox material, the production of solar fuels could be described as follows:

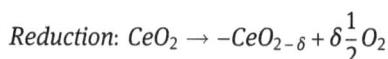

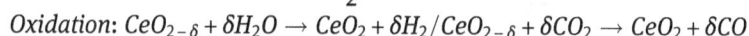

$$Reduction: \; CeO_2 \rightarrow -CeO_{2-\delta} + \delta \frac{1}{2} O_2$$

$$Oxidation: \; CeO_{2-\delta} + \delta H_2 O \rightarrow CeO_2 + \delta H_2 / CeO_{2-\delta} + \delta CO_2 \rightarrow CeO_2 + \delta CO$$

The amount of produced hydrogen or carbon monoxide, respectively, depends on the amount of released oxygen in the preceding reduction step. By introducing doping elements into the fluorite-type structure of CeO_2, the oxygen storage capacity can be influenced (see, e.g., [5]) through the incorporation of 4+ valence dopants such as Zr^{4+} the reduction extent of CeO_2 could be increased [6–11]. Different studies show that the reduction extent increases with $[Zr^{4+}]$ concentrations up to 15–25 mol% [6, 11, 12]. The improvement is often attributed to the small ionic radius of Zr^{4+} in comparison to Ce^{4+}/Ce^{3+} which could reduce strain during lattice expansion in consequence of reduction [13–15]. Moreover, 7-fold coordination of Zr^{4+} with oxygen accounts for facilitated oxygen vacancy formation with respect to eight-fold coordinated Ce^{4+} [10, 16, 17]. Different thermodynamic studies have been performed at varying temperatures and oxygen partial pressures [9, 10, 18–21]. All previous studies indicate higher non-stoichiometry parameters δ for $(Ce,Zr)O_{2-\delta}$ as compared to $CeO_{2-\delta}$, which depends on dopant concentration and changes both, enthalpy and entropy of redox reaction. To compare the scattering data of previous studies, the Gibbs free energy change for the reaction $Ce_{1-x}Zr_xO_{1.95} \leftrightarrow Ce_{1-x}Zr_xO_2$ as function of temperature is shown in Fig. 6.3. In summary, it can be stated that the enthalpy and entropy of the redox-reaction is smaller than for undoped cerium oxide. For low $[Zr^{4+}]$-concentration (x = 0.05) only small deviation between different studies are recorded. But scattering of reported data rises up with increasing $[Zr^{4+}]$-concentration. Thereby the maximum water-splitting temperature varies over a broad range.

Beside thermodynamic studies, water and carbon dioxide splitting tests were also performed by several groups [6, 8, 12, 22]. In comparison to pure ceria rather sluggish re-oxidation kinetics was observed. Beside changes in thermodynamic parameters, sinter effects are discussed as a reason for slower oxidation kinetics [8, 23]. According to Le Gal et al. [12] even with high $[Zr^{4+}]$ concentrations, for example, of x = 0.5 hydrogen

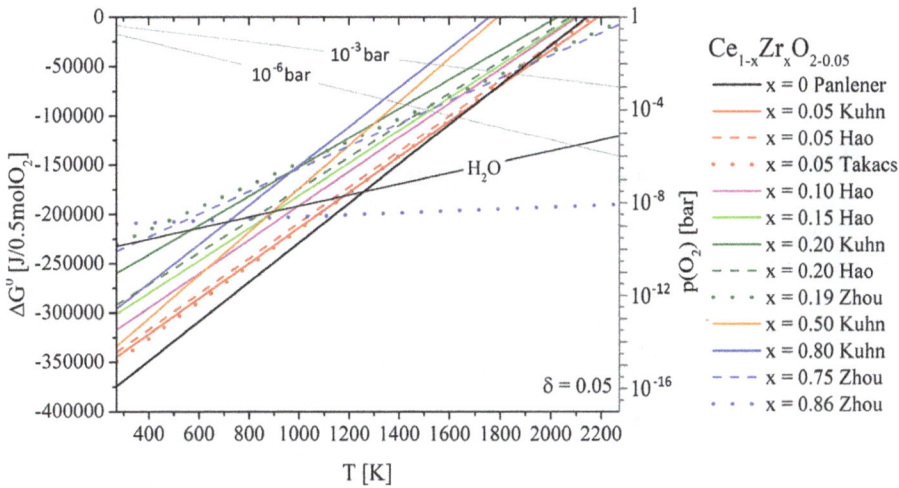

Fig. 6.3: Gibbs free energy change versus temperature for the oxidation of CeO$_{2-\delta}$ (black [19]), Ce$_{0.95}$Zr$_{0.05}$O$_{2-\delta}$ (red) [9, 10, 18], Ce$_{0.9}$Zr$_{0.1}$O$_{2-\delta}$ (pink) [18], Ce$_{0.85}$Zr$_{0.15}$O$_{2-\delta}$ (bright green) [18], Ce$_{0.80}$Zr$_{0.20}$O$_{2-\delta}$ (green) [10, 18].

can be produced by water-splitting at 1,323 K. This result, however, is inconsistent with the discussed thermodynamic studies. A coherent study of redox thermodynamics, water-splitting behavior and morphology/microstructure is still missing. There is, however, some evidence that thermal aging of Zr-doped cerium oxide is a crucial step, especially for higher [Zr^{4+}]-concentration and in reducing atmospheres [24–27]. But the influence of thermal pre-treatment at high temperature and morphology of lower Zr-doped cerium in ambient pressure were not investigated so far. In the present study, we investigate the thermochemical performance of Ce$_{0.85}$Zr$_{0.15}$O$_2$ thermally pre-treated at 1,923 K in comparison to 1,473 K pre-treated samples as well as their water splitting behavior. The Zr concentration of 15 mol% corresponds to the optimum described by Call et al. [11]. Furthermore, we study changes of morphology and crystal structure as a result of thermal pre-treatment. In combination with re-oxidation experiments in air and dilatometric studies an appropriate defect model is developed to describe changes in chemical expansions during reduction.

6.2 Results and discussion

6.2.1 Sample characterization

The compositions of Ce$_{1-x}$Zr$_x$O$_2$ samples synthesized via Pechini-method [28] and thermally pre-treated at different temperatures were analyzed by energy-dispersive X-ray spectroscopy (EDS, INCA-Software, Oxford Instruments, Abingdon, UK). Figure 6.4 shows a typical EDS spectrum of a $Ce_{1-x}Zr_xO_2$ specimen fired at 1,473 K confirming the target composition of 15 mol% ZrO_2.

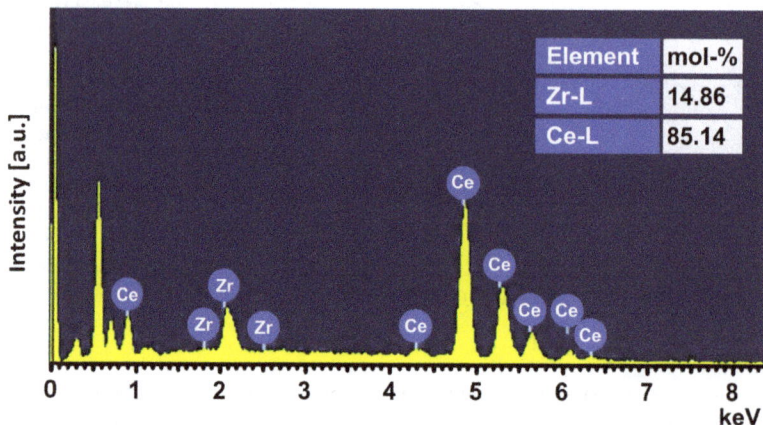

Element	mol-%
Zr-L	14.86
Ce-L	85.14

Fig. 6.4: EDS analysis of $Ce_{1-x}Zr_xO_2$ pellet sample.

The Zr-doped cerium oxide samples appear in different color depending on thermal heat treatment. With increasing temperature, the color shifts from yellow to blue. The blue coloration is well known for reduced pure cerium oxide [29]. The bluish color indicates the formation of Ce^{3+}-ions, which is known to be favored in the presence of Zr^{4+} (e.g., [10, 14, 16]). Accordingly, the reduction temperature of cerium oxide could be shifted to lower temperature by addition of Zr^{4+} (see Fig. 6.3). The results of SEM-analyses (Ultra 55 FEG, Carl Zeiss, Germany) show a decrease of porosity and an increase of grain size with increasing temperature. After heat treatment at 1,923 K more than 95% of theoretical density is reached.

Beside optical and morphological changes also small changes of the crystal structure are detected by X-ray diffraction (HT-XRD D8 Advance A25, Bruker, Germany). Figure 6.6 shows the X-ray diffraction patterns after heat treatment at different temperatures.

$Ce_{0.85}Zr_{0.15}O_2$ XRD patterns reveal decreasing peak width and a gradual peak shift toward higher diffraction angles with increasing thermal treatment. No other phase than fluorite-type solid solution can be detected. The development of lattice constants is shown in Fig. 6.7 together with apparent Zr concentrations calculated according to Varez et al. [30]. Data suggest that a regular solid solution with the target concentration of 15 mol% ZrO_2 is reached only at temperatures above 1,600 K. Obvious smaller lattice constants occurring in the temperature range from 1,000 to 1,600 K are interpreted as deviation from Vegard's law. This is in line with the observation that zirconia-doped ceria tends to separate into CeO_2-rich and ZrO_2-rich

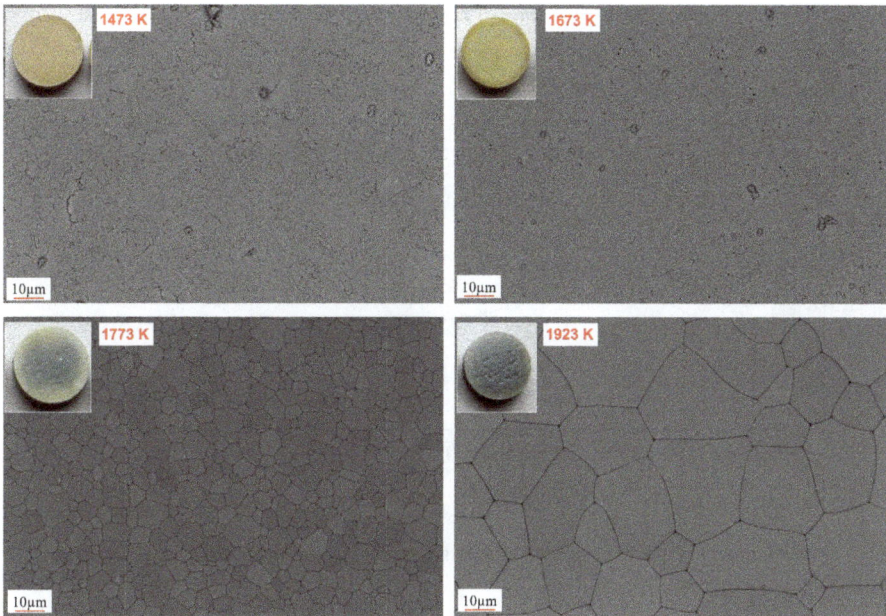

Fig. 6.5: SEM-image and pictures of $Ce_{0.85}Zr_{0.15}O_2$ pellet samples after heat treatment at 1,473 K, 1,673 K, 1,773 K and 1,923 K for 2 h.

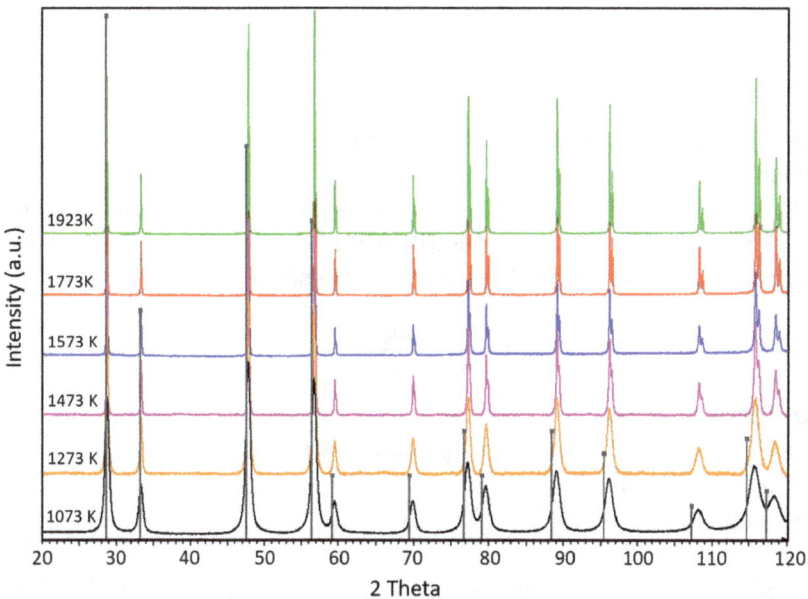

Fig. 6.6: XRD patterns of $Ce_{0.85}Zr_{0.15}O_2$ pellet samples after thermal pre-treatment at 1,073 K, 1,273 K, 1,473 K, 1,573 K,1,773 K and 1,923 K (2 h) in comparison to pattern of CeO_2 (ICDD PDF-00-034-039423).

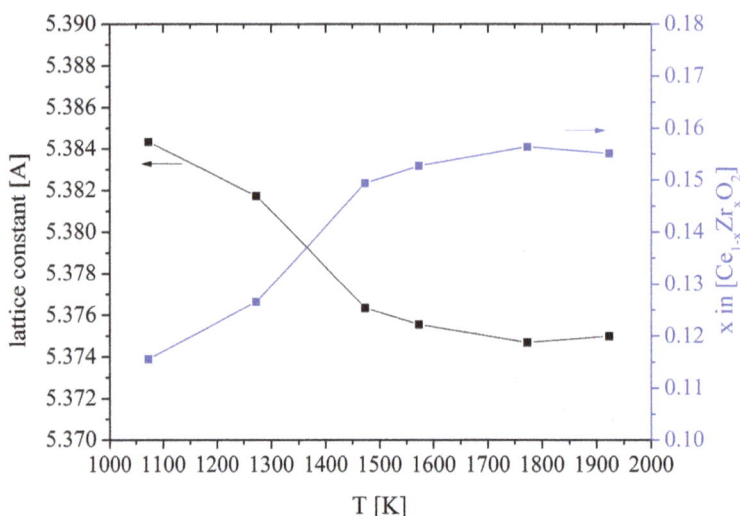

Fig. 6.7: Development of lattice constants and corresponding Zr-concentrations of $Ce_{1-x}Zr_xO_2$ with increasing temperature.

domains at lower temperature as previously described (e.g., [31]). At 1,923 K a slight increase of the crystal cell is observed. This effect can be attributed to incipient reduction which is accompanied by formation of bigger Ce^{3+}-ions [32].

6.2.2 Thermodynamic calculations

The following diagrams show the determined δ-values (eq. (6.1)) as function of reduction temperature and pO_2 for two $Ce_{0.85}Zr_{0.15}O_2$ samples thermally pre-treated at different temperatures. Data are derived from mass loss measurements obtained by thermogravimetric analyses (TGA, Model STA 449 F3 Jupiter, Germany, equipped with oxygen ion pump and oxygen detector (Nernst Setnag, France)).

$$\delta = \frac{M \cdot \Delta meq}{M_O \cdot m_s} \tag{6.1}$$

M = molar mass of $Ce_{1-x}Zr_xO_2$,
Δmeq = equilibrium value of the mass change during the redox-reaction,
M_o = molar mass of oxygen
m_s = sample mass

The Zr-doped samples clearly show higher δ-values compared to undoped cerium oxide. Moreover, the thermal pre-treatment has an influence on δ(T,pO_2) function.

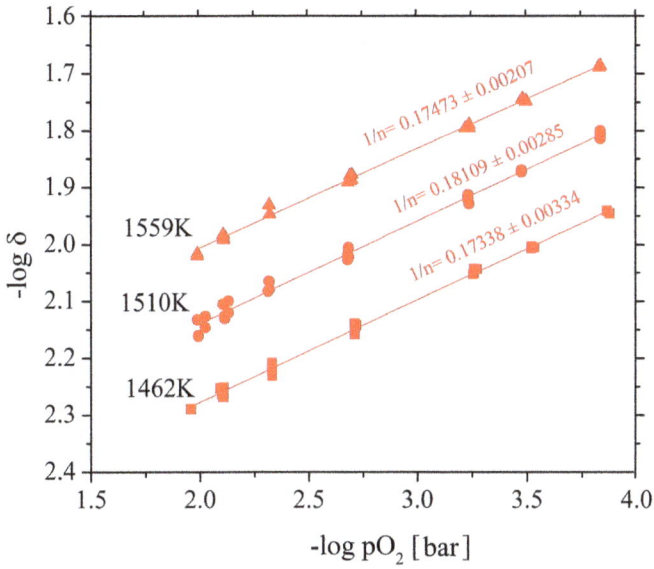

Fig. 6.8: Measured δ values of $Ce_{0.85}Zr_{0.15}O_2$ pre-treated at 1,473 K (red) as function of temperature and pO_2 [33].

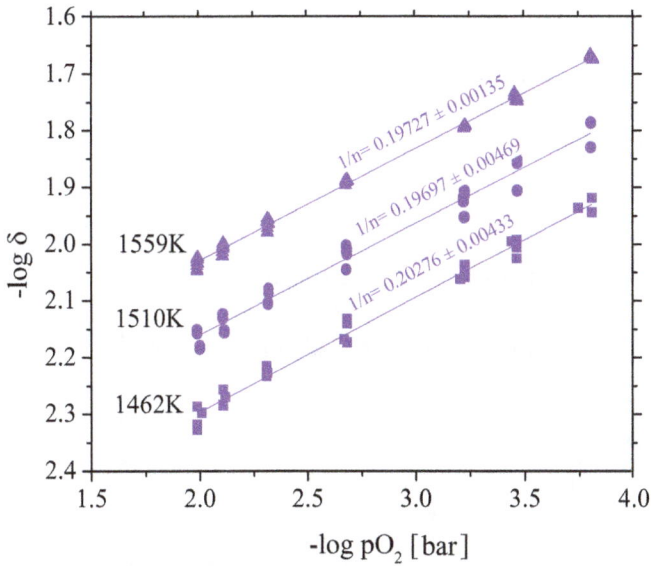

Fig. 6.9: Measured δ values of $Ce_{0.85}Zr_{0.15}O_2$ pre-treated at 1,923 K (purple) as function of temperature and pO_2 [33].

Samples sintered at low temperatures release more oxygen during reduction at higher oxygen partial pressures than samples sintered at higher temperatures. At low oxygen partial pressures a reverse behavior is observed.

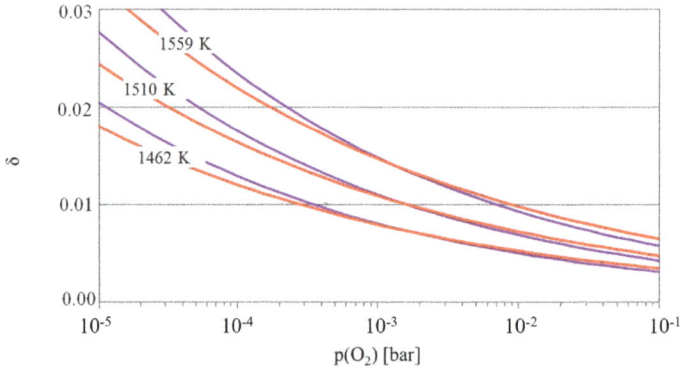

Fig. 6.10: Experimental equilibrium δ values vs. pO_2, obtained isothermally with in graph labels showing the temperature in K, for $Ce_{0.85}Zr_{0.15}O_2$ pre-treated at 1,473 K (red) and $Ce_{0.85}Zr_{0.15}O_2$ pre-treated at 1,923 K (purple) plotted with the best fit for model (eq. (6.2)) [33].

The results recorded by variation of temperature and pO_2 conditions were fitted by an analytical model of Bulfin et al. [34] modified by Call [35] and Knoblauch [36], (eq. (6.2)).

$$\delta(p(O_2), T) = \frac{\delta_{max} \cdot \frac{A_{Red}}{A_{Ox}} \cdot p^{-1/n}(O_2) \cdot \exp\left(\frac{-\Delta E}{RT}\right)}{1 + \frac{A_{Red}}{A_{Ox}} \cdot p^{-1/n}(O_2) \cdot \exp\left(\frac{-\Delta E}{RT}\right)} \tag{6.2}$$

δ_{max} = 0.425 which corresponds to complete reduction of Ce^{4+} to Ce^{3+} for the composition $Ce_{0.85}Zr_{0.15}O_{1.575}$

Curve fittings result in the following parameters listed in Tab. 6.1:

Tab. 6.1: Resulting parameter A_{red}/ A_{Ox}, 1/n, ΔE of function δ(pO_2,T) (eq. (6.2)) using data of Figs. 6.8 and 6.9. δ_{max} is fixed to 0.425.

Sample	δ_{max}	A_{red}/ A_{Ox} [bar^n]	1/n	ΔE [kJ/mol]	R^2
CeZr-1650	0.425	101 ± 11	0.2084 ± 0.0019	121.6 ± 1.4	0.9893
CeZr-1200	0.425	133 ± 11	0.1817 ± 0.0012	122.8 ± 1.1	0.9979
CeO$_2$	0.5	12,575 ± 7550	0.2320 ± 0.0079	206.7 ± 8.2	0.9754

Compared to undoped cerium oxide, the effective activation energy ΔE is reduced from 207 kJ/mol to 122 kJ/mol by introducing 15 mol% ZrO_2. The 1/n values characterizing the T/pO_2 dependence have also been slightly influenced by Zr^{4+} incorporation.

Here, the oxygen partial pressure exponents $1/n$ are in the range of $1/5$ to $1/6$, reflecting the tendency to isolated oxygen vacancies [37]. Likewise, Kuhn et al. [10] determine a dependency of $1/n \approx 1/6$ for $Ce_{0.8}Zr_{0.2}O_2$ in the range of $\delta = 0.0001\text{–}0.1$. Thus, according to the obtained $1/n$ results in the present δ range of $0.005\text{–}0.02$ interactions between oxygen vacancies and the Ce^{3+} ions are not presumed. On the other hand, Shah et al. [21] suggest that, Zr^{4+} is preferentially localized near Ce^{3+} ions and $V_O^{\cdot\cdot}$ According to Yang et al. [16], the formation of clusters between Ce^{3+}, Zr^{4+}, and $V_O^{\cdot\cdot}$ would also reduce the energy required for reduction. In case of a cluster with pyrochlore configuration (see Fig. 6.11), the oxygen vacancy at position 8a is encompassed by Zr^{4+} ions [38]. Such stabilization of the oxygen vacancies in the form of a pyrochlore configuration could also explain the observed reduction of the activation energies.

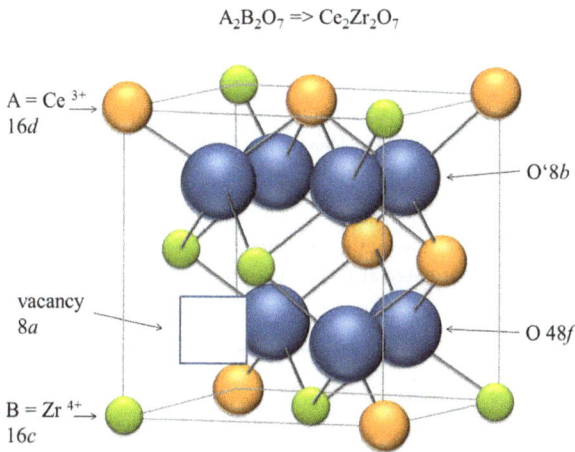

$$A_2B_2O_7 => Ce_2Zr_2O_7$$

A = Ce $^{3+}$
16d

O'8b

vacancy
8a

O 48f

B = Zr $^{4+}$
16c

Fig. 6.11: 1/8 unit cell of $Ce_2Zr_2O_7$.

Based on the resulting δ-pO_2 function, the thermodynamic eqs. (6.3) and (6.4) enable the calculation of ΔG and subsequently a Ellingham-type presentation (Fig. 6.12)

$$\Delta G^0 = 0.5 \cdot R \cdot T \cdot \ln \frac{pO_2}{p^0} \qquad (6.3)$$

$$\Delta G^0 = \Delta H^0 - T \cdot \Delta S^0 \qquad (6.4)$$

The Ellingham-Richardson diagram (Fig. 6.12) depicts the corresponding equilibrium curves of re-oxidations of the differently pre-treated $Ce_{0.85}Zr_{0.15}O_{2\text{-}\delta}$ with δ-values of 0.01, 0.05, and 0.1, respectively. In addition, Fig. 6.13 shows the Gibbs free energy of the redox reaction of undoped CeO_2 and of the water-splitting reaction. According to eq. (6.4), the intercept and slope correspond to redox enthalpy and entropy. The comparison of the results reveals that apart from doping, the thermal pre-treatment influences enthalpy and entropy of redox-reaction. The entropy change of

1,923 K pre-treated samples is strikingly reduced. As a consequence the tendency to water splitting decreased.

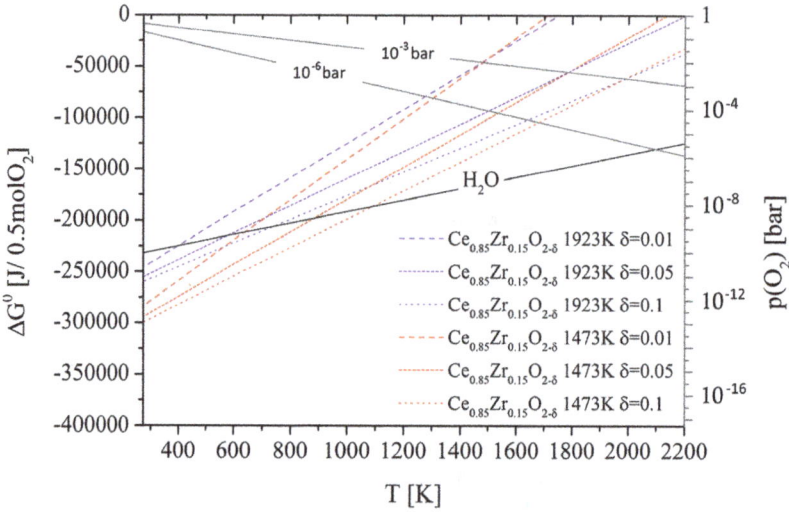

Fig. 6.12: Gibbs free energy change versus temperature for the reduction $Ce_{0.85}Zr_{0.15}O_2$ pre-treated at 1,473 K (red) and $Ce_{0.85}Zr_{0.15}O_2$ pre-treated at 1,923 K (purple) for different δ values.

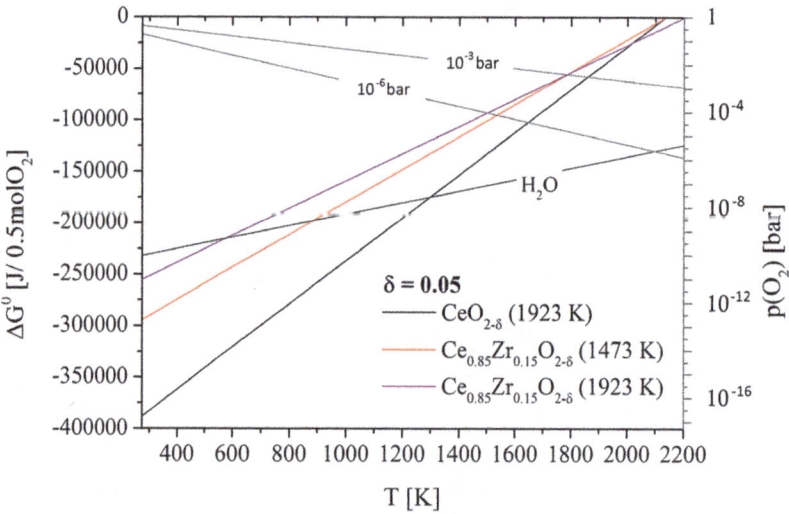

Fig. 6.13: Gibbs free energy change versus temperature for the re-oxidation of $Ce_{0.85}Zr_{0.15}O_{2-\delta}$ (δ = 0.05) pre-treated at 1,473 K (red), $Ce_{0.85}Zr_{0.15}O_{2-\delta}$ pre-treated at 1,923 K (purple), $CeO_{2-\delta}$ (black) as for the water splitting reaction.

Redox entropies result from oxygen release, vibration, and configuration entropy (eq. (6.5)).

$$\Delta S^0 = \frac{1}{2}S^0_{O_2} + \Delta S^0_{vib} + \Delta S_{con} \qquad (6.5)$$

The entropy change due to oxygen release $S^0_{O_2}$ is the same for all samples depicted in Fig. 6.12. The amount of vibration entropy ΔS^0_{vib} is small in comparison to $S^0_{O_2}$ and ΔS_{con} [19]. As a consequence ΔS_{con} must be the decisive factor to explain the observed redox entropy differences between CeO_2 and $Ce_{0.85}Zr_{0.15}O_{2-\delta}$ of ≈ 76 J/K per 0.5 mol O_2. So in first approximation it can be assumed that:

$$\Delta S_{con}(CeO2) - \Delta S_{con}(Ce0.85Zr0.15O2,1923K) \cong 76\frac{J}{0.5molO_2K} \cong 3.8\frac{J}{\delta molK} \qquad (6.6)$$

The difference in configuration entropy can be related to a decrease of oxygen vacancies sites in the postulated pyrochlore environment (see Fig. 6.11):

$$\Delta S = R[\ln(N!) - ln(n!) - \ln((N-n)!)] \qquad (6.7)$$

In comparison to the fluorite structure with eight symmetric equivalent oxygen vacancy sites only one oxygen vacancy position may exists in the pyrochlore structure (under consideration of unit cells Ce_4O_8 and $Ce_2Zr_2O_7$). This results in the following configuration entropies:

$$\Delta S_{con}(Ce4O8) = R[N \cdot \ln(N) - n \cdot \ln(n) - (N-n) \cdot \ln(N-n)]$$

$$= R[8 \cdot \ln(8) - (4\delta) \cdot \ln(4\delta) - (8 - 4\delta) \cdot \ln(8 - 4\delta)] = 7.8\frac{J}{\delta molK} \qquad (6.8)$$

$$\Delta S_{con}(Ce2Zr2O8) = R[N \cdot \ln(N) - n \cdot \ln(n) - (N-n) \cdot \ln(N-n)]$$

$$= R[1 \cdot \ln(1) - (4\delta) \cdot \ln(4\delta) - (1 - 4\delta) \cdot \ln(1 - 4\delta)] = 4.2\frac{J}{\delta molK} \qquad (6.9)$$

The difference amounts to:

$$\Delta S_{con}(Ce4O8) - \Delta S_{con}(Ce2Zr2O8) \cong 3.6\frac{J}{\delta molK} \qquad (6.10)$$

Thus, the estimated difference in configuration entropy between CeO_2 and $Ce_{0.85}Zr_{0.15}O_{2-\delta}$ corresponds well to the observed data.

Likewise, Shah et al. [21] obtained an entropy change during the reduction of $Ce_{0.81}Zr_{0.19}O_2$ to $Ce_{0.81}Zr_{0.19}O_{1.9}$ and justified it by the formation of pyrochlore-type clusters (here $Ce_{0.81}Zr_{0.19}O_{1.9}$ is approximately equal to $(CeO_2)_{0.62}(Ce_2Zr_2O_7)_{0.095}$ or $(CeO_2)_{0.62}(Ce_{0.5}Zr_{0.5}O_{1.75})_{0.38}$), respectively. In contrast to the observations of Shah et al. and other authors [24–27], our data (Fig. 6.13) suggest that the pyrochlore-ordering transition occurs also during mild reduction under ambient pressure during pre-treatment at 1,923 K and lower $[Zr^{4+}]$ concentration of $x = 0.15$.

According to the thermal pre-treatment/reduction conditions of this work (1,923 K, 0.2 bar) for $Ce_{0.85}Zr_{0.15}O_2$ a reduction value of $\delta = 0.028$ is obtained (calculation is based on the obtained parameter of Tab. 6.1). This condition results in a Ce^{3+} concentration of 5.6%. Assuming fractional pyrochlore-type ordering, the following composition can be taken into account:

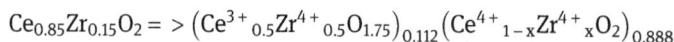

$$Ce_{0.85}Zr_{0.15}O_2 = > \left(Ce^{3+}_{0.5}Zr^{4+}_{0.5}O_{1.75}\right)_{0.112}\left(Ce^{4+}_{1-x}Zr^{4+}_xO_2\right)_{0.888}$$

During cooling down of the reduced sample under ambient pressure re-oxidation can be expected. However, the pellet sample remains slightly blue (see Fig. 6.5) suggesting the re-oxidation during cooling down is incomplete. On the other hand, the diffraction pattern after cooling down (see Fig. 6.6) does not show any pyrochlore phase like $Ce_2Zr_2O_7$ or $Ce_2Zr_2O_8$ [39], which in general consists of main pattern closely related to that of ceria plus additional superstructure peaks due to cation and oxygen vacancy ordering (Fig. 6.8). The resulting lattice constant after cooling (5.375 Å, Fig. 6.7) and the corresponding difference to the fully oxidized state indicates a small remaining reduction value of $\delta = 0.001$, based on $\Delta a/a = (0.0543 \pm 0.00157) \cdot \delta$ [36] .

The composition $Ce_{0.85}Zr_{0.15}O_{1.999}$ produced by high-temperature sintering and subsequent cooling down in air, however, may retain its high-temperature structural ordering provided that cation-ordered pyrochlore-type phases such as $Ce_2Zr_2O_{7.5}$ or $Ce_2Zr_2O_8$ are formed by oxidation of $Ce_2Zr_2O_7$ as suggested by Achary et al. [39]. Assuming the 11.2% fraction of $Ce^{3+}_{0.5}Zr^{4+}_{0.5}O_{1.75}$ existing at 1,923 K ($\delta = 0.0279$) transforms into $Ce^{3+}_{0.5}Zr^{4+}_{0.5}O_2$ by re-oxidation ($\delta = 0.001$) composition

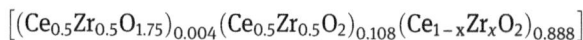

$$\left[(Ce_{0.5}Zr_{0.5}O_{1.75})_{0.004}(Ce_{0.5}Zr_{0.5}O_2)_{0.108}(Ce_{1-x}Zr_xO_2)_{0.888}\right]$$

can be envisaged after cooling down rather than a true solid solution $Ce_{0.85}Zr_{0.15}O_{1.999}$.

To sum up, lower redox entropy of $Ce_{0.85}Zr_{0.15}O_2$ after 1,923 K pre-treatment accounts for higher structural ordering after cooling down and re-oxidation. In contrast to $Ce_{0.85}Zr_{0.15}O_2$ ceramics fired at 1,923 K, the sample pre-treated at only 1,473 K ($\delta \approx 0.0029$) becomes fully oxidized after cooling down as derived from coloring (pale yellow, Fig. 6.5) and lattice constants. Due to the low initial reduction condition and/or lower temperature obviously no significant cation ordering has been developed.

To shed more light on the question of structural ordering of $Ce_{0.85}Zr_{0.15}O_2$ fired at 1,923 K, dilatometric studies were performed since pyrochlore ordering phenomena do affect cell parameters and sample volumes, respectively.

6.2.3 Chemical expansion during reduction

With increasing temperature Ce^{4+} will be gradually reduced to Ce^{3+} going along with the formation of oxygen vacancies. These changes induce a volume change of the crystal lattice. In the case of undoped cerium oxide the lattice variation is caused by the change of ionic radii of Ce^{4+} −97 pm- versus Ce^{3+} −114 pm- [32], by the formation of oxygen vacancies which provides charge neutrality and by repulsion of defects and their atomic neighbors [15, 36, 40, 41]. The cause of resulting chemical expansion is controversially discussed. According to Bishop [15] and Marrocchelli et al. [40], the lattice contraction induced by the formation of oxygen vacancies is overcompensated by the formation of bigger Ce^{3+} ions thus leading to expansion of $CeO_{2-\delta}$ lattice constants with increasing δ value. In contrast to Bishop and Marrocchelli et al., Muhich [41] shows by electronic structure calculations (DFT + U) that repulsion of defects and their atomic neighbors are crucial effects.

If Zr^{4+} ions are present, in a first approximation the same relative chemical expansion during reduction of Ce^{4+} can be taken into account (mechanism A, see below). However, this assumption does not include any interaction between Zr^{4+} and Ce^{3+} ions, which may lead to pyrochlore-cluster formation (mechanism B, C, see below). Here, the concentrations of $[Zr^{4+}]$ and $[Ce^{3+}]$ are the crucial parameters. For the given Zr-concentration x = 0.15, the pyrochlore-cluster concentration is limited to $\delta = x/2$. In the following some possible pyrochlore-cluster-formation mechanisms are discussed and related chemical expansions are estimated:

A) $Ce^{4+}_{0.85}Zr_{0.15}O_2 \rightarrow (Ce^{4+}_{1-2\delta},Ce^{3+}_{2\delta})_{0.85}Zr_{0.15}O_{2\,-\delta}$ (solid solution, no interaction between Zr^{4+} and Ce^{3+}-ions)

B) $Ce^{4+}_{0.85}Zr_{0.15}O_2 \rightarrow (Ce^{3+}_{0.5}Zr_{0.5}O_{1.75})_a + (Ce^{4+}_{1-x}Zr_xO_2)_b$; x < 0.15
    (solid solution in the starting state gradually transforms into pyrochlore structure. Zr-enrichment of pyrochlor with respect to starting composition is compensated by gradual Zr depletion in the coexisting solid solution. At $\delta = 0.075$, all available Zr is incorporated into pyrochlore which then coexists with pure CeO_2.)

C) $(Ce_{0.5}Zr_{0.5}O_2)_{0.112} + (Ce_{1-x}Zr_xO_2)_{0.888} \rightarrow (Ce_{0.5}Zr_{0.5}O_{1.75})_c + (Ce_{0.5}Zr_{0.5}O_2)_d + (Ce_{1-x}Zr_xO_{2-x})_e$
    ("oxidized" pyrochlore phase (amount limited by $[Ce^{3+}]$ concentration) of previous high-temperature treatment ($\delta = 0.028$) gradually transforms into "regular" pyrochlore phase. For reduction values above $\delta = 0.028$, all $Ce_{0.5}Zr_{0.5}O_2$ is consumed and hence cerium reduction in the coexisting solid solution $Ce_{1-x}Zr_xO_2$ has to occur which leads to higher chemical expansion. As a consequence, the expansion curve of reaction C shows a discontinuity at $\delta = 0.028$.)

The corresponding amounts a-e are results of the Zr-concentration and the effective reduction value δ. The resulting chemical expansion depends on the assumed mechanism. The chemical expansion of undoped cerium oxide is given by:

$$\Delta l/l = (0.0726 \pm 0.00335) \cdot \delta \qquad (6.11)$$

This correlation is based on lattice constants as function of δ published by Hull et al. [42]. It is also used for the chemical expansion of the ideal $Ce_{1-x}Zr_xO_{2-\delta}$ solid solution. For the calculation of the starting lattice constants of ceria/zirconia solid solution, data published by Varez et al. [30] are used:

$$(-0.2362 \pm 0.00733) \cdot x + \mathbf{a}(CeO_2) = \mathbf{a}(Ce_{1-x}Zr_xO_2) \qquad (6.12)$$

Furthermore, the lattice constants of $Ce_2Zr_2O_7$ (a = 10.6924(3)Å) and $Ce_2Zr_2O_8$ (a = 10.5443(3)Å) [39] are from literature data. Based on these values and Vegard's rule [43], all resulting lattice constants and chemical expansions could be estimated for given reduction states. The comparison of calculated expansion data with experimental ones (Fig. 6.14) indicates that the experimental chemical expansion of $Ce_{0.85}Zr_{0.15}O_{2-\delta}$ during isothermal reduction recorded by a horizontal dilatometer (TA Instruments, DIL 803, SiC furnace, UK, see also [36]) is significantly lower than estimated for a solid solution (Fig. 6.14, red vs. black line). A similar finding of lower chemical expansion by addition of zirconia was already mentioned by Bishop et al. [15] for $Ce_{0.5}Zr_{0.5}O_2$ and was explained by a stronger displacement of Zr^{4+} toward neighboring vacancies as compared to Ce^{4+}. In our study, the lower chemical expansion was also observed at lower Zr-concentration, that is $Ce_{0.85}Zr_{0.15}O_2$. The observed expansion characteristics can be interpreted by pyrochlore-type zones since the formation of pyrochlore coordination leads to a lattice contraction [38]. As mentioned before, thermodynamic data suggest that the pyrochlore-coordination is formed during thermal pre-treatment. This assumption could be encouraged by the measured chemical expansion. Since the observed chemical expansion is higher than calculated for in situ pyrochlore-formation from initial ceria/zirconia solid solution (reaction B, green curve in Fig. 6.14), a starting condition comprising pyrochlore-type coordination ($Ce_{0.5}Zr_{0.5}O_2$, mechanism C) is most obvious. In case of reaction C, the amount of oxidized pyrochlore $Ce_{0.5}Zr_{0.5}O_2$ [39] in the starting state (11.2%) corresponds to the amount of $Ce_{0.5}Zr_{0.5}O_{1.75}$ which is supposed to exists at 1,923 K ($\delta = 0.028$). Thus, for reduction values above $\delta = 0.028$, cerium reduction in the coexisting solid solution $Ce_{1-x}Zr_xO_2$ has to occur which leads then to higher chemical expansion. As a consequence, the expansion curve of reaction C shows a discontinuity around $\delta = 0.028$. Actually, our experimental dilatometry data indicate such discontinuity in the range of $\delta = 0.03$. Thus, the present dilatometry data support the idea of pre-existing pyrochlore-coordination which develops during thermal pre-treatment.

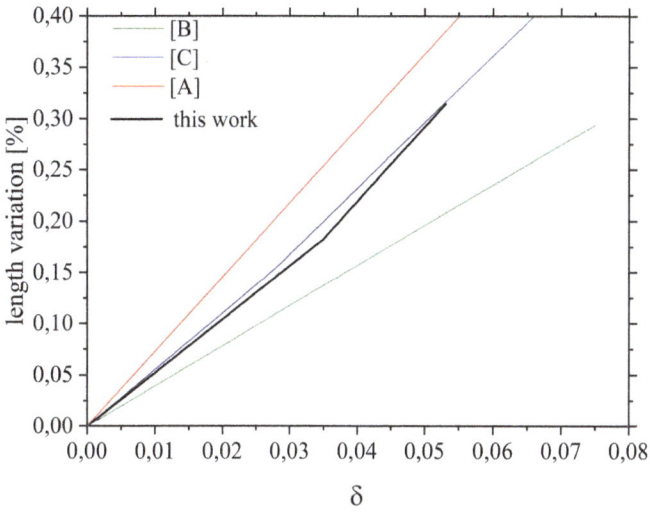

Fig. 6.14: Experimental (black) and calculated dilatometric data of $Ce_{0.85}Zr_{0.15}O_2$ during isothermal reduction. Modeled curves refer to mechanism [A] (solid solution without Ce^{3+}/Zr interactions), [B] ($Ce_{0.5}Zr_{0.5}O_{1.75}$ pyrochlore zones gradually forming from $Ce_{0.85}Zr_{0.15}O_2$), [C] (reduction of pre-existing $Ce_{0.5}Zr_{0.5}O_2$ to $Ce_{0.5}Zr_{0.5}O_{1.75}$). For details see text.

6.2.4 Water-splitting tests

The water-splitting tests are performed in a lab rector described in Fig. 6.15.

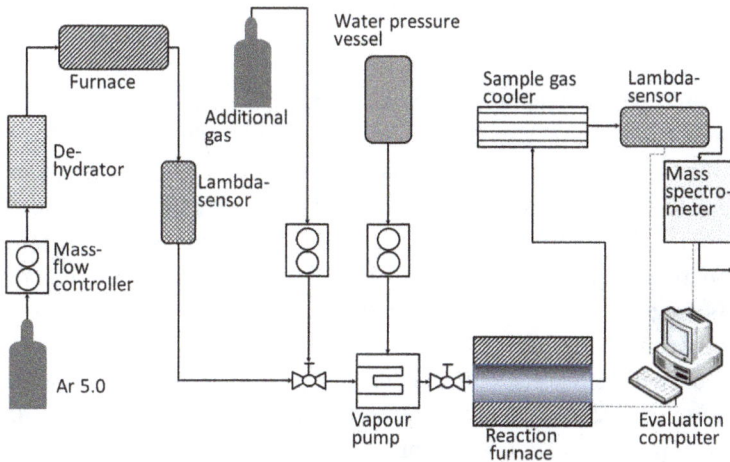

Fig. 6.15: Water-splitting test setup, for more details see [44].

To evaluate the results also of pure cerium oxide samples, thermally pre-treated at 1,473 K and 1,923 K, are investigated. Figure 6.16 shows the hydrogen development of an exemplary water-splitting test using a CeO_2 pellet sample thermally pre-treated at 1,473 K. The sample is reduced at 1,573 K under Ar (pO_2 = 1E-3 atm) and oxidized at 1,150 K by water steam. As a result of oxidation (Fig. 6.16, 135 min), the hydrogen concentration increases. In addition, the oxygen concentration in the process gas, which is required for the transport of the water vapor, decreases. In the following reduction step (Fig. 6.16, 170 min), the oxygen concentration rises again.

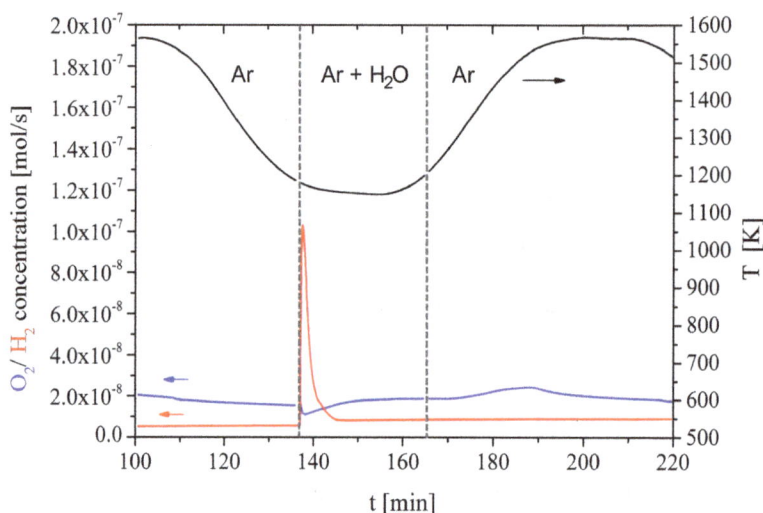

Fig. 6.16: Temperature and O_2/H_2 evolution during water-splitting experiment with CeO_2 (thermally pre-treated at 1,473 K).

Figures 6.17–6.20 depict the hydrogen evolution of CeO_2 and $Ce_{0.85}Zr_{0.15}O_2$ pellet samples after thermal pre-treatment at 1,473 K or 1,923 K and subsequently reduction at 1,573 K.

Except from the $Ce_{0.85}Zr_{0.15}O_2$ sample thermally pre-treated at 1,923 K, the hydrogen evolution starts as soon as the sample gets contact to the water steam. Figure 6.21 depicts the amount of released hydrogen [µmol/g] during water-splitting test of each material. The amount of hydrogen depends on doping and in particular on thermal pre-treatment. As long as pre-treatment temperature is 1,473 K, Zr-doping leads to an increase of hydrogen release. Nevertheless this increase is not as high as the improved reduction behavior suggests. This phenomenon was reported before, for example, by Le Gal et al. [22] but is not fully understood. Hydrogen release data suggest that thermal pre-treatment is more important than doping. Through a thermal pre-treatment at 1,923 K, the water-splitting activity of $Ce_{0.85}Zr_{0.15}O_2$ decreases dramatically. At first, the decrease of porosity has to be considered. Hence, for full oxidation, the oxidation time has to be extended. Nonetheless pre-treatment at 1,923 K of

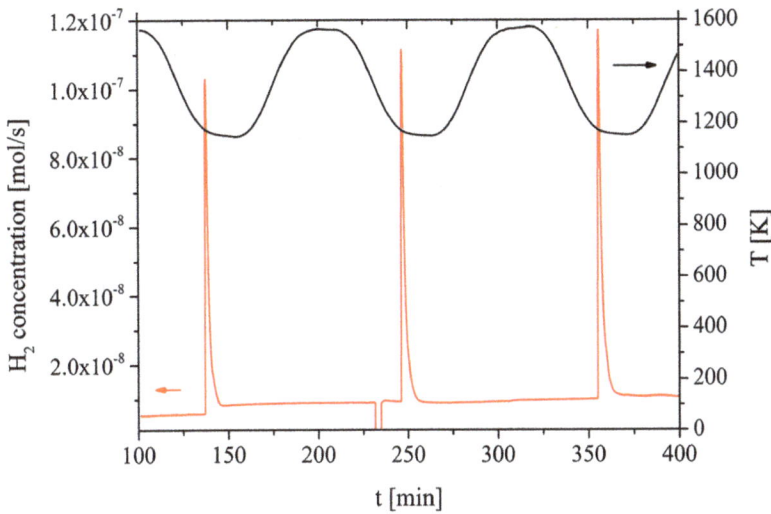

Fig. 6.17: H_2 evolution during water-splitting experiment with CeO_2 thermally pre-treated at 1,473 K and subsequently reduced at 1,573 K under Ar [45].

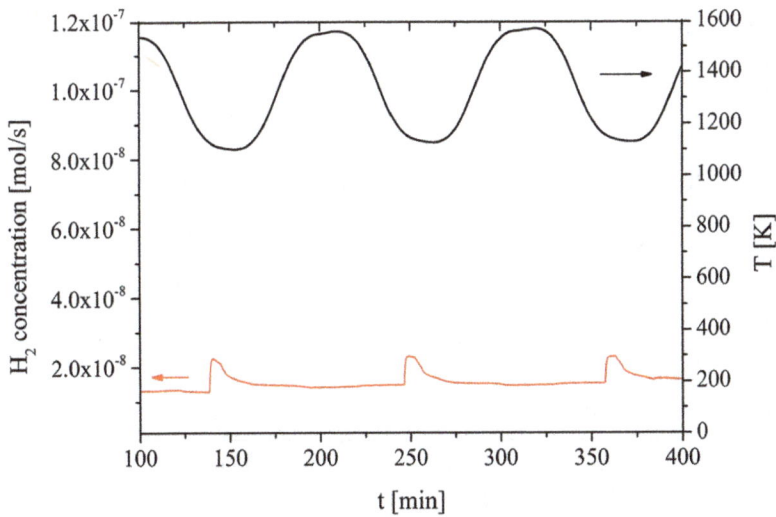

Fig. 6.18: H_2 evolution during water-splitting experiment with CeO_2 thermally pre-treated at 1,923 K and subsequently reduced at 1,573 K under Ar [45].

Zr-doped cerium oxide pellets results in complete suppression of any water-splitting activity. Neither by a decreased oxidation temperature which implies favorable thermodynamic conditions nor by milling the pellets into powder hence increasing the surface area drastically, the water-splitting activity could be reactivated.

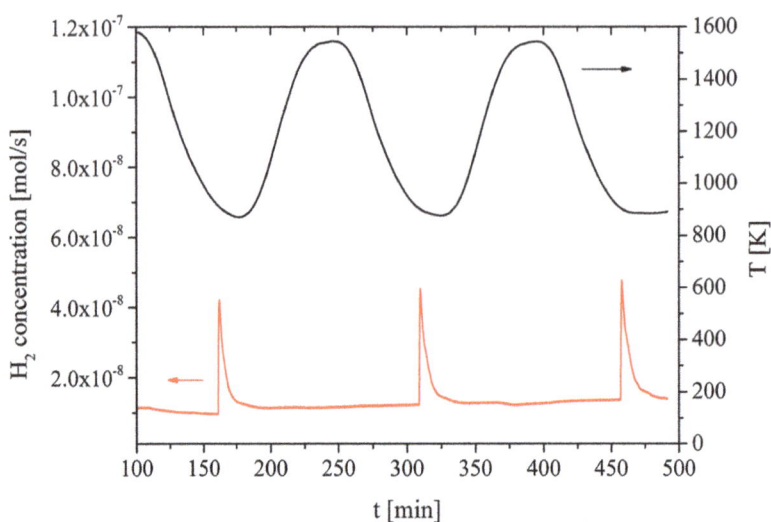

Fig. 6.19: H_2 evolution during water-splitting experiment with $Ce_{0.85}Zr_{0.15}O_2$ thermally pre-treated at 1,473 K and subsequently reduced at 1,573 K under Ar [45].

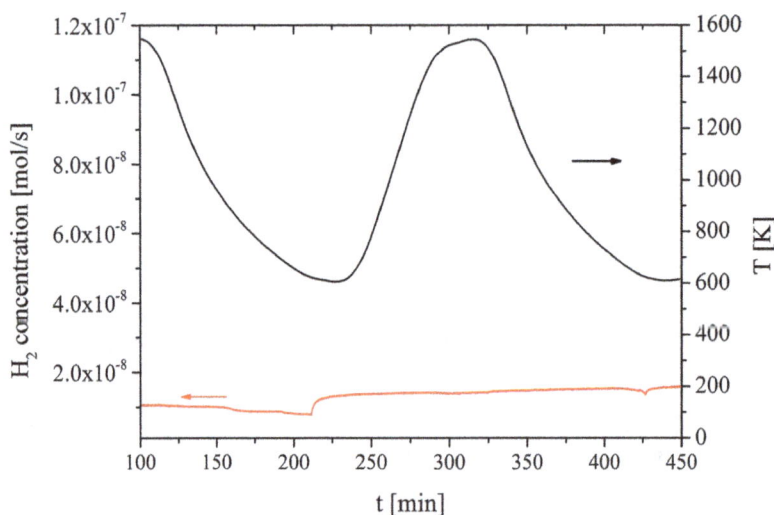

Fig. 6.20: H_2 evolution during water-splitting experiment with $Ce_{0.85}Zr_{0.15}O_2$ thermally pre-treated at 1,923 K subsequently reduced at 1,573 K under Ar [45].

High-temperature pre-treatment of zirconia-doped ceria obviously leads to a stabilization of the reduced state. This finding can be explained in terms of interactions between oxygen vacancies, Ce^{3+} and Zr^{4+} ions, forming pyrochlore-type coordination as discussed before. The pyrochlore coordination implies a decrease of redox entropy and water-splitting temperature (see Fig. 6.13). The following

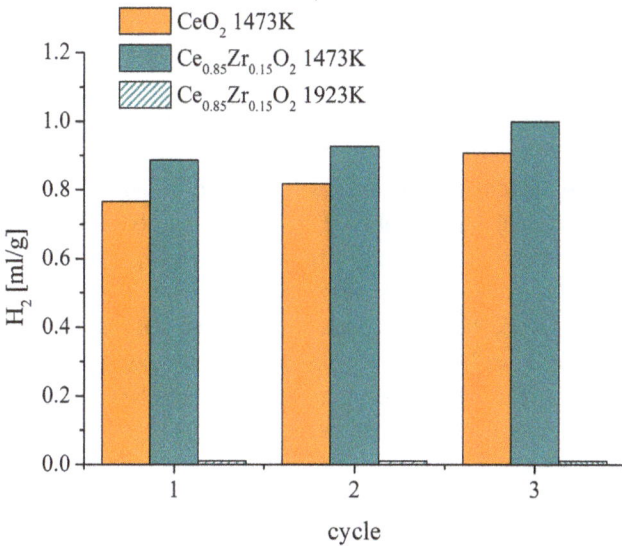

Fig. 6.21: Amount of H_2 released during water-splitting test of CeO_2 and $Ce_{0.85}Zr_{0.15}O_2$ with thermal pre-treatments of 1,473 K and 1,923 K, respectively [45].

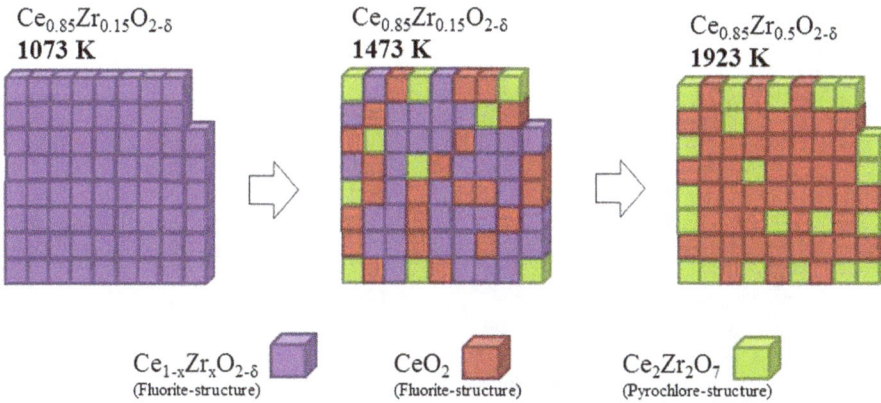

Fig. 6.22: Schematic diagram of the proposed pyrochlore-cluster formation as result of thermal pre-treatment.

schematic diagram depicts the increase of postulated pyrochlore coordination with higher temperature during thermal pre-treatment.

According to Fig. 6.22, considerable amounts of Zr^{4+} ions are coordinated in pyrochlore-type clusters after thermal pre-treatment at 1,923 K. To compensate the Zr-enrichment in the pyrochlore clusters cerium rich zones will coexist. Due to general lowering of redox enthalpy with increasing Zr^{4+} content (see Fig. 6.3), oxygen vacancies are assumed to be preferentially located in the Zr-rich pyrochlore-clusters. Both, the stabilization of oxygen vacancies in the pyrochlore zones and the low oxygen vacancy concentration in the coexisting CeO_2-rich regions affect the mobility of oxygen ions. Thus, especially at low reaction temperatures, the water splitting reaction is kinetically suppressed. At lower pre-treatment temperatures (e.g., 1,473 K), on the other hand, the formation of pyrochlore-clusters is believed to be less favored and hence mainly a solid solution with disordered fluorite-structure exists. Thereby, not only the thermodynamic conditions for the redox reaction are modified but also oxygen diffusion will be facilitated due to less localized oxygen vacancies. Thus, both effects will result in an improved re-oxidation. In order to identify the critical temperature between 1,473 K and 1,923 K, the thermal pre-treatment temperature was successively lowered. In the course of this study, it turned out that Zr doped pellet samples pre-treated at 1,773 K could be re-oxidized with water but only after grinding (Fig. 6.23 vs. Fig. 6.24).

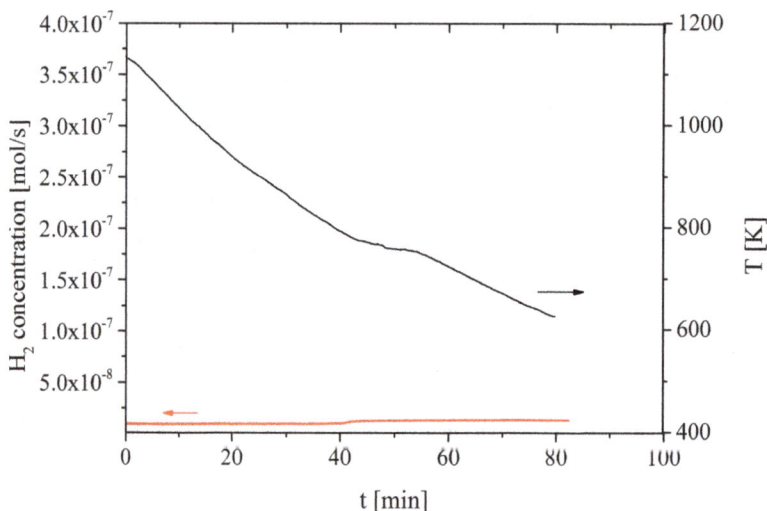

Fig. 6.23: Mass flow of hydrogen (red) during the oxidation with steam of a reduced $Ce_{0.85}Zr_{0.15}O_2$ pellet (thermal pre-treatment at 1,773 K for 2 h).

After pre-treatment at 1,773 K obviously an interesting intermediate structural stage appears. Here, $Ce_{0.85}Zr_{0.15}O_{2-\delta}$ pellet samples behave inert to water vapor (Fig. 6.23) while pulverized samples are re-oxidized within a short time. In general, this observation could be explained by larger active surfaces and shorter diffusion paths of the powder samples, but with a view to the extreme differences and due to the fact that after a thermal pre-treatment at 1,923 K grinding the pellet has no effect in view of water-splitting activity; this interpretation is rather unsatisfactory. However, an alternative explanation could be found in the observations that under low partial

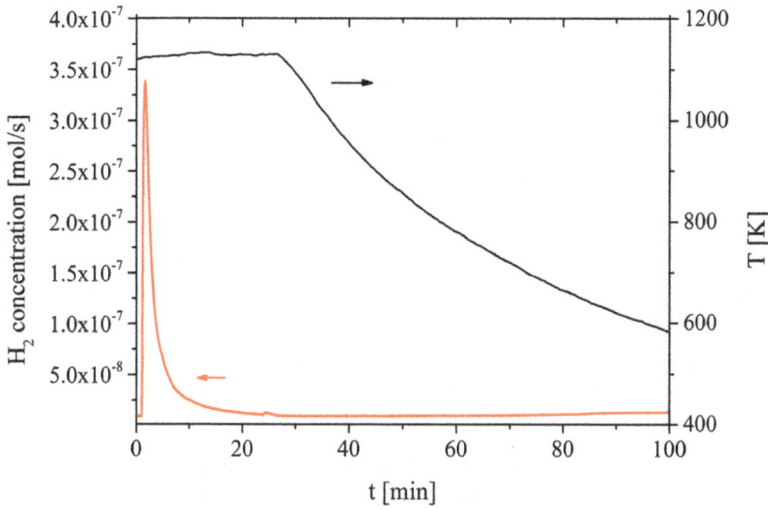

Fig. 6.24: Mass flow of hydrogen (red) during the oxidation with steam of a reduced $Ce_{0.85}Zr_{0.15}O_2$ powder sample (thermal pre-treatment at 1,773 K for 2 h).

pressure of oxygen a pyrochlore phase enrichment takes place at the surface [46]. It would therefore be quite conceivable that, even in the case of a pre-treatment at 1,773 K, the pyrochlore-like short-range-order tends to be present at the sample surface rather than inside the sample and thus suppress the water-splitting reaction. Only by grinding the pellet sample the still existing cerium rich areas of the sample interior are exposed to water vapor and hence a water-splitting reaction becomes possible (Fig. 6.25).

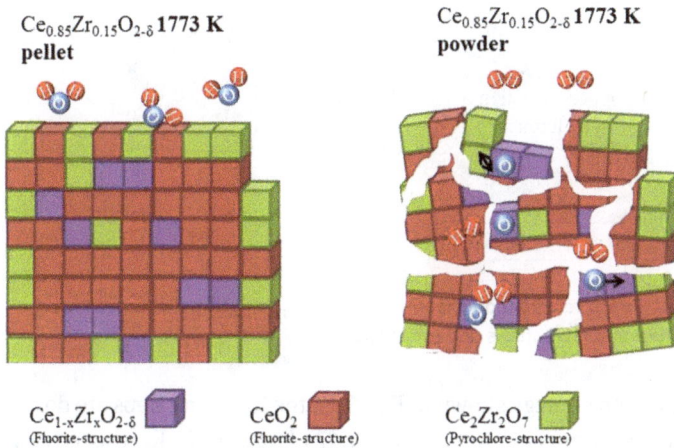

$Ce_{0.85}Zr_{0.15}O_{2-\delta}$ 1773 K
pellet

$Ce_{0.85}Zr_{0.15}O_{2-\delta}$ 1773 K
powder

$Ce_{1-x}Zr_xO_{2-\delta}$
(Fluorite-structure)

CeO_2
(Fluorite-structure)

$Ce_2Zr_2O_7$
(Pyrochlore-structure)

Fig. 6.25: Supplement of Fig. 6.22; proposed pyrochlore-cluster formation as a consequence of thermal pre-treatment at 1,773 K and the effect of powderization.

6.2.5 Oxidation in air

So far, thermodynamic data, water-splitting experiments, and dilatometric studies provide evidence for the formation of $Ce_2Zr_2O_7$ pyrochlore-type zones forming at high-temperature (1,923 K) sintering. $Ce_2Zr_2O_8$ zones (oxidized pyrochlore) are formed during cooling down in air but no pyrochlore-type ordering seems to develop during 1,473 K sintering. In other words, reduction (Ce^{3+}, oxygen vacancies) is a necessary but not sufficient condition for pyrochlore-type ordering in $Ce_{0.85}Zr_{0.15}O_{2-\delta}$ while high-temperature treatment is the strict pre-requirement. However, the role of temperature as necessary condition for structural ordering is not fully clear; moreover, data interpretation is difficult since thermal pre-treatment does not only affect possible structural ordering on nanometer scale but also has strong influence on porosity, grain size, internal surfaces, and oxygen diffusion paths. To shed more light on these issues, additional re-oxidation experiments in air using the strategy shown in Fig. 6.26 were carried out. Re-oxidation is considered as sensitive measure of pyrochlore-type ordering because of related oxygen vacancy stabilization which in turn affects the re-oxidation kinetics.

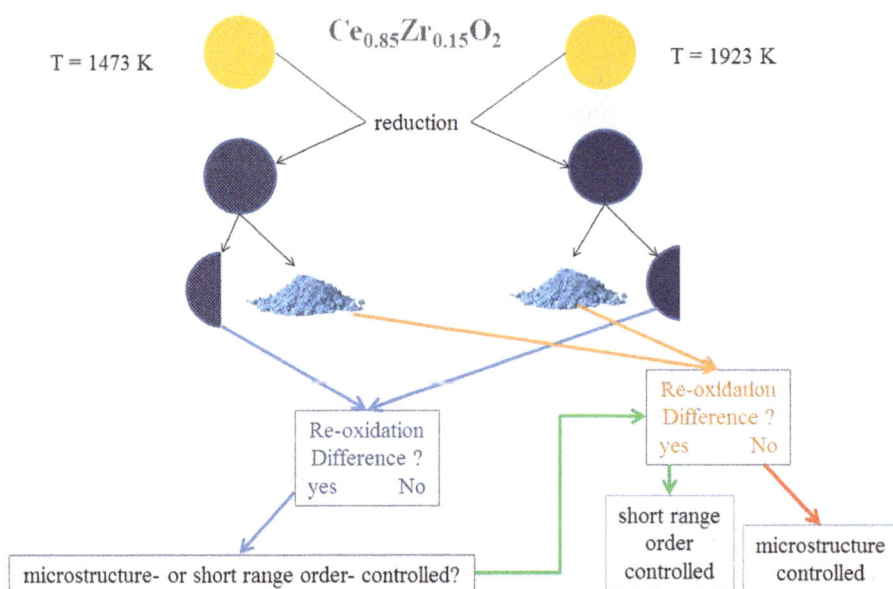

Fig. 6.26: Schematic illustration of re-oxidation experiments of $Ce_{0.85}Zr_{0.15}O_2$ pellet and powder samples pre-treated at different temperatures and pre-reduced.

According to the re-oxidation strategy shown in Fig. 6.26, the following steps are done:
1. Pellets of $Ce_{0.85}Zr_{0.15}O_2$ were sintered in air at 1,473 K and 1,923 K, respectively. Reduction values δ during these high-temperature conditions are 0.002 and 0.03, respectively.

2. Both pellet samples were re-oxidized by slow cooling down resulting in δ values < 0.001.
3. Both pellet samples were reduced in vacuum ($pO_2 \approx$ E-6 mbar, T = 1,473 K) to δ values of ≈ 0.04.
4. A part of pellet samples was powdered by milling.
5. Pellets and powders were re-oxidized under ambient air. The mass changes were detected by thermogravimetric analysis.

Since pyrochlore-type ordering is assumed to stabilize the reduced state, re-oxidation kinetics should reflect structural differences between 1,473 K and 1,923 K pre-treatment. Moreover, possible differences in microstructure (porosity, grain size) are leveled off when ground samples were employed and compared to each other. Data on re-oxidation kinetics of pre-reduced pellet samples under dynamic (STA) conditions actually show an effect of thermal pre-treatment. The re-oxidation of pellet samples thermally pre-treated at 1,473 K starts at temperatures as low as 400 K (Fig. 6.27, black curve) while pellet samples thermally pre-treated at 1,923 K start to re-oxidize at considerably higher temperatures (650 K, Fig. 6.27, red curve).

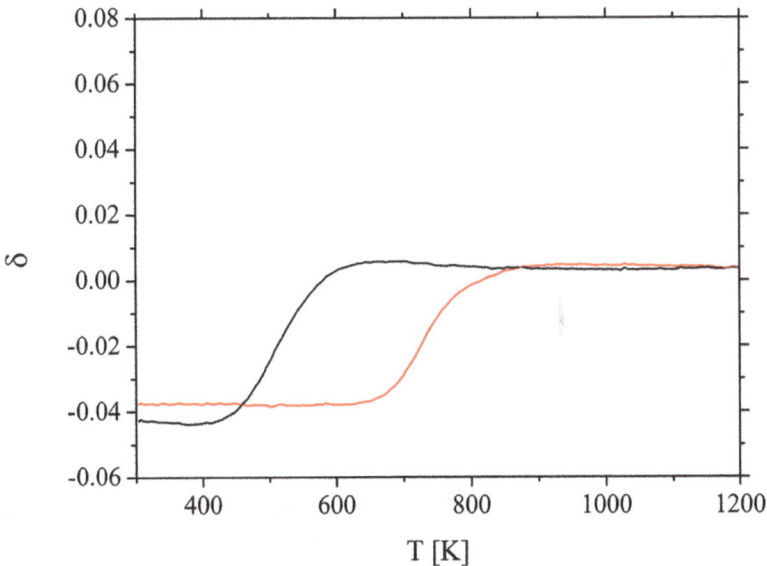

Fig. 6.27: Re-oxidation of $Ce_{0.85}Zr_{0.15}O_2$ pellet samples after thermal pre-treatment at 1,473 K (black), 1,923 K (red) and subsequent reduction at 1,473 K/ pO_2 = 2E-6 mbar.

Thus, it actually turns out that the thermal pre-treatment has strong influence on re-oxidation kinetics. This might be due to the postulated pyrochlore-type ordering, but it cannot be ruled out that the different microstructures (porosity, grain size, internal surfaces, and oxygen diffusion paths) also affect re-oxidation kinetics. To exclude possible microstructural reasons, the pellet samples were finely ground to powder before re-oxidation under dynamic conditions. Figure 6.28 shows the re-oxidation kinetics of the corresponding powder samples which behave very similar to the pellet samples (Fig. 6.27). We assume therefore that microstructural differences between samples pre-treated at 1,473 and 1,923 K, respectively, not play a crucial role. Instead, kinetic data suggest that sluggish re-oxidation of the 1,923 K treated samples is caused by pyrochlore-type zones which favor the reduced condition by stabilization of oxygen vacancies.

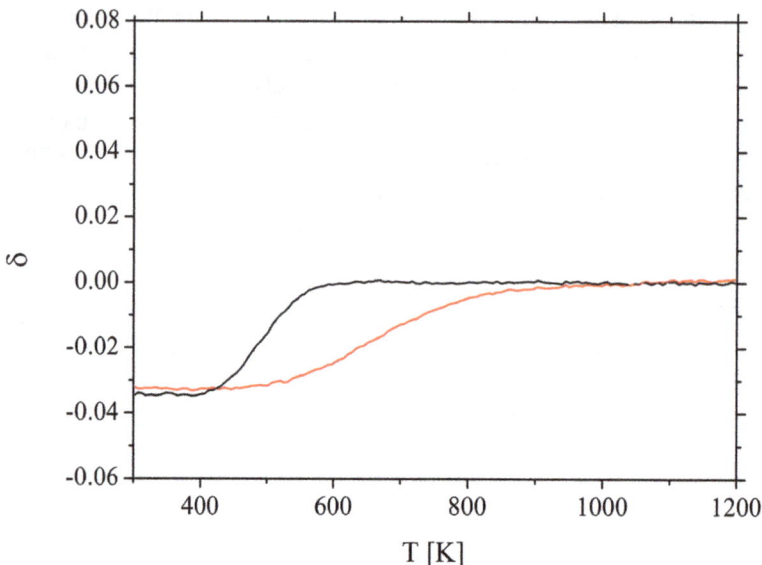

Fig. 6.28: Re-oxidation of $Ce_{0.85}Zr_{0.15}O_2$ powder samples after thermal pre-treatment at 1,473 K (black), 1,923 K (red) and subsequent reduction at 1,473 K/ pO_2 = 2E-6 mbar.

To shed more light on the degree of pyrochlore-type ordering, further re-oxidation experiments on higher reduced samples were carried out. As already mentioned by interpretation of dilatometry results, the pyrochlore development during thermal pre-treatment at 1,923 K in air is believed to be limited by the existing Ce^{3+} concentration, thus resulting in 11% pyrochlore:

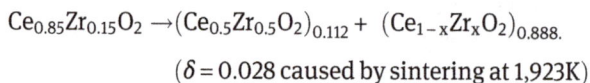

$$Ce_{0.85}Zr_{0.15}O_2 \rightarrow (Ce_{0.5}Zr_{0.5}O_2)_{0.112} + (Ce_{1-x}Zr_xO_2)_{0.888.}$$

$$(\delta = 0.028 \text{ caused by sintering at } 1{,}923K)$$

If the oxygen vacancy concentration evolving from the second reduction step exceeds half of the Ce^{3+} ion concentration in pyrochlore configuration, gradually oxygen vacancies have to form within the fluorite-type solid solution. The mobility of these oxygen vacancies is higher than in pyrochlore configuration due to lower solute-vacancy interaction (see also eqs. (6.8) and **(6.9)**) and therefore higher re-oxidation kinetics can be expected.

The comparison of the re-oxidation behavior of $Ce_{0.85}Zr_{0.15}O_2$ thermal pre-treated at 1,473 K and 1,923 K and subsequently reduced to $\delta \approx 0.1$ reveal no clear shift of re-oxidation start temperature as observed before (Fig. 6.29/6.30). This finding supports the idea that fast moving oxygen vacancies did form in fluorite domains of samples thermally pre-treated at 1,923 K, which were reduced afterward to a degree beyond the Ce^{3+} ion concentration given by the pre-formed pyrochlore ($\delta = 0.028$).

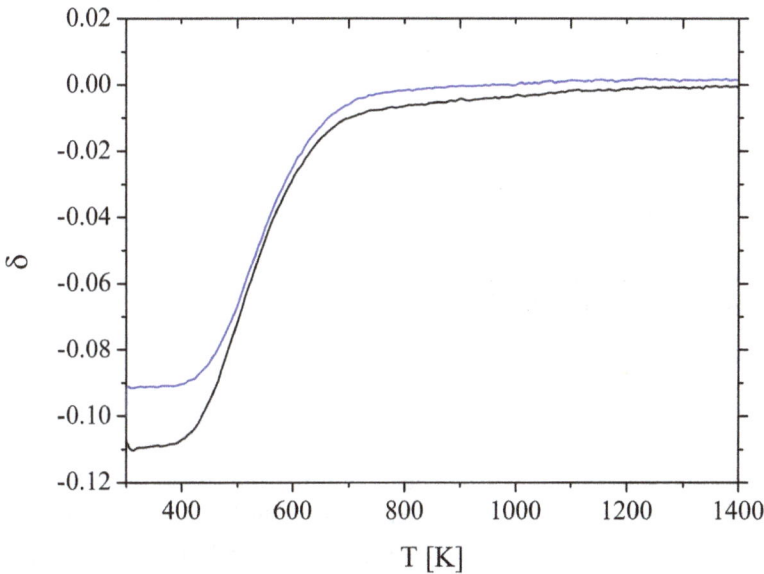

Fig. 6.29: Re-oxidation of $Ce_{0.85}Zr_{0.15}O_2$ pellet samples after thermal pre-treatment at 1,473 K and subsequent reduction at 1,473 K/ $pO_2 \approx$ 2E-6 mbar.

Closer inspection of the re-oxidation curves in Fig. 6.30 suggests a change in reaction kinetics in the range of $\delta = 0.05$–0.06. At first, re-oxidation proceeds rapidly and then slows down when a critical δ-value of = 0.05–0.06 is reached. Thus, the rapid oxidation regime can be interpreted as re-oxidation of defect fluorite-type solid solution while the gradually fading re-oxidation in the temperature range >650 K rather corresponds to sluggish pyrochlore oxidation.

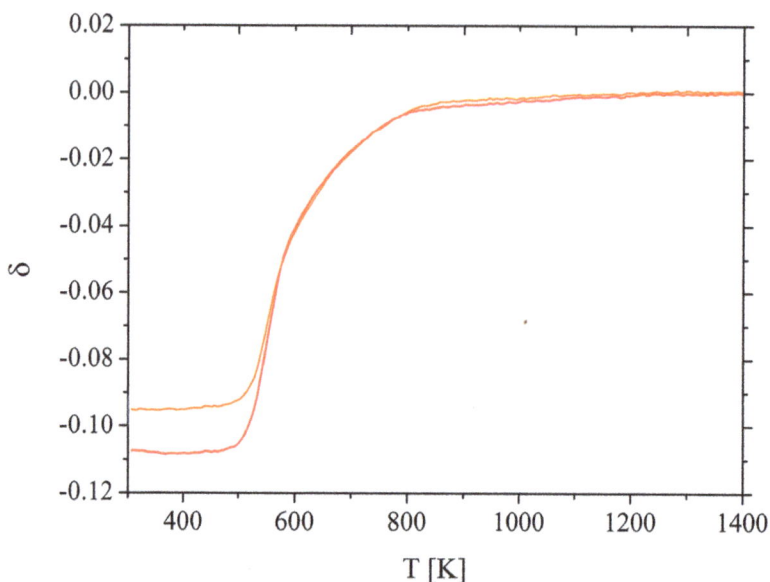

Fig. 6.30: Re-oxidation of $Ce_{0.85}Zr_{0.15}O_2$ pellet samples after thermal pre-treatment at 1,923 K and subsequent reduction at 1,473 K/ $pO_2 \approx$ 2E-6 mbar.

6.2.6 Temperature-controlled pyrochlore formation

Thermodynamic data, water-splitting experiments, dilatometric studies and re-oxidation experiments in air can be interpreted consistently with the idea of pyrochlore-type ordering of $Ce_{0.85}Zr_{0.15}O_2$ occurring at temperatures above 1,673 K–1,773 K. Thus, besides reducing conditions (formation of Ce^{3+} and oxygen vacancies), the high-temperature treatment is obviously the necessary condition for the postulated pyrochlore-type ordering. This finding, however, is a contradictory to widespread observation that ordering in solid solutions typically develops at low temperature whereas structural disorder is favored at high temperature because of the increasing significance of entropy.

A tentative explanation of high-temperature ordering in $Ce_{0.85}Zr_{0.15}O_{2-\delta}$ under mild reducing conditions can be derived from the constitution of $(Ce,Zr)O_2$ mixed crystals. Figure 6.7 reveals that a true solid solution with the target concentration of 15 mol% ZrO_2 is reached only at temperatures above 1,600 K. The smaller lattice constants occurring in the temperature range from 1,000 to 1,600 K are interpreted as deviation from Vegard's rule which is in line with the observation that zirconia-doped ceria tends to separate into CeO_2-rich and ZrO_2-rich domains at lower temperature (e.g., [31]. The temperature-dependent development of a solid solution with ZrO_2-rich domains toward a regular solid solution with random ZrO_2 distribution is assumed to be the key factor of subsequent pyrochlore-type ordering: Only if random solute distribution is reached, or, in other words, if Zr ions have a sufficient number of cerium-neighbors, pyrochlore-type ordering may develop, provided a sufficient number of Ce^{3+} ions, and oxygen vacancies are formed by simultaneous reduction. This explanation approach is visualized in Fig. 6.31.

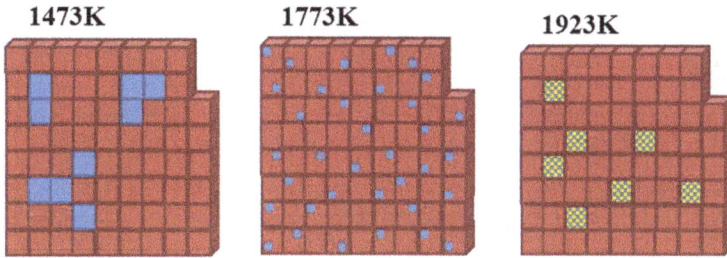

Fig. 6.31: Sketch of elemental distribution in $Ce_{0.85}Zr_{0.15}O_2$ as function of temperature. 1,473 K: solid solution with ZrO_2-rich domains; 1,773 K: solid solution with random ZrO_2 distribution increasing the number of neighboring Ce and Zr ions; 1,923 K: incipient reduction from Ce^{4+} to Ce^{3+} resulting in pyrochlore-type ordering.

Once formed, cation distribution of $Ce_2Zr_2O_7$ zones is believed to be retained after re-oxidation, thus leading to clusters of $Ce_2Zr_2O_8$ after cooling down. It should be emphasized that pyrochlore-type ordering stabilizes the reduced condition in $Ce_{0.85}Zr_{0.15}O_{2-\delta}$ by interactions between oxygen vacancies, Ce^{3+} and Zr^{4+} ions, and hence being detrimental in view of water splitting. Therefore, for re-oxidation of $Ce_{0.85}Zr_{0.15}O_{2-\delta}$ by water vapor pyrochlore-type ordering should be avoided which means that processing temperatures should not exceed $\approx$1,673 K.

6.3 Conclusion

Thermodynamic data, dilatometric studies, and re-oxidation experiments in water vapor and air are interpreted consistently with the idea of pyrochlore-type ordering of $Ce_{0.85}Zr_{0.15}O_2$ occurring at temperatures above 1,673 K–1,773 K. The formation of pyrochlore-type clusters enriched in Zr with respect to the starting composition goes along with the evolution of fluorite-type ceria-rich domains $(Ce_{1-x}Zr_xO_2)$ with x < 0.15.

In contrast to the fluorite structure, the pyrochlore structure offers a lower number of potential oxygen vacancy sites and shows cation ordering. Thus, with increasing amounts of pyrochlore the overall redox entropy decreases. In case of thermal pre-treatment at 1,923 K, the redox entropy is reduced to 132 J/molK ($CeO_{2-0.05}$ = 208 J/molK) while a pre-treatment of only 1,473 K results in a redox entropy of 158 J/mol K. Furthermore the interaction of Zr^{4+} cations with oxygen vacancies leads to a contraction of the crystal lattice which counteracts the $Ce^{4+} \rightarrow Ce^{3+}$ cation expansion or repulsion of defects by their atomic neighbors upon reduction. The Zr^{4+}/oxygen vacancy interaction results in a smaller chemical expansion in comparison to pure ceria. The stabilization of oxygen vacancies in pyrochlore domains counteracts the re-oxidation kinetics. The re-oxidation in air of samples pre-treated at 1,923 K needs significantly higher temperatures than the re-oxidation of 1,473 K pre-treated samples. This observation is independent of sample microstructure and morphology as evidenced by samples milled to powder before re-oxidation. Beside the re-oxidation behavior in air also the water-splitting behavior is influenced by thermal pre-treatment. It turns out that $Ce_{0.85}Zr_{0.15}O_2$ samples pre-treated at 1,923 K are not able to split water regardless of whether they were tested as powder or pellets. In contrast, samples pre-treated at 1,473 K/1,573 K could be re-oxidized by water vapor with H_2 yields higher than obtained by pure ceria. All data show that the efficiency of redox materials of the system CeO_2–ZrO_2 is strongly influenced by thermal pre-treatment. These effects may counteract the improvement of CeO_2 by Zr doping. As consequence of the achieved results, the reduction temperature needs to be low as possible but still high enough to reach an adequate oxygen vacancy concentration for good water splitting performance.

Although cerium oxide has some drawbacks as a redox material for thermochemical cycle processes, it is promising yet compared to other metal oxides such as ZnO, SnO_2, Fe_3O_4, and perovskites, investigated so far (e.g., 2). First experiments on water-splitting behavior of perovskites showed a higher oxygen release during the reduction as compared with doped ceria, but the entropy change during the redox reaction is relatively low. Therefore, the subsequent oxidation step is thermodynamically less favored, resulting in moderate hydrogen release (e.g., [47, 48]). Nonetheless, perovskite phases still have potential for further optimization and hence, according to the authors' estimation, they may compete in future with doped ceria-based redox materials.

References

[1] Siegel, N.P. et al. Factors affecting the efficiency of solar driven metal oxide thermochemical cycles. Ind. Eng. Chem. Res. 2013, 52, 3276–3286.
[2] Scheffe, J.R., Steinfeld, A. Oxygen exchange materials for solar thermochemical splitting of H_2O and CO_2: A review. Mater. Today 2014, 17, 341–348.
[3] Nakamura, T. Hydrogen production from water utilizing solar heat at high temperatures. Solar Energy 1977, 19, 467–475.
[4] Ihara, S. On the study of hydrogen production from water using solar thermal energy. Int. J. Hydrogen Energy 1980, 5, 527–534.
[5] Meng, Q.L. et al. Reactivity of CeO_2-based ceramics for solar hydrogen production via a two-step water-splitting cycle with concentrated solar energy. Int. J. Hydrogen Energy 2011, 36, 13435–13441.
[6] Scheffe, J.R. et al. Synthesis, characterization, and thermochemical redox performance of Hf^{4+}, Zr^{4+}, and Sc^{3+} doped ceria for splitting CO_2. J. Phys. Chem. C 2013, 117, 24104–24114.

[7]   Pappacena, A. et al. New insights into the dynamics that control the activity of ceria–zirconia solid solutions in thermochemical water splitting cycles. J. Phys. Chem. C 2017, 121, 17746–17755.

[8]   Abanades, S. et al. Cerium based oxides for H_2 production by thermochemical Two Step water-splitting. J. Mater. Sci. 2010, 45, 4163–4173.

[9]   Takacs, M., Scheffe, J.R., Steinfeld, A. Oxygen nonstoichiometry and thermodynamic characterization of Zr doped ceria in the 1573–1773 K temperature range. Phys. Chem. Chem. Phys. 2015, 17, 7813–7822.

[10] Kuhn, M. et al. Structural characterization and oxygen nonstoichiometry of ceria-zirconia ($Ce_{1-x}Zr_xO_{2-\delta}$) solid solutions. Acta Mater. 2013, 61, 4277–4288.

[11] Call, F. et al. Thermogravimetric Analysis of Zirconia-Doped Ceria for Thermochemical Production of Solar Fuel. Am. J. Analyt Chem. 2013, 4, 37–45.

[12] LeGal, A. et al. Reactivity of doped ceria-based mixed oxides for solar thermochemical hydrogen generation via two-step water-splitting cycles. Energy Fuels 2013, 27, 6068–6078.

[13] Shannon, R.D. Revised effective ionic radii and systematic studies of interatomic distances in halides and chalcogenides. Acta Crystallographia 1976, A 32, 751–767.

[14] Kaneko, H., Taku, S., Tamaura, Y. Reduction reactivity of CeO_2–ZrO_2 oxide under high O_2 partial pressure in two-step water splitting process. Solar Energy 2011, 85, 2321–2330.

[15] Bishop, S.R. et al. Reducing the chemical expansion coefficient in ceria by addition of zirconia. Energy Environ. Sci. 2013, 6, 1142–1146.

[16] Yang, Z., Woo, T.K., Hermansson, K. Effects of Zr doping on stoichiometric and reduced ceria: A first-principles study. J. Chem. Phys. 2006, 124, 224704.

[17] Rodriguez, J.A. et al. Properties of CeO_2 and $Ce_{1-x}ZrxO_2$ Nanoparticles: X-ray Absorption Near-Edge Spectroscopy, Density Functional, and Time-Resolved X-ray Diffraction Studie. J. Phys. Chem. B 2003, 107, 3535–3543.

[18] Hao, Y., Yang, C.K., Haile, S.M. Ceria–Zirconia Solid Solutions ($Ce_{1-x}Zr_xO_{2-\delta}$, $x \leq 0.2$) for Solar Thermochemical Water Splitting: A Thermodynamic Study. Chem. Mater. 2014, 26, 6073–6082.

[19] Panlener, R.J., Blumenthal, R.N., Garnier, J.E. A thermodynamic study of nonstoichiometric cerium dioxide. J. Phys. Chem. Solids 1975, 36, 1213–1222.

[20] Zhou, G. et al. Oxidation entropies and enthalpies of ceria-zirconia solid solutions. Catalysis Today 2007, 123, 86–93.

[21] Shah, P.R. et al. Evidence for Entropy Effects in the Reduction of Ceria–Zirconia Solutions. Chem. Mater. 2006, 18, 5363–5369.

[22] LeGal, A., Abanades, S., Flamant, G. CO_2 and H_2O splitting for thermochemical production of solar fuels using nonstoichiometric ceria and Ceria/Zirconia solid solutions. Energy Fuels 2011, 25, 4836–4845.

[23] LeGal, A., Abanades, S. Dopant incorporation in ceria for enhanced water-splitting activity during solar thermochemical hydrogen generation. J. Phys. Chem. C 2012, 116, 13516–13523.

[24] Baker, R.T. et al. Reversible changes in the redox behaviour of a $Ce_{0.68}Zr_{0.32}O_2$ mixed oxide: Effect of alternating the re-oxidation temperature after reduction at 1223 K. ChemComm. 1999, 149–150.

[25] Fornasiero, P. et al. Effects of thermal pretreatment on the redox behaviour of $Ce_{0.5}Zr_{0.5}O_2$: Isotopic and spectroscopic studies. Phys. Chem. Chem. Phys. 2002, 4, 149–159.

[26] Montini, T. et al. Promotion of reduction in $Ce_{0.5}Zr_{0.5}O_2$: The pyrochlore structure as effect rather than cause?. Phys. Chem. Chem. Phys. 2002, 4, 1–3.

[27] Yeste, M.P. et al. Redox behavior of thermally aged ceria–zirconia mixed oxides. Role of their surface and bulk structural properties. Chem. Mater. 2006, 18, 2750–2757.

[28] Pechini, M.P. Method of preparing lead and alkaline earth titanates and niobates. US Patent 3.330.697. 1967.

[29] Bevan, D.J.M. Ordered intermediate phases in the system CeO_2-Ce_2O_3. J. Inorg. Nucl. Chem. 1955, 1, 49–59.

[30] Varez, A., Garcia-Gonzales, E., Sanz, J. Cation miscibility in CeO_2–ZrO_2 oxides with fluorite structure. A combined TEM, SAED and XRD Rietveld analysis. J. Mater. Chem. 2006, 16, 4249–4256.

[31] Lee, T.A. et al. Enthalpy of formation of the cubic fluorite phase in the ceria-zirconia system. J. Mater. Res. Technol. 2008, 23, 1105–1112.

[32] Ahrens, L.H. The use of ionization potentials Part 1. Ionic radii of the elements. Geochim. Cosmochim. Acta 1952, 2, 155–169.

[33] Hoffmann, L. Thermodynamische Untersuchung von Cer-basierten Redoxmaterialien zur solaren Wasserstoffherstellung. Master Thesis, Ruhr Universität Bochum. 2015.

[34] Bulfin, B. et al. Analytical model of CeO_2 oxidation and reduction. J. Phys. Chem. C 2013, 117, 24129–24137.

[35] Call, F. Investigation of ceria-based redox materials for thermochemical solar fuel production. Ph. D. Thesis, RWTH Aachen. 2014.

[36] Knoblauch, N. Synthese, Charakterisierung und Untersuchung zum Redoxverhalten von ceroxidbasierten Materialien zur Erzeugung solarer Brennstoffe. Ph. D. Thesis, TU Clausthal, 2014.

[37] Toft-Soerensen, O. Thermodynamic studies of the phase relationships of nonstoichiometric cerium oxides at higher temperatures. J. Solid State Chem. 1976, 18, 217–233.

[38] Subramanian, M.A., Aravamudan, G., Subba Rao, G.V. Oxide pyrochlores – A review. Prog. Solid State Chem. 1983, 15, 55–143.

[39] Achary, S.N. et al. Intercalation/deintercalation of oxygen: A sequential evolution of phases in Ce_2O_3/CeO_2–ZrO_2 pyrochlores. Chem. Mater. 2009, 21, 5848–5859.

[40] Marrocchelli, D. et al. Understanding chemical expansion in non-stoichiometric oxides: Ceria and zirconia case studies. Adv. Funct. Mater. 2012, 22, 1958–1965.

[41] Muhich, C.L. re-evaluating CeO_2 expansion upon reduction: Noncounterpoised forces, not ionic radius effects, are the cause. J. Phys. Chem. C 2017, 121, 8052–8059.

[42] Hull, S. et al. Oxygen vacancy ordering within anion-deficient Ceria. J. Solid State Chem. 2009, 182, 2815–2821.

[43] Vegard, L. Die Konstitution der Mischkristalle und die Raumfüllung der Atome. Zeitschrift für Physik 1921, 5, 17–26.

[44] Neises, M., et al., Investigations of the Regeneration Step Of A Thermochemical Cycle Using Mixed Iron Oxides Coated on SiC substrates. Proc. ASME 2011, 5th Int. Conference on Energy Sustainability, Paper 2011-54193. 2011.

[45] Esser, C. Thesis, TH Köln, 2016.

[46] Knoblauch, N. et al. Ceria: Recent results on dopant-induced surface phenomena. Inorganics 2017, 5, 76–102.

[47] McDaniel, A.H. et al. Sr- and Mn-doped $LaAlO_{3-\delta}$ for solar thermochemical H2 and CO production. Energy Environ. Sci. 2013, 6, 2424–2428.

[48] Scheffe, J.R. et al. Lanthanum–strontium–manganese perovskites as redox materials for solar thermochemical splitting of H_2O and CO_2. Energy & Fuels 2013, 27(8), 4250–4257.

Christophe Didier, Wei Kong Pang, and Vanessa K. Peterson

7 The influence of electrode material crystal structure on battery performance

Abstract: Energy storage demands are increasing as a result of the growing use of electric vehicles and intermittent energy sources. Lithium-ion batteries, commercialized over two decades ago, have enabled the widespread use of portable electronics, and these are being implemented both in electric vehicles and to store intermittently available energy. This type of battery operates with the so-called "rocking-chair" mechanism, where lithium ions are reversibly exchanged between two electrodes. The mechanism relies on the reversible insertion of lithium into sites within the electrode material crystal structure, with the repeated processes of lithium insertion and extraction generating atomic-level change that the bulk electrode material must accommodate. These are both the short- and long-range structural changes within electrodes that dictate, to a large degree, the performance of the whole battery. Hence, the characterization of electrode crystal structure, particularly during battery cycling, is key to understanding and improving energy storage in batteries. In this chapter, we explore the relationship between the crystal structure of electrode materials and their performance in lithium-ion batteries, with a focus on structure types that have found commercial applications. The discussion identifies what crystallographic features are useful for electrode materials and how crystal structure influences electrochemical behavior.

Keywords: battery, powder diffraction, in situ, in operando, phase transition, diffusion pathway, electrode

7.1 Concepts

7.1.1 The rocking-chair mechanism

Since the first lithium-ion battery (LIB) was introduced and commercialized by Sony in 1991 [1], LIBs have become the most promising energy storage for portable electronic devices, as well as new electric transportation applications, because they have the highest gravimetric and volumetric energy densities compared to other available battery technologies [2]. The importance of LIBs in our daily life is recognized by the Nobel Prize in Chemistry 2019, which was jointly awarded to John B. Goodenough, M. Stanley Whittingham, and Akira Yoshino for their significant contribution to their development.

Christophe Didier, Vanessa K. Peterson, ANSTO – Sydney, New Illawarra Rd, Lucas Heights NSW 2234, Australia and Institute for Superconducting and Electronic Materials, University of Wollongong
Wei Kong Pang, Institute for Superconducting and Electronic Materials, University of Wollongong

https://doi.org/10.1515/9783110674910-007

In a LIB, there are two electrodes, separated by an insulating but porous separator surrounded by ion-conducting electrolyte (Fig. 7.1). Electricity is stored in chemical form until the battery is discharged, where the stored electrons and Li^+ ions mobilize. The separator prevents internal short-circuiting, and the electrolyte within the pores of the separator ensures the migration of charge carriers such as Li^+ between these electrodes while electrons are forced around the external circuit to create a current. During battery use the electrons flow from the most reducing (lowest potential) to the most oxidizing (highest potential) electrode, the difference of potentials being the driving force of the reaction. The discharge proceeds until electrodes reach the same potential, or one of the electrodes is unable to accept/release additional Li^+/ e^-, or the external circuit is disconnected. When charging, an electrical current (galvanostatic charge) or voltage (potentiostatic charge) is applied to the battery via the external circuit, increasing the potential difference between the two electrodes.

If the Li^+ insertion and extraction processes are repeatable (reversible), the battery is rechargeable. One repetition means one cycle, where increased repetitions without a significant decrease in the number of mobile Li^+ yields better life cycle. The concept of this rocking-chair mechanism has been further developed to encompass other battery chemistries such as sodium-ion batteries (SIBs), potassium-ion batteries (PIBs), and other metal-ion rechargeable batteries.

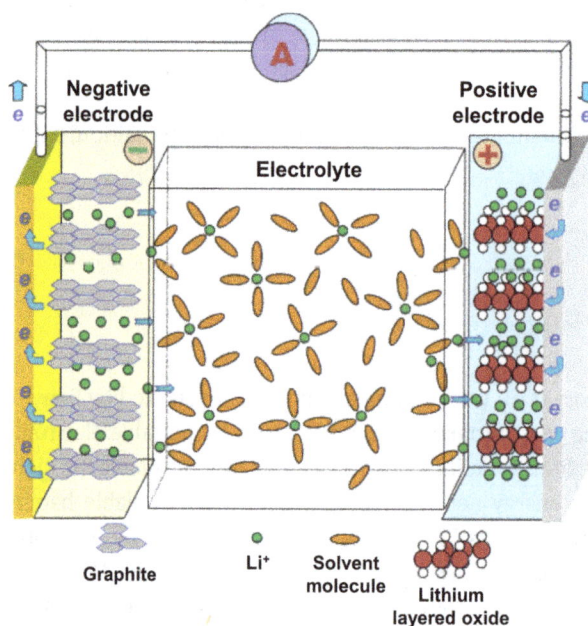

Fig. 7.1: Working principle of a LIB. Arrows symbolize the electron movement in the outer circuit and the Li^+ movement within the battery during discharge. Adapted with permission from [3]. Copyright 2004 *American Chemical Society*.

7.1.2 Principles of intercalation in electrode materials

Electrode materials typically used in commercial LIBs operate with an intercalation mechanism where the crystal structure can reversibly accommodate lithium ions in empty crystallographic sites with slight structural deformations. Other mechanisms such as alloying/conversion (e.g., lithium, silicon), where the material structure changes dramatically or adsorption (e.g., carbon blacks) where charge carriers stay at the surface of the electrode, have also been investigated as electrodes. Each mechanism has advantages and challenges; however, intercalation materials are the most successful commercially applied and so this discussion will focus on these materials.

The specific capacity of an intercalation electrode material is given by the number of Li^+ that can be inserted or removed per unit of volume or mass. It is directly related to the crystal structure of the material as it depends on the number of crystallographic sites available to Li^+. Hence, the crystal structure of an electrode material largely determines how much energy can ultimately be stored in an LIB.

The crystal structure of an electrode material is also responsible for other performance-related properties: the size, shape, and degree of interconnectivity between crystallographic sites available to Li^+ impact the ionic conductivity, that is, how easily ions can move within the structure. A low conductivity will limit the current rate of the battery, restricting how fast it can be cycled. An electrode material with high electronic conductivity is also desirable; however, this may be circumvented by the addition of conductive additives. The electrode should contain an element, often a transition metal, with variable oxidation state to enable the material to accept or release electrons (be reduced or oxidized) via the external circuit. Finally, the insertion and removal of Li^+ should proceed with moderate and reversible structural changes, allowing repeated charge/discharge cycles.

A high battery voltage is desirable, leading to higher energy stored per electron. The battery voltage is given by the difference in electrochemical potentials of the two electrodes, and varies with the battery state-of-charge (the amount of Li stored in each electrode). Thus, pairings of high potential (more oxidizing) and low potential (more reducing) electrode materials are preferred, with those measured against a common electrochemical couple, typically Li^+/Li. Finally, good chemical stability is desirable for long-term battery use. The electrochemical potential shown against the specific capacity of selected commercial electrode materials is given in Fig. 7.2, noting that other parameters such as stability leading to high rate capability are important battery performance characteristics.

The first electrode materials used in an LIB were $LiCoO_2$ (LCO) and graphite [1, 12]. These two materials satisfy the previously identified requirements of high ionic and electronic conductivity and reversible transformations during cycling, arising from their layered crystallographic structures. As the LCO electrode limited the battery capacity (140 mAh/g for LCO vs. 372 mAh/g for graphite), electrode research has focused on the development of materials to be used against the graphite electrode. Of the materials

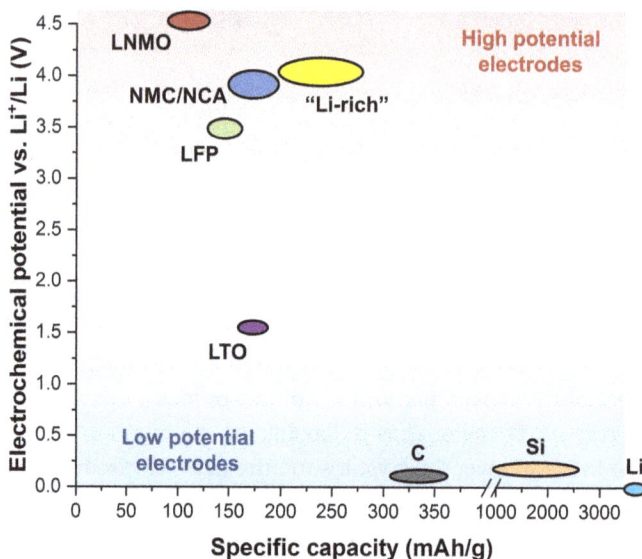

Fig. 7.2: Typical nominal voltage and specific capacity of the commercially available LIB electrodes $LiNi_{0.5}Mn_{1.5}O_4$ (LNMO) [4], $LiNi_xMn_{(1-x)/2}Co_{(1-x)/2}O_2$ (NMC) [5], $LiNi_{0.8}Co_{0.05}Al_{0.15}O_2$ (NCA) [6], x $Li_2MnO_3 \cdot$ (1-x) $LiMO_2$ (Li-rich) [7], $LiFePO_4$ (LFP) [8], $Li_4Ti_5O_{12}$ (LTO) [9], graphite (C) [10], silicon (Si) [11], and lithium (Li).

studied, only three groups of materials are used commercially (specific capacity in parenthesis): layered oxides such as LCO (140 mAh/g), $LiNi_xMn_{(1-x)/2}Co_{(1-x)/2}O_2$ (165–200 mAh/g), and $LiNi_{0.8}Co_{0.05}Al_{0.15}O_2$ (190 mAh/g) °C [12–14], spinel oxides such as $LiMn_2O_4$ (120 mAh/g) [15], and olivine-like poly-anion oxides such as $LiFePO_4$ (170 mAh/g) [16, 17]. These are all intercalation-type electrodes used in commercial LIBs, mainly due to their exceptional cycling stability and high energy/power density [18–20]. Understanding the relationship between the structure and electrochemical properties of a material that make it useful as an electrode is key to addressing many battery performance issues. Electrode materials are often structurally complex, with the variation of charge balancing ions over the operational range as well as the chemical engineering of the materials through substantial ion substitution resulting in compositional variation and complex multiphase systems with short-range order that sometimes differs from the structural average.

A number of tools are available to the crystallographer in order to determine the crystal structure of electrode materials, with structural variations during cycling the most valuable information for understanding the (de)lithiation mechanism. Because electrode materials are relatively difficult to extract and handle out of the battery, and the structure at equilibrium may not be representative of the structure under load, it is preferable to perform measurements in situ or in operando on the materials inside the battery during cycling. The most frequently used technique for the investigation of long-range structural ordering is in situ diffraction, with synchrotron X-rays and neutrons predominantly employed thanks to their high penetration, although more readily available lab X-rays have also been successfully used. Different cell designs have been engineered to improve the observed signal while maintaining normal electrochemical behavior [21–24]. Other notable techniques include pair distribution function (PDF) from total scattering and nuclear magnetic resonance (NMR), with those providing short-range structural information averaged over the probed volume, which can be performed in situ under appropriate conditions [25–27]. Local structural information can be determined using scanning transmission electron microscopy (STEM) and selected area electron diffraction (SAED), noting that they are often performed ex situ, and the local structure may not be representative of the overall sample.

We note that the practical performance is also strongly influenced by other parameters such as the size of any crystallite domains that exists within macroscopic particles, the morphology and size of those particles, as well as any coatings or secondary phases that may be present. The characterization of these parameters is important and can sometimes be done using crystallographic techniques. When relevant, these important features will also be discussed in the following paragraphs, although the main focus will be on the intrinsic properties conferred by the crystal structure of the electrode. In this chapter, we therefore focus on the influence of average crystal structure on battery performance, discussing this for each structure type.

7.2 Crystal structures of commercial electrode materials and their limitations

7.2.1 Layered materials: graphite

Graphite was one of the first materials studied as an LIB electrode [28] and is the most used electrode today in light of its excellent electrochemical properties [29]. The crystal structure of graphite is well-known in its most stable form [30] called 2H (for 2 layers per hexagonal unit cell). The conventional unit cell is hexagonal ($P6_3/mmc$ spacegroup). The 2H structure is composed of flat layers of covalently bonded C atoms held together by weak Van der Waals interactions in a staggered stacking configuration (commonly noted as ABAB . . . stacking), with the stacking direction along the c-axis of the hexagonal unit cell [31]. The space between layers can accommodate intercalants resulting in strongly anisotropic conductivity parallel to the layers [32]. Many elements and molecules intercalate in graphite, including lithium [33].

The galvanostatic voltage curve of graphite against lithium metal shows several voltage variations (Fig. 7.3). The phase transitions were initially characterized using in situ X-ray diffraction [34, 35]. The largest achievable capacity in the material is 372 mAh/g, corresponding to the composition LiC_6, where all spaces between carbon

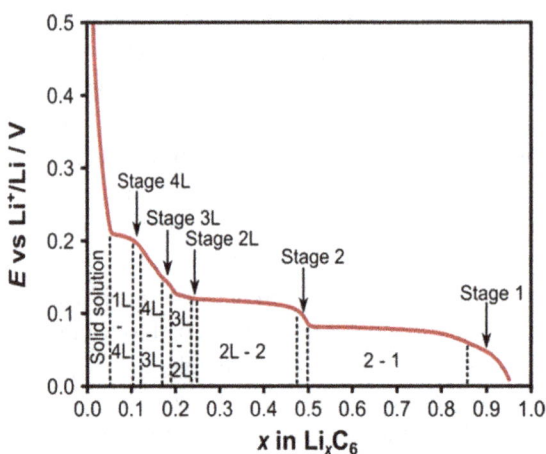

Fig. 7.3: Typical voltage variation during galvanostatic cycling of graphite against lithium metal. Proposed stages during cycling are annotated. Adapted with permission from [39]. Copyright 2018 *American Chemical Society.*

layers are occupied with lithium. The plateau at 0.05 V vs. Li⁺/Li corresponds to a two-phase transformation between LiC_6 and LiC_{12} where in LiC_{12} ($Li_{1/2}C_6$), only half the layers are alternatively occupied by lithium. A consideration of the system in which intercalation layers are either occupied or empty was proposed, with the number of occupied layers changing with the intercalant amount. The consequential long-range stacking direction ordering between full and empty layers, known as staging [31], is common in graphite intercalation compounds [33]. A stage n compound contains $1/n$ occupied layers, with LiC_6 a stage 1 compound and LiC_{12} a stage 2 compound. Other potential variations are observed at higher voltage in the galvanostatic curve and a second stage 2, as well as stage 3 and stage 4 phases have been observed [34–38].

The crystal structures of stage 1 LiC_6 and stage 2 LiC_{12} are accepted. In LiC_6, all interlayers are occupied by Li with each facing the center of hexagons on each side with carbon layers stacked in an eclipsed configuration (A/A/A/ . . . stacking), with the intercalation provoking a shift between carbon layers away from that in graphite where carbon layers are staggered (ABAB . . .). Li⁺ are ordered within the (a,b) plane, forming a √3a × √3a supercell (Fig. 7.4) [40]. The structure of the stage 2 LiC_{12} material is similar, with Li in-plane ordering conserved and carbon layers stacked in an eclipsed configuration (/AA/AA/AA/ stacking) [40, 41]. The structure of other stage phases is still debated. The majority agree that no in-plane Li ordering is observed for these compounds, often denoted with an L for liquid. The most accurate structural characterization of stage 2L and 3L lithium-intercalated graphite used in situ X-ray diffraction of highly oriented pyrolytic graphite platelets with the incident beam either parallel or perpendicular to the platelets, emphasizing different reflections in each configuration [36, 37]. It was found that in the stage 2L compound layers are

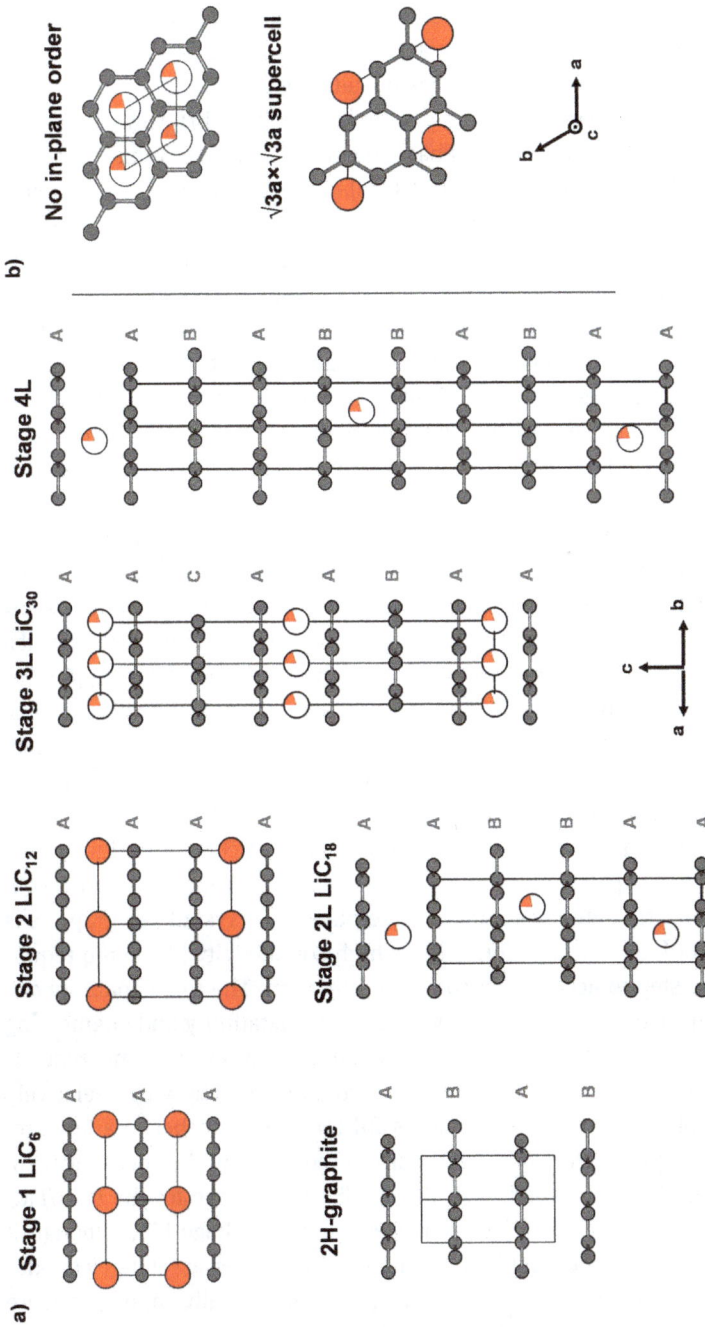

Fig. 7.4: a) Representation of the crystal structures of lithium-intercalated graphite for different stage organizations. b) Lithium in-plane ordering in dilute stages (2L, 3L, 4L) and fully ordered stages (1 and 2). Li and C are red and gray circles, respectively. Reprinted with permission from [38]. Copyright 2020 American Chemical Society.

eclipsed around Li as in stage 1 and 2 compounds, but staggered around empty layers, as in parent graphite, resulting in /AB/BA/stacking (Fig. 7.4) [36]. The structure of the stage 3L material suggests an /ABA/ACA/ stacking, although some /ABA/ABA/ may be simultaneously present, perhaps as stacking faults [37]. An /ABAB/BABA/ stacking is proposed for the stage 4L material [42.]. Phase diagram predictions from theoretical calculations are challenging for this system [43–45], and calculated structures sometimes contradict experimental results. In operando neutron diffraction [38] supported the structures proposed by Billaud et al. (Fig. 7.4) and phase evolution determined by Dahn [34]; however, the structure adopted for stages > 2 is still contested [46, 47].

An interesting question arises where, due to negligible Li conduction perpendicular to carbon layers [32], the transition between stage 2 and 3 or stage 3 and 4 requires complex Li reordering expected to be kinetically difficult. The Daumas–Herold model was proposed where the stage ordering could be maintained if Li occupied islands between carbon layers, and the stage transitions were accomplished by the migration of these islands across layers (Fig. 7.5) [48].

This model implies slight distortions of the carbon layers which have been observed using transmission electron microscopy in graphite intercalated with $FeCl_3$ and $SbCl_5$ [50, 51], but not in lithium-intercalated graphite. First-principle calculations show that these distortions are thermodynamically favorable in stage > 2 lithium-intercalated graphite, and Monte-Carlo simulations have predicted their formation (Fig. 7.5) [49, 52–55]. The determined coherence length of the stacking axis is consistent with such distortions during phase transitions [38]. These distortions should impact intercalation kinetics, with the phase evolution sensitive to the degree of crystallinity, temperature, and cycling conditions [34, 56–58].

The unit cell formula of the richest lithium-intercalated compound LiC_6 corresponds to a theoretical capacity of 372 mAh/g. Although such capacities may be obtained at low current, the rate capability of large plate well-crystallized graphite is notably lower [59, 60], in contradiction with its high conductivity [32], often requiring a potentiostatic step to accomplish complete lithiation. The rate capability can be dramatically improved by controlling the degree of crystallinity and engineering the microstructure of graphitized particles, with spherical mesocarbon microbeads (MCMB), ~ 10 μm diameter particles combining high surface area with a relatively high degree of graphitization, the most successful form [61]. Further to a large reversible capacity and high current rate, MCMB also presents a low electrochemical potential, close to that of lithium metal, with the equilibrium voltage for the LiC_{12}/LiC_6 redox couple being 20 mV versus Li^+/Li. This yields a high voltage LIB, although it also presents safety issues where lithium plating can occur over graphite electrodes [62], and current collectors that are stable at this potential, typically copper, increase substantially the overall cost and weight of the battery.

A disadvantage of graphite electrodes are secondary surface reactions that occur with the electrolyte. Some solvents, such as propylene carbonate, can co-intercalate between the layers leading to the fast defoliation of graphite and negligible reversibility

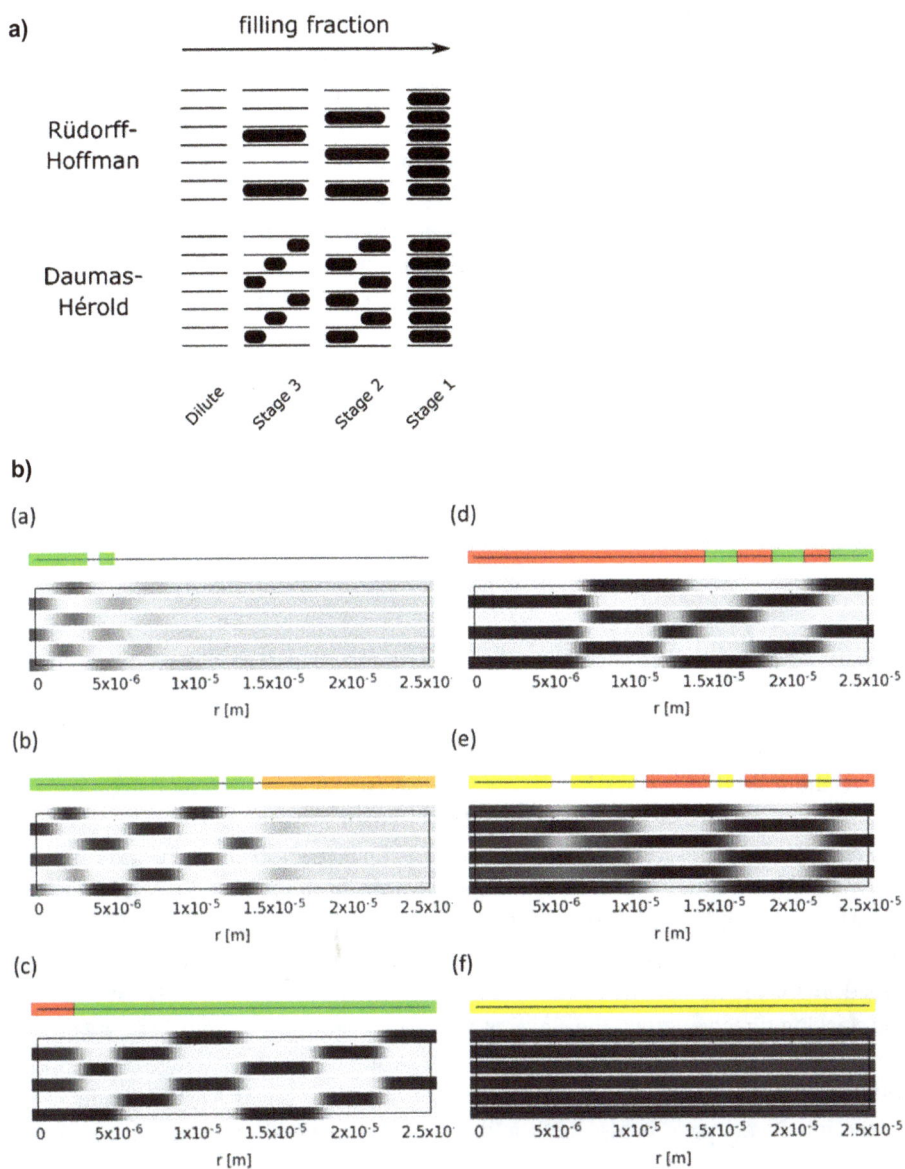

Fig. 7.5: a) Comparison between the classical staging (Rudorff–Hoffman) and Daumas–Herold model of Li intercalation in graphite. In the classical model, simultaneous complete emptying and filling of some layers is required. In the Daumas–Herold model, staging can proceed via migration of Li within the same layer. b) Monte-Carlo simulation of the formation of Daumas–Herold domains in graphite during Li intercalation. Adapted with permission from [49]. Copyright 2019 *American Chemical Society*.

[63]. Even in conventional electrolyte solutions, an initial capacity loss of ~ 15% is observed during the first cycle, due to electrolyte decomposition resulting in the formation of a thick solid electrolyte interphase layer and release of gas [64]. The solid electrolyte interphase plays a crucial role in determining electrochemical properties and so-called formation cycling is required to prime batteries containing MCMB, a serious bottleneck in their preparation [65, 66].

In summary, graphite is one of the first electrode materials used in a lithium-ion battery and remains the most commonly used commercial material, as a result of its high capacity and low electrochemical potential. Upon lithiation, the crystal structure of graphite transforms into several phases where lithium-filled and empty layers are ordered in the stacking direction; this mechanism is common in graphite intercalation compounds and is called staging. Despite the widespread use of graphite in lithium-ion batteries, the phase diagram is not completely understood, with its determination complicated likely as a result of the mechanism of transformation between the different stages. The low electrochemical potential of lithiated graphite electrodes also results in some chemical instability that can be problematic in practical applications.

7.2.2 Layered materials: lithium oxides, $LiMO_2$

The materials that we will refer to as lithium layered oxides with $LiMO_2$ formula, M being a transition metal, crystallize in the α-$NaFeO_2$ structure type. This structure type derives from the NaCl structure type, where two cations with different sizes (e.g., Li and M) are present. The size mismatch results in their ordering into layers, distorting the cubic lattice of rock salt into a trigonal one. The conventional unit cell for lithium layered oxides is hexagonal with space-group symmetry $R\bar{3}m$. Their structure consists of MO_6 octahedra linked by their edges and forming layers in the (a,b) plane of the hexagonal unit cell, separated by lithium ions in octahedral coordination in the so-called interlayer space. The resulting structure alternates "MO_2" and Li layers, with the stacking direction along the c-axis of the hexagonal unit cell (Fig. 7.6). The transition metals M have variable oxidation states allowing the release/storage of electrons, with electron delocalization usually high within the "MO_2" layers. The Li^+ conductivity in the interlayer space is also high and Li^+ can be inserted/removed with minimal structural rearrangement, confirming the suitability of this structure type as an electrode material.

$LiCoO_2$ (LCO) and graphite were the first ever commercialized electrode materials for use in LIBs [12, 28], with LCO still the focus of intense study aimed at improving its performance in full commercial batteries. LCO adopts the α-$NaFeO_2$ structure type and delivers approximately 140 mAh/g, equivalent to the removal of 0.55 Li, over the in-use voltage window 3.0–4.2 V (vs. Li^+/Li). Li_1CoO_2 is the composition with the highest lithium content as the interlayer lacks further sites available for lithium. The structural changes in LCO during lithium insertion and extraction were first studied using

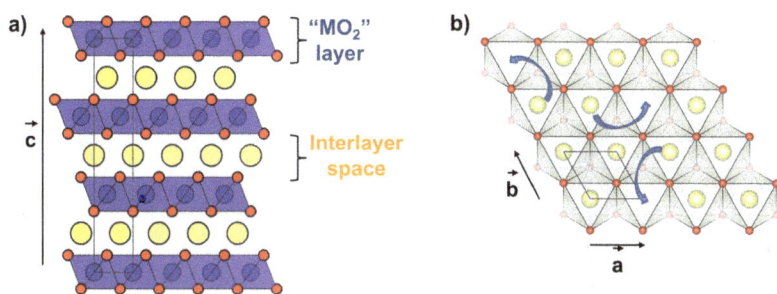

Fig. 7.6: a) Representation of lithium layered oxide crystal structure [67] with Li, Co, and O as yellow, blue, and red circles, respectively, and unit cell in black. b) Representation of the lithium conduction mechanism where Li hops between octahedral sites of the interlayer.

in situ X-ray diffraction [68, 69]. The observed phase transitions are directly related to the voltage variations observed during galvanostatic cycling (Fig. 7.7). The delithiaton of Li_xCoO_2 shows a voltage plateau between $0.95 > x > 0.75$ associated with a biphasic transformation between semiconductor $Li_{0.95}CoO_2$ and metallic $Li_{0.75}CoO_2$ [70, 71]. The metallic behavior of $Li_{0.75}CoO_2$ (electronic resistivity increases with temperature) is observed for $T > 175$ K and is supported by ^7Li Magic Angle Spinning (MAS) NMR spectroscopy [70] and theoretical considerations [72], noting a complex behavior at lower temperatures [70]. Further delithiation proceeds as a solid-solution with associated linear unit cell changes and voltage response. At $x = 0.5$, a voltage step is associated with a superstructure where Li^+/vacancy ordering occurs within the interlayer and charge ordering between Co^{3+} and Co^{4+} in the transition metal layer, resulting in a monoclinic distortion of the structure [73, 74]. A second solid-solution transformation is observed for $0.5 > x > 0.4$. Single phases are observed when the voltage is sloped $(dV/dx \neq 0)$ whereas two phase transformations occur around voltage plateaus $(dV/dx \approx 0)$. LCO exemplifies the relationship between the voltage response of the electrode and crystal structure, as seen for other materials in subsequent sections.

The cycling of Li_xCoO_2 for $1 > x > 0.4$ (3.0–4.2 V) is reversible; however further delithiation results in capacity fading due to substantial structural changes. Completely deintercalated Li_0CoO_2 can be obtained with the CdI_2 structure type (Fig. 7.7) [75, 77], where the CoO_2 layers collapse onto each other with ~ 10% reduction of the unit cell volume. This transformation leads to particle cracking and is only partially reversible, leading to fast capacity fading [77, 78].

Due to the limitations of LCO and the cost and toxicity associated with the use of cobalt, other isostructural materials have been investigated. $LiNiO_2$ has a high capacity (220 mAh/g or 0.8 Li removed) [79, 80] but suffers from major stability issues. The first issue with $LiNiO_2$ is the tendency for Ni to be initially present in the interlayer due to the comparable ionic radius of Ni^{2+} and Li^+ (0.69 and 0.76 Å, respectively) [81], noting that samples are commonly non-stoichiometric with composition closer to $[Li_{1-z}Ni_z]NiO_2$ [82, 83]. The degree of cation mixing between Li and Ni has

a)

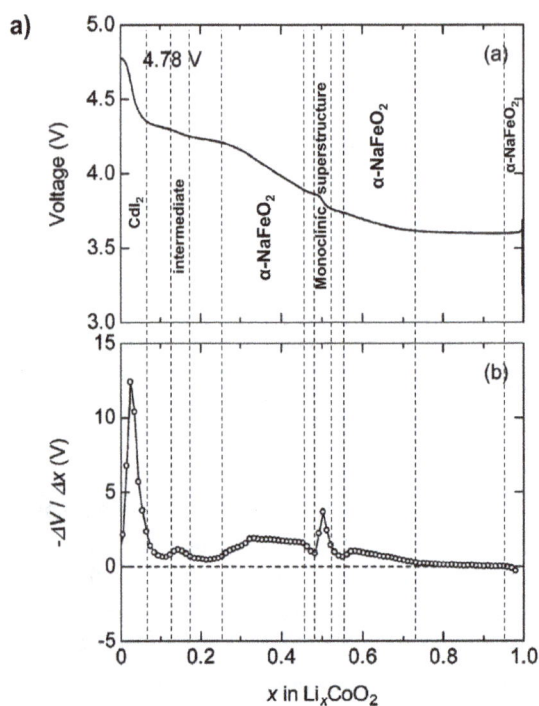

b)

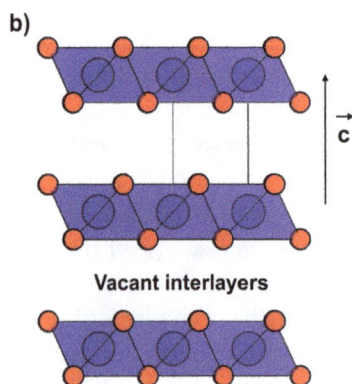

Vacant interlayers

Fig. 7.7: a) Galvanostatic voltage curve of the delithiation of LCO and its derivative dV/dx. The structure adopted by the material during cycling is denoted for regions where a single phase is observed, with those separated by two-phase regions where both end phases are present. The "intermediate" structure is thought to incorporate a mixture of occupied and empty interlayers, despite not being well characterized due to disorder. Adapted with permission from [75]. Copyright 2007 *American Chemical Society*. b) Representation of the crystal structure of completely deintercalated Li_0CoO_2 [76] with the CdI_2 structure type. Co atoms form columns along the stacking direction as opposed to the staggered configuration in LCO with the α-NaFeO$_2$ structure type.

been found to negatively affect electrochemical properties. The presence of Ni in the interlayer decreases the amount of sites available for Li, reducing capacity [82, 83] and hindering ionic diffusion within the interlayer resulting in lower rate capability. A second issue arises at low Li content, with Ni migrating into the interlayer and forming NiO on the surface, and the excess oxygen reacting with the electrolyte [79, 84]. This transformation is irreversible and particle cracking arises from the volume change, resulting in fast capacity fading [79, 80, 84].

A large number of elemental substitutions have been used to modify the electrochemical properties of these two lithium layered oxides, with Ni, Mn, and Co the most common [85–88]. The bulk of the capacity is provided by Ni with both Ni^{3+}/Ni^{2+} and Ni^{4+}/Ni^{3+} couples active in the investigated voltage range, with higher Ni content usually resulting in higher capacity. Other dopants such as Co^{3+}, Mn^{4+}, and Al^{3+}, although electrochemically inactive, are regarded as stabilizers of the layered structure. Their small size compared to that of Li^+ enhances their segregation into separate layers, reducing cation mixing between Li^+ and Ni^{2+}, and stabilizing the layered structure during cycling [89, 90].

Despite considerable studies concerning mixed transition metal layered oxides, the influence of these substitutions on transition metal ordering and how they impact electrochemical behavior is still debated. Long-range ordering as shown by superstructure features in diffraction data can be difficult to observe. Probes of short-range ordering such as nuclear magnetic resonance (NMR) spectroscopy and pair distribution function (PDF) analysis of total scattering data have shown a non-random organization within MO_2 layers [91–93], with Mn^{4+} and Ni^{2+} found to form local clusters that minimize electrostatic repulsions within MO_2 layers [93]. This has important implications for the electrochemical behavior of the material as Li vacancy ordering within the interlayer is intimately related to charge ordering in the transition metal layer, and a correlation between the metal-insulator transition (as seen in undoped LCO) and the degree of transition metal ordering exists [93]. Two materials have been particularly successful in commercial applications: $LiNi_{1/3}Mn_{1/3}Co_{1/3}O_2$ (NMC333, 150 mAh/g, 2.5–4.2 V vs. Li^+/Li) [94] and $LiNi_{0.8}Co_{0.15}Al_{0.05}O_2$ (NCA, 190 mAh.g^{-1}, 3.0–4.4 V vs. Li^+/Li) [13], both with high electrochemical potential, excellent cyclability, and high rate capability. A wide number of applied and fundamental studies have attempted to understand and improve the performance of these materials, with the focus shifting toward Ni-rich oxides for their reduced Co content, higher voltage, and higher capacity [95, 96]. Comparable Ni and Co content yield similar cycling performance and stability in NMC and NCA, noting slightly better long-term capacity retention in NCA [6].

Despite the enhanced stability provided by substitutions, Ni-rich layered oxides still suffer from Li/Ni cation mixing, similar to $LiNiO_2$. Quantification of the degree of cation mixing is thus critical, with powder diffraction data the most used method for this [79, 90, 97, 98]. Careful constraints in the structural model during refinement are often required as the presence of several cations on the same crystallographic site increases correlations between parameters. The use of combined X-ray and neutron data is preferred where the different scattering for the two radiation types (as well as additional independent observations) facilitates a unique solution when constrained by additional information from measurements such as compositional analysis and magnetism [98]. The NMC and NCA oxides also suffer from irreversible transformations at high voltage and instability at elevated temperature [99–101], with a correlation between the amount of Ni and the decomposition temperature and oxygen gas released [5]. A compromise between capacity and stability is often made (Fig. 7.8), and attempts to resolve this conflict using material doping, surface coating, or electrolyte formulation are underway [95, 96, 102–104].

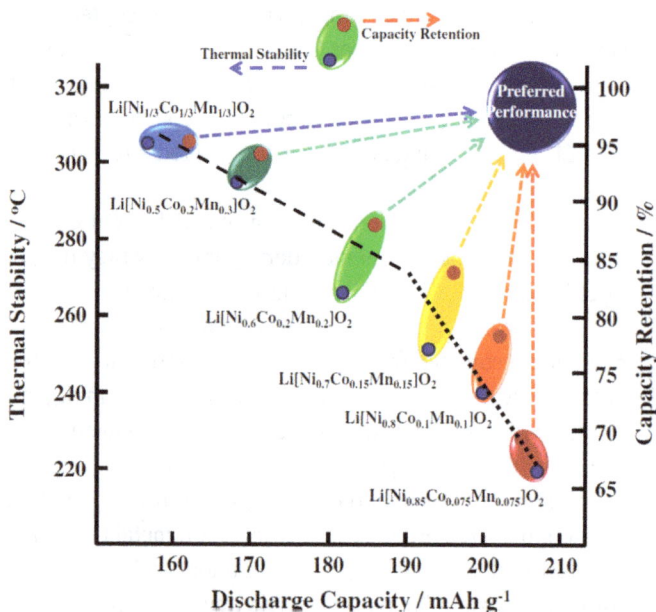

Fig. 7.8: Correlation between the Ni content in NMC electrodes and their initial capacity, thermal stability, and cyclability. Reprinted from [5], Copyright 2013, with permission from *Elsevier*.

Lithium layered oxides have been studied and engineered extensively since their early implementation in lithium-ion batteries. As the amount of lithium between the layers of the structure is varied, several phase transitions are observed as a result of in-plane ordering of lithium ions and/or the charge carried by transition metals. The overall capacity is often limited by instability at low lithium concentration where large structural changes occur, often with the irreversible migration of transition metal ions into the lithium layers. Metal doping has been employed to enhance the separation of lithium and transition metals resulting in appreciable increases of the capacity and long-term cyclability.

7.2.3 Layered materials: "lithium-rich" oxides and Li_2MnO_3

The crystal structure of Li_2MnO_3 is closely related to that of the lithium layered oxides $LiMO_2$ (Fig. 7.9). The same stratified organization is observed, where "MO_2" and Li layers alternate along the stacking direction; however, in Li_2MnO_3, the M is split across two distinct positions occupied by either Li or Mn. The resulting unit cell is monoclinic with $C2/m$ space-group symmetry in which each LiO_6 octahedron is surrounded by 6 MnO_6 octahedra. The chemical formula can be written as $Li[Li_{1/3}Mn_{2/3}]O_2$, emphasizing the overall ratio of 1/3 Li and 2/3 Mn within "MO_2" layers.

As expected from the structural similarity to lithium layered oxides, Li^+ are mobile within Li_2MnO_3 and the material can be used as an LIB electrode. Similar to layered oxides, Li_2MnO_3 is the composition with the highest Li content with full interlayers. The theoretical capacity, assuming complete Li removal from Li_2MnO_3, is 458 mAh/g; however, the initial capacity varies over 50–300 mAh/g (up to 1.3 Li removed), with a strong dependence on synthesis and cycling conditions [106, 107]. A capacity of 300 mAh/g corresponds to the removal of 1.3 Li per Li_2MnO_3, suggesting that some lithium in "MO_2" layers contributes to the capacity, however this is complicated by the participation of oxygen in the redox process, as explained later. Li_2MnO_3 electrodes suffer from large initial capacity loss and fast capacity decay [108].

Voltage variations during the first charge of Li_2MnO_3 are clearly different from subsequent cycles (Fig. 7.9), revealing distinct electrochemical mechanisms. During the first charge, a sloped voltage plateau is observed around 4.5 V and is associated with a loss of oxygen gas [106]. X-ray absorption spectroscopy suggests that Mn^{4+} is electrochemically inactive during this step, and it is widely accepted that oxygen participates in the redox process instead [107, 109]. The lack of reversibility may be explained by the further reaction and migration of oxygenated species away from Li_2MnO_3 [106]. The following discharge shows a significant reduction of the capacity, by around 30%, with a complex voltage profile in the 2.0–4.7 V range, dissimilar from the first charge and conserved in following cycles (Fig. 7.9) [106]. Subsequent cycles have no further gas evolution and Mn becomes electrochemically active [106, 107]. Fast capacity fading is observed for this material, with negligible capacity after only 20 cycles, making Li_2MnO_3 not suitable for practical application.

It is possible to prepare layered materials with the formula $(1-x)Li_2MnO_3 \cdot x\ LiMO_2$ (M often a mixture of Mn, Ni, and Co) where both Li_2MnO_3-like layers and $LiMO_2$-like layers are intergrown within the same structure, as evidenced by scanning transmission electron microscopy [110]. These materials are sometimes called "lithium-rich," "lithium-excess," or "overlithiated" in light of their overall $Li_{1+x}M_{1-x}O_2$ composition analogously to the formula $LiMO_2$ of layered oxides. Interest for this family of materials was brought on by the very large observed specific capacity at relatively elevated voltage (> 250 mAh/g in the 3.0–4.5 V range) [111–114], noting that performance is remarkably sensitive to synthesis conditions [113, 115].

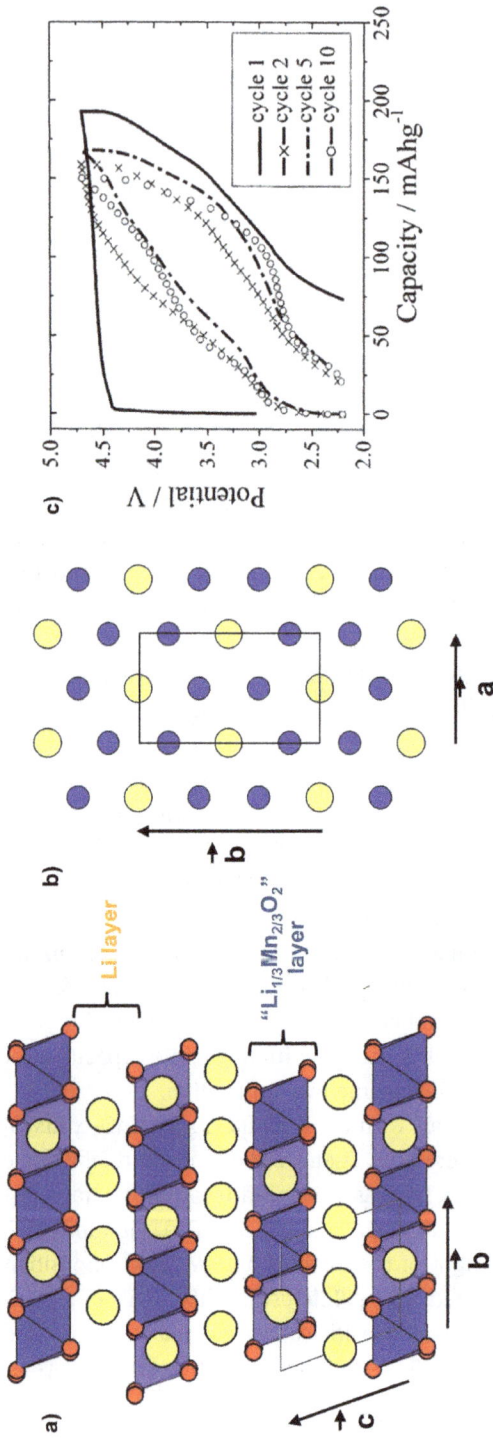

Fig. 7.9: a) Representation of the crystal structure of Li_2MnO_3; [105] Li, Mn, and O are yellow, blue, and red circles, respectively. Similarity to layered oxides can be appreciated by comparison with Fig. 7.6. b) In-plane Li/Mn ordering within "$Li_{1/3}Mn_{2/3}O_2$" layers, O atoms omitted for clarity. c) Typical voltage variations of batteries containing Li_2MnO_3 during galvanostatic cycling. Adapted with permission from [106]. Copyright 2003 *American Chemical Society*.

Due to the structural and compositional similarity of Li_2MnO_3 and $LiMO_2$, their diffraction patterns overlap almost perfectly [112], and thus it is relatively difficult to distinguish between the two layer types using conventional powder diffraction. The only non-overlapping reflections are those observed between 4.9 and 2.8 Å d-spacing, which arise from in-plane Li/Mn ordering within Li_2MnO_3-like layers. These superstructure reflections are often broad, following a Warren fall peak shape as a result of stacking disorder [116, 117]. Conventional Rietveld refinements are unable to properly model their shape and intensity, and as a result, it is common practice to limit the characterization of the structure to an average simplified model, often excluding those reflections. More specialist programs, such as FAULTS [118], can model stacking disorder between Li_2MnO_3-type layers and Li/Mn site mixing within those layers. Each contributes differently to the broadening and intensity of the superstructure reflections, (Fig. 7.10) and it is possible to quantify the composition within each layer [119, 120], with a correlation between the electrochemical behavior and degree of disorder noted [119].

The cycling behavior of lithium-rich layered oxides is intermediate between that of $LiMO_2$ and Li_2MnO_3. Two distinct regimes can be seen in galvanostatic curves, corresponding to the contribution of each layer type (Fig. 7.10) [7, 113]. During the first charge, between 3.0 and 4.5 V, $LiMO_2$-like layers contribute to capacity, after which a plateau associated with Li_2MnO_3-like layers is observed at around 4.6 V, with a concomitant release of oxygen as a gas [121, 122], as similar to Li_2MnO_3. After the first charge, both layers contribute to the complex voltage curve between 3.0 and 4.5 V, with subsequent cycles mostly reversible. The $LiMO_2$-like layers are regarded as stabilizers of the structure, as evidenced by the vastly improved cyclability compared to that of Li_2MnO_3, however, the structure of "active" Li_2MnO_3-like layers is still ambiguous, despite many investigations of the transformation mechanism. Theoretical calculations suggest that Mn ions migrate irreversibly into the Li layers during the first charge, with the structure of "activated" Li_2MnO_3-like layers closer to that of disordered rock salt [123], as supported by transmission electron microscopy [124] and in situ synchrotron [125] measurements and consistent with the large voltage hysteresis of the material. The mechanism involves the oxidation of Mn^{4+} to Mn^{7+} during the first charge [123], in contradiction to X-ray absorption spectroscopy measurements [109, 122, 126], with the instability of Mn^{7+} thought possibly responsible for the contradiction.

Performance-wise, lithium-rich oxides have a high reversible specific capacity (Fig. 7.10), up to 300 mAh/g in the 2.0–5.0 V range [113, 122]. The larger capacity required during the first charge is advantageous against a graphite counter-electrode, with graphite consuming excess Li during the first charge, enabling full use of the capacity of both electrodes through load balancing [127]. Although the cyclability of lithium-rich electrodes is improved compared to Li_2MnO_3, they still undergo significant capacity fading that limits their commercial application.

In conclusion, lithium-rich layered oxides can be regarded as a composite of two layered structures with both Li_2MnO_3 and $LiMO_2$ –like layers; how these layers are distributed within crystallites is the subject of intense debate, with the type and degree of

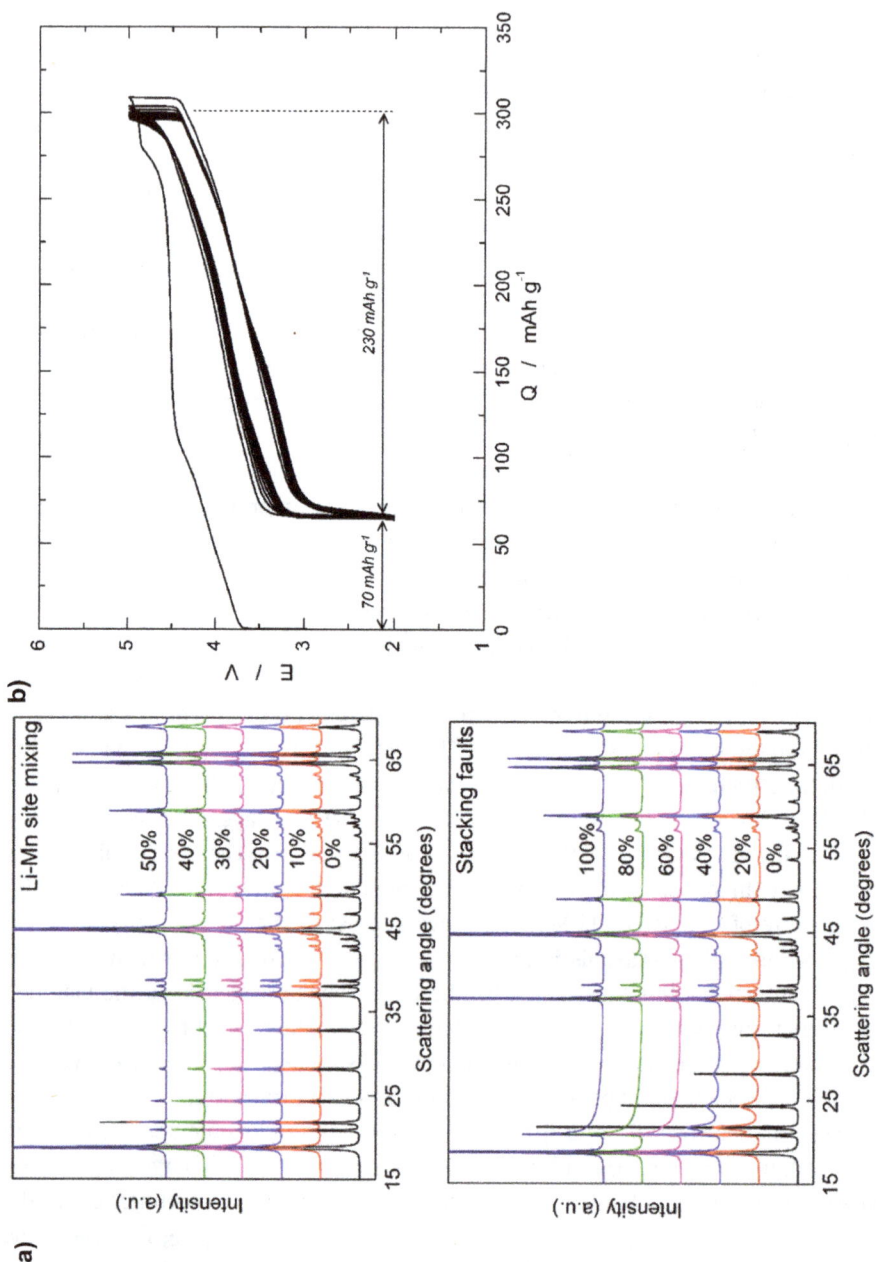

Fig. 7.10: a) Simulation using DIFFaX of X-ray diffraction patterns of lithium-rich layered oxides with varying Li/Mn mixing and stacking faults between Li_2MnO_3 layers. Adapted with permission from [119]. Copyright 2020 *American Chemical Society*. b) Typical voltage variations during the galvanostatic cycling of lithium-rich layered oxide. Republished with permission of *Royal Society of Chemistry*, from [113]; permission conveyed through Copyright Clearance Center, Inc.

disorder related to electrochemical behavior. The performance of the composite is intermediate between that of the two pure compounds, benefitting from larger capacity, but also suffering from the inherent instability of Li_2MnO_3-like layers, which undergo an irreversible transformation during the first delithiation which is poorly understood. Overall the material is promising but requires improvement before it can be considered for practical application.

7.2.4 Layered materials: sodium layered oxides

The higher natural abundance of sodium compared to lithium offers the possibility for more affordable energy storage. Consequently, interest for sodium-ion batteries has been steadily growing. The larger size and mass of Na^+ may result in lower specific capacity (e.g., the maximum theoretical capacities of LCO and $NaCoO_2$ are 274 mAh/g and 235 mAh/g, respectively) making these more suitable for stationary energy storage as required for intermittent energy sources. Research of sodium-ion battery electrode materials began with materials similar to those used in LIBs, with layered oxides heavily studied [128].

The general structure of layered oxides A_xMO_2, with A an alkali metal ion and M a transition metal, has been described for A = Li, where layers of edge-sharing MO_6 octahedra are separated by the interlayer space occupied by A^+ alkali metal ions. The larger size of Na^+ enables both octahedral and prismatic coordination with oxygen ions on both sides of the interlayer, resulting in polymorphism where different stackings can be obtained depending on the sample composition and history. The notation commonly employed to describe each stacking uses a letter to symbolize the coordination geometry of the alkali metal ion, O for octahedral and P for prismatic, and a number to indicate the number of layers in one unit cell [129]. The most common stacking types are O3, P3, P2, and O2, with respective oxygen positions /AB/CA/BC/, /AB/BC/ CA/, /AB/BA/, and /AB/CB/ (Fig. 7.11). O3 corresponds to the α-$NaFeO_2$ structure type. All structures are conventionally described using a hexagonal unit cell, with c the stacking axis, and space-group symmetry $R\bar{3}m$, $R3m$, $P6_3/mmc$, and $P6_3mc$ for O3, P3, P2, and O2 phases, respectively.

The P3 and O3 stacking structures are related to each other via a gliding of "MO_2" layers and the same relationship exists between the P2 and O2 stacking structures (Fig. 7.11). This relatively low-energy transformation is frequently observed during electrochemical cycling at room temperature [130–136], although the transformation is sometimes kinetically limited [131]. Preference for octahedral or prismatic coordination depends on electrostatic repulsions [137]. O–O repulsions are stronger in prismatic coordination as oxygen ions face each other on each side of the interlayer, however, a higher density of Na crystallographic sites (Fig. 7.12) is available in P3 and P2 compared to O3 and O2, driving longer Na–Na distances in structures where Na is in prismatic coordination. As a result of competing interactions, octahedral coordination is usually

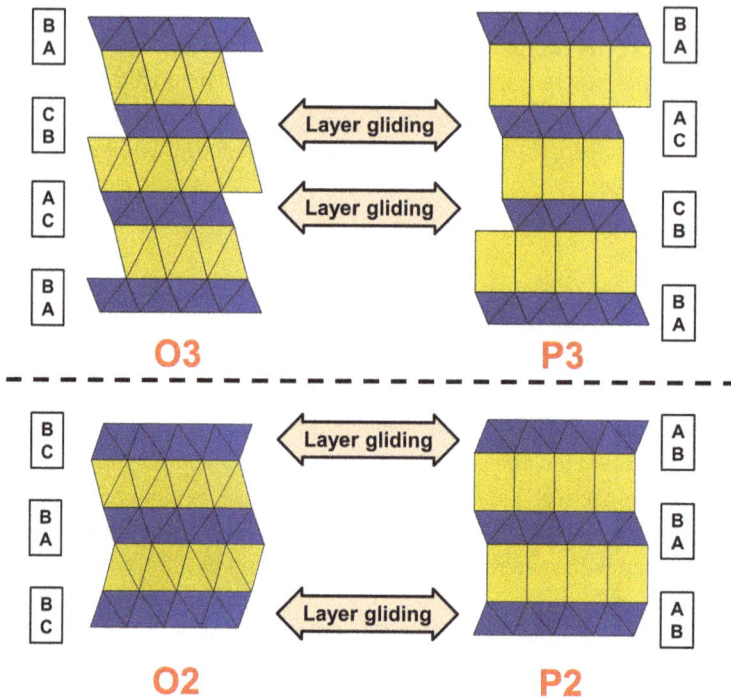

Fig. 7.11: Representation of different stacking structures in Na_xMO_2. MO_6 octahedra and NaO_6 polyhedra are represented in blue and yellow, respectively, with prisms depicted as rectangles and octahedra as crossed rhombuses. Letters indicate the position occupied by oxygen ions.

favored at high and low sodium content when the interlayer space is small, and prismatic coordination at intermediate compositions [137]. The larger number of crystallographic sites and more accessible migration pathways also results in generally faster ionic diffusion in the P phases compared to O phases [138]. The transformation between 2-layered (O2 and P2) and 3-layered (O3 and P3) stacking sequences requires the breaking of M-O bonds, and so only occurs at high temperature, with the sequence usually determined after the initial synthetic step. O3, P3, and P2 phases can be obtained using conventional high temperature synthetic routes [130, 139] whereas O2 phases are not stable at high temperature [140] but rather metastable at room temperature when obtained from a P2 phase [134, 140, 141].

Na–Na interactions are stronger than Li–Li, particularly in P3 and P2 stacking structures, where NaO_6 prismatic sites share faces [137, 142]. The reduction of Na distances results in Na-vacancy ordering within the interlayer. The density and organization of crystallographic sites differ between P and O stacking types, resulting in distinct in-plane organizations depending on the amount of sodium (Fig. 7.12). This has consequences for the electrochemical potential, with steeper voltage curves and complex variations related to the relative stability of each ordering. The position of second-

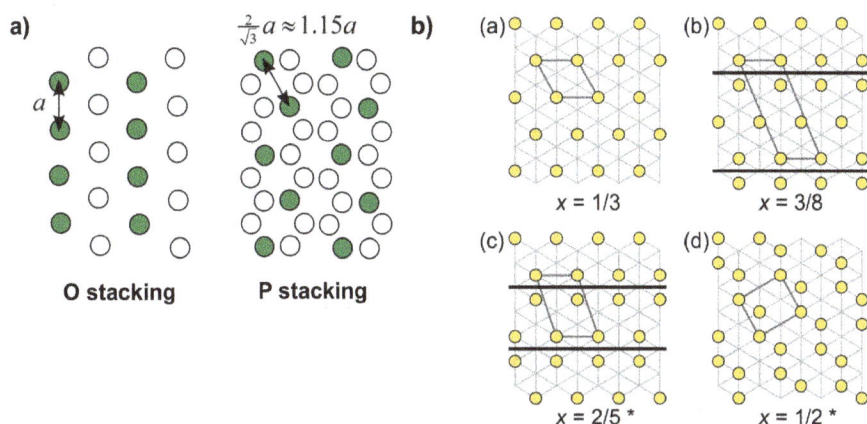

Fig. 7.12: a) Representation of crystallographic sites (circles) available for Na within the interlayer of O3/O2 and P3/P2 structures. The predicted Na ordering (occupied sites in green) is given for the $Na_{1/2}MO_2$ composition, emphasizing larger Na–Na distances in the P3 than the O3 structure. Adapted with permission from [137]. Copyright 2016 *American Chemical Society*. b) Predicted stable Na ordering in the P3 structure at different x in Na_xMO_2 from ab initio calculations. Reprinted figure with permission from [142]. Copyright 2019 by *American Physical Society*.

neighbor transition metal M^{n+} in the "MO$_2$" layers also influence the sodium ordering, slightly destabilizing NaO$_6$ sites that share faces with MO$_6$ octahedra; this explains the distinct order observed in P3 and P2 stacking phases despite sharing identical interlayer organization. Based on these considerations, the Na-vacancy order can be predicted using first-principle calculations [142, 143] and many of the predicted orders have been experimentally observed using X-ray or electron diffraction.

As a result of variable stacking sequence, interlayer sodium order, and symmetry-lowering distortion due to Jahn–Teller or charge order, the crystal chemistry of sodium layered oxides is particularly rich and a wide range of superstructures have been observed during cycling [128, 132, 144–148]. Na_xVO_2 phases present interesting crystallography and will be given as an example. Three polytypes can be prepared, P2, O3, and P3, with the stacking dependent on the sample history [131, 149, 150]. The conversion between O3 and P3 stacking via layer gliding is possible; however, unlike other known sodium layered oxides, the O3 to P3 transformation in Na_xVO_2 requires thermal activation, with both O3 and P3 phases stable (or metastable) at room temperature over an overlapping range of composition [131]. Each polytype exhibits a distinct voltage curve during cycling, with slopes and plateaus related to transformations between phases of different interlayer sodium order (Fig. 7.13) [131, 150, 151]. A superstructure occurs at $Na_{1/2}VO_2$ in the three polytypes, solved for the O3 and P2 phases using X-ray powder diffraction. The refined structures show Na order in the interlayer and a shift of V atoms in the "VO$_2$" layers from their ideal position, forming relatively short V–V distances (< 2.6 Å) in the shape of dimers in the O3 phase or

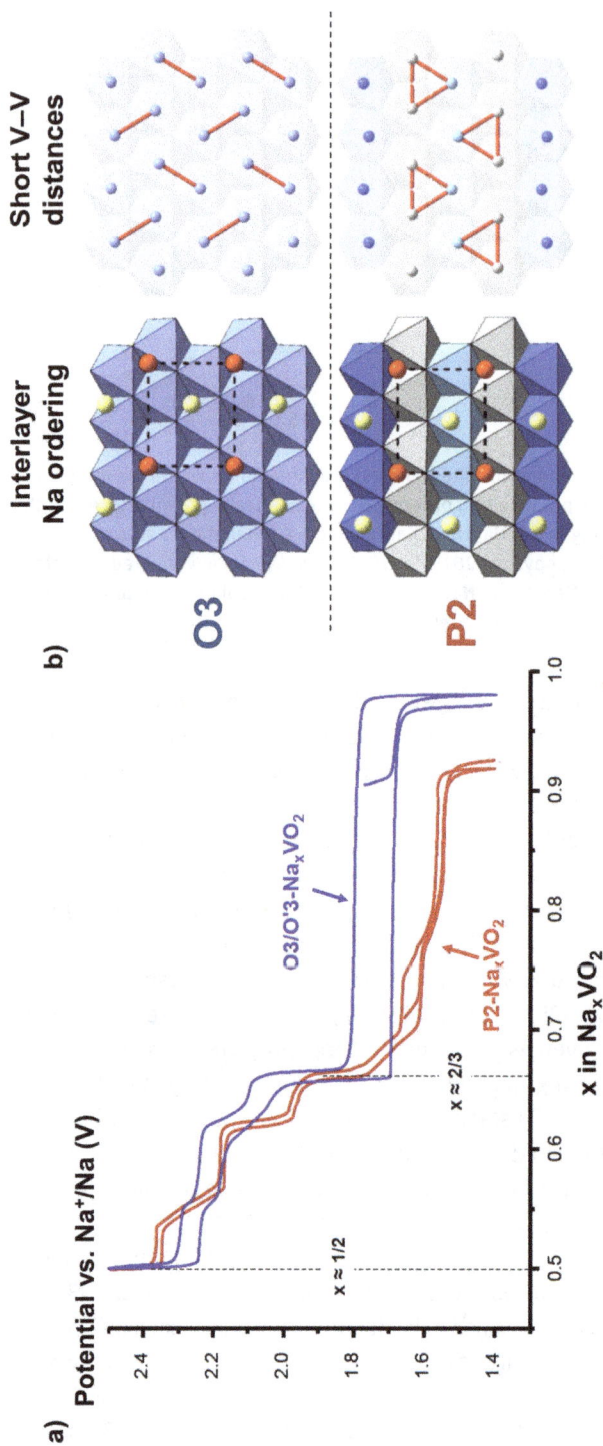

Fig. 7.13: a) Galvanostatic voltage curves of O3-Na_xVO_2 and P2-Na_xVO_2 exhibiting phase transitions between phases with different Na order. b) Representations of the in-plane ordering of Na ions in the interlayer and V–V distances in the "VO_2" layers in $Na_{1/2}VO_2$ for the O3 and P2 stackings. Adapted from [158].

trimers in the P2 phase (Fig. 7.13) [150, 151]. Bond valence sum calculations suggest partial charge order, consistent with the sharp increase of electronic conductivity when the trimers disappear at 320 K in P2-$Na_{1/2}VO_2$ [150, 152]. The resulting magnetic interactions are relatively complex, coupling local spin order between dimers/trimers within the magnetically frustrated triangular network of the "VO_2" layers. Another interesting crystallographic aspect occurs in P3-Na_xVO_2 where a single phase is observed during cycling in the $0.55 < x < 0.57$ composition and the corresponding superstructure reflections can be indexed by a supercell that is incommensurate with the subcell of the average structure, the modulation vector q directly proportional to the amount of sodium x in Na_xVO_2 [131, 153]. Incommensurate order has also been observed in the Na_xCoO_2 layered oxides [154–157], and they show a similar relation between the modulation vector and Na content. Although the structure of those compounds remains to be solved using the incommensurate formalism, it is hypothesized that the supercell of the Na order in the interlayer being incommensurate with the unit cell of "MO_2" layers provides a large number of long-range arrangements that minimize electrostatic repulsions between Na^+, with commensurate order special cases that occur at discrete compositions, for example at $x = 1/2$.

Although fundamentally interesting, stacking sequence changes and Na order are often undesirable for the purpose of energy storage due to the sharp volume changes they induce and the lower diffusion rate around ordered compositions [159–161]. Partial substitution at the M site is commonly employed to destabilize sodium order while benefitting properties such as higher voltage or improved stability. One of the most promising sodium-layered oxides for energy storage is doped P2-$Na_xNi_{1/3}Mn_{2/3}O_2$, which exhibits a specific capacity of 150 mAh/g in the voltage range 2.0–4.5 V [134]. The material does not contain any Co, resulting in a relatively more affordable and environmentally friendlier material compared to lithium oxides; however, the capacity and voltage are comparatively lower.

Undoped P2-$Na_xNi_{1/3}Mn_{2/3}O_2$ shows fast capacity fading when cycled at low sodium content ($x < 0.4$). In situ X-ray diffraction has revealed a two-phase transformation at high voltage and an abrupt reduction of 20% of the c unit cell parameter in the stacking direction upon desodiation [134]. The desodiated phase adopts the O2 structure as a result of stronger O–O interactions in the collapsed interlayer. It was found that chemical substitution could induce a continuous change of unit cell parameters instead, resulting in improved cyclability [162–164]. The same transformation was observed in P2-$Na_xMn_{1/2}Fe_{1/2}O_2$ [165]; however, the complex broadening of reflections at high voltage has prevented complete structural analysis of the desodiated "Z" phase. An elegant structural model has been proposed for this Z phase (Fig. 7.14), where randomly arranged sodium-filled P2 and empty O2 layers are intergrown [162, 164], as supported by annular dark-field scanning transmission electron microscopy observations [164].

The stacking fault model reproduces remarkably well the anisotropically broadened reflections observed using in operando X-ray diffraction [162, 164]. Not unlike the staging mechanism in graphite, a commensurate order alternating empty (O2) and

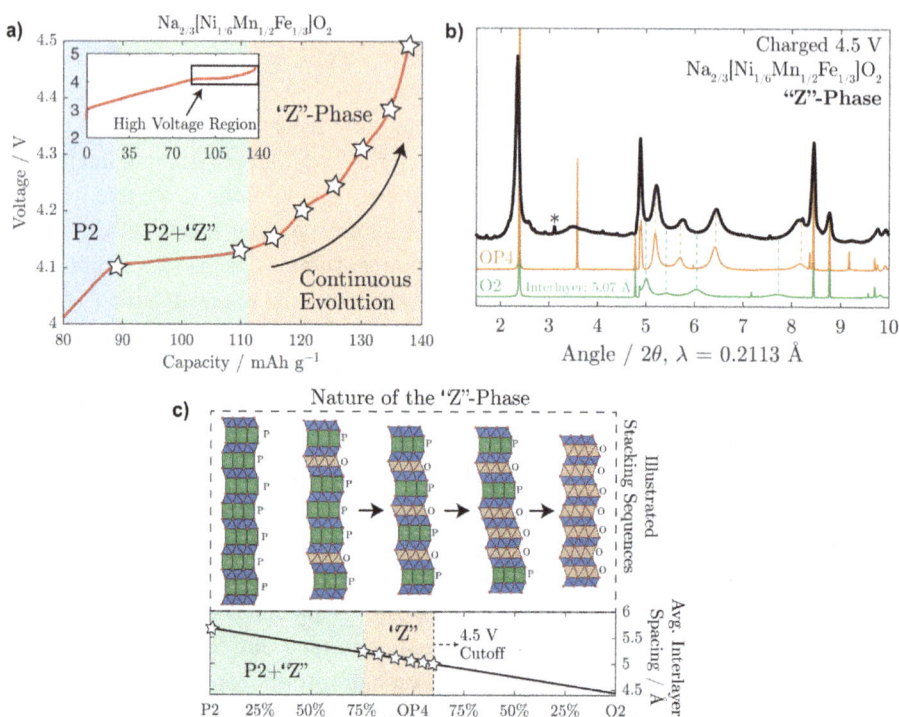

Fig. 7.14: a) Galvanostatic voltage curve of $Na_xNi_{1/6}Mn_{1/2}Fe_{1/6}O_2$ and corresponding phase diagram where the structure of the layered oxide is P2 or Z phase. b) Diffraction pattern of $Na_xNi_{1/6}Mn_{1/2}Fe_{1/6}O_2$ extracted from a battery at 4.5 V and simulation of the pattern using a mixed P2-O2 model for the Z phase. 3) Proposed structural evolution for the Z phase at high voltage from P2 to O2 via OP4. Adapted from [164] under the Creative Commons CC BY License. Published by the *Royal Society of Chemistry*.

sodium-occupied (P2) layers in the stacking direction is sometimes observed when half the layers are empty, with the structure named OP4 [164, 166, 167]. Although the P2–O2 transformation is smoother when the material is doped, the fundamental mechanism of the Z phase formation is poorly understood. Further understanding of this could potentially lead to better Na-ion battery performance, given that sodium layered oxides are already used as electrodes in several commercial applications.

In summary, sodium layered oxides may be suitable electrode materials for use in sodium-ion batteries that follow a similar rocking-chair mechanism as lithium-ion batteries. Structural changes during (de)sodiation can be rather complex due to the size of sodium ions, resulting in variable coordination geometry, stacking axis arrangements, and sodium ordering within layers, with the resulting structure dependent on the sample history. The successive transitions between phases of different stacking or in-plane ordering results in a staircase electrochemical potential. It is sometimes possible to predict these orderings for a given layered oxide using first principles calculations. At low sodium content, the sharp collapse of the interlayer results in capacity degradation, limiting the overall capacity of sodium layered oxide materials. Elemental substitution can potentially smooth out the change of interlayer distances, resulting in cyclability improvement.

7.2.5 Spinel-type materials: LiMn$_2$O$_4$

LiMn$_2$O$_4$ crystallizes with a spinel-type face-centered cubic AB$_2$O$_4$ structure in which A and B cations occupy some of the four octahedral sites and eight tetrahedral sites, their occupancy depending on the relative stability of the two cations in each environment. In the case of LiMn$_2$O$_4$, it is said to be a direct spinel as half of the octahedral sites contain Mn and one-eighth of tetrahedral sites contain Li (Fig. 7.15). Within the $Fd\bar{3}m$ space-group symmetry Li, Mn, and O occupy 8a, 16d, and 32e sites, respectively. MnO$_6$ octahedra are linked by their edges and Li$^+$ migrate via unoccupied octahedral 16c interstitial sites, forming a network of interconnected tunnels along [110] (12 equivalents in the cubic system).

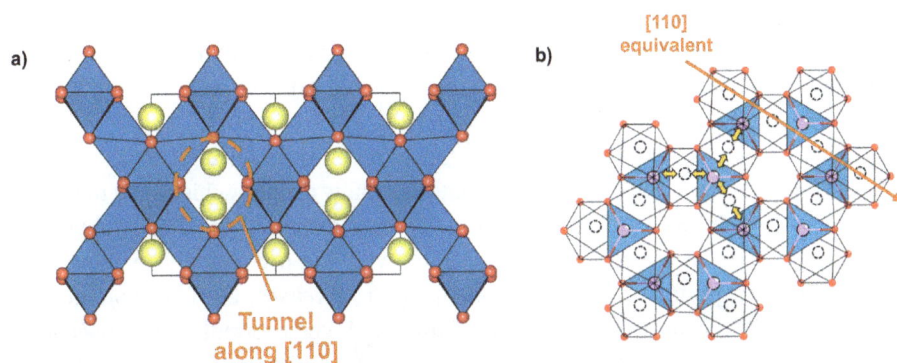

Fig. 7.15: a) Representation of the crystal structure of spinel LiMn$_2$O$_4$ [174] with MnO$_6$ octahedra in blue and Li and O as yellow and red spheres, respectively. b) Representation of Li migration pathways between 8a tetrahedral sites (blue) via unoccupied 16c octahedral sites. Adapted from [175] under the Creative Commons CC BY License.

The electrochemical extraction of lithium from LiMn$_2$O$_4$ was demonstrated as early as 1983 [15, 168]. The three-dimensional connectivity between lithium interstitial sites provides good ionic diffusion, which results in excellent rate capability. Combined with an elevated electrochemical potential (4.2 V vs. Li$^+$/Li) and moderate specific capacity (148 mAh/g theoretical, 120 mAh/g practical equivalent to 0.8 Li removal), LiMn$_2$O$_4$ has been employed as an electrode in high-power battery applications such as electric vehicles and power tools [169]. Compared to cobalt-based layered oxides, the material also benefits from lower cost and toxicity. One of the main disadvantages of LiMn$_2$O$_4$ is fast capacity fading, especially at moderate temperatures (> 50 °C) due to irreversible Mn dissolution into the electrolyte at high voltage [170] upon the dismutation of Mn^{3+} [171]. This problem is compounded in commercial batteries that are commonly stored in a charged state. It is possible to intercalate Li into $LiMn_2O_4$ as not all interstitial sites of the spinel structure are occupied in the pristine material. Li occupies both $8a$ and $16c$ sites in $Li_{1+x}Mn_2O_4$ within the spinel structure with tetragonal distortion due to co-operative Jahn–Teller effects of Mn^{3+} [168, 172]. In practice, "overlithiation" is avoided due to poor cyclability, the strong Jahn–Teller distortion partly responsible for particle cracking [173]. Doping strategies have aimed at overcoming this issue, with research focused on Ni-doped $LiNi_{0.5}Mn_{1.5}O_2$, where Ni is the active redox center instead of Mn, increasing electrochemical potential (4.7 V vs. Li^+/Li) while maintaining excellent rate capability, low cost, low toxicity, and moderate specific capacity [4].

The large difference in charge and ionic radius between Mn^{4+} and Ni^{2+} results in cation ordering, with a long-range ordering occurring when Ni : Mn = 1 : 3 and $LiNi_{1/2}Mn_{3/2}O_2$ with $P4_332$ space-group symmetry, compared with the $Fd\bar{3}m$ space-group symmetry achieved for a random distribution. The degree of ordering is influenced by the synthesis temperature, ordering obtained after annealing at 700 °C and the disordered phase obtained at higher temperature [176, 177]. Loss of long-range order is driven by the partial reduction of manganese due to the formation of oxygen vacancies at high temperature, where Ni : Mn is no longer 1 : 3 due to the formation of a rock salt $Ni_{1-x}Li_xO$ secondary phase, with the composition of the disordered phase closer to $LiNi_{0.5-\varepsilon}Mn_{1.5+\varepsilon}O_{4-\delta}$ [176–178].

Important to electrode function, the degree of order affects electrochemical performance (Fig. 7.16), with better capacity and rate capability usually achieved by the disordered material [179, 180], although this is sometimes disputed [181, 182]. The voltage curve of the material is influenced by ordering, with the ordered phase showing a single plateau around 4.7 V whereas, as disorder increases, the high voltage plateau splits into two sloped plateaus at 4.6 and 4.7 V corresponding to consecutive Ni^{3+}/Ni^{2+} and Ni^{4+}/Ni^{3+} redox processes, respectively, and a smaller third plateau is observed at 4 V that corresponds to the Mn^{4+}/Mn^{3+} redox couple [4, 177]. The phase transformation behavior is altered as well, with a two-phase transformation for the ordered phase at high voltage as opposed to an apparently single-phase transformation in the disordered phase, although two-voltage plateaus are visible [176, 183]. The order related reflections are barely distinguishable using X-ray diffraction; however, they are easily observed using neutron diffraction thanks to the significant difference in neutron scattering lengths of Ni and Mn (+10.3 and –3.73 fm, respectively). The degree of Ni/Mn mixing at $16d$ sites, related to the degree of long-range transition metal ordering, is routinely characterized using neutron diffraction, also allowing the extraction of the overall Ni/Mn stoichiometry, the oxygen non-stoichiometry, the amount of $Ni_{1-x}Li_xO$, and structural distortion [178].

Although clearly distinct average long-range ordering is observed in $LiNi_{0.5}Mn_{1.5}O_4$, average short-range ordering is also observed. Pair distribution function analysis from

a)

b)

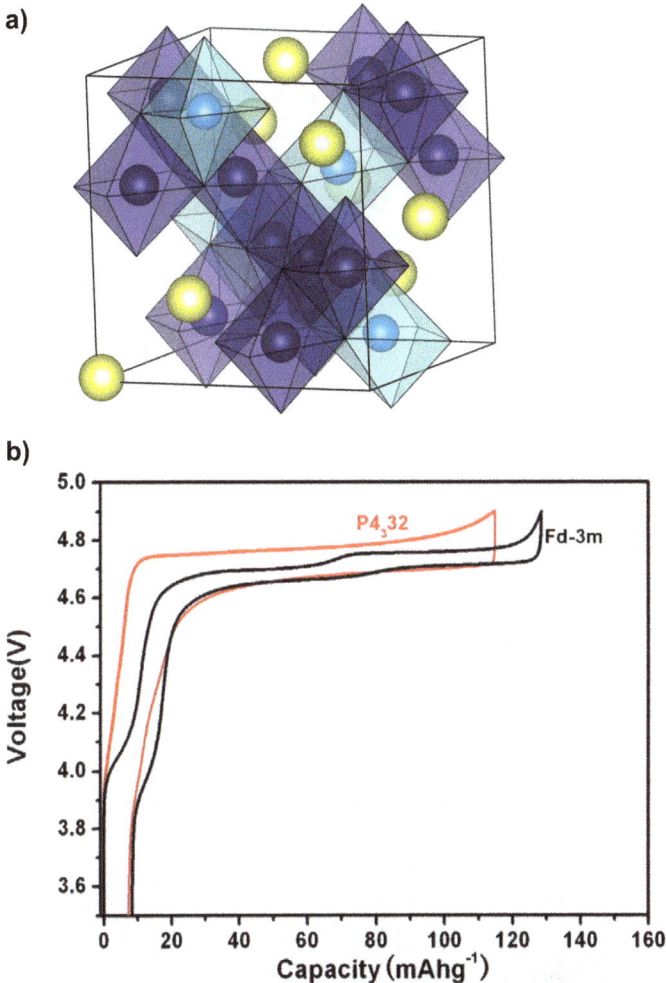

Fig. 7.16: a) Representation of the crystal structure of cation-ordered $LiNi_{1/2}Mn_{3/2}O_4$ [184] with $P4_332$ space-group symmetry. Ni, Mn, and Li are light blue, dark blue, and yellow spheres, respectively. Only MO_6 octahedra (M = Ni, Mn) are shown and oxygen ions omitted, for clarity. b) Typical galvanostatic voltage curve of cation ordered ($P4_332$ space-group symmetry) and disordered ($Fd\bar{3}m$) space-group symmetry $LiNi_{0.5}Mn_{1.5}O_4$. Adapted from [180], Copyright 2011, with permission from *Elsevier*.

neutron total scattering revealed the same short-range Ni/Mn ordering in both average long-range ordered and disordered forms, with disorder in the $LiNi_{0.5}Mn_{1.5}O_4$ with $Fd\bar{3}m$ space-group symmetry only apparent from the third or higher nearest neighbors (> 15 Å, Fig. 7.17) [185]. This order is a result of the strong tendency for Ni^{2+} to be surrounded only by Mn^{4+} first neighbors to lower electrostatic repulsions, not dissimilarly to previously discussed NMC materials and as supported by ab initio calculations

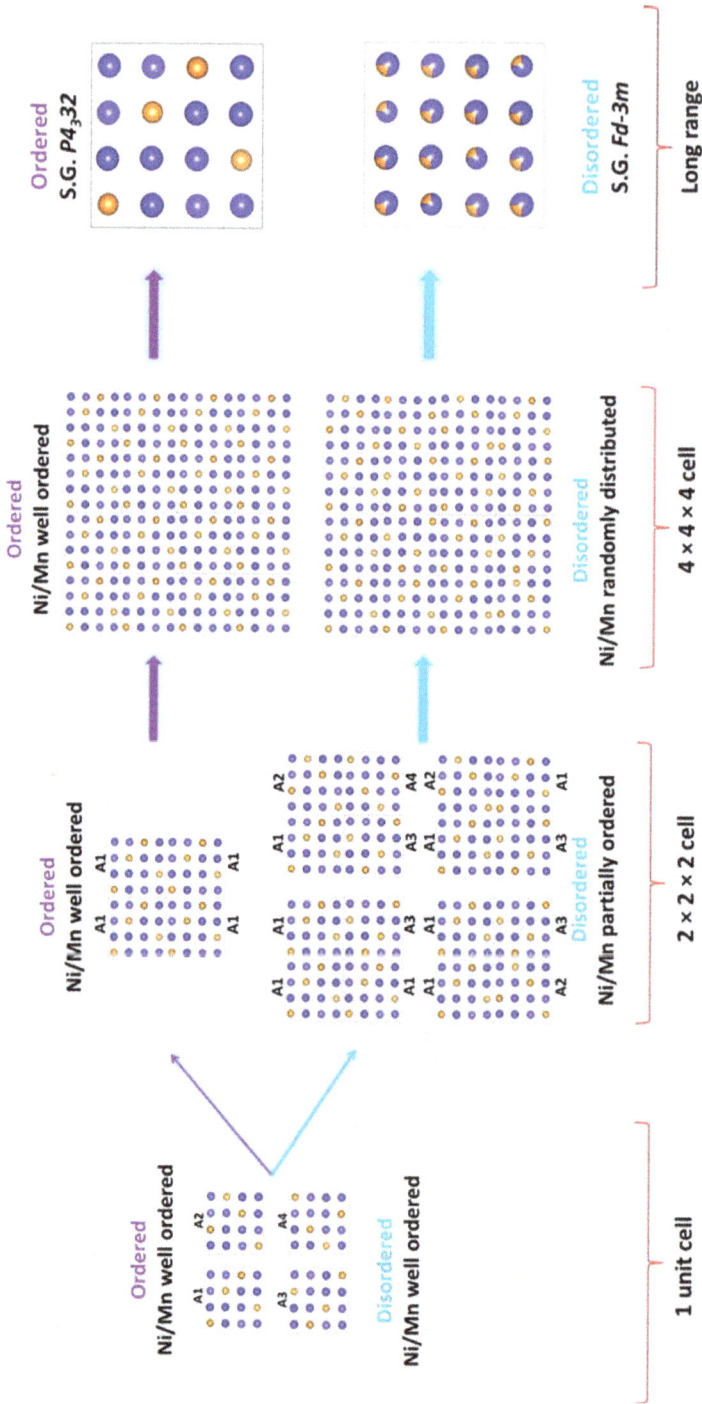

Fig. 7.17: Proposed short- and long-range order of Ni and Mn at 16d sites of TM disordered ($Fd\bar{3}m$ space-group) and ordered ($P4_{3}32$ space-group) LiNi$_{0.5}$Mn$_{1.5}$O$_{4}$. Reprinted with permission from [185]. Copyright 2016 American Chemical Society.

[186]. Liu et al. suggest that this often uncharacterized feature may explain the variable performance of seemingly similar LNMO materials. Theoretical calculations show a complex correlation between the Ni/Mn order and lithium intercalation mechanisms in LNMO, which may explain the different electrochemical behavior [186, 187].

The inactivity of Mn in $LiNi_{0.5}Mn_{1.5}O_4$ greatly reduces Mn dissolution compared to parent $LiMn_2O_4$; however a small amount of unstable Mn^{3+} is still commonly present as revealed by the plateau at 4 V and some Mn is often observed on the surface of the other electrode in full cell configurations [188]. Despite improvement of the cyclability of $LiNi_{0.5}Mn_{1.5}O_4$ in half-cells against Li, full batteries containing $LiNi_{0.5}Mn_{1.5}O_4$ against graphite are plagued by capacity fading and spontaneous discharge at high voltage [189, 190]. The reasons for the high-voltage instability are under debate, with dissolved Mn in the electrolyte thought to disturb the solid-electrolyte interphase at the graphite surface [190]. Excellent stability is obtained against an alternative counter electrode such as $Li_4Ti_5O_{12}$, at the cost of battery voltage [191]. Strategies such as doping, surface modification, separator choice, and solid-state electrolyte, show promising stability improvement [4, 191–195], giving this material strong potential for commercial application, which is said to have been used commercially in the electric car industry [169].

Lithium manganese spinels $LiMn_2O_4$ and their nickel counterpart $LiNi_{0.5}Mn_{1.5}O_4$ present enticing properties for battery applications, including a high chemical potential, excellent rate capability, and low cost/toxicity. The three-dimensional connectivity of tunnels in the spinel structure confers high mobility to Li^+ beneficial for high current applications. The cyclability of $LiMn_2O_4$ is hindered by Mn dissolution at high voltage, with this effect reduced in $LiNi_{0.5}Mn_{1.5}O_4$ where Ni is the active transition metal. In $LiNi_{0.5}Mn_{1.5}O_4$, complex ordering of Ni and Mn occurs at both short and long range, which seems to affect the electrochemical behavior; however, the relationship is poorly understood. Despite the noticeable improvement in stability compared to $LiMn_2O_4$, some dissolution of Mn persists which disturbs long-term cyclability against graphite counter electrodes. Several strategies have been explored to counteract this effect, making $LiNi_{0.5}Mn_{1.5}O_4$ a promising candidate for practical application.

7.2.6 Spinel-type materials: $Li_4Ti_5O_{12}$

$Li_4Ti_5O_{12}$ adopts a spinel-type structure as described for $LiMn_2O_4$. The AB_2O_4 equivalent formula is $Li^{8a}[Li_{1/3}Ti_{5/3}]^{16d}O_4$ where 1/8 of 8a tetrahedral sites are occupied by Li and 1/2 of 16d octahedral sites are shared by 1/3 Li and 5/3 Ti in the cubic unit cell with $Fd\bar{3}m$ space-group symmetry. No long-range order of Li and Ti is observed. TiO_6 octahedra are distorted (83 and 97° O-Ti-O angles) [196] due to second-order Jahn–Teller effects of Ti^{4+} (d^0). 8a tetrahedral sites are linked by empty octahedral 16c sites, forming a three-dimensional network of tunnels for Li migration.

The cycling of $Li_4Ti_5O_{12}$ was first reported in 1989 [197]. Unlike isostructural $LiMn_2O_4$, Li cannot be extracted from $Li_4Ti_5O_{12}$ due to the difficulty in further

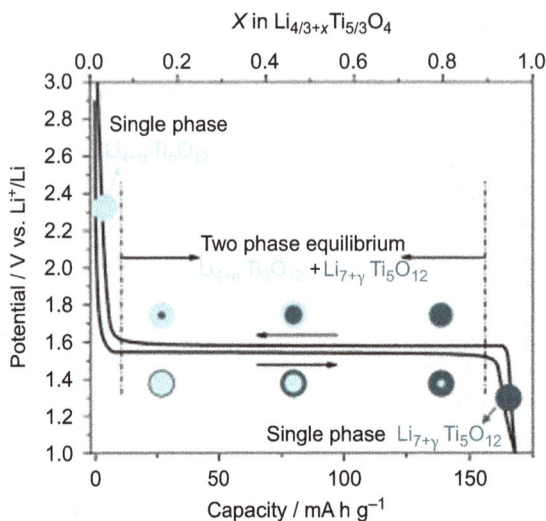

Fig. 7.18: Typical galvanostatic cycling curve of $Li_4Ti_5O_{12}$ against lithium metal. Reprinted from [9], Copyright 2015, with permission from *Elsevier*.

oxidizing Ti^{4+}; however insertion is possible thanks to vacant $8a$ and $16c$ interstitial sites of the spinel structure. The galvanostatic curve in the 1.0–2.0 V range vs. Li^+/Li shows a single plateau at 1.55 V (Fig. 7.18), characteristic of a two-phase transformation, with a specific capacity of 170 mAh/g which corresponds to the composition $Li_7Ti_5O_{12}$ for the lithiated end member. $Li_4Ti_5O_{12}$ and $Li_7Ti_5O_{12}$ have a similar volume, with only ~ 0.2% lattice volume change during cycling [198, 199]. The transformation between these structures is so-called "zero-strain" behavior, and used to explain the excellent reversibility [200] and remarkable long-term stability of $Li_4Ti_5O_{12}$ electrodes [201]. In $Li_7Ti_5O_{12}$, Li occupies $16c$ octahedral sites instead of $8a$ tetrahedral sites as in $Li_4Ti_5O_{12}$ (Fig. 7.19) [198]. The structure of $Li_7Ti_5O_{12}$ resembles that of the rock salt structure type, where oxygen ions form a face-centered cubic arrangement and all octahedral sites are occupied with cations. It is possible to insert further Li into vacant $8a$ crystallographic sites of $Li_7Ti_5O_{12}$, increasing capacity and lowering the potential to ~ 10 mV vs. Li^+/Li [202]. The transformation is reversible; however, the first cycle requires extra capacity for the formation of a solid electrolyte interphase at these potentials, as similar to graphite.

Ex situ X-ray diffraction suggests the presence of the two phases at equilibrium, noting that the almost perfect overlap of $Li_4Ti_5O_{12}$ and $Li_7Ti_5O_{12}$ reflections renders the distinction between the two phases difficult [198–200]. Consequently, structural characterization is usually performed with an averaged single-phase model. Some argue that the presence of two phases during cycling represents a metastable state [203], in contradiction to ab initio calculations suggesting a phase of intermediate composition such as "$Li_{5.5}Ti_5O_7$" is thermodynamically less stable than a mixture of

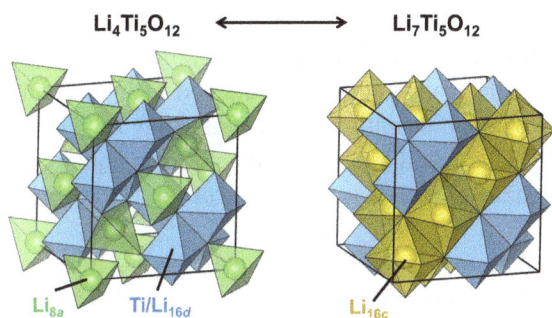

Fig. 7.19: Representation of the crystal structures of $Li_4Ti_5O_{12}$ and $Li_7Ti_5O_{12}$ [198]. The 8a, 16d, and 16c sites are in green, blue, and yellow, respectively.

$Li_4Ti_5O_{12}$ and $Li_7Ti_5O_{12}$ [199]. The presence of two phases at equilibrium was confirmed using transmission electron microscopy and electron energy loss spectroscopy, with phase boundaries parallel to the [110] or equivalent directions observed [204]. In operando neutron powder diffraction revealed a nonlinear lattice response to lithiation, showing that Li is initially inserted at 8a sites in the average structure, causing lattice expansion, and later at 16c sites as $Li_7Ti_5O_{12}$ forms, causing contraction [205]. Octahedral distortion was observed during this transition as driven by a repositioning of the oxygen atoms initiated by Li insertion at the 16c sites. The initial structure is not completely recovered after the first cycle where some 16c sites remain occupied by some lithium [205].

$Li_4Ti_5O_{12}$ electrodes show remarkably high rate capability, in contrast with the very low ionic conductivity measured for the as-made material [206], with Li diffusion being composition dependent [207]. A kinetically activated mechanism of intercalation at elevated currents has been suggested using combined in situ electron energy loss spectroscopy and density functional theory calculations [208]. At the boundary between $Li_4Ti_5O_{12}$ and $Li_7Ti_5O_{12}$, Li occupies both 8a tetrahedral and 16c octahedral sites, at a small energy cost as they share faces. Once this activation energy, on the order of 200 meV, is reached under high current, the energy landscape for Li diffusion becomes relatively flat thanks to migration facilitated by local distortions of neighboring O and Li, allowing fast diffusion of the boundary across the grain [208].

Combining excellent stability, high-rate capability, and appreciable specific capacity, there is considerable interest in using $Li_4Ti_5O_{12}$ as an electrode. Similar to graphite, $Li_4Ti_5O_{12}$ is nontoxic, relatively inexpensive and thermally stable. No formation cycle is necessary and no Li plating occurs at the electrochemical potential of $Li_7Ti_5O_{12}$. One of the main barriers to its wide adoption is the spontaneous formation of gas (H_2, CO_2) from electrolyte decomposition at the surface of $Li_4Ti_5O_{12}$ electrodes, particularly at elevated temperature, leading to safety issues [209, 210]. The process is not completely understood, but promising results have been obtained using surface coating or different electrolyte formulations [9, 211]. Although the increased electrochemical potential and reduced capacity of $Li_4Ti_5O_{12}$ leads to lower voltage batteries (2.5 V against NMC333) with lower capacity than graphite, there is considerable interest in this material which has already found a few commercial applications. The "zero-strain" property is also compatible with solid-state electrolytes, a desirable replacement for the currently used volatile, flammable, and toxic liquid electrolytes, where electrode volume changes often lead to the loss of contact against a rigid electrolyte [212].

The spinel $Li_4Ti_5O_{12}$ can be used as an electrode material in a lithium-ion battery. Upon lithiation, $Li_7Ti_5O_{12}$ is formed where lithium ions migrate from tetrahedral interstitial sites to octahedral sites within the tunnels of the spinel structure. A higher capacity can potentially be reached, at compositions $Li_xTi_5O_{12}$ with $x > 7$; however this is often not realized due to the instability associated with the low electrochemical potential of this phase. Diffraction data show a first-order transformation between $Li_4Ti_5O_{12}$ and $Li_7Ti_5O_{12}$, although in situ data suggest a more complex transformation, noting that X-ray diffraction analysis is difficult. A remarkable aspect of the transformation is the almost identical unit cell parameters of the two limit phases, resulting in excellent long-term cyclability. Combined with the high current rates enabled by the good mobility of lithium ions via interconnected tunnels of the spinel structure, there is considerable interest in using this material for practical applications despite the higher electrochemical potential, cost, and lower capacity compared to graphite.

7.2.7 Olivine-type materials: LiFePO$_4$

The report of the reversible electrochemical (de)intercalation of Li in $LiFePO_4$ in 1997 [16] took the battery community by surprise. This compound crystallizes with an olivine-type structure, which can be described as a distorted hexagonal close-packed arrangement of oxygen where half of the octahedral sites contain Fe or Li and one-eighth of tetrahedral sites contain P. The conventional unit cell is orthorhombic with *Pnma* space-group symmetry. Due to the strong covalent P–O bonds, the PO_4 tetrahedra can be regarded as PO_4^{3-} surrounding Li^+ and Fe^{2+}. Each FeO_6 octahedron shares four corners with others, forming Fe layers in the (b,c) plane, and shares two edges with PO_4 tetrahedra that link Fe layers in the a direction (Fig. 7.20). Li occupies octahedra next to the PO_4 "pillars," in-between the Fe layers. Due to the sharing of edges with small inflexible PO_4 tetrahedra, LiO_6 and FeO_6 octahedra are strongly distorted. Li movement is only possible along the b unit cell axis [213], resulting in highly anisotropic ionic conductivity, with Li pathways experimentally observable from scattering density maps obtained from neutron diffraction, Li being the only element with negative scattering length in $LiFePO_4$ [214]. Despite the presence of vacant crystallographic sites, the highest attainable Li concentration is $LiFePO_4$, due to the difficult further reduction of Fe^{2+} without major structural change.

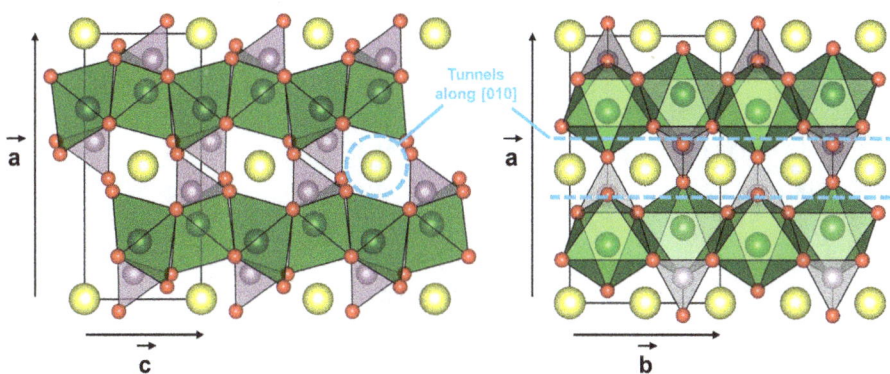

Fig. 7.20: Crystal structure of LiFePO$_4$ [215], Li, Fe, P, and O represented as yellow, green, purple, and red spheres, respectively.

The reduced connectivity between FeO$_6$ octahedra where only corners are shared results in poor electrical conductivity, and carbon additives are used to improve this. This is regarded as one of the reasons for the relatively late demonstration of the electrochemistry of LiFePO$_4$, which was achieved by surface carbon coating [16]. In optimized preparations, carbon is often added early during the synthesis of LiFePO$_4$ to promote intimate mixing but also to induce a slightly reducing environment that stabilizes Fe^{2+}, less stable than Fe^{3+} in air [8]. Although first studies report specific capacities of only 50–80% of theoretical capacity [16], optimized C-LiFePO$_4$ electrodes exhibit capacities close to the theoretical 170 mAh/g in the 2.0–4.0 V range, corresponding to the complete removal of Li from LiFePO$_4$ to FePO$_4$ [8]. Despite reduced specific capacity and voltage compared to other electrode materials, C-LiFePO$_4$ shows excellent reversibility, high current rate capability and long-term stability particularly at elevated temperatures, making this material very well suited for electromotive applications [169].

The electrochemical curve of the material has a single plateau at 3.45 V versus Li$^+$/Li, typical of a two-phase transformation (Fig. 7.21), and end-members exhibit small solid-solution domains, that is, Li$_{1-y}$FePO$_4$ and Li$_z$FePO$_4$ with small y and z [216]. The degree of miscibility is correlated to particle size, with a smaller gap (larger y and z) in smaller particles [217, 218]; however, a significant miscibility gap subsists even at 20 nm particle size, in agreement with X-ray diffraction and transmission electron microscopy data that show a majority of either fully lithiated or delithiated particles at equilibrium [219].

Despite the closely related structures of LiFePO$_4$ and FePO$_4$, with identical connectivity of FeO$_6$ octahedra and PO$_4$ tetrahedra, there is a substantial lattice mismatch between the two, with a volume change of ~ 6.6% and the smallest misfit in the (b,c) plane [222]. The induced strain results in a coherent but unstable boundary parallel to (b,c) that propagates quickly across the grain in the a direction, analogously to falling

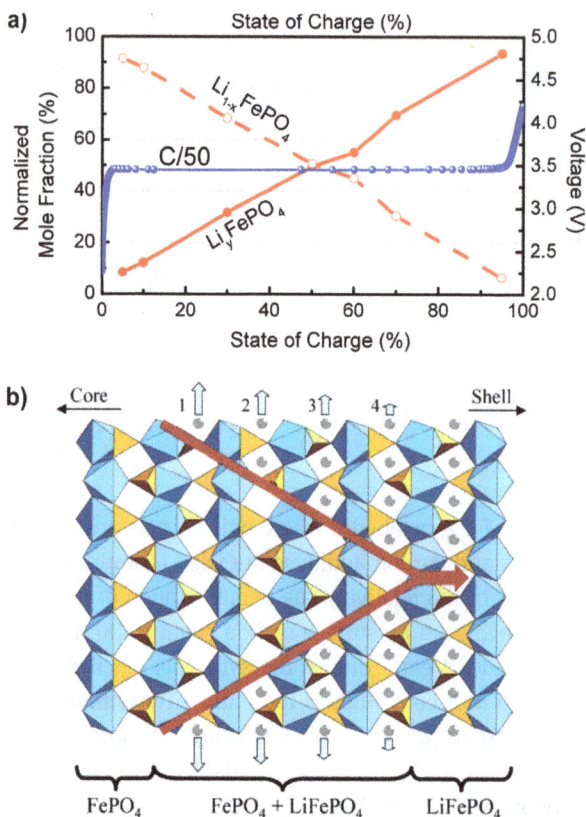

Fig. 7.21: a) Typical voltage curve during galvanostatic discharge of LiFePO$_4$ and phase fractions of the end-members obtained from in situ X-ray diffraction. Adapted with permission from [220]. Copyright 2008 *American Chemical Society*. b) Illustration of the domino cascade mechanism during discharge, with the front between FePO$_4$ and LiFePO$_4$ in red moving in the a direction. The lattice mismatch is not represented. Arrows symbolize Li extraction in the b direction. Longer arrows correspond to more energetically favored extractions. Adapted with permission from [221]. Copyright 2006 *American Chemical Society*.

dominos [223]. This so-called domino-cascade mechanism is facilitated by the fast migration of lithium in the b direction, with both ionic and electronic conductivities thought to be locally exacerbated by deformations at the interface [219, 223]. This mechanism explains the high rate capability of LiFePO$_4$ in contradiction to the low conductivity of the pristine material. These considerations led to work investigating the influence of particle size and shape on performance. Larger boundaries, particularly in the (b,c) plane, in larger particles experience higher strain and should travel faster; however, the increased strain may lead to particle cracking over repeated cycling [219], which was experimentally confirmed by scanning and transmission electron microscopy of > 200 nm particles [224, 225]. The easier relaxation of structural constraints in smaller grains is consistent with a smaller miscibility gap and improved cyclability [218, 220]. There may be a critical size below which the domino-cascade is not the driving mechanism [220] and a critical current above which phase separation is kinetically suppressed [226, 227]. Research aimed at controlling $LiFePO_4$ particle morphology in conjunction with optimization of the carbon-additive microstructure has produced C-$LiFePO_4$ electrodes with remarkable performance [8, 228–230].

$LiFePO_4$ with an olivine structure was the first material with so-called polyanion units (PO_4^{3-} in this case) to be used as an electrode in a lithium-ion battery. It is possible to delithiate the material completely, with the structure of $FePO_4$ similar to that of $LiFePO_4$. Li^+ are mobile across tunnels along one direction of the olivine structure, resulting in a highly anisotropic and limited ionic conductivity. Despite the limited conductivity, the transformation between the two phases proceeds remarkably fast, in a process called domino-cascade, as a result of the lattice mismatch while maintaining coherence between the two extreme phases. As a result of this mechanism, the phase transformation and electrochemical behavior are sensitive to the size and shape of particles in the electrode. The discovery of $LiFePO_4$ has sparked the study of a vast array of polyanionic materials presenting similar structural features, such as $LiMnPO_4$, $LiFeP_2O_7$, $LiFeSiO_4F$, or $LiVPO_4F$, and the reader is invited to read the review by Masquelier and Croguennec for more details [8].

7.3 Concluding remarks

The crystal structure of materials has a major influence on properties, and crystalline battery electrode materials, that reversibly accommodate charge balancing ions and electronic structure changes, are an important class of materials where this is very well demonstrated. Broadly speaking, several crystal structure types have emerged as useful for application as electrode materials, notably materials with a layered structure, as well as spinel- and olivine-type structures feature predominantly. It may be tempting to identify specific structural features that are beneficial to battery performance. Although layered materials such as layered oxides and graphite have been used extensively as electrodes, for historical reasons perhaps, research on spinels and olivines show that a layered structure is not a necessary prerequisite. An essential feature is the presence of interstitial sites that can accommodate lithium ions, with those sites interconnected at long-range, forming tunnels or layers. A high connectivity is indeed necessary to enable the migration of lithium from the surface to the bulk of crystallites. In some cases, it is possible to predict the performance of materials using theoretical calculations, and a few electrode materials have been successfully discovered this way [231, 232].

Importantly, these functional materials must undergo change to accommodate the continuously varying composition during their use, a remarkable material demand where the performance of the whole battery is influenced heavily by the change in properties of the electrode material during this process. These structural changes generate stress that can lead to particle cracking or chemical instability, a major issue in some intercalation materials. These problems have also prevented the use of conversion/alloying materials, such as Si or FeF_3, despite their high capacity [169, 233]. A stable framework seems to be an important feature of successful electrode materials used in rechargeable batteries, with research trending toward $L_4Ti_5O_{12}$ or $LiFePO_4$ in light of excellent stability during cycling. Importantly, structural modifications during (de)lithiation can often only be determined experimentally, using in situ diffraction for example. Understanding the link between the material crystallography and properties such as ionic conductivity and stress/strain of the material during use is key to overcoming existing limitations of electrode materials and will drive the strategic chemical engineering of new electrode materials with superior function to meet the increasing demands of the technologies powered by rechargeable batteries.

References

[1]   Yoshino, A. Development of lithium-ion battery. Mol. Cryst. Liq. Cryst. 2000, 340, 425–429.
[2]   Armand, M., Tarascon, J.-M. Building better batteries. Nature 2008, 451, 652–657.
[3]   Xu, K. Nonaqueous liquid electrolytes for lithium-based rechargeable batteries. Chem. Rev. 2004, 104(10), 4303–4418.
[4]   Manthiram, A., Chemelewski, K., Lee, E.-S. A perspective on the high-voltage $LiMn_{1.5}Ni_{0.5}O_4$ spinel cathode for lithium-ion batteries. Energy Environ. Sci. 2014, 7, 1339.
[5]   Noh, H.-J., Youn, S., Yoon, C.S., Sun, Y.-K. Comparison of the structural and electrochemical properties of layered $Li[Ni_xCo_yMn_z]O_2$ (x = 1/3, 0.5, 0.6, 0.7, 0.8 and 0.85) cathode material for lithium-ion batteries. J. Power Sources 2013, 233, 121–130.
[6]   Li, W., Liu, X., Celio, H., Smith, P., Dolocan, A., Chi, M., Manthiram, A. Mn versus Al in layered oxide cathodes in lithium-ion batteries: A comprehensive evaluation on long-term cyclability. Adv. Energy Mater. 2018, 8, 1703154.
[7]   Yu, H., Zhou, H. High-energy cathode materials (Li_2MnO_3–$LiMO_2$) for lithium-ion batteries. J. Phys. Chem. 2013, 4, 1268–1280.
[8]   Masquelier, C., Croguennec, L. Polyanionic (phosphates, silicates, sulfates) frameworks as electrode materials for rechargeable Li (or Na) batteries. Chem. Rev. 2013, 113, 6552–6591.
[9]   Zhao, B., Ran, R., Liu, M., Zongping, S. A comprehensive review of $Li_4Ti_5O_{12}$-based electrodes for lithium-ion batteries: The latest advancements and future perspectives. Mater. Sci. Eng., R 2015, 98, 1–71.
[10]   Ridgway, P., Honghe Zheng, A.F., Bello, X.S., Xun, S., Chong, J., Battaglia, V. Comparison of cycling performance of lithium ion cell anode graphites. J. Electrochem. Soc. 2012, 159(5), A520–A524.
[11]   Zuo, X., Zhu, J., Muller-Buschbaum, P., Cheng, Y.-J. Silicon based lithium-ion battery anodes: A chronicle perspective review. Nano Energy 2017, 31, 113–143.
[12]   Mizushima, K., Jones, P.C., Wiseman, P.J., Goodenough, J.B. Li_xCoO_2 (0<x<1): A new cathode material for batteries of high energy density. Mater Res. Bull. 1980, 15(6), 783–789.

[13]   Lee, K.K., Yoon, W.S., Kim, K.B., Lee, K.Y., Hong, S.T. Characterization of $LiNi_{0.85}Co_{0.10}M_{0.05}O_2$
       (M = Al, Fe) as a cathode material for lithium secondary batteries. J. Power Sources 2001,
       97–98, 308–312.
[14]   Yabuuchi, N., Ohzuku, T. Novel lithium insertion material of $LiCo_{1/3}Ni_{1/3}Mn_{1/3}O_2$ for advanced
       lithium-ion batteries. J. Power Sources 2003, 119–121, 171–174.
[15]   Thackeray, M.M., Johnson, P.J., De Picciotto, L.A., Bruce, P.G., Goodenough,
       J.B. Electrochemical extraction of lithium from $LiMn_2O_4$. Mater Res. Bull. 1984, 19(2),
       179–187.
[16]   Padhi, A.K., Nanjundaswamy, K.S., Goodenough, J.B. Phospho-olivines as positive-electrode
       materials for rechargeable lithium batteries. J. Electrochem. Soc. 1997a, 144(4), 1188–1194.
[17]   Padhi, A.K., Nanjundaswamy, K.S., Masquelier, C., Okada, S., Goodenough, J.B. Effect of
       structure on the Fe^{3+}/Fe^{2+} redox couple in iron phosphates. J. Electrochem. Soc. 1997b, 144,
       1609.
[18]   Fergus, J.W. Recent developments in cathode materials for lithium ion batteries. J. Power
       Sources 2010, 195(4), 939–954.
[19]   Nitta, N., Feixiang, W., Lee, J.T., Yushin, G. Li-Ion battery materials: Present and future. Mater.
       Today 2015, 18(5), 252–264.
[20]   Xu, B., Qian, D., Wang, Z., Meng, Y.S. Recent progress in cathode materials research for
       advanced lithium ion batteries. Mater. Sci. Eng., R 2012, 73(5–6), 51–65.
[21]   Bak, S.-M., Shadike, Z., Lin, R., Xiqian, Y., Yang, X.-Q. In situ/operando synchrotron-based
       X-ray techniques for lithium-ion battery research. NPG Asia Mater. 2018, 10, 563–580.
[22]   Liang, G., Didier, C., Guo, Z., Pang, W.K., Peterson, V.K. Understanding rechargeable battery
       function using in operando neutron powder diffraction. Adv. Mater. 2019, 32(18), 1904528.
[23]   Sharma, N., Pang, W.K., Guo, Z., Peterson, V.K. In situ powder diffraction studies of electrode
       materials in rechargeable batteries. ChemSusChem 2015, 8, 2826–2853.
[24]   Xia, M., Liu, T., Zheng, R., Cheng, X., Zhu, H., Haoxiang, Y., Shui, M., Shu, J. Lab-scale in situ
       X-Ray diffraction technique for different battery systems: Designs applications, and
       perspectives. Small Methods 2019, 3(7), 1900119.
[25]   Blanc, F., Leskes, M., Grey, C.P. In situ solid-state NMR spectroscopy of electrochemical cells:
       Batteries, supercapacitors, and fuel cells. Acc. Chem. Res. 2013, 46(9), 1952–1963.
[26]   Chapman, K.W. Emerging operando and X-ray pair distribution function methods for energy
       materials development. MRS Bull. 2016, 41(3), 231–240.
[27]   Trease, N.M., Zhou, L., Chang, H.J., Zhu, B.Y., Grey, C.P. In situ NMR of lithium ion batteries:
       Bulk susceptibility effects and practical considerations. Solid State Nucl. Magn. Reson. 2012,
       42, 62–70.
[28]   Goodenough, J.B., Mizuchima, K. 1981. Electrochemical cell with new fast ion conductors US.
       Patent US4302518A.
[29]   Tarascon, J.M., Armand, M. Issues and challenges facing rechargeable lithium batteries.
       Nature 2001, 414, 359–367.
[30]   Trucano, P., Chen, R. Structure of graphite by neutron diffraction. Nature 1975, 258, 136–137.
[31]   Boehm, H.-P., Setton, R., Stumpp, E. Nomenclature and terminology of graphite intercalation
       compounds (IUPAC Recommendations 1994). Pure Appl. Chem. 1994, 66(9), 1893–1901.
[32]   McRae, E., Billaud, D., Mareche, J.-F., Herold, A. Basal plane resistivity of alkali metal-
       graphite compounds. Physica. B 1980, 99, 489–493.
[33]   Dresselhaus, M.S., Dresselhaus, G. Intercalation compounds of graphite. Adv. Phys. 2010,
       51(1), 1–186.
[34]   Dahn, J.R. Phase diagram of Li_xC_6. Phys. Rev. B 1991, 44(17), 9170–9177.

[35] Ohzuku, T., Iwakoshi, Y., Sawai, K. Formation of lithium-graphite intercalation compounds in nonaqueous electrolytes and their application as a negative electrode for a lithium ion (Shuttlecock) Cell. J. Electrochem. Soc. 1993, 140(9), 2490–2498.

[36] Billaud, D., Henry, F.-X., Lelaurain, M., Willmann, P. Revisited structures of dense and dilute stage II lithium-graphite intercalation compounds. J. Phys. Chem. Solids 1996, 57(6–8), 775–781.

[37] Billaud, D., Henry, F.X. Structural studies of the stage III lithium-graphite intercalation compound. Solid State Commun. 2002, 124, 299–304.

[38] Didier, C., Pang, W.K., Guo, Z., Schmidt, S., Peterson, V.K. Phase evolution and intermittent disorder in electrochemically lithiated graphite determined using in operando neutron diffraction. Chem. Mater. 2020, 32(6), 2518–2531.

[39] Schweidler, S., De Biasi, L., Schiele, A., Hartmann, P., Brezesinski, T., Janek, J. Volume changes of graphite anodes revisited: A combined operando X-ray diffraction and in situ pressure analysis study. J. Phys. Chem. C 2018, 122(16), 8829–8835.

[40] Guerard, A., Herold, D. Intercalation of lithium into graphite and other carbons. Carbon 1975, 13, 337–345.

[41] Billaud, D., McRae, E., Mareche, J.-F., Herold, A. New results concerning the lithium-pyrographite system. Synth. Met. 1981, 3, 21–26.

[42] Konar, S., Häusserman, U., Svensson, G. Intercalation compounds from LiH and graphite: Relative stability of metastable stages and thermodynamic stability of dilute stage I_d. Chem. Mater. 2015, 27, 2566–2575.

[43] Hazrati, E., De Wijs, G.A., Brocks, G. Li intercalation in graphite: A Van der Waals density functional study. Phys. Rev. B 2014, 90, 155448.

[44] Imai, Y., Watanabe, A. Energetic evaluation of possible stacking structures of Li-intercalation in graphite using a first-principle pseudopotential calculation. J. Alloys Compd. 2007, 439, 258–267.

[45] Persson, K., Hinuma, Y., Meng, Y.S., Van Der Ven, A., Ceder, G. Thermodynamic and kinetic properties of the Li-graphite system from first-principles calculations. Phys. Rev. B 2010, 82, 125416.

[46] Matsunaga, T., Takagi, S., Shimoda, K., Okazaki, K.-I., Ishikawa, Y., Yonemura, M., Ukyo, Y., Fukunaga, T., Matsubara, E. Comprehensive elucidation of crystal structures of lithium-intercalated graphite. Carbon 2018, 142, 513–517.

[47] Senyshyn, A., Dolotko, O., Muhlbauer, M.J., Nikolowski, K., Fuess, H., Ehrenberg, H. Lithium intercalation into graphitic carbons revisited: Experimental evidence for twisted bilayer behaviour. J. Electrochem. Soc. 2013, 160(5), A3198–A3205.

[48] Daumas, N., Herold, A. Sur les relations entre la notion de stade et les mecanismes reactionnels dans les composes d'insertion du graphite. On relations between the notion of stage and reaction mechanisms in graphite insertion compounds. Comptes Rendus de l'Academie des Sciences de Paris C 1969, 268, 373–375.

[49] Chandesris, M., Caliste, D., Jamet, D., Pochet, P. Thermodynamics and related kinetics of staging in intercalation compounds. J. Phys. Chem. C 2019, 123, 23711–23720.

[50] Thomas, J.M., Millword, G.R., Schlogl, R.F., Boehm, H.P. Direct imaging of a graphite intercalate: Evidence of interpenetration of 'stages' in graphite: Ferric chloride. Mater Res. Bull. 1980, 15(5), 671–676.

[51] Timp, G., Dresselhaus, M.S. The ultramicrostructure of commensurate graphite intercalation compounds. J. Phys. C: Solid State Phys. 1984, 17, 2641–2645.

[52] Hawrylak, P., Subbaswamy, K.R. Kinetic model of stage transformation and intercalation in graphite. Phys. Rev. Lett. 1984, 53(22), 2098–2101.

[53] Krishnan, S., Brenet, G., Nachado-Charry, E., Caliste, D., Genovese, L., Deutsch, T., Pochet, P. Revisiting the domain model for lithium intercalated graphite. Appl. Phys. Lett. 2013, 103, 251904.

[54] Safran, S.A., Hamann, D.R. Long-range elastic interactions and staging in graphite intercalation compounds. Phys. Rev. Lett. 1979, 42(21), 1410–1413.

[55] Smith, R.B., Khoo, E., Martin Bazant, Z. Intercalation kinetics on multiphase-layered materials. J. Phys. Chem. C 2017, 121, 12505–12523.

[56] Taminato, S., Yonemura, M., Shiotani, S., Kamiyama, T., Torii, S., Nagao, M., Ishikawa, Y., Mori, K., Fukunaga, T., Onodera, Y., Naka, T., Morishima, M., Ukyo, Y., Adipranoto, D.S., Arai, H., Uchimoto, Y., Ogumi, Z., Suzuki, K., Hirayama, M., Kanno, R. Real-time observations of lithium battery reactions – operando neutron diffraction analysis during practical operation. Sci. Rep. 2016, 6, 28843.

[57] Zheng, T., Dahn, J.R. Effect of turbostratic disorder on the staging phase diagram of lithium-intercalated graphitic carbon hosts. Phys. Rev. B 1996, 53(6), 3061–3071.

[58] Zinth, V., Von Lüders, C., Hofmann, M., Hattendorff, J., Buchberger, I., Erhard, S., Rebelo-Kornmeier, J., Jossen, A., Gilles, R. Lithium plating in lithium-ion batteries at sub-ambient temperatures investigated by in situ neutron diffraction. J. Power Sources 2014, 271, 152–159.

[59] Hess, M., Novak, P. Shrinking annuli mechanism and stage-dependent rate capability of thin-layer graphite electrodes for lithium-ion batteries. Electrochim. Acta 2013, 106, 149–158.

[60] Takami, N., Satoh, A., Hara, M., Ohsaki, T. Structural and kinetic characterization of lithium intercalation into carbon anodes for secondary lithium batteries. J. Electrochem. Soc. 1995, 142(2), 371.

[61] Dokko, K., Nakata, N., Suzuki, Y., Kanamura, K. High-rate lithium deintercalation from lithiated graphite single-particle electrode. J. Phys. Chem. C 2010, 114, 8646–8650.

[62] Harris, S.J., Timmons, A., Baker, D.R., Monroe, C. Direct in situ measurements of Li transport in Li-ion battery negative electrodes. Chem. Phys. Lett. 2010, 485, 265–274.

[63] Buqa, H., Wursig, A., Goers, D., Hardwick, L.J., Holzapfel, M., Novak, P., Krumeich, F., Spahr, M.E. Behaviour of highly crystalline graphites in lithium-ion cells with propylene carbonate containing electrolytes. J. Power Sources 2005, 146(1–2), 134–141.

[64] Buqa, H., Wursig, A., Vetter, J., Spahr, M.E., Krumeich, F., Novak, P. SEI film formation on highly crystalline graphitic materials in lithium-ion batteries. J. Power Sources 2006, 153(2), 385–390.

[65] An, S.J., Jianlin, L., Daniel, C., Debasish, M., Nagpure, S., Wood, D.L. The state of understanding of the lithium-ion-battery graphite solid electrolyte interphase (SEI) and its relationship to formation cycling. Carbon 2016, 105, 52–76.

[66] Wood, D.L., Jianlin, L., An, S.J. Formation challenges of lithium-ion battery manufacturing. Joule 2019, 3, 1–5.

[67] Holzapfel, M., Haak, C., Ott, A. Lithium conductors of the system $LiCo_{(1-x)}Fe_{(x)}O_2$, preparation and structural investigation. J. Solid State Chem. 2001, 156, 470–479.

[68] Ohzuku, T., Ueda, A. Solid-state redox reactions of $LiCoO_2$ ($R\bar{3}m$) for 4 volt secondary lithium cells. J. Electrochem. Soc. 1994, 141(11), 2972–2977.

[69] Reimers, J.N., Dahn, J.R. Electrochemical and in situ X-ray diffraction studies of lithium intercalation in Li_xCoO_2. J. Electrochem. Soc. 1992, 139(8), 2091–2097.

[70] Menetrier, M., Saadoune, I., Levasseur, S., Delmas, C. The insulator–metal transition upon lithium deintercalation from $LiCoO_2$: Electronic properties and ^7Li NMR study. J Mater Chem 1999, 9, 1135–1140.

[71] Nishizawa, M., Yamamura, S., Itoh, T., Uchida, I. Irreversible conductivity change of $Li_{1-x}CoO_2$ on electrochemical lithium insertion/extraction, desirable for battery applications. Chem. Commun. 1998, 16, 1631–1632.

[72] Marianetti, C.A., Kotliar, G., Ceder, G. A first-order Mott transition in Li_xCoO_2. Nat. Mater. 2004, 3, 627–631.

[73] Shao-Horn, Y., Levasseur, S., Weill, F., Delmas, C. Probing lithium and vacancy ordering in O3 layered Li_xCoO_2 ($x \approx 0.5$). An electron diffraction study. J. Electrochem. Soc. 2003, 150(3), A366–A373.

[74] Takahashi, Y., Kijima, N., Tokiwa, K., Watanabe, T., Akimoto, J. Single-crystal synthesis, structure refinement and electrical properties of $Li_{0.5}CoO_2$. J. Phys.: Condens. Matter 2007, 19, 436202.

[75] Motohashi, T., Katsumata, Y., Ono, T., Kanno, R., Karppinen, M., Yamauchi, H. Synthesis and properties of CoO_2, the x = 0 end member of the Li_xCoO_2 and Na_xCoO_2 systems. Chem. Mater. 2007, 19, 5063–5066.

[76] Tarascon, J.M., Vaughan, G., Chabre, Y., Seguin, L., Anne, M., Strobel, P., Amatucci, G.G. In situ structural and electrochemical study of $Ni_{1-x}Co_xO_2$ metastable oxides prepared by soft chemistry. J. Solid State Chem. 1999, 147, 410–420.

[77] Amatucci, G.G., Tarascon, J.M., Klein, L.C. CoO_2, The end member of the Li_xCoO_2 solid solution. J. Electrochem. Soc. 1996, 143(3), 1114–1123.

[78] Seong, W.M., Yoon, K., Lee, M.H., Jung, S.-K., Kang, K. Unveiling the intrinsic cycle reversibility of a $LiCoO_2$ electrode at 4.8V cutoff voltage through subtractive surface modification for lithium-ion batteries. Nano Lett. 2018, 19, 29–37.

[79] Bianchini, M., Roca-Ayats, M., Hartmann, P., Brezesinski, T., Janek, J. There and back again – The journey of $LiNiO_2$ as a cathode active material. Angewandte Chemie International Edition. 2018, 58(31), 10434–10458.

[80] Dahn, J.R., Von Sacken, U., Michal, C.A. Structure and electrochemistry of $Li_{1\pm y}NiO_2$ and a new Li_2NiO_2 phase with the $Ni(OH)_2$ structure. Solid State Ionics 1990, 44(1–2), 87–97.

[81] Shannon, R.D. Revised effective ionic radii and systematic studies of interatomic distances in halides and chalcogenides. Acta Crystallographica A 1976, 32, 751.

[82] Delmas, C.J., Peres, P., Alain Rougier, A., Weill, D.F., Chadwick, A., Broussely, M., Perton, F., Biensan, P., Willmann, P. On the behavior of the Li_xNiO_2 system: An electrochemical and structural overview. J. Power Sources 1997, 68(1), 120–125.

[83] Rougier, A., Gravereau, P., Delmas, C. Optimization of the composition of the $Li_{1-z}Ni_{1+z}O_2$ electrode materials: Structural, magnetic, and electrochemical studies. J. Electrochem. Soc. 1996, 143(4), 1169–1175.

[84] Yoon, C.S., Jun, D.-W., Myung, S.-T., Sun, Y.-K. Structural stability of $LiNiO_2$ cycled above 4.2 V. ACS Energy Lett. 2017, 2(5), 1150–1155.

[85] Delmas, C., Saadoune, I., Rougier, A. The cycling properties of the lithium nickel cobalt oxide ($Li_xNi_{1-y}Co_yO_2$) electrode. J. Power Sources 1993, 44, 595–602.

[86] MacNeil, D.D., Lu, Z., Dahn, J.R. Structure and electrochemistry of $Li[Ni_xCo_{1-2x}Mn_x]O_2$ ($0 \leq x \leq 1/2$). J. Electrochem. Soc. 2002, 149(10), A1332–A1336.

[87] Rossen, E., Jones, C.D.W., Dahn, J.R. Structure and electrochemistry of $Li_xMn_yNi_{1-y}O_2$. Solid State Ionics 1992, 57(3–4), 311–318.

[88] Saadoune, I., Delmas, C. $LiNi_{1-y}Co_yO_2$ positive electrode materials: Relationships between the structure, physical properties and electrochemical behavior. J. Mater. Chem. 1996, 6(2), 193–199.

[89] Myung, S.-T., Maglia, F., Park, K.-J., Yoon, C.S., Lamp, P., Kim, S.-J., Sun, Y.-K. Nickel-rich layered cathode materials for automotive lithium-ion batteries: Achievements and perspectives. ACS Energy Lett. 2017, 2(1), 196–223.

[90] Katana, N.J., Chernova, N.S., Ma, M., Mamak, M., Zavalij, P.Y., Wittingham, M.S. The synthesis, characterization and electrochemical behavior of the layered $LiNi_{0.4}Mn_{0.4}Co_{0.2}O_2$ compound. J. Mater. Chem. 2003, 14, 214–220.

[91] Breger, J., Nicolas Dupre, P.J., Chupas, P.L., Lee, T.P., Parise, J.B., Grey, C.P. Short- and long-range order in the positive electrode material, $Li(NiMn)_{0.5}O_2$: A Joint X-ray and neutron diffraction, pair distribution function analysis and nmr study. J. Am. Chem. Soc. 2005, 127(20), 7529–7537.

[92] Yoon, W.-S., Steven Iannopollo, C.P., Grey, D.C., Gorman, J., Reed, J. Local structure and cation ordering in O3 lithium nickel manganese oxides with stoichiometry $Li[Ni_xMn_{(2-x)/3}Li_{(1-2x)/3}]O_2$: NMR studies and first principles calculations Electrochem. Solid-State Lett. 2004, 7(7), A167–A171.

[93] Zeng, D., Cabana, J., Breger, J., Yoon, W.-S., Grey, C.P. Cation ordering in $Li[Ni_xMn_xCo_{(1-2x)}]O_2$-layered cathode materials: a nuclear magnetic resonance (NMR), pair distribution function, X-ray absorption spectroscopy, and electrochemical study. Chem. Mater. 2007, 19, 6277–6289.

[94] Ohzuku, T., Makimura, Y. Layered lithium insertion material of $LiCo_{1/3}Ni_{1/3}Mn_{1/3}O_2$ for lithium-ion batteries. Chem. Lett. 2001, 7, 642.

[95] Liu, L., Meicheng, L., Chu, L., Jiang, B., Ruoxu, L., Xiaopei, Z., Cao, G. Layered ternary metal oxides: Performance degradation mechanisms as cathodes, and design strategies for high-performance batteries. Prog. Mater. Sci. 2020, 111, 100655.

[96] Wang, X., Yuan-Li, D., Deng, Y.-P., Chen, Z. Ni-rich/Co-poor layered cathode for automotive Li-ion batteries: promises and challenges. Adv. Energy Mater. 2020, 10, 1903864.

[97] Hirano, A., Kanno, R., Kawamoto, Y., Takeda, Y., Yamaura, K., Takano, M., Ohyama, K., Ohashi, M., Yamaguchi, Y. Relationship between non-stoichiometry and physical properties in $LiNiO_2$. Solid State Ionics 1995, 78(1–2), 123–131.

[98] Ma, M., Chernova, N.A., Toby, B.H., Zavalij, P.Y., Stanley Wittingham, M. Structural and electrochemical behavior of $LiMn_{0.4}Ni_{0.4}Co_{0.2}O_2$. J. Power Sources 2006, 165, 517–334.

[99] Bak, S.-M., Enyuan, H., Zhou, Y.-N., Xiqian, Y., Senanayake, S.D., Cho, S.-J., Kim, K.-B. Structural changes and thermal stability of charged $LiNi_xMn_yCo_zO_2$ cathode materials studied by combined in situ time-resolved XRD and mass spectroscopy. ACS Appl. Mater. Interfaces 2014, 6(24), 22594–22601.

[100] Belharouak, I., Wenquan, L., Vissers, D., Amine, K. Safety characteristics of $Li(Ni_{0.8}Co_{0.15}Al_{0.05})O_2$ and $Li(Ni_{1/3}Co_{1/3}Mn_{1/3})O_2$. Electrochem. commun. 2006, 8, 329–335.

[101] Ma, L., Nie, M., Xia, J., Dahn, J.R. A systematic study on the reactivity of different grades of charged $Li[Ni_xMn_yCo_z]O_2$ with electrolyte at elevated temperatures using accelerating rate calorimetry. J. Power Sources 2016, 327, 145–150.

[102] Kim, J., Lee, H., Cha, H., Yoon, M., Park, M., Cho, J. Prospect and reality of Ni-rich cathode for commercialization. Adv. Energy Mater. 2017, 8(6), 1702028.

[103] Liu, W., Pilgun, O., Liu, X., Lee, M.-J., Cho, W., Chae, S., Kim, Y., Cho, J. Nickel-rich layered lithium transitional-metal oxide for high-energy lithium-ion batteries. Angew. Chem. Int. Ed. 2015, 54, 4440–4458.

[104] Zhang, S.S. Problems and their origins of Ni-rich layered oxide cathode materials. Energy Storage Mater. 2020, 24, 247–254.

[105] Matsunaga, T., Komatsu, H., Shimoda, K., Minato, T., Yonemura, M., Kamiyama, T., Kobayashi, S., Kato, T., Hirayama, T., Ikuhara, Y., Arai, H., Ukyo, Y., Uchimoto, Y., Ogumi, Z. Structural understanding of superior battery properties of partially Ni-doped Li_2MnO_3 as cathode material. J. Phys. Chem. 2016, 7, 2063–2067.

[106] Robertson, A.D., Bruce, P.G. Mechanism of electrochemical activity in Li_2MnO_3. Chem. Mater. 2003, 15, 1984–1992.

[107] Yu, D.Y.W., Yanagida, K., Kato, Y., Nakamura, H. Electrochemical activities in Li_2MnO_3. J. Electrochem. Soc. 2009, 156(6), A417–A424.

[108] Rana, J., Papp, J.K., Lebens-Higgins, Z., Mateusz, Z., Kaufman, L. A., Goel, A., Schmuch, R., Winter, M., Wittingham, M.S., Yang, W., McCloskey, B.D., Piper, L.F.J. Quantifying the capacity contributions during activation of Li_2MnO_3. ACS Energy Lett. 2020, 5(2), 634–641.

[109] Kubobuchi, K., Mogi, M., Ikeno, H., Tanaka, I., Imai, H., Mizoguchi, T. Mn $L_{2,3}$-edge X-ray absorption spectroscopic studies on charge-discharge mechanism of Li_2MnO_3. Appl. Phys. Lett. 2014, 104, 053906.

[110] Yu, H., Ishikawa, R., Shibata, N., Kudo, T., Zhou, H., Ikuhara, Y. Direct atomic-resolution observation of two phases in the $Li_{1.2}Mn_{0.567}Ni_{0.166}Co_{0.067}O_2$ cathode material for lithium-ion batteries. Angew. Chem. Int. Ed. 2013, 53(23), 5969–5973.

[111] Johnson, C.S., Kim, J.-S., Lefief, C., Li, N., Vaughey, J.T., Thackeray, M.M. The significance of the Li_2MnO_3 component in composite x Li_2MnO_3·(1-x)$LiMn_{0.5}Ni_{0.5}O_2$ electrodes. Electrochem. commun. 2004, 6, 1085–1091.

[112] Lu, Z.L., Beaulieu, Y., Donaberger, R.A., Thomas, C.L., Dahn, J.R. Synthesis, structure, and electrochemical behavior of $Li[Ni_xLi_{1/3-2x/3}Mn_{2/3-x/3}]O_2$. J. Electrochem. Soc. 2002, 149(6), A778–A791.

[113] Ohzuku, T., Nagayama, M., Tsuji, K., Ariyoshi, K. High-capacity lithium insertion materials of lithium nickel manganese oxides for advanced lithium-ion batteries: Toward rechargeable capacity more than 300 mAh.g^{-1}. J. Mater. Chem. 2011, 21, 10179.

[114] Thackeray, M.M., Kang, S.-H., Johnson, C.S., Vaughey, J.T., Benedek, R., Hackney, S.A. Li_2MnO_3-stabilized $LiMO_2$ (M = Mn, Ni, Co) electrodes for lithium-ion batteries. J. Mater. Chem. 2007, 17, 3112–3125.

[115] Zheng, J.M., Wu, X.B., Yang, Y. A comparison of preparation method on the electrochemical performance of cathode material $Li[Li_{0.2}Mn_{0.54}Ni_{0.13}Co_{0.13}]O_2$ for lithium ion battery. Electrochim. Acta 2011, 56, 3071–3078.

[116] Boulineau, A., Croguennec, L., Delmas, C., Weill, F. Reinvestigation of Li_2MnO_3 structure: Electron diffraction and high resolution TEM. Chem. Mater. 2009, 21, 4216–4222.

[117] Boulineau, A., Croguennec, L., Delmas, C., Weill, F. Structure of Li_2MnO_3 with different degrees of defects. Solid State Ionics 2010, 180, 1652–1659.

[118] Casas-Cabanas, M., Reynaud, M., Rikarte, J. Horbach, P., Rodriguez-Carvajal, J. FAULTS: a program for refinement of structures with extended defects. J. Appl. Crystallogr. 2016, 49, 2259–2269.

[119] Menon, A. S., Ojwang, D. O., Willhammar, T., Peterson, V. K., Edström, K., Gomez, C. P., Brant, W. R. Influence of synthesis routes on the crystallography, morphology, and electrochemistry of Li_2MnO_3. ACS Appl. Mater. Interfaces 2020, 12, 5939–5950.

[120] Shunmugasundaram, R., Arumugam, R.S., Dahn, J.R. A study of stacking faults and superlattice ordering in some Li-rich layered transition metal oxide positive electrode materials. J. Electrochem. Soc. 2016, 163(7), A1394–A1400.

[121] Armstrong, R.A., Holzapfel, M., Novak, P., Johnson, C.S., Kang, S.-H., Thackeray, M.M., Bruce, P.G. Demonstrating oxygen loss and associated structural reorganization in the lithium battery cathode $Li[Ni_{0.2}Li_{0.2}Mn_{0.6}]O_2$. J. Am. Chem. Soc. 2006, 128, 8694–8698.

[122] Yabuuchi, N., Yoshii, K., Myung, S.-T., Nakai, I., Komaba, S. Detailed studies of a high-capacity electrode material for rechargeable batteries, Li_2MnO_3–$LiCo_{1/3}Ni_{1/3}Mn_{1/3}O_2$. J. Am. Chem. Soc. 2011, 133(12), 4404–4419.

[123] Radin, M.D., Vinckeviciute, J., Seshadri, R., Van der Van, A. Manganese oxidation as the origin of the anomalous capacity of Mn-containing Li-excess cathode materials. Nat. Energy 2019, 4, 639–646.

[124] Genevois, C., Koga, H., Croguennec, L., Menetrier, M., Delmas, C., Weill, F. Insight into the atomic structure of cycled lithium-rich layered oxide $Li_{1.20}Mn_{0.54}Co_{0.13}Ni_{0.13}O_2$ using HAADF STEM and electron nanodiffraction. J. Phys. Chem. C 2015, 119, 75–83.

[125] Kleiner, K., Strehle, B., Baker, A.R., Day, S.J., Tang, C.C., Buchberger, I., Chesneau, F.-F., Gasteiger, H.A., Piana, M. Origin of high capacity and poor cycling stability of Li-Rich layered oxides: A long-duration in situ synchrotron powder diffraction study. Chem. Mater. 2018, 30, 3656–3667.

[126] Luo, K., Roberts, M.R., Hao, R., Guerrini, N., Pickup, D.M., Liu, Y.-S., Edstrom, K., Guo, J., Chadwick, A.V., Duda, L.C., Bruce, P.G. Charge-compensation in 3d-transition-metal-oxide intercalation cathodes through the generation of localized electron holes on oxygen. Nat Chem 2016, 8, 684–691.

[127] Crompton, K.R., Staub, J.W., Hladky, M.P., Landi, B.J. Lithium rich cathode/graphite anode combination for lithium ion cells with high tolerance to near zero volt storage. J. Power Sources 2017, 343, 109–118.

[128] Kubota, K., Kumakura, S., Yoda, Y., Kuroki, K., Komaba, S. Electrochemistry and solid-state chemistry of $NaMeO_2$ (Me = 3d transition metals). Adv. Energy Mater. 2018, 8, 1703415.

[129] Delmas, C., Fouassier, C., Hagenmuller, P. Structural classification and properties of the layered oxides. Physica. B 1980, 99, 81–85.

[130] Delmas, C., Braconnier, J.-J., Fouassier, C., Hagenmuller, P. Electrochemical intercalation of sodium in Na_xCoO_2 bronzes. Solid State Ionics 1981, 3/4, 165–169.

[131] Didier, C., Guignard, M., Suchomel, M. R., Carlier, D., Darriet, J., Delmas, C. Thermally and electrochemically driven topotactical transformations in sodium layered oxides Na_xVO_2. Chem. Mater. 2016, 28(5), 1426–1471.

[132] Han, M.H., Gonzalo, E., Casas-Cabanas, M., Rojo, T. Structural evolution and electrochemistry of monoclinic $NaNiO_2$ upon the first cycling process. J. Power Sources 2014, 258, 266–271.

[133] Jung, Y.H., Christensen, A.S., Johnsen, R.E., Norby, P., Kim, D.K. In Situ X-Ray diffraction studies on structural changes of a P2 layered material during electrochemical desodiation/sodiation. Adv. Funct. Mater. 2015, 25, 3227–3237.

[134] Lu, Z., Dahn, J.R. In situ X-Ray diffraction Study of P2-$Na_{2/3}[Ni_{1/3}Mn_{2/3}]O_2$. J. Electrochem. Soc. 2001, 148(11), A1225–A1229.

[135] Schacklette, L.W. Rechargeable electrodes from sodium cobalt bronzes. J. Electrochem. Soc. 1988, 135, 2669.

[136] Zhou, Y. N., Ding, J. J., Nam, K.-W., Yu, X., Bak, S.-M., Hu, E., Liu, J., Bai, J., Li, H., Fu, Z.-W., Yang, X.-Q. Phase transition behavior of $NaCrO_2$ during sodium extraction studied by synchrotron-based Xray diffraction and absorption spectroscopy. J. Mater. Chem. A 2013, 1, 11130–11134.

[137] Radin, M.D., Van Der Ven, A. Stability of prismatic and octahedral coordination in layered oxides and sulfides intercalated with alkali and alkaline-earth metals. Chem. Mater. 2016, 28, 7898–7904.

[138] Mo, Y., Ong, S.P., Ceder, G. Insights into diffusion mechanisms in P2 layered oxide materials by first-principles calculations. Chem. Mater. 2014, 26(18), 5208–5214.

[139] Lei, Y., Xin, L., Liu, L., Ceder, G. Synthesis and stoichiometry of different layered sodium cobalt oxides. Chem. Mater. 2014, 26, 5288–5296.

[140] Carlier, D., Saadoune, I., Croguennec, L., Menetrier, M., Suard, E., Delmas, C. On the metastable O_2-type $LiCoO_2$. Solid State Ionics 2001, 144, 263–276.

[141] Carlier, D., Van Der Ven, A., Delmas, C., Ceder, G. First-principles investigation of phase stability in the O_2-$LiCoO_2$ system. Chem. Mater. 2003, 15, 2651–2660.

[142] Kaufman, J.L., Van Der Ven, A. Na_xCoO_2 phase stability and hierarchical orderings in the O3/P3 structure family. Phys. Rev. Mater. 2019, 3, 015402.

[143] Vinckeviciute, J., Radin, M.D., Van der Ven, A. Stacking-sequence changes and Na ordering in layered intercalation materials. Chem. Mater. 2017, 28, 8640–8650.

[144] Berthelot, R., Carlier, D., Delmas, C. Electrochemical investigation of the P2–Na_xCoO_2 phase diagram. Nat. Mater. 2011, 10, 74.

[145] Shou-Hang, B., Li, X., Toumar, A.J., Ceder, G. Layered-to-rock-salt transformation in desodiated Na_xCrO_2 (x<0.4). Chem. Mater. 2014, 28, 1419–1429.

[146] Chen, X., Wang, Y., Wiaderek, K., Sang, X., Borkiewicz, O., Chapman, K., LeBeau, J., Lynn, J., Li, X. Super charge separation and high voltage phase in Na_xMnO_2. Adv. Funct. Mater. 2018, 28(50), 1805105.

[147] Lee, E., Brown, D.E., Alp, E.E., Ren, Y., Lu, J., Woo, J.-J., Johnson, C.S. New insights into the performance degradation of Fe-based layered oxides in sodium-ion batteries: Instability of Fe $^{3+}$/Fe^{4+} Redox in α-$NaFeO_2$. Chem. Mater. 2015, 27(19), 6755–6764.

[148] Wu, D., Li, X., Xu, B., Twu, N., Liu, L., Ceder, G. $NaTiO_2$: A layered anode material for sodium-ion batteries. Energy Environ. Sci. 2015, 8, 195–202.

[149] Didier, C., Guignard, M., Denage, C., Szajwaj, O., Ito, S., Saadoune, I., Darriet, J., Delmas, C. Electrochemical Na-deintercalation from $NaVO_2$. Electrochem. Solid-State Lett. 2011, 14(5), A75.

[150] Guignard, M., Didier, C., Darriet, J., Bordet, P., Elkaim, E., Delmas, C. P2-Na_xVO_2 system as electrodes for batteries and electron-correlated materials. Nat. Mater. 2013, 12, 74.

[151] Didier, C., Guignard, M., Darriet, J., Delmas, C. O'3–Na_xVO_2 System: a superstructure for $Na_{1/2}VO_2$. Inorg. Chem. 2012, 51, 11007–11017.

[152] Guignard, M., Carlier, D., Didier, C., Suchomel, M. R., Elkaïm, E., Bordet, P., Decourt, R., Darriet, J., Delmas, C. Vanadium clustering/declustering in P2–$Na_{1/2}VO_2$ layered oxide. Chem. Mater. 2014, 26, 1538–1548.

[153] Szajwaj, O., Gaudin, E., Weill, F., Darriet, J., Delmas, C. Investigation of the new P'3-$Na_{0.60}VO_2$ phase: Structural and physical properties. Inorg. Chem. 2009, 48, 9147–9154.

[154] Feyerherm, R., Dudzik, E., Valencia, S., Wolter, A.U.B., Milne, C.J., Landsgesell, S., Alber, D., Argyriou, D.N. Transition from a phase-segregated state to single-phase incommensurate sodium ordering in gamma-Na_xCoO_2 (x=0.53). Phys. Rev. 2010, B 82, 024103.

[155] Lang, G., Bobroff, J., Alloul, H., Collin, G., Blanchard, N. Spin correlations and cobalt charge states: Phase diagram of sodium cobaltates. Phys. Rev. B 2008, 78, 155116.

[156] Viciu, L., Bos, J.W.G., Zandbergen, H.W., Huang, Q., Foo, M.L., Ishiwata, S., Ramirez, A.P., Lee, M., Ong, N.P., Cava, R.J. Crystal structure and elementary properties of Na_xCoO_2 (x = 0.32, 0.51, 0.6, 0.75, and 0.92) in the three-layer $NaCoO_2$ family. Phys. Rev. 2006, B 73, 174104.

[157] Zandbergen, H.W., Foo, M., Xu, Q., Kumar, V., Cava, R.J. Sodium ion ordering in Na_xCoO_2: Electron diffraction study. Phys. Rev. B 2004, 70, 024101.

[158] Didier, C. 2013. Etude des oxydes lamellaires Na_xVO_2 : électrochimie, structure et propriétés physiques.Study of Na_xVO_2 layered oxides: Electrochemistry, structure and physical properties. Universite Sciences et Technologies – Bordeaux I.

[159] Shu, G.J., Chou, F.C. Sodium-ion diffusion and ordering in single-crystal P2-Na_xCoO_2. Phys. Rev. B 2008, 78, 052101.

[160] Van Der Ven, A., Ceder, G., Asta, M., Tepesch, P.D. First-principles theory of ionic diffusion with nondilute carriers. Phys. Rev. 2001, B 64, 184307.

[161] Van Der Ven, A., Deng, Z., Banerjee, S., Ong, S.P. Rechargeable Alkali-ion battery materials: theory and computation. Chem. Rev. 2020, 120(14), 6977–7019.

[162] Kubota, K., Yoda, Y., Komaba, S. Origin of enhanced capacity retention of P2-type $Na_{2/3}Ni_{1/3-x}Mn_{2/3}Cu_xO_2$ for Na-ion batteries. J. Electrochem. Soc. 2017, 164(12), A2368-A2373.

[163] Singh, G., Tapia-Ruiz, N., Lopez Del Amo, J.M., Maitra, U., Somerville, J.W., Armstrong, R.A., Martinez de Ilarduya, J., Rojo, T., Bruce, P.G. High voltage Mg-doped $Na_{0.67}Ni_{0.3-x}Mg_xMn_{0.7}O_2$

(x = 0.05, 0.1) Na-ion cathodes with enhanced stability and rate capability. Chem. Mater. 2016, 28, 5087–5094.

[164] Somerville, J.W., Sobkowiak, A., Tapia-Ruiz, N., Billaud, J., Lozano, J. G., House, R. A., Gallington, L. C., Ericsson, T., Häggström, L., Roberts, M.R., Maitra, U., Bruce, P.G. Nature of the "Z"-phase in layered Na-ion battery cathodes. Energy Environ. Sci. 2019, 12, 2223.

[165] Mortemard de Boisse, B., Carlier, C., Guignard, M., Bourgeois, L., Delmas, C. P2-Na$_x$Mn$_{1/2}$Fe$_{1/2}$O$_2$ phase used as positive electrode in Na batteries: structural changes induced by the electrochemical (de)intercalation process. Inorg. Chem. 2014, 53, 11197–11205.

[166] Kumakura, S., Tahara, Y., Kubota, K., Chihara, K., Komaba, S. Sodium and manganese stoichiometry of P2-Type Na$_{2/3}$MnO$_2$. Angew. Chem. Int. Ed. 2016, 55, 1–5.

[167] Yabuuchi, N., Kajiyama, M., Iwatate, J., Nishikawa, H., Hitomi, S., Okuyama, R., Usui, R., Yamada, Y., Komaba, S. P2-type Na$_x$[Fe$_{1/2}$Mn$_{1/2}$]O$_2$ made from earth-abundant elements for rechargeable Na batteries. Nat. Mater. 2012, 11, 512.

[168] Thackeray, M.M., David, W.I.F., Bruce, P.G., Goodenough, J.B. Lithium insertion into manganese spinels. Mater. Res. Bull. 1983, 18, 461–472.

[169] Andre, D., Kim, S.-J., Lamp, P., Lux, S.F., Maglia, F., Paschos, O., Stiaszny, B. Future generations of cathode materials: An automotive industry perspective. J. Mater. Chem. A 2015, 3, 6709.

[170] Gummow, R., De Kock, A., Thackeray, M.M. Improved capacity retention in rechargeable 4 V lithium/lithium manganese oxide (spinel) cells. Solid State Ionics 1994, 69(1), 59–67.

[171] Hunter, J.C. Preparation of a New Crystal Form of Manganese Dioxide: λ-MnO$_2$. J. Solid State Chem. 1981, 39, 142–147.

[172] David, W.I.F., Thackeray, M.M., De Picciotto, L.A. Structure refinement of the spinel related phases Li$_2$Mn$_2$O$_4$ and Li$_{0.2}$Mn$_2$O$_4$. J. Solid State Chem. 1987, 67(2), 316–323.

[173] Kosilov, V.V., Potapenko, A.V., Kirillov, S.A. Effect of overdischarge (overlithiation) on electrochemical properties of LiMn$_2$O$_4$ samples of different origin. J. Solid State Electrochem. 2017, 21, 3269–3279.

[174] Berg, H., Thomas, J.O., Wen, L., Farrington, G.C. A neutron diffraction study of Ni substituted LiMn$_2$O$_4$. Solid State Ionics 1998, 112, 165–168.

[175] Manthiram, A. A reflection on lithium-ion battery cathode chemistry. Nat. Commun. 2020, 11, 1550.

[176] Kim, J.-H., Myung, S.-T., Yoon, C.S., Kang, S.G., Sun, Y.-K. Comparative study of LiNi$_{0.5}$Mn$_{1.5}$O$_{4-\lambda}$ and LiNi$_{0.5}$Mn$_{1.5}$O$_4$ cathodes having two crystallographic structures: $Fd\bar{3}m$ and P4$_3$32. Chem. Mater. 2004, 16, 906–914.

[177] Takahashi, K., Saitoh, M., Sano, M., Fujita, M., Kifune, K. Electrochemical and structural properties of a 4.7 V-class LiNi$_{0.5}$Mn$_{1.5}$O$_4$ positive electrode material prepared with a self-reaction method. J. Electrochem. Soc. 2004, 151(1), A173–A177.

[178] Cabana, J., Casas-Cabanas, M., Omenya, F. O., Chernova, N. A., Zeng, D., Wittingham, M. S., Grey, C.P. Composition-structure relationships in the Li-ion battery electrode material LiNi$_{0.5}$Mn$_{1.5}$O$_4$. Chem. Mater. 2012, 24, 2952–2964.

[179] Kundaraci, M., Amatucci, G.G. The effect of particle size and morphology on the rate capability of 4.7V LiMn$_{1.5+\delta}$Ni$_{0.5-\delta}$O$_4$ spinel lithium-ion battery cathodes. Electrochim. Acta 2008, 53, 4193–4199.

[180] Wang, L., Hong, L., Huang, X., Baudrin, E. A comparative study of $Fd\bar{3}m$ and P4$_3$32 "LiNi$_{0.5}$Mn$_{1.5}$O$_4$". Solid State Ionics 2011, 193, 32–38.

[181] Ma, X., Kang, B., Ceder, G. High rate micron-sized ordered LiNi$_{0.5}$Mn$_{1.5}$O$_4$. J. Electrochem. Soc. 2010, 157, A925.

[182] Shin, D.W., Bridges, C.A., Huq, A., Paranthaman, M.P., Manthiram, A. Role of cation ordering and surface segregation in high-voltage spinel $LiMn_{1.5}Ni_{0.5-x}M_xO_4$ (M = Cr, Fe, and Ga) cathodes for lithium-ion batteries. Chem. Mater. 2012, 24(19), 3720–3731.

[183] Ariyoshi, K., Iwakoshi, Y., Nakayama, N., Ohzuku, T. Topotactic two-phase reactions of $Li[Ni_{1/2}Mn_{3/2}]O_4(P4_332)$ in nonaqueous lithium cells. J. Electrochem. Soc. 2004, 151(2), A296–A303.

[184] Gryffroy, D., Vandenberghe, R.E., Legrand, E. A neutron diffraction study of some spinel componds containing octahedral Ni and Mn at a 1:3 ratio. Mater. Sci. Forum. 1991, 79, 785–790.

[185] Liu, J., Huq, A., Moorhead-Rosenberg, Z., Manthiram, A., Page, K. Nanoscale Ni/Mn ordering in the high voltage spinel cathode $LiNi_{0.5}Mn_{1.5}O_4$. Chem. Mater. 2016, 28(19), 6817–6821.

[186] Lee, E., Persson, K.A. Revealing the coupled cation interactions behind the electrochemical profile of $Li_xNi_{0.5}Mn_{1.5}O_4$. Energy Environ. Sci. 2012, 5, 6047.

[187] Lee, E., Persson, K.A. Solid-solution Li intercalation as a function of cation order/disorder in the high-voltage $Li_xNi_{0.5}Mn_{1.5}O_4$ Spinel. Chem. Mater. 2013, 25, 2885–2889.

[188] Fang, X., Ding, N., Feng, X., Lu, Y., Chen, C.H. Study of $LiNi_{0.5}Mn_{1.5}O_4$ synthesized via a chloride-ammonia co-precipitation method: Electrochemical performance, diffusion coefficient and capacity loss mechanism. Electrochim. Acta 2009, 54, 7471.

[189] Aurbach, D., Markovsky, B., Talyossef, Y., Salitra, G., Kim, H.-J., Choi, S. Studies of cycling behavior, ageing, and interfacial reactions of $LiNi_{0.5}Mn_{1.5}O_4$ and carbon electrodes for lithium-ion 5-V cells. J. Power Sources 2006, 162, 780.

[190] Kim, J.-H., Pieczonka, N. P. W., Li, Z., Wu, Y., Harris, S., Powell, B. R. Understanding the capacity fading mechanism in $LiNi_{0.5}Mn_{1.5}O_4$/graphite Li-ion batteries. Electrochim. Acta 2013, 90, 556–562.

[191] Kim, J.-H., Pieczonka, N.P.W., Yang-Kook, S., Powell, B.R. Improved lithium-ion battery performance of $LiNi_{0.5}Mn_{1.5-x}Ti_xO_4$ high voltage spinel in full-cells paired with graphite and $Li_4Ti_5O_{12}$ negative electrodes. J. Power Sources 2014, 262, 62–71.

[192] Arunkumar, T.A., Manthiram, A. Influence of lattice parameter differences on the electrochemical performance of the 5 V Spinel $LiMn_{1.5-y}Ni_{0.5-z}M_{y+z}O_4$ (M = Li, Mg, Fe, Co, and Zn). Electrochem. Solid-State Lett. 2005, 8(8), A403.

[193] Leitner, K.W., Wolf, H., Garsuch, A., Chesneau, F., Dobrick-Schulz, M. Electroactive separator for high voltage graphite/$LiNi_{0.5}Mn_{1.5}O_4$ lithium ion batteries. J. Power Sources 2013, 244, 548–551.

[194] Liang, G., Zhibin, W., Didier, C., Zhang, W., Cuan, J., Baohua, L., Ko, K.-Y., Hung, P. Y., Cheng Zhang, L., Chen, Y., Leniec, G., Kaczmarek, S. M., Johanessen, B., Thomsen, L., Peterson, V. K., Pang, W. K., Guo, Z. A long cycle-life high-voltage spinel lithium-ion battery electrode achieved by site-selective doping. Angew. Chem. Int. Ed. 2020, 59, 10594.

[195] Liu, J., Manthiram, A. Improved electrochemical performance of the 5 V spinel cathode $LiMn_{1.5}Ni_{0.42}Zn_{0.08}O_4$ by surface modification. J. Electrochem. Soc. 2009, 156, A66–A72.

[196] Kataoka, K., Takahashi, Y., Kijima, N., Akimoto, J., Ohshima, K.-I. Single crystal growth and structure refinement of $Li_4Ti_5O_{12}$. J. Phys. Chem. Solids 2008, 69, 1454–1456.

[197] Colbow, K.M., Dahn, J.R., Haering, R.R. Structure and electrochemistry of the spinel oxides $LiTi_2O_4$ and $Li_{4/3}Ti_{5/3}O_4$. J. Power Sources 1989, 26, 397–402.

[198] Ariyoshi, K., Yamato, R., Ohzuku, T. Zero-strain insertion mechanism of $Li[Li_{1/3}Ti_{5/3}]O_4$ for advanced lithium-ion (shuttlecock) batteries. Electrochim. Acta 2005, 51, 1125–1129.

[199] Kataoka, K., Takahashi, Y., Kijima, N., Hayakawa, H., Akimoto, J., Ohshima, K.-I. A single-crystal study of the electrochemically Li-ion intercalated spinel-type $Li_4Ti_5O_{12}$. Solid State Ionics 2009, 180, 631–635.

[200] Ohzuku, T., Ueda, A., Yamamoto, N. Zero-strain insertion material of $Li[Li_{1/3}Ti_{5/3}]O_4$ for rechargeable lithium cells. J. Electrochem. Soc. 1995, 142(5), 1431.

[201] Jansen, A.N., Kahaian, A.J., Kepler, K.D., Nelson, P.A., Amine, K., Dees, D.W., Vissers, D.R., Thackeray, M.M. Development of a high-power lithium-ion battery. J. Power Sources 1999, 81–82, 902–905.

[202] Liu, H., Zhu, Z., Huang, J., Xin, H., Chen, Y., Zhang, R., Lin, R., Yejing, L., Sicen, Y., Xing, X., Yan, Q., Xiangguo, L., Frost, M.J., An, K., Feng, J., Kostecki, R., Xin, H., Ong, S.P., Liu, P. Elucidating the limit of Li insertion into the spinel $Li_4Ti_5O_{12}$. ACS Mater. Lett. 2019, 1, 96–102.

[203] Wagemaker, M., Simon, D.R., Kelder, E.M., Schoonman, J., Ringpfeil, C., Haake, U., Lutzenkirchen-Hecht, D., Frahm, R., Mulder, F.M. A kinetic two-phase and equilibrium solid solution in Spinel $Li_{4+x}Ti_5O_{12}$. Adv. Mater. 2006, 18, 3169–3173.

[204] Kitta, M., Akita, T., Tanaka, S., Kohyama, M. Two-phase separation in a lithiated spinel $Li_4Ti_5O_{12}$ crystal as confirmed by electron energy-loss spectroscopy. J. Power Sources 2014, 257, 120–125.

[205] Pang, W.K., Peterson, V.K., Sharma, N., Shiu, -J.-J., Wu, S.-H. Lithium migration in $Li_4Ti_5O_{12}$ studied using in situ neutron powder diffraction. Chem. Mater. 2014, 26, 2318–2326.

[206] Wilkening, M., Amade, R., Iwaniak, W., Heitjans, P. Ultraslow Li diffusion in spinel-type structured $Li_4Ti_5O_{12}$ – A comparison of results from solid state NMR and impedance spectroscopy. Phys. Chem. Chem. Phys. 2007, 9, 1239–1246.

[207] Schmidt, W., Bottke, P., Sternad, M., Gollob, P., Hennige, V., Wilkening, M. Small change-great effect: Steep increase of Li Ion dynamics in $Li_4Ti_5O_{12}$ at the early stages of chemical Li insertion. Chem. Mater. 2015, 27, 1740–1750.

[208] Zhang, W., Seo, D.-H., Chen, T., Wu, L., Topsakal, M., Zhu, Y., Lu, D., Ceder, G., Wang, F. Kinetic pathways of ionic transport in fast-charging lithium titanate. Science 2020, 367, 1030–1034.

[209] Bernhard, R., Meini, S., Gasteiger, H.A. On-line electrochemical mass spectrometry investigations on the gassing behavior of $Li_4Ti_5O_{12}$ electrodes and its origins. J. Electrochem. Soc. 2014, 161, A497–A505.

[210] Yuan, T., Tan, Z., Ma, C., Yang, J., Ma, Z.-F., Zheng, S. Challenges of spinel $Li_4Ti_5O_{12}$ for lithium-ion battery industrial applications. Adv. Energy Mater. 2017, 1601625.

[211] Li, W., Li, X., Chen, M., Xie, Z., Zhang, J., Dong, S., Qu, M. AlF_3 modification to suppress the gas generation of $Li_4Ti_5O_{12}$ anode battery. Electrochim. Acta 2014, 139, 104–110.

[212] Manthiram, A., Yu, X., Wang, S. Lithium battery chemistries enabled by solid-state electrolytes. Nat. Rev. Mater. 2017, 2, 16103.

[213] Morgan, D., Van Der Ven, A., Ceder, G. Li conductivity in Li_xMPO_4 (M = Mn, Fe, Co, Ni) olivine materials. Electrochem. Solid-State Lett. 2004, 7(2), A30–A32.

[214] Nishimura, S.-I., Kobayashi, G., Ohoyama, K., Kanno, R., Yashima, M., Yamada, A. Experimental visualization of lithium diffusion in Li_xFePO_4. Nat. Mater. 2008, 7, 707–711.

[215] Andersson, A.S., Kalska, B., Häggström, L., Thomas, J.O., Huang, Y., Shan, B., Chen, R. Lithium extraction/insertion in $LiFePO_4$: An X-ray diffraction and Mossbauer spectroscopy study. Solid State Ionics 2000, 130(41), 52.

[216] Yamada, A., Koizumi, H., Nishimura, S.-I., Noriyuki, S., Kanno, R., Yonemura, M., Nakamura, T., Kobayashi, Y. Room-temperature miscibility gap in Li_xFePO_4. Nat. Mater. 2006, 5, 357–360.

[217] Meethong, N., Huang, H.-Y., Carter, W. C., Chiang, Y.-M. Size-dependent lithium miscibility gap in nanoscale $Li_{1-x}FePO_4$. Electrochem. Solid-State Lett. 2007a, 10(5), A134–A138.

[218] Wagemaker, M., Singh, D. P., Borghols, W. J. H., Lafont, U., Haverkate, L., Peterson, V. K., Mulder, F. M. Dynamic solubility limits in nanosized olivine $LiFePO_4$. J. Am. Chem. Soc. 2011, 133, 10222–10228.

[219] Brunetti, G., Robert, D., Bayle-Guillemaud, P., Rouviere, J.-L., Rauch, E.F., Martin, J.-F., Colin, J.-F., Bertin, F., Cayron, C. Confirmation of the domino-cascade model by $LiFePO_4/FePO_4$ precession electron diffraction. Chem. Mater. 2011, 23, 4515–4524.

[220] Meethong, N., Kao, Y.-H., Tang, M., Huang, H.-Y., Carter, W. C., Chiang, Y.-M. Electrochemically induced phase transformation in nanoscale olivines $Li_{1-x}MPO_4$ (M = Fe, Mn). Chem. Mater. 2008, 20, 6189–6198.

[221] Laffont, L., Delacourt, C., Gibot, P., Wu, M.Y., Kooyman, P., Masquelier, C., Tarascon, J.-M. Study of the $LiFePO_4/FePO_4$ Two-phase system by high-resolution electron energy loss spectroscopy. Chem. Mater. 2006, 18, 5520–5529.

[222] Meethong, N., Huang, H.-Y., Speakman, S. A., Carter, W. C., Chiang, Y.-M. Strain accommodation during phase transformations in olivine-based cathodes as a materials selection criterion for high-power rechargeable batteries. Adv. Funct. Mater. 2007b, 17, 1115–1123.

[223] Delmas, C., Maccario, M., Croguennec, L., Le Cras, F., Weill, F. Lithium deintercalation in $LiFePO_4$ nanoparticles via a domino-cascade model. Nat. Mater. 2008, 7, 665.

[224] Chen, G., Song, X., Richardson, T.J. Electron microscopy study of the $LiFePO_4$ to $FePO_4$ phase transition. Electrochem. Solid-State Lett. 2006, 9(6), A295–A298.

[225] Wang, D., Xiaodong, W., Wang, Z., Chen, L. Cracking causing cyclic instability of $LiFePO_4$ cathode material. J. Power Sources 2005, 140, 125–128.

[226] Bai, P., Cogswell, D.A., Bazant, M. Z. Suppression of phase separation in $LiFePO_4$ nanoparticles during battery discharge. Nano Lett. 2011, 11, 4890–4896.

[227] Liu, H., Strobridge, F.C., Borkiewicz, O.J., Wiaderek, K.M., Chapman, K.W., Chupas, P.J., Grey, C.P. Capturing metastable structures during high-rate cycling of $LiFePO_4$ nanoparticle electrodes. Science 2014, 344, 1252817–1252811.

[228] Li, H., Peng, L., Wu, D., Wu, J., Zhu, Y.-J., Hu, X. Ultrahigh-capacity and fire-resistant $LiFePO_4$-based composite cathodes for advanced lithium-ion batteries. Adv. Energy Mater. 2019, 9, 1802930.

[229] Lu, Z., Chen, H., Robert, R., Zhu, B.Y.X., Deng, J., Wu, L., Chung, C.Y., Grey, C.P. Citric acid- and ammonium-mediated morphological transformations of olivine $LiFePO_4$ particles. Chem. Mater. 2011, 23, 2848.

[230] Nan, C., Lu, J., Li, L., Li, L., Peng, Q., Li, Y. Size and shape control of $LiFePO_4$ nanocrystals for better lithium ion battery cathode materials. Nano Res. 2013, 6, 469–477.

[231] Harper, A.F., Evans, M.L., Darby, J.P., Karasulu, B., Kocer, C.P., Nelson, J.R., Morris, A.J. Ab initio structure prediction methods for battery materials. Johnson Matthey Technol. Rev. 2020, 64(2), 103–118.

[232] Jain, A., Shin, Y., Persson, K.A. Computational predictions of energy materials using density functional theory. Nat. Rev. Mater. 2016, 1, 1–13.

[233] Cabana, J., Monconduit, L., Larcher, D., Palacin, R. Beyond intercalation-based Li-Ion batteries: The state of the art and challenges of electrode materials reacting through conversion reactions. Adv. Energy Mater. 2010, 22, E170–E192.

Arndt Remhof and Radovan Černý

8 Hydroborates as novel solid-state electrolytes

Abstract: Hydroborates are a promising class of superionic conductors for lithium and sodium all-solid-state batteries. They combine high ionic conductivity with (electro-)chemical and thermal stability and favorable mechanical properties. In this chapter, we elucidate their conductivity mechanism based on their structural chemistry and their crystal lattice dynamics, and discuss selected compounds that were successfully employed in proof-of-concept batteries.

Keywords: solid-state battery, solid-state electrolyte, hydroborates

8.1 State-of-the-art lithium-ion batteries

Batteries play a key role in the energy supply of modern society. Due to their high energy and power density, Li-ion batteries (LIBs) dominate the market for portable electronic devices and for electric vehicles. In 2019, the development of LIBs has been awarded by the Nobel Prize in Chemistry.[1] LIBs are electrochemical storage devices that deliver (and store) electrical energy in a chemical form by redox reactions. In a LIB, this is the reversible (de-)intercalation of Li from a graphite anode into a transition metal oxide, acting as a cathode. Examples for the cathode active materials are layered oxides such as lithium cobalt oxide (LCO), olivines such as lithium iron phosphate (LFP), or oxides with a tunneled structure such as lithium manganese oxide (LMO). Despite its high price, toxicity, and limited availability, cobalt is still used in state-of-the-art cathode materials such as lithium nickel manganese cobalt oxide (NMC) or lithium nickel cobalt aluminum oxide (NCA) as they offer higher energy densities as compared to cobalt-free materials. The flexible valence of the transition metal is a necessary requirement to accommodate varying amounts of lithium.

The electrodes are composites, consisting of the electrochemically active material (i.e., the transition metal oxide), an electrolyte soaked in the electrodes, a binder, and an electrically conductive additive to enhance electrical contact with the external circuit via the current collectors. A graphitic anode is used in rechargeable LIB instead of a metallic anode, as lithium electro-deposition during charge is not homogeneous,

---

[1] (https://www.nobelprize.org/prizes/chemistry/2019/press-release/)

Arndt Remhof, Empa, Überlandstrasse 129, 8600 Dübendorf, Switzerland
Radovan Černý, University of Geneva, 24 Quai Ernest Ansermet, 1211 Genève 4, Switzerland

https://doi.org/10.1515/9783110674910-008

leading to the formation of metallic needles called dendrites. They grow upon charge-discharge cycles, ultimately short-circuiting the cell and thereby limiting its lifetime. Intercalating lithium into graphite circumvents this problem. As graphite is electrically conducting, there is no need for an additional conductive additive.

An electrically insulating, but ionically conducting, *separator* prevents the direct contact between the two electrodes. The separator is typically a glassy membrane or a polymer, soaked with an electrolyte solution, separating the tasks of electrical insulation and ion conduction. Due to the reactivity and limited electrochemical stability of water, organic solvents are used. To ensure a homogeneous and fast supply (and removal) with electrons, the electrodes are contacted via metallic *current collectors*. On the cathode side, aluminum foils are used for that purpose. Due to alloy formation between lithium and aluminum at low voltages, copper foils have to be used on the anode side. Figure 8.1 depicts the principle of operation of an LIB on the cell level.

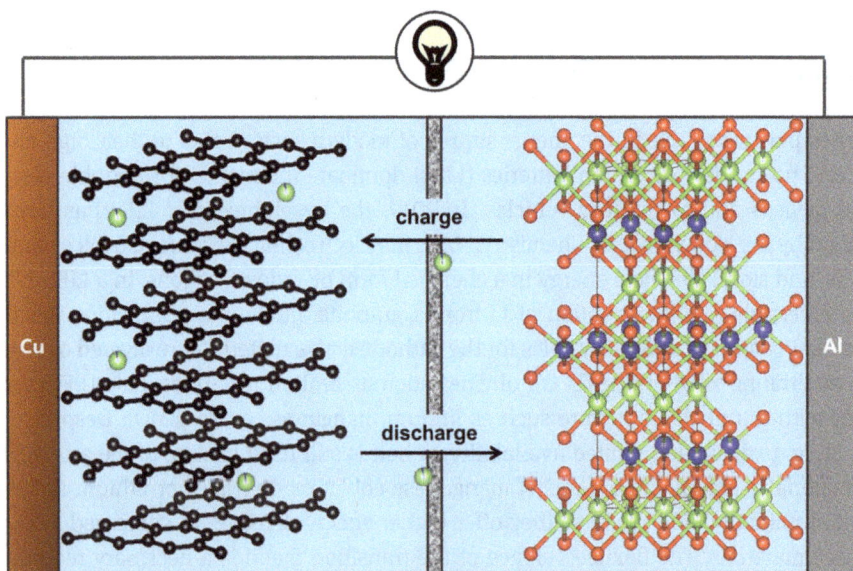

Fig. 8.1: Principle of operation of a lithium-ion battery. Electrical energy is stored as chemical energy by reversible (de)intercalation of Li^+ from a graphite anode (left) into a transition metal oxide (right), acting as a cathode. Lithium is depicted in green, carbon in black, oxygen in red and transition metals in dark blue. Anode and cathode are soaked with a liquid electrolyte (light blue), to facilitate ionic transport. The electrodes are contacted via a copper foil (on the anode side) and an aluminum foil (cathode side) as current collectors to the external circuit. To prevent unwanted chemical reactions and short-circuiting, anode and cathode are kept apart by an electrically insulating, but ionically conducting separator.

8.2 All-solid-state batteries

LIB technology is a mature technology nowadays. Current research on LIBs focuses on high-voltage cathode materials without critical materials such as Co, higher-capacity anodes, and safer and cheaper electrolytes, such as polymers. However, abrupt improvements are not to be expected. Therefore, alternatives to LIBs are discussed, so-called post-LIB technologies, such as lithium-sulfur and lithium-air batteries as well as sodium and magnesium batteries, and all-solid-state batteries.

Solid-state batteries employing a solid-state electrolyte promise to overcome the challenges associated with the liquid organic electrolyte conventionally used in LIBs:
- Organic liquid electrolytes are typically flammable, which poses a safety risk, especially in the case of overcharging and thermal runaway.
- The thermal and (electro-)chemical stability is in many cases limited by the solvent.
- The counter-ions are also mobile within the liquid electrolyte, but they do not undergo an oxidation or reduction reaction during battery charge or discharge. Subsequently, a concentration gradient is built up, reducing the voltage of the battery, especially at fast discharge rates.
- The dendrite formation during electro-deposition of Li is not prevented, inhibiting the use of a high-capacity lithium metal anode.
- The liquid electrolyte may leach material from the electrodes and subsequently serve as a carrier, enabling diffusion across the battery. Prominent examples are transition metal leaching in Li-ion batteries or the "poly-sulfide shuttle" in lithium-sulfur batteries, limiting the cycling stability of the battery.

A suitable solid-state electrolyte should primarily exhibit an ionic conductivity above 1 mS cm^{-1} at room temperature to deliver comparable performance in a battery to its liquid counterparts. Negligible electronic conduction is necessary to avoid short-circuiting and self-discharge of the battery. Furthermore, solid-state electrolytes need to be compatible with high-energy electrodes, which require chemical stability between electrode and electrolyte. Solid-state electrolytes with large electrochemical stability windows facilitate the use of high-voltage cathodes and of metallic anodes that increase the energy density of the cell. In addition to maximizing the gravimetric energy density, it is desirable to use metallic anodes. To minimize interface and grain boundary resistance, the solid-state electrolyte needs favorable mechanical properties to guarantee intimate contact with the electrode materials.

Simultaneously meeting all the above-mentioned requirements for a solid-state electrolyte is still a major challenge. Currently, a few material classes that show sufficiently high ionic conductivities are under consideration as potential candidates for solid-state electrolytes in LIBs. Ion conducting oxides, such as garnets, show moderate to high Li-ion conductivities about 10^{-3} to 1 mS cm^{-1} at room temperature [1]. They are hard materials that require sintering (and thermally stable electrode materials) to reduce interface resistance. Some conductive oxides such as the perovskite-type $Li_{3x}La_{2/3-x}TiO_3$ are unstable against metallic Li. Thiophosphates, such as $Li_{10}GeP_2S_{12}$ and $Li_2S-P_2S_5$ with conductivities of 12 mS cm^{-1} and 17 mS cm^{-1}, respectively, have recently emerged as a promising alternative with conductivities at room temperature comparable to organic liquid electrolytes [2]. Despite the high bulk conductivity of thiophosphates, their limited chemical and electrochemical interface stability with anode and cathode, and the risk of potential release of toxic H_2S, have hindered their integration into competitive batteries [3]. The current state-of-the-art of oxide- and sulfide-based solid conductors has been presented by Zhang et al. [4].

Apart from the inorganic, crystalline materials discussed above, solid polymer electrolytes composed of a Li salt dissolved in a polymer matrix are another class of solid-state Li-ion conductors. The dielectric polymer matrix enables dissociation of the Li salt and transport of the Li-ions while being electronically insulating. Up to now, conductivities in the order of 10^{-2} mS cm^{-1} were reached at room temperature, requiring elevated operating temperatures of ~60 °C (~333 K) at which the conductivity reaches values of ~1 mS cm^{-1}. Despite the limited thermal range of operation and low conductivity at room temperature for solid polymer electrolytes, they are still in the focus of many research activities due to their low interfacial resistance, easy processability, and low cost.

To address the issues of today's solid-state electrolytes, different strategies have been developed (i) to protect the electrolyte from the electrodes to avert stability issues or (ii) to improve the poor contact between electrolyte and electrode as in the case of oxide materials [4]. An example of a protective interphase is the use of a $LiNbO_3$ cathode coating to allow for the assembly of $Li_{9.54}Si_{1.74}P_{1.44}S_{11.7}Cl_{0.3}$- and $Li_{9.6}P_3S_{12}$-based high-power all-solid-state batteries, however, at the expense of an elevated operating temperature of 100 °C (373 K) [5]. The insertion of a soft interlayer [6] or the use of a hybrid ceramic/polymer electrolyte [7] is an example of strategies to reduce the contact resistance.

As ionic conduction in a solid usually proceeds via thermally activated hopping, a high charge density of the ions, leading to strong electrostatic interaction, seems to be disadvantageous. Classical examples of superionic conduction therefore comprise Na$^+$ conducting β-alumina, the Ag$^+$ conductors AgI and Ag_2S, or the F$^-$conducting CaF_2. All these examples comprise medium-sized ions with a lower charge density than the Li-ion. Na β-alumina-based electrolytes are commercially used in molten salt batteries, operating typically at 300 °C (573 K). At this temperature, the Na metal anode is liquid, guaranteeing a large contact area and a low interfacial resistance and avoiding dendrite formation.

NASICON ("**Na S**uper **I**onic **CON**ductor")-type conductors are another class of ceramic Na$^+$ electrolytes with the chemical formula $Na_{1+x}Zr_2Si_xP_{3-x}O_{12}$ ($0 < x < 3$) and ionic conductivities on the order of 1 mS cm^{-1} at room temperature. Unlike β-alumina, NASICON is not stable versus a liquid Na anode. Both materials are hard ceramics that require sintering and thermally stable electrodes, as in the case of the ceramic Li-conductors to increase the contact to the active material in composite electrodes and to minimize interface resistance. Furthermore, the structural integrity of a hard, sintered composite electrode is challenged when the active material experiences strong volume changes upon sodiation and desodiation, while the electrolyte and conducting additives stay unchanged.

Complex metal hydrides and in particular metal *closo*-hydroborates (salts with polyhedral $B_nH_n^{2-}$ ($n = 6$–12) anions), carborates (with monovalent $CB_{n-1}H_n^-$ anions), and mixtures thereof recently emerged as a new class of favorable Li- and especially Na-solid-state electrolytes [8–15]. Within the closo-borate family, the salts of the $B_{10}H_{10}^{2-}$ and $B_{12}H_{12}^{2-}$ anions stand out due to their high thermal and chemical stability [16]. They combine liquid-like ionic conductivity, electrochemical stability >3 V [11, 17–19], high thermal stability, favorable mechanical properties [20], and they are solution processable, enabling roll-to-roll processes [21]. They owe their fast ionic mobility to a high density of accessible vacant sites for ion hopping. Also, the absence of site preference and structural frustration, resulting from different anion and lattice symmetry, supports cation diffusivity. Dynamical disorder due to re-orientational anion motion averages the energetic landscape and prevents cation ordering [22]. *Closo*-borate and carborate salts exhibit structural transitions to highly conductive phases above room temperature, characterized by dynamical disorder [23, 24]. Proof-of-concept all-solid-state batteries were built, using lithium borohydride [25, 26] and lithium carborate [13] as solid-state electrolytes, operated at temperatures above their structural phase transition (>100 °C (373 K)). Among the strategies to stabilize the superionic phases at room temperature and below, anion substitution has been most successful so far [18] and stable cycling in all-solid-state cells has been demonstrated using TiS_2, $NaCrO_2$ and $Na_3(VOPO_4)_2F$ as cathode materials [19, 27–29].

The current literature agrees that in the case of Li and Na "the bottleneck of all solid-state battery development is no longer maximizing the ionic conductivity, but has instead shifted toward the integration of the solid electrolyte and the electrodes" [30]. With the discovery of superionic conductivity in the mixed *closo*-borates, also this material class has matured to this point.

In this chapter, we elucidate their conductivity mechanism based on their structural chemistry and their crystal lattice dynamics and discuss selected compounds that were successfully employed in proof-of-concept all-solid-state batteries.

8.3 Anion packing in hydroborates

The crystal chemistry of metal hydroborates based on the simplest hydroborate anion BH_4^- (commonly addressed as *borohydride*) has been recently analyzed, and the similarity to known structure types of oxides recognized [31]. Contrary to the borohydrides, the currently known hydroborates based on boron clusters are limited to alkali-metals and alkali-earth, that is, to ionic salts, with only a few examples of transition metal hydroborates [32, 33]. No hydroborate-based double-cation boron clusters containing cations with strongly different electronegativity are currently known [34].

The most stable hydroborates, which form also the compounds mostly used in various applications, are *closo*-dodecaborate salts based on the anion $B_{12}H_{12}^{2-}$ and *closo*-decaborate salts based on $B_{10}H_{10}^{2-}$ (Fig. 8.2). While the boron cluster in the first has the form of an ideal icosahedron with the symmetry I_h, the latter is the gyro-elongated square bipyramid (bi-capped square antiprism) with the symmetry D_{4d}. Both polyhedra belong to deltahedra (all faces are equilateral triangles), and their non-crystallographic symmetry is at the origin of their orientational disorder in most of the disordered phases.

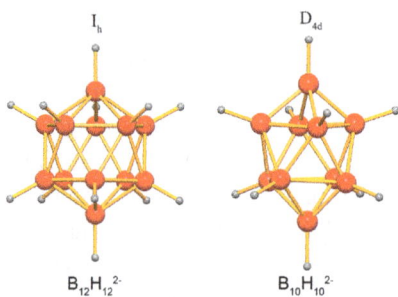

Fig. 8.2: Two most stable hydroborate anions: closo-dodecaborate (left) and closo-decaborate (right) showing their point group symmetry.

The high symmetry of the two anions is also responsible for building the crystal structures according to Pauling's rules as packing of spherical anions with cations occupying the available interstitial sites [35]. The empirical parameters constraining the structure type of the single cation hydroborates are the same as those determining the structure types of metal oxides: (i) chemical formula (i.e., cation oxidation state), (ii) preferred coordination polyhedron of the cation, and (iii) degree of covalency in the cation-anion bond. The constraints may be simplified as follows: First, the preferred cation coordination number is evaluated by means of the ionic size ratio (anion/cation) following Pauling's 1st rule [35]. The structural prototype is then selected according to the chemical formula and the degree of covalency, in agreement with the 2nd and 3rd Pauling's rules. In borohydrides all types of inorganic solid compounds can be found, ranging from salts based on anions packing up to 2D and 3D coordination polymers [31]. The higher hydroborates on the other hand usually form salts where the large hydroborate anions pack mostly according to one of the three common packing types: two close packings, cubic close packing (*ccp*) and hexagonal close packing (*hcp*), and a less dense packing in body centered cubic cell (*bcc*) [34]. The exception are hydroborates of beryllium, which form molecular compounds, and hydroborates of transition metals, which are based on anion packing, but the metal-anion interaction is covalent. These compounds should be classified as coordination polymers.

Strictly speaking, *ccp* and *bcc* packing only apply for structures with cubic symmetry, and *hcp* only for hexagonal or trigonal crystals with lattice parameters ratio of c/a = √8/3 = 1.633. Very often, the true arrangement of anions in hydroborates deviates slightly from the ideal packing, and the crystal symmetry is lowered from the ideal symmetry of hard spheres packing. We will allocate the *ccp* and *hcp* type of packing to a given hydroborate if the anion-anion first coordination sphere is close to the cuboctahedron or anti-cuboctahedron, respectively. The *bcc* packing will be allocated if the first coordination sphere is close to a cubic shape and the second coordination sphere close to an ideal octahedron. The shape of the anion-anion coordination spheres may slightly deviate from the ideal form, but the number and type of the interstitial sites, as well as their connectivity, should not change with respect to the ideal packing. The typical interstitial sites occupied by the cations have the shape of a tetrahedron (T-site) and octahedron (O-site).

It is not always easy to detect the packing type in hydroborates, as it is a packing of poly-anions, which usually deviates from the ideal packing, and the hexagonal layers of the close packing are not always parallel to simple crystallographic planes. In some studies, the anion packing is analyzed with the algorithms developed for molecular dynamics simulation [36], which gives the results as mixtures of three basic packings, and gives their fractions in the analyzed structure. This is justified for the analysis of molecular dynamics simulations as the atomic positions are given as a distribution of atoms, and therefore the distribution of anions packing is justified. It makes, however, no sense to use such algorithm for the analysis of a periodic structure, even if it is disordered, where the atoms are located on Wyckoff sites in the unit cell.

Packing of BH_4^- usually deviates more from the three basic packings compared to the more complex hydroborate anions due to the less spherical shape of the former. It is interesting to note that no salt based on BH_4^- anion crystallizes in the *bcc* packing, while most of the best ionic conductors among *closo*-hydroborate salts are based on *bcc* anion packing [34]. In this context, we want to remind the reader that in the close packing of anions there exist n O-sites for n anions, and $2n$ T-sites, while for *bcc* packing there exists $3n$ O-sites and $6n$ T-sites. While in the ideal *ccp* and *hcp*, the O- and T-sites have ideal shape, this is not true even in the ideal *bcc* packing. This may explain why the *bcc* packing is common among mono-atomic structures, that is, metals, but is rare in ionic structures where the O- and T-sites do not have favorable shape to be occupied by cations.

The number of available interstitial sites for mobile cations is not the only important parameter for the cation conduction. The connectivity between the sites and the dimensionality of so created conduction pathways as well as the number of charge carriers (cations), number of available empty sites for cation jumps and energy barriers between two sites are the parameters determining more importantly the observed ionic conductivity values. From this point of view, it has been suggested since early times of solid electrolytes that the *bcc* anions packing has an advantage over *ccp* and *hcp* in having T-sites (often the preferred sites for alkali metals in hydroborates) percolating in 3D frameworks without involving the O-sites (Fig. 8.3). Not involving the O-sites in the conduction pathways eliminates higher energy barriers of

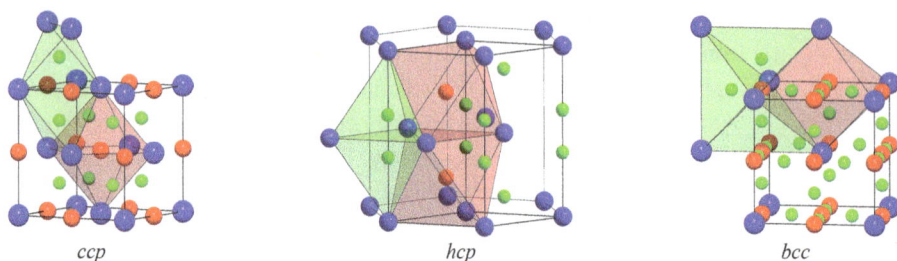

ccp   *hcp*   *bcc*

Fig. 8.3: Octahedral (red) and tetrahedral (green) sites in three basic anions (blue) packings. Note the percolation of tetrahedral sites in bcc, while in ccp and hcp, the 3D conduction pathways based on tetrahedral sites require involving also octahedral sites. The connectivity between all sites (bottleneck for cation mobility) is always a triangular face as high-energy barriers penalize cations jumps over an edge.

the T-O jumps. The bottleneck for cations mobility in all three packings is always a triangular face, the jumps over and edge being penalized by high-energy barriers. Recent *ab-initio* calculations have confirmed this observation for thiophosphate-based solid electrolytes [37], but experimental results on *closo*-hydroborates solid electrolytes do not exclude *ccp* and *hcp* from high performing solid electrolytes [11, 12].

An interesting concept of ideal anion packing for high cation mobility has been proposed in [38]: Cations are expected to diffuse well in hosts where they are in non-preferred but stable sites and migrate through favored but unstable sites.

8.4 Conduction pathways in hydroborates

The cation conductions pathways can be analyzed by using three approaches:
1. Geometric approach using the information on cation and anion size.
2. Energy approach based on the concept of bond valence sums.
3. Energy approach based on *ab-initio* calculations.

8.4.1 Geometric approach

Pauling's 1st rule predicts the preferred cation coordination in iono-covalent compounds from the ratio of cation and anion radii [35]. The cation radii in such calculations are usually taken according to [39]. As only very few estimations of hydroborate radii are available in the literature (for BH_4^- see [31]), best estimations are the radii from [10] where the diameter of the anion was calculated as the maximal distance between two terminal hydrogens. Evidently, these radii correspond to the lower limit of the estimation. The upper limit of the hydroborate anions diameter is estimated from solid crystal structures: 3.18 Å for $B_{10}H_{10}^{2-}$ [40] or 3.28 Å for $B_{12}H_{12}^{2-}$ [41] or from calculation of electrostatic potential surface maps [42].

The interstitial sites available in the anion packing and channels connecting them are considered as suitable pathways for cation conduction according to their size that should be bigger than the mobile cation size with usual tolerance of 10–15% (softness of the ions). The program TOPOS[2] simplifies these calculations by using the Dirichlet-Voronoi tessellation of the free space between the atoms in a crystal structure [43].

8.4.2 Energy approach based on the concept of bond valence sums

The concept of bond valence sum [44] has been further related to bond energy scale introducing the bond valence force field containing attractive and repulsive terms described by empirical potentials [45]. It allows searching for pathways where the cation-anion interaction is minimized, and provides the values of energy barriers for cation hopping between available interstitial sites. The calculations can be easily performed using the program SoftBV [46].

8.4.3 Energy approach based on ab-initio calculations

The analysis of conduction pathways using the *ab-initio* calculations is usually performed by a constant volume molecular dynamics approach [47], and by the climbing image nudged elastic band (NEB) method to calculate the energy barriers [48].

8.5 Controlling anion packing

As we have seen above, the type of anion packing may play important role in mobility of cations. While the packing type for monoatomic structures such as metals is simply controlled by space-filling efficiency without any directional bonding, it is not as simple for poly-atomic structures such as ionic compounds. In the case of metal *closo*-hydroborates, the polyanions are only approximatively spherical and their orientation is given by directional metal-hydrogen bonding (less pronounced for alkali metals). It means that the anion packing may be controlled by cation-anion interaction. The packing type depends on the temperature and pressure, as,

---

2 https://topospro.com/

for example, in AgI [49] or $Na_2B_{12}H_{12}$ [20, 41]. We will focus only on packing control at ambient conditions, where the all-solid-state batteries will mostly operate.

8.5.1 Mixing cations

One of the possibilities is adding a second cation in the structure. The control parameters are the size of the second cation, its preferred coordination type, and the number of interstitial sites with such preferred coordination in various packing types. A good example among solid electrolytes is the stabilization of highly conducting high-temperature *bcc* phase of AgI at room temperature by replacing 1/5 of Ag cation by bigger Rb in $RbAg_4I_5$ [50]. As a result, not the *bcc* packing has been stabilized at room temperature, but a new type of anion packing was obtained that contains conduction pathways made of face-sharing tetrahedra providing high Ag^+ mobility (Fig. 8.4).

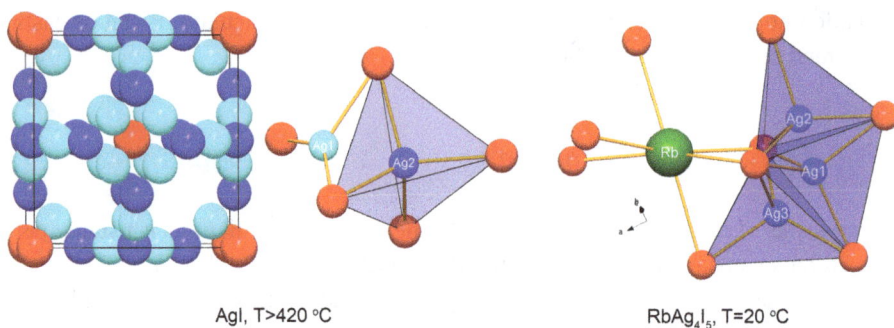

AgI, T>420 °C $RbAg_4I_5$, T=20 °C

Fig. 8.4: Stabilization of high-temperature superionics AgI with bcc anionic packing at room temperature by cation mixing as $RbAg_4I_5$ with a new type of anion packing. While in AgI the silver cations are disordered over tetrahedral (dark blue) and triangular (light blue) sites, in $RbAg_4I_5$ they are disordered only on tetrahedral sites which create 3D conduction pathways by sharing the faces.

An analogical approach has been attempted in Li and Na *closo*-hydroborates with the aim to stabilize the disordered conducting high-temperature phases $Li_2B_{12}H_{12}$ and $Na_2B_{12}H_{12}$ down to rt. Small and mobile alkali metals were mixed with bigger alkali metals (Na, K, and Cs) in the molar ratio 1:1 [51]. Indeed, the anion packing is controlled by the size of the bigger cation that occupies O-site. Starting from the K^+ cation, the packing turns to be *hcp* (Fig. 8.5). Contrary to $RbAg_4I_5$, no new type of packing has been obtained. Moreover, in the three obtained packing types, the presence of bigger cation on the O-site blocks the mobility of the smaller cation on the T-site even in the case of *bcc* (i.e., neighboring O- and T-sites cannot be occupied simultaneously).

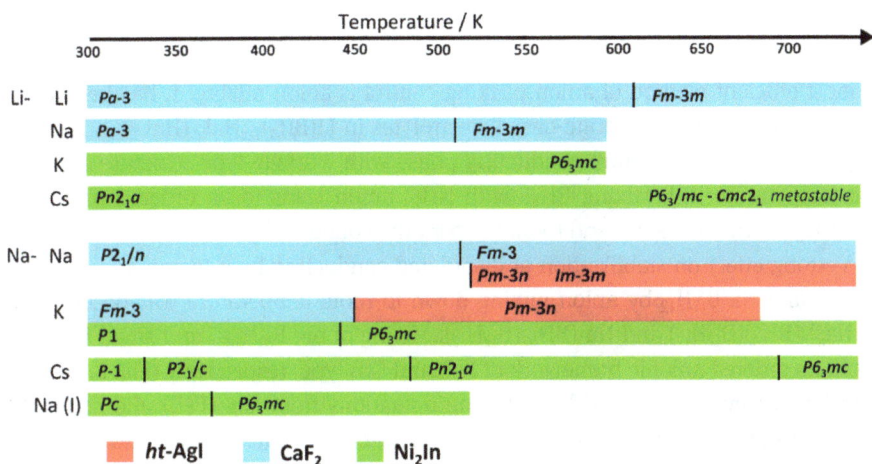

Fig. 8.5: Anion packing (in color code) and polymorphism of alkali metal *closo*-hydroborates (from [51]).

8.5.2 Anion modification

Modifying the chemistry of the anion is anoher way of controlling the anion packing. Partial halogenation of hydroborates, for example $Na_2(B_{12}H_{12-x}I_x)$ with $x \sim 0.1$ leads to anions *hcp* and high Na^+ conductivity close to 10 mS cm^{-1} at 87 °C (360 K) with very low activation energy of 140 meV [51]. The anion packing stay *hcp* at room temperature, but the structure deforms into a monoclinic symmetry and conduction pathways are closed. The *ab-initio* calculations have attributed the changes in anion packing to the volume increase of 1%. The modification of charge distribution on the partially halogenated *closo*-borate is another reason of *hcp* preference. Decreasing the sphericity of the *closo*-hydroborate anion by its partial halogenation, by replacing $B_{12}H_{12}{}^{2-}$ with $B_{10}H_{10}{}^{2-}$ or hydroborates with *carba*-hydroborates or *nido*-hydroborates, increases the probability of *hcp* [34]. It can be understood that in *hcp* (12 anions around the central one), the two triangles above and below the hexagonal plane have the same orientation (contrary to *ccp* where they have the opposite orientation), which may be preferred by aspherical anions due to Coulombic repulsion. It is interesting to note that the full halogenation of the *closo*-hydroborate anion leads to *ccp* and an increase of order-disorder transition to higher temperatures resulting in low conductivity at room temperature [42].

8.5.3 Anion mixing

The most efficient method of anion packing control is anion mixing. It has been used for the first time for hydroborate-based electrolytes in $Li(BH_4)_{1-x}Hal_x$ (Hal = Cl, Br, I) to stabilize the high-temperature conducting phase with wurtzite-type structure down to room temperature [52]. Mixing BH_4^- with NH_2^- anion leads to an ordered phase Li_4 $(BH_4)(NH_2)_3$ with anions *ccp* and room temperature conductivity of 0.2 mS cm^{-1} [53].

A strong effect on stabilization of disordered conducting high-temperature phases $Li_2B_{12}H_{12}$ and $Na_2B_{12}H_{12}$ by anion mixing down to room temperature was observed in $(Li_{1-x}Na_x)_3(BH_4)(B_{12}H_{12})$ and $Na_3(BH_4)(B_{12}H_{12})$. Mixing of big $B_{12}H_{12}^{2-}$ and small BH_4^- results in an ordered anionic framework of FeB and CrB type, respectively (Fig. 8.6). Both packing types may be derived by twinning operations from *hcp* (FeB) and *ccp* (CrB) stacking of the larger units, that is, Fe, Cr, or *closo*-hydroborates. Li^+ and Na^+ are then highly mobile at room temperature without any order-disorder transition [8]. Unfortunately, a low thermal stability of this mixed anion framework penalizes the compounds for the all-solid-state batteries application.

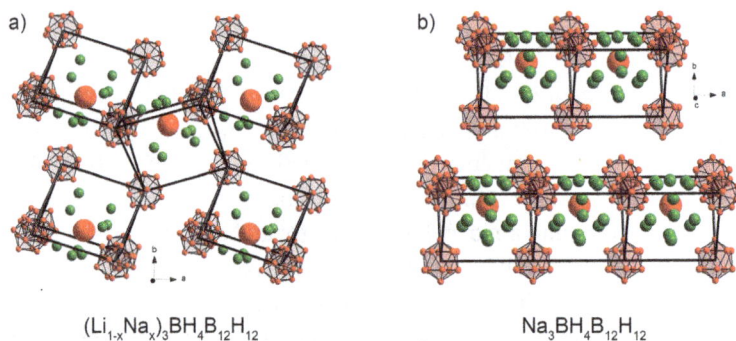

a)   b)

$(Li_{1-x}Na_x)_3BH_4B_{12}H_{12}$ $Na_3BH_4B_{12}H_{12}$

Fig. 8.6: Crystal structures of mixed anion hydroborates $(Li_{1-x}Na_x)_3(BH_4)(B_{12}H_{12})$ with FeB type and $Na_3(BH_4)(B_{12}H_{12})$ with CrB type anion packing showing the connectivity and stacking of basic building units, trigonal prisms of closo-hydroborates. Only B atoms (red) of the closo-hydroborates are shown; the borohydrides are shown simplified as big red spheres and disordered positions of cations as green spheres. (from [8]).

An even more convincing example is mixing of sodium salts of dodeca- and deca-*closo*-hydroborates and dodeca- and deca-*closo*-carba-hydroborates of sodium. Cubic Na_4 $(B_{10}H_{10})(B_{12}H_{12})$ stabilizes the *ccp* without any monoclinic deformation at room temperature contrary to both precursors. The dodeca- and deca-*closo*-hydroborates mixing does not create any anion ordering, and results in room temperature conductivity of 0.9 mS cm^{-1}, thermal stability up to 300 °C, and a large electrochemical stability window of 3 V including stability toward sodium metal anodes [11]. Better performance, room temperature conductivity of 2 mS cm^{-1} (Fig. 8.7), thermal stability up to more than

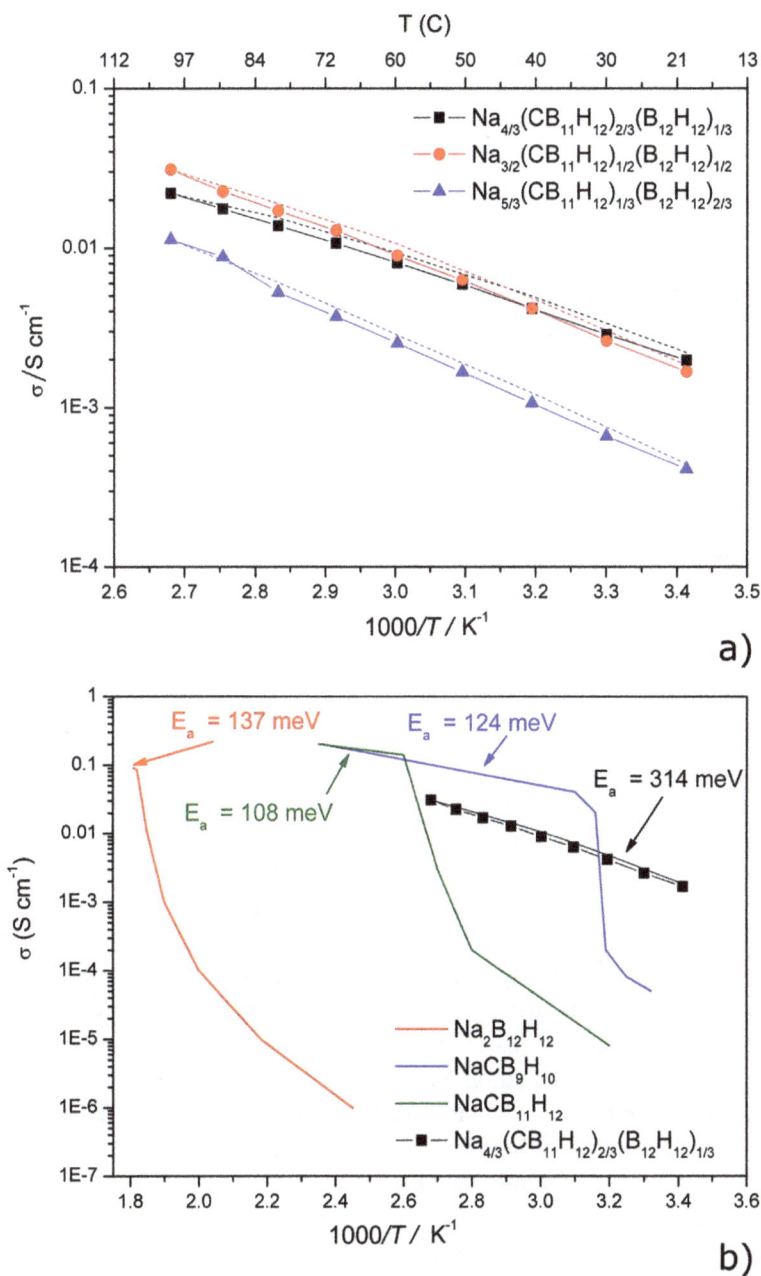

Fig. 8.7: Arrhenius plots for a) Na$^+$ conductivity of Na$_{2-x}$(CB$_{11}$H$_{12}$)$_x$(B$_{12}$H$_{12}$)$_{1-x}$ in the temperature range 20–100 °C (293–373 K) and b) different closo- and carba-closo-borates Na$^+$ conductors, with their respective activation energies in the superionic regime (from [12]).

540 °C, and a stability window of 4.1 V, is provided by hydroborate-carba-hydroborate mixing in $Na_{2-x}(CB_{11}H_{12})_x(B_{12}H_{12})_{1-x}$ [12]. This mixing stabilizes the *bcc* anion packing at room temperature, which is the packing of conducting high-temperature $Na_2B_{12}H_{12}$. For the equimolar mixture ($x = 0.5$), this compound is therefore an equivalent of a superionic α-Ag_3SI, which, however, orders into the anti-perovskite type below 245 °C (519 K) with 2 orders of magnitude lower conductivity [54]. Neutron powder diffraction on isotopically labelled samples has shown no ordering of the two hydroborate anions at room temperature. No slope change in the thermal expansion, signalizing a possible anion ordering, has been observed by X-ray diffraction down to −103 °C (170 K) [55].

8.6 Anion dynamics in hydroborates

As outlined in the introduction, dynamical disorder in hydroborate salts favor cation mobility, leading to liquid-like ionic conductivity at elevated temperature and in various mixed anion compounds. In the following, we elucidate the conduction mechanism of two specific compounds which exemplify the interplay between structure and dynamics of these compounds.

8.6.1 LiBH$_4$

$LiBH_4$ is an ionic compound consisting of BH_4^- anions and Li^+ cations. Within the covalently bound polyanion, the four hydrogen atoms form a regular tetrahedron surrounding the central boron atom. The tetrahedral point group T_d or $\bar{4}3m$ is compatible with a cubic symmetry. The BH_4 tetrahedron can be envisioned as being embedded into a cube of edge length $a = \sqrt{\frac{2}{4}}d_{B-H}$, where d_{B-H} is the boron-hydrogen distance. Each edge of the tetrahedron is a diagonal of one of the cube's faces and each vertex of the tetrahedron is a vertex of the cube. As a cube possesses twice as many vertices as a tetrahedron, that is, eight compared to four, there are two orientations to fit a tetrahedron into a cube. The tetrahedron has two sets of high-symmetry axes: three twofold axes called C2 and four threefold axes called C3 as shown in Fig. 8.8a. A 180° rotation about a C2 axis or a 120° rotation about a C3 axis map the tetrahedron to itself, a 90° rotation about a C2 axis changes the orientation of the tetrahedron.

At ambient conditions, $LiBH_4$ crystallizes in the orthorhombic *Pnma* structure (space group no. 62). At around 110 °C (~380 K), $LiBH_4$ undergoes a first-order structural phase transition to a hexagonal $P6_3mc$ structure (space group 186) [56, 57]. In both cases, the symmetry of the Wyckoff site where the polyanion is located does not match the symmetry of the tetrahedron and the C2 axis of the tetrahedron does not coincide with any low-index direction of the crystal. Therefore, the BH_4-moiety is structurally frustrated.

Moreover, due to the electrostatic attraction of the positively charged cation and the negatively charged anion, a tridentate configuration, that is, a configuration in which three hydrogen atoms point toward the metal ion, would be the most favorable [58]. However, this configuration cannot be achieved simultaneously for all Li^+ in neither the low-temperature phase nor the high-temperature phase of $LiBH_4$, adding to the structural frustration and leading to a mixed bidentate/tridentate configuration. The high ionic conduction in the high-temperature phase of $LiBH_4$ has been attributed to a split of the position of the Li^+ ions along the hexagonal c-axis separated by 0.9 Å [59]. A low energy barrier of 0.02 eV has been calculated between the two sites, allowing the frequent Li^+ hopping between them. The fast reorientations of the anion support the cation diffusion (i) in flattening the overall average potential energy landscape and (ii) in breaking the orientation between the anion and the cation, leading to dynamical frustration.

Figure 8.8b and 8.8c display the potential energy surfaces for (b) the low-temperature and (c) the high-temperature phase, respectively [60]. The calculation assumes rotation around a C2 axis, followed by a C3 axis rotation. As that group of rotations is not commutative, a rotation in a different order, that is, the first around C3 and the second around C2, would result in a different representation of the energy landscape. In the low-temperature phase (Fig. 8.8b) the BH_4 moieties are localized within well-defined energy minima, separated by 120° and 180° for rotations about the C3 and the C2 axis, respectively. The lowest barriers involve a combined motion about both axes and have a height of >0.6 eV. The overall potential energy landscape in the high-temperature phase is much flatter and the equilibrium positions are less well defined. A rotating BH_4 moiety experiences very shallow energy valleys that pass through the whole energy landscape with barrier heights of about 0.2 eV.

Experimentally, the rotational diffusion of the BH_4 moieties has been observed by quasi-elastic neutron scattering (QENS). As neutrons and protons have a similar mass, neutron scattering is a technique that is especially suited for hydrogen-containing compounds. Inelastic and quasi-elastic neutron scattering exploits the energy and momentum transfer between the sample and the incident neutrons. Thereby inelastic neutron scattering is comparable to other spectroscopic techniques such as infrared or Raman spectroscopy, where the discrete vibrational spectrum of the sample is seen as a set of discrete signals, often referred to as "lines" that are clearly separated from the elastic line, that is, the signal at zero energy transfer. QENS refers to energy transfers that are small compared to the incident energy and lead to a broadening of the elastic line. In the spectrometer, the energy E_0 and the resolution ΔE of the incident neutrons are defined by the instrumental settings, mainly by its optics and velocity selectors. Due to the energy-time uncertainty principle, a ΔE is inversely proportional to Δt, the temporal extension of the neutron's wave package. To maintain the given energy resolution during the scattering event, the target atom that scatters the neutron has to be at rest on a time-scale exceeding Δt. If the scattering event is interrupted by a translational or rotational jump of the scattering atom, Δt is shortened and consequently ΔE becomes

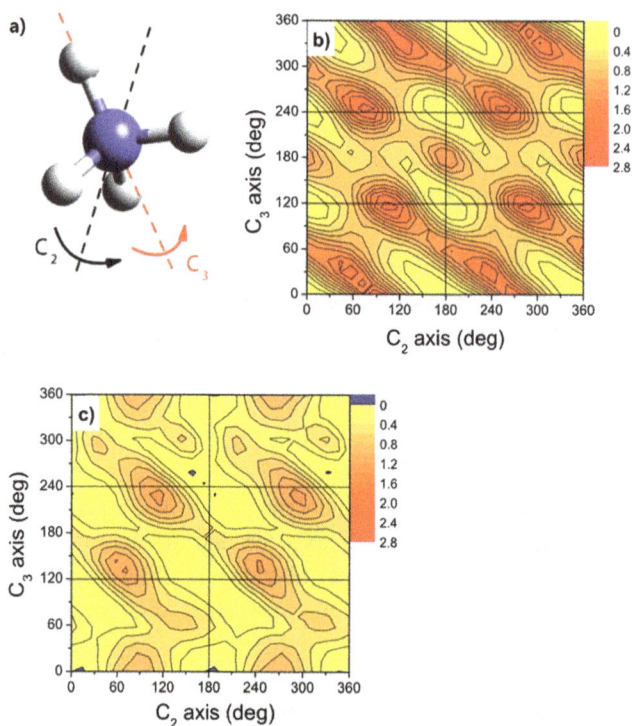

Fig. 8.8: Calculation of the rotational energy barriers of the BH$_4$ tetrahedra in LiBH$_4$. (a) Rotation axis C2 and C3 used for the description of all the orientations of the tetrahedra. Potential energy surface calculated along the two axes C2 and C3 (b) for Pnma space group, (c) for P63mc space group [60].

larger. In other words, the dynamics within the sample disturbs the energy definition of the scattered neutrons, symmetrically broadening the elastic line. The broadening directly yields the time scale of the dynamic event and temperature-dependent measurements yield the apparent activation energies. Further details such as the determination of the jump direction, jump length, and rotational axis require the simultaneous determination of the momentum transfer and the fitting of the data against often complex model systems. Especially in the case of overlapping motions, such as rotations around different axes, a deconvolution is challenging and consequently details of the dynamics are still debated in the literature [41, 61, 62].

8.6.2 Li(BH$_4$)$_{1/4}$(NH$_2$)$_{3/4}$

Li(BH$_4$)$_{1/4}$(NH$_2$)$_{3/4}$ is synthesized by ball milling and subsequently heating (120 °C (393 K), 12 h) of a mixture of LiBH$_4$ and LiNH$_2$ in a 1:3 molar ratio. Li(BH$_4$)$_{1/4}$(NH$_2$)$_{3/4}$ crystallizes with a cubic lattice (space group $I2_1 3$) with three inequivalent Li sites

[63]. Each Li-site is tetrahedrally coordinated by four NH_2^- anions or by a combination of NH_2^- and BH_4^- anions as shown in Fig. 8.9. The polyanions form a *ccp*, in which Li occupies T-sites. For charge neutrality, half of the T-sites remain unoccupied, similar to the sphalerite structure type, but the distribution of filled (blue) and empty (yellow) T-sites is novel. The occupied tetrahedra are slightly contracted as compared to the unoccupied ones due to the electrostatic attraction between the cation and the anions.

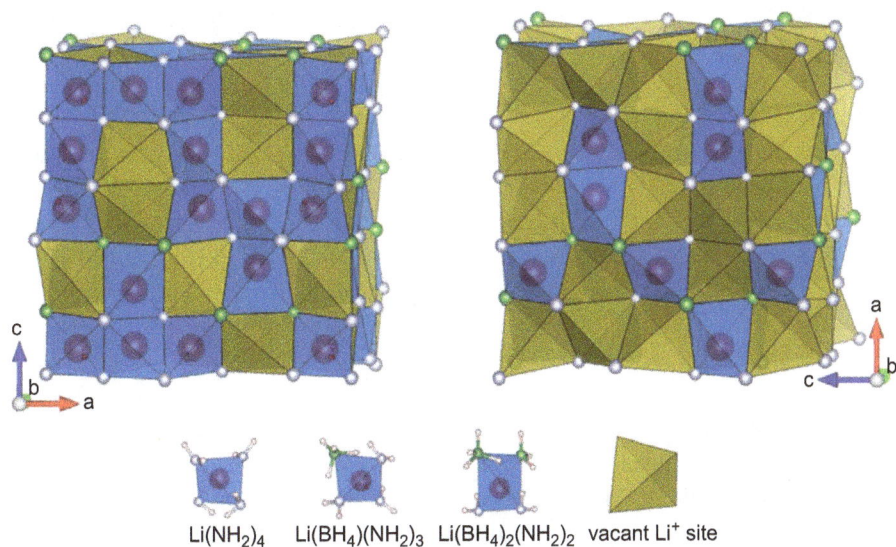

$Li(NH_2)_4$ $Li(BH_4)(NH_2)_3$ $Li(BH_4)_2(NH_2)_2$ vacant Li^+ site

Fig. 8.9: Polyhedral representation of $Li(BH_4)_{1/4}(NH_2)_{3/4}$ (left) and its symmetry-related dual structure (right) along with the three inequivalent lithium sites occurring in the structure. Anion tetrahedra filled with lithium (empty) are shown in blue (yellow). Li, B, and N atoms are colored red, green, and grey, respectively (From [64]).

As shown in Fig. 8.10, conduction involves first the excitation of a Li^+-ion into an unoccupied site (from A to B). This requires the intermediate occupation of the unfavorable octahedrally coordinated site, resulting in a high energy barrier of about 1 eV. Hopping events within the unoccupied sites (B to C) are jumps between two edge-sharing tetrahedral sites with energy barriers of about 0.3 to 0.4 eV [64].

As in the case of $LiBH_4$, the reorientation of the anions lowers the energy barrier for migration significantly and facilitates efficient Li^+ conduction in the presence of an electric field.

The reorientational dynamics of the anions were studied by QENS. In order to distinguish the two anions, deuterium labelling was used, revealing that the quasi-elastic broadening is mainly related to the BH_4^- reorientations [65].

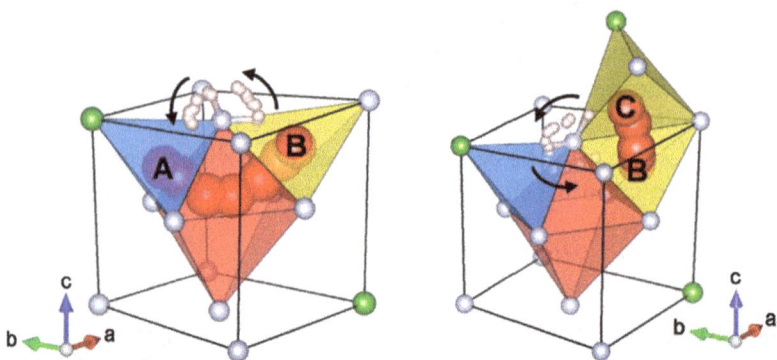

Fig. 8.10: Lithium migration paths: left excitation of a Li$^+$-ion into an unoccupied site (position A to B). The right panel shows migration between unoccupied sites (B to C). The intermediate octahedral site employed in transition is shown in red. The rotation of the anion (here a NH$_2^-$) is indicated by arrows (From [64]).

8.7 Closo-hydroborate-based solid electrolyte batteries

The salts of the B$_{10}$H$_{10}^{2-}$ and B$_{12}$H$_{12}^{2-}$ anions and their carborate-analogues were used for assembling all-solid-state batteries. A Lithium-sulphur battery with 0.7Li(CB$_9$H$_{10}$)-0.3Li(CB$_{11}$H$_{12}$) as solid electrolyte provided a reversible capacity of 1,017 mAhg^{-1} at room temperature with a Coulombic efficiency of ~100% retained after 100 cycles with a working potential ~2.1 V [13]. No formation of Li dendrites has been observed.

Sodium-metal batteries have been assembled with two different mixed anion electrolytes Na$_4$(B$_{10}$H$_{10}$)(B$_{12}$H$_{12}$) and Na$_{2-x}$(CB$_{11}$H$_{12}$)$_x$(B$_{12}$H$_{12}$)$_{1-x}$, respectively [28, 29]. In both cases, NaCrO$_2$ has been used as positive electrode (Fig. 8.11). The impregnation of the cathode has considerably improved the capacity retention (Fig. 8.12).

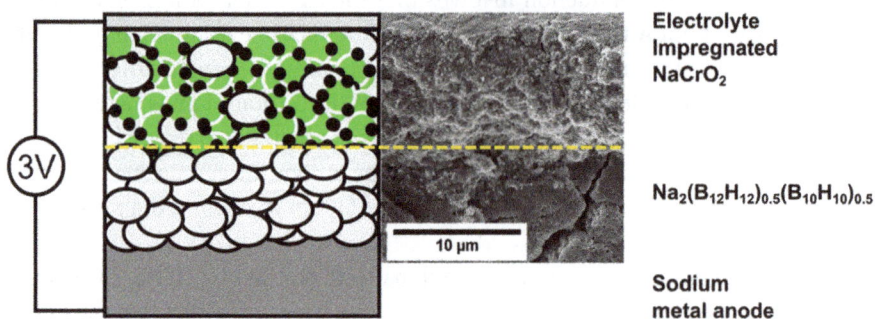

Fig. 8.11: Schematic and SEM cross-section of the device investigated in this study showing the different components of the cell (From [28]).

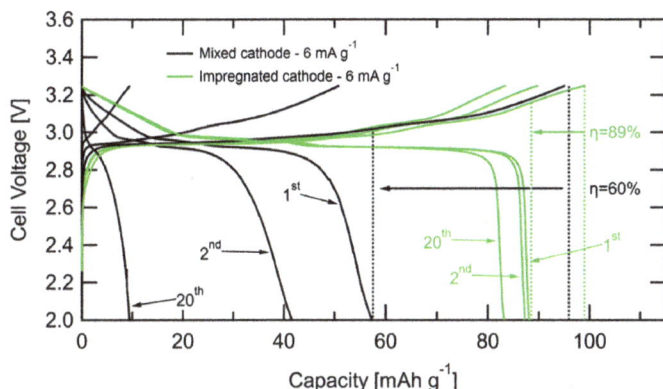

Fig. 8.12: Charge/discharge profiles for the 1st, 2nd and 20th cycle of Na|Na($B_{12}H_{12}$)($B_{10}H_{10}$)|NaCrO$_2$ cells with mixed and impregnated cathode mixtures operating at 60 °C (From [28]).

The higher conductivity of Na$_4$(CB$_{11}$H$_{12}$)$_2$(B$_{12}$H$_{12}$) electrolyte allowed testing of the battery at room temperature. Galvanostatic measurements of a symmetrical Na|Na$_4$ (CB$_{11}$H$_{12}$)$_2$(B$_{12}$H$_{12}$)|Na cell are shown in Fig. 8.13. They match well those measured with Na$_4$(B$_{10}$H$_{10}$)(B$_{12}$H$_{12}$) electrolyte at 60 °C (Fig. 8.12). After an open circuit voltage (OCV) period of 5 h, a current density of ±25.5 µAcm^{-2} was alternatively applied for 30 min. Repeated plating/stripping occurs with a steady and limited polarization, starting from ±8.5 mV that further decreases, stabilizing around ±6 mV vs. Na$^+$/Na. The electrolyte stays therefore stable toward Na metal for >500 operating hours. To investigate the evolution of the interfacial resistance, *in-situ* electrochemical impedance spectroscopy (EIS) was carried out during both the OCV period and after each sweep. Three distinct contributions were detected in the spectra. The first one was attributed to the SE with a resulting conductivity of σ ~2 mS cm^{-2}. The others two events feature a remarkable evolution in the first 100 cycles (Fig. 8.13b). One is nearly stable between 20 and 30 Ω cm^2 (R1 in Fig. 8.13b) for the whole measurement, while the second shows an important drop from 100 to 55 Ω cm^2 in the first 70 cycles (R2 in Fig. 8.13b). Their resonance frequencies amount to 3.55 MHz for R1 while for R2 it lies in the kHz region (Fig. 8.13b inset). Both events likely correspond to interfacial resistance. The R1 was associated with the interfacial resistance at the Na/ solid electrolyte boundary. The nature of R2 contribution points to a chemical/mechanical instability as the resonance peak moves toward lower frequencies with cycling.

So far, cell assembly was achieved by powder processes, that is, by powder stacking, mixing, and pressing. The ability of the mixed anion system Na$_4$(B$_{10}$H$_{10}$)(B$_{12}$H$_{12}$) to crystallize in its highly conducting phase from a solution of its precursors, Na$_2$(B$_{10}$H$_{10}$) and Na$_2$(B$_{12}$H$_{12}$) enables solution impregnation of a porous Sn metal anode and of a NaCrO$_2$ cathode casted on an aluminum current collector. This process can be integrated into today's cell production lines and high-throughput roll-to-roll process

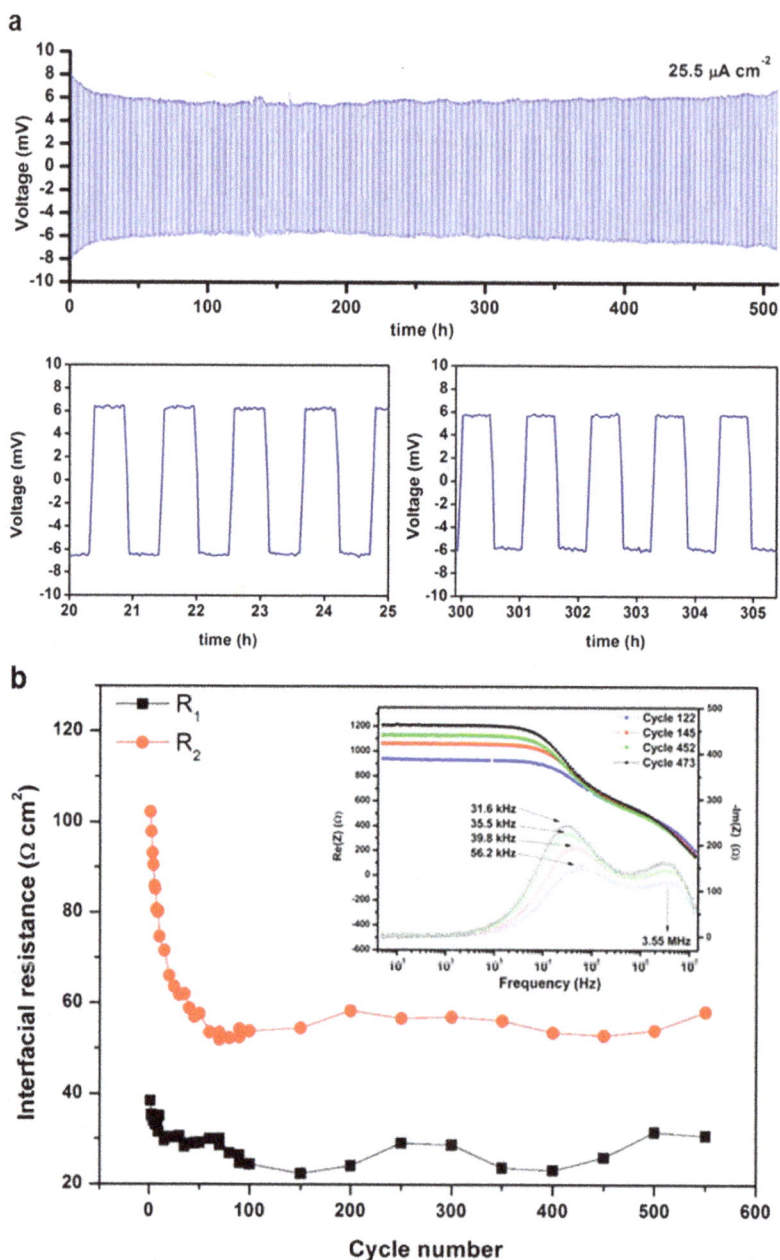

Fig. 8.13: a) Galvanostatic profile of the Na|Na$_4$(CB$_{11}$H$_{12}$)$_2$(B$_{12}$H$_{12}$)|Na cell, RT, 30-min sweeps b) Interfacial resistance evolution obtained by in situ EIS. A rapid decrease is observed while the battery is cycling for R2 (red dots), until stabilizing for the whole operating time. Less pronounced variation of R1 (black squares). Inset: frequency dependence of real and imaginary impedance for selected cycles (from [29]).

becomes possible, facilitating the adoption of all-solid-state battery technology at a larger scale [21].

Based on this method, NaCrO$_2$|Na$_4$(B$_{12}$H$_{12}$)(B$_{10}$H$_{10}$)|Na-Sn half-cells with mass loadings of up to 7 mg cm^{-2} were assembled and room temperature cycling at rates up to 5 C (4.5 mAcm^{-2}) was demonstrated. The Na-Sn anode was used to circumvent the Na dendrite formation occurring at high current densities and/or at lower temperatures. Capacity retention >80% of the initial capacity after 100 cycles at 0.5 C respectively was reached at an applied pressure of 3 MPa. So far, high rate capability requires over-dimensioned anodes. Capacity-balanced full cells show stronger capacity fading and only ~50% of the initial capacity was observed in a balanced NaCrO$_2$|Na$_4$(B$_{12}$H$_{12}$)(B$_{10}$H$_{10}$)|Na-Sn full cell at a rate of C/5 after 50 cycles [21].

Recently, a self-passivation of the cathode/solid electrolyte interface was observed in Na$_3$(VOPO$_4$)$_2$F|Na$_4$((CB$_{11}$H$_{12}$)2)(B$_{12}$H$_{12}$)|Na cells, enabling the stable room temperature operation of a 4 V-class all-solid-state battery without using artificial protective layers. This cell achieved the highest reported average discharge cell voltage and specific energy per cathode active material among all-solid-state sodium batteries [19].

8.8 Conclusion and outlook

In the past, the scientific community mainly considered insufficient ionic conductivity of solid-state electrolytes as the key bottleneck to develop all-solid-state batteries. Today, several classes of materials, including thiophosphates- and hydroborate-based compounds, show lithium- and sodium-ion conductivities in the range of 1 to 10 mS cm^{-1} at room temperature, which is comparable to typical conductivities of commercial liquid electrolytes. Prototypes of all-solid-state batteries clearly show that the (electro-)chemical stability, system integration, and the challenge to maintain the structural integrity of an all-solid-state cell during battery cycling are the main challenges to develop reliable all-solid-state batteries.

The emerging all-solid-state battery technology will compete with more mature lithium-ion technology that is aiming for higher energy and power densities, while avoiding the use of rare or critical raw materials. Besides all-solid-state lithium and sodium batteries, metal-sulfur and metal-air batteries are also in development as possible post-lithium-ion technologies. However, all-solid-state batteries, especially the hydroborate-based ones, are a young field and, despite the challenges, surprises are to be expected. The discovery of a means to increase the operating voltage beyond the electrochemical stability window, for example, by the use of protective coatings or by avoiding dendrite formation in conjunction with metallic anodes, could be a game changer. The race is open for a competitive all-solid-state battery technology to fulfill the promise of a safe, high energy density battery, exclusively made out of earth-abundant materials.

References

[1] Thangadurai, V., Narayanan, S., Pinzaru, D. Garnet-type solid-state fast Li ion conductors for Li batteries: Critical review. Chem. Soc. Rev. 2014, 43, 4714.

[2] Kamaya, N., Homma, K., Yamakawa, Y., Hirayama, M., Kanno, R., Yonemura, M., Kamiyama, T., Kato, Y., Hama, S., Kawamoto, K., Mitsui, A. A lithium superionic conductor. Nat. Mater. 2011, 10, 682.

[3] Wenzel, S., Randau, S., Leichtweiß, T., Weber, D.A., Sann, J., Zeier, W.G., Janek, J. Direct observation of the interfacial instability of the fast ionic conductor $Li_{10}GeP_2S_{12}$ at the lithium metal anode. Chem. Mater. 2016, 28, 2400.

[4] Zhang, Z., Shao, Y., Lotsch, B., Hu, Y.S., Li, H., Janek, J., Nazar, L.F., Nan, C., Maier, J., Armand, M., Chen, L. New horizons for inorganic solid state ion conductors. Energy Environ. Sci. 2018, 11, 1945.

[5] Kato, Y., Hori, S., Saito, T., Suzuki, K., Hirayama, M., Mitsui, A., Yonemura, M., Iba, H., Kanno, R. High-power all-solid-state batteries using sulfide superionic conductors. Nat. Energy. 2016, 1, 16030.

[6] Gao, H., Xue, L., Xin, S., Park, K., Goodenough, J.B. A plastic–crystal electrolyte interphase for all-solid-state sodium batteries. Angew. Chemie – Int. Ed. 2017, 129, 5633.

[7] Kim, J.-K., Lim, Y.J., Kim, H., Cho, G.B., Kim, Y. A hybrid solid electrolyte for flexible solid-state sodium batteries. Energy Environ. Sci. 2015, 8, 3589.

[8] Sadikin, Y., Brighi, M., Schouwink, P., Černý, R. Superionic conduction of sodium and lithium in anion-mixed hydroborates $Na_3BH_4B_{12}H_{12}$ and $(Li_{0.7}Na_{0.3})_3BH_4B_{12}H_{12}$. Adv. Energy Mater. 2015, 5, 1501016.

[9] Mohtadi, R., Remhof, A., Jena, P. Complex metal borohydrides: Multifunctionalmaterials for energy storage and conversion. J. Condens. Matter 2016, 28, 353001.

[10] Hansen, B.R.S., Paskevicius, M., Li, H.W., Akiba, E., Jensen, T.R. Metal boranes: Progress and applications. Coord. Chem. Rev. 2016, 323, 60–70.

[11] Duchêne, L., Kühnel, R.-S., Rentsch, D., Remhof, A., Hagemann, H., Battaglia, C. A highly stable sodium solid-state electrolyte based on a dodeca/deca-borate equimolar mixture. Chem. Commun. 2017, 53, 4195.

[12] Brighi, M., Murgia, F., Łodziana, Z., Schouwink, P., Wołczyk, A., Černý, R. A mixed anion hydroborate/carba-hydroborate as a room temperature Na-ion solid electrolyte. J. Power Sources 2018, 404, 7–12.

[13] Kim, S., Oguchi, H., Toyama, N., Sato, T., Takagi, S., Otomo, T., Arunkumar, D., Kuwata, N., Kawamura, J., Orimo, S.-I. A complex hydride lithium superionic conductor for high-energy-density all-solid-state lithium metal batteries. Nat. Comm 2019, 10, 1081.

[14] Hagemann, H. Boron hydrogen compounds for hydrogen storage and as solid ionic conductors. Chimia 2019, 73, 868–873.

[15] Payandeh, S., Asakura, R., Avramidou, P., Rentsch, D.L., Łodziana, Z., Černý, R., Remhof, A., Battaglia, C. Nido-Borate/closo-borate mixed-anion electrolytes for all-solid-state batteries. Chem. Matr. 2020, 32, 1101–1110.

[16] Muetterties, E.L., Balthis, J.H., Chia, Y.T., Knoth, W.H., Miller, H.C. Chemistry of boranes. VIII. Salts and acids of $B_{10}H_{10}{}^{-2}$ and $B_{12}H_{12}{}^{-2}$. Inorg. Chem. 1964, 3, 444.

[17] Asakura, R., Duchêne, L., Kühnel, R.-S., Remhof, A., Hagemann, H., Battaglia, C. Electrochemical oxidative stability of hydroborate-based solid-state electrolytes. ACS Appl. Energy Mater. 2019, 2, 6924–6930.

[18] Brighi, M., Murgia, F., Černý, R. Closo-hydroborate sodium salts as an emerging class of room-temperature solid electrolytes. Cell Rep. Phys. Sci 2020, 1, 100217.

[19] Asakura, R., Reber, D., Duchêne, L., Payandeh, S., Remhof, A., Hagemann, H., Battaglia, C. 4 V room-temperature all-solid-state sodium battery enabled by a passivating cathode/hydroborate solid electrolyte interface. Energy Environ. Sci., in press. 2020, Doi: 10.1039/d0ee01569e.

[20] Moury, R., Łodziana, Z., Remhof, A., Duchêne, L., Roedern, E., Gigante, A., Hagemann, H. Pressure-induced phase transitions in $Na_2B_{12}H_{12}$, structural investigation on a candidate for solid state electrolyte. Acta Cryst. B 2019, 75, 406–413.

[21] Duchêne, L., Kim, D.H., Song, Y.B., Jun, S., Moury, R., Remhof, A., Hagemann, H., Jung, Y.S., Battaglia, C. Crystallization of closo-borate electrolytes from solution enabling infiltration into slurry-casted porous electrodes for all-solid-state batteries. Energy Storage Matr. 2020, 26, 543–549.

[22] Kweon, K.E., Varley, J.B., Shea, P., Adelstein, N., Mehta, P., Heo, T.W., Udovic, T.J., Stavila, V., Wood, B.C. Structural, Chemical, and Dynamical Frustration: Origins of Superionic Conductivity in closo-Borate Solid Electrolytes Chem. Matr. 2017, 29, 9142.

[23] Udovic, T.J., Matsuo, M., Unemoto, A., Verdal, N., Stavila, V., Skripov, A.V., Rush, J.J., Takamura, H., Orimo, S. Sodium superionic conduction in $Na_2B_{12}H_{12}$. Chem. Commun. 2014, 50, 3750.

[24] Tang, W.S., Unemoto, A., Zhou, W., Stavila, V., Matsuo, M., Wu, H., Orimo, S.-I., Udovic, T.J. Unparalleled lithium and sodium superionic conduction in solid electrolytes with large monovalent cage-like anions. Energy Environ. Sci. 2015, 8, 3637.

[25] Takahashi, K., Hattori, K., Yamazaki, T., Takada, K., Matsuo, M., Orimo, S.-I., Maekawa, H., Takamura, H. All-solid-state lithium battery with LiBH4 solid electrolyte. J. Power Sources 2013, 226, 61.

[26] Unemoto, A., Ikeshoji, T., Yasaku, S., Matsuo, M., Stavila, V., Udovic, T.J., Orimo, S.-I. Stable interface formation between TiS_2 and $LiBH_4$ in bulk-type all-solid-state lithium batteries. Chem. Mater. 2015, 27, 5407.

[27] Yoshida, K., Sato, T., Unemoto, A., Matsuo, M., Ikeshoji, T., Udovic, T.J., Orimo, S.-I. Fast sodium ionic conduction in $Na_2B_{10}H_{10}$-$Na_2B_{12}H_{12}$ pseudo-binary complex hydride and application to a bulk-type all-solid-state battery. Appl. Phys. Lett. 2017, 110, 103901.

[28] Duchêne, L., Kühnel, R.-S., Stilp, E., Reyes, E.C., Remhof, A., Hagemann, H., Battaglia, C. A stable 3 V all-solid-state sodium–ion battery based on a closo-borate electrolyte. Energy Environ. Sci 2017, 10, 2609–2615.

[29] Murgia, F., Brighi, M., Černý, R. Room-temperature-operating Na-ion solid state-battery with complex hydride as electrolyte. Electrochem. Comm. 2019, 106, 106534.

[30] Tian, Y., Shi, T., Richards, W.D., Li, J., Kim, J.C., Bo, S.-H., Ceder, G. Compatibility issues between electrodes and electrolytes in solid-state batteries. Energy Environ. Sci. 2017, 10, 1150.

[31] Černý, R., Schouwink, P. The crystal chemistry of inorganic metal borohydrides and their relation to metal oxides. Acta Cryst. 2015, B 71, 619–640.

[32] Didelot, E., Sadikin, Y., Łodziana, Z., Černý, R. Hydrated and anhydrous dodecahydro closo-dodecaborates of 3d transition metals and of magnesium. Solid State Sci. 2019, 90, 86–94.

[33] Didelot, E., Łodziana, Z., Murgia, F., Černý, R. Ethanol- and methanol-coordinated and solvent-free dodecahydro closo-dodecaborates of 3d transition metals and of magnesium. Crystals 2019, 9, 372.

[34] Černý, R., Brighi, M., Murgia, F. The crystal chemistry of inorganic hydroborates. Chemistry 2020, 2, 805–826.

[35] Pauling, L. The principles determining the structure of complex ionic crystals. Am. Chem. Soc. 1929, 51, 1010–1026.

[36] Larsen, P.M., Schmidt, S., Schiøtz, J. Robust structural identification via polyhedral template matching. Modelling Simul. Mater. Sci. Eng. 2016, 24, 055007.

[37] Wang, Y., Richards, W.D., Ong, S.P., Miara, L.J., Kim, J.C., Mo, Y., Ceder, G. Design principles for solid-state lithium superionic conductors. Nat. Mater. 2015, 14, 1026–1031.

[38] Gautam, G.S., Canepa, P., Malik, R., Liu, M., Persson, K., Ceder, G. First-principles evaluation of multi-valent cation insertion into orthorhombic V_2O_5. Chem. Commun. 2015, 51, 13619–13622.

[39] Shannon, R.D. Revised effective ionic radii and systematic studies of interatomic distances in halides and chalcogenides. Acta Cryst. A 1976, 32, 751.

[40] Wu, H., Tang, W.S., Stavila, V., Zhou, W., Rush, J.J., Udovic, T.J. Structural Behavior of $Li_2B_{10}H_{10}$. J. Phys. Chem. C 2015, 119, 6481–6487.

[41] Verdal, N., Her, J.-H., Stavila, V., Soloninin, A.V., Babanova, O.A., Skripov, A.V., Udovic, T.J., Rush, J.J. Complex high-temperature phase transitions in $Li_2B_{12}H_{12}$ and $Na_2B_{12}H_{12}$. J. Solid State Chem. 2014, 212, 81–91.

[42] Hansen, B.R.S., Paskevicius, M., Jørgensen, M., Jensen, T.R. Halogenated Sodium-closo-Dodecaboranes as Solid-State Ion Conductors. Chem. Mater. 2017, 29, 3423–3430.

[43] Anurova, N.A., Blatov, V.A., Ilyushin, G.D., Blatova, O.A., Ivanov-Schitz, A.K., Demyanets, L.N. Migration maps of Li^+ cations in oxygen-containing compounds. Solid State Ion 2008, 179, 2248.

[44] Donnay, G., Allmann, R. How to recognize O^{2-}, OH^-, and H_2O in crystal structures determined by x-rays. Am. Mineral. 1970, 55, 1003–1015.

[45] Adams, S., Rao, R.P. Transport pathways for mobile ions in disordered solids from the analysis of energy-scaled bond-valence mismatch landscapes. Phys. Chem. Chem. Phys. 2009, 11, 3210–3216.

[46] Chen, H., Wong, L.L., Adams, S. SoftBV – a software tool for screening the materials genome of inorganic fast ion conductors. Acta Cryst. 2019, B 75, 18–33.

[47] Nose, S. A unified formulation of the constant temperature molecular dynamics methods. J. Chem. Phys. 1984, 81, 511–519.

[48] Henkelman, G., Uberuaga, B.P., Jonsson, H.A. A climbing image nudged elastic band method for finding saddle points and minimum energy paths. J. Chem. Phys. 2000, 113, 9901–9904.

[49] Burley, G. Polymorphism of silver iodide. Amer. Min. 1963, 48, 1266–1276.

[50] Bradley, J.N., Green, P.D. Relationship of structure and ionic mobility in solid MAg_4I_5. Trans. Faraday Soc. 1967, 63, 2516–2521.

[51] Sadikin, Y., Brighi, M., Łodziana, Z., Černý, R. Modified Anion Packing of Na2B12H12 in Close to Room Temperature Superionic Conductors. Inorg. Chem. 2017, 56, 5006.

[52] Maekawa, H., Matsuo, M., Takamura, H., Ando, M., Noda, Y., Karahashi, T., Orimo, S.-I. Halide-stabilized LiBH4, a Room-temperature lithium fast-ion conductor. J. Am. Chem. Soc. 2009, 131, 894–895.

[53] Matsuo, M., Remhof, A., Martelli, P., Caputo, R., Ernst, M., Miura, Y., Sato, T., Oguchi, H., Maekawa, H., Takamura, H., Borgschulte, A., Züttel, A., Orimo, S.-I. Complex Hydrides with $(BH_4)^-$ and $(NH_2)^-$ Anions as New Lithium Fast-Ion Conductors. J. Am. Chem. Soc. 2009, 131, 16389–16391.

[54] Hull, S., Keen, D.A., Gardner, N.J.G., Hayes, W. The crystal structures of superionic Ag3SI. J. Phys.: Condens. Matter 2001, 13, 2295.

[55] Brighi, M., Metal hydroborates as solid electrolyte for Li- and Na-ion batteries. Thesis no. 5257, University of Geneva (2018).

[56] Soulié, J.P., Renaudin, G., Černý, R., Yvon, K. Lithium boro-hydride LiBH4: I. Crystal structure. J. Alloys Compd. 2002, 346, 200.

[57] Hartman, M., Rush, J., Udovic, T., Bowman, R. Jr, Hwang, S.-J. Structure and vibrational dynamics of isotopically labeled lithium borohydride using neutron diffraction and spectroscopy. J. Solid State Chem. 2007, 180, 1298.

[58] Łodziana, Z., Van Setten, M.J. Binding in alkali and alkaline-earth tetrahydroborates: Special position of magnesium tetrahydroborate. Phys. Rev. B 2010, 81, 024117.

[59] Ikeshoji, T., Tsuchida, E., Ikeda, K., Matsuo, M., Li, H.-W., Kawazoe, Y., Orimo, S.-I. Diffuse and doubly split atom occupation in hexagonal LiBH$_4$. Appl. Phys. Lett. 2009, 95, 221901.

[60] Buchter, F., Łodziana, Z., Ph. Mauron, A., Remhof, O.F., Borgschulte, A., Züttel, A., Sheptyakov, D., Th. Strässle, A., Ramirez-Cuesta, J. Dynamical properties and temperature induced molecular disordering of LiBH$_4$ and LiBD$_4$. Phys. Rev. B 2008, 78, 094302.

[61] Remhof, A., Züttel, A., Ramirez-Cuesta, T., García-Sakai, V., Frick, B. Hydrogen dynamics in the low temperature phase of LiBH4 probed by quasielastic neutron scattering. Chem. Phys. 2013, 427, 18–21.

[62] Silvi, L., Zhao-Karger, Z., Röhm, E., Fichtner, M., Petry, W., Lohstroh, W. A quasielastic and inelastic neutron scattering study of the alkaline and alkaline-earth borohydrides LiBH$_4$ and Mg(BH$_4$)$_2$ and the mixture LiBH$_4$ + Mg(BH$_4$)$_2$. Phys. Chem. Chem. Phys. 2019, 21, 718–728.

[63] Filinchuk, Y.E., Yvon, K., Meisner, G.P., Pinkerton, F.E., Balogh, M.P. On the composition and crystal structure of the new quaternary hydride phase Li$_4$BN$_3$H$_{10}$. Inorg. Chem. 2006, 45, 1433–1435.

[64] Yan, Y., Kühnel, R.S., Remhof, A., Duchêne, L., Cuervo Reyes, E., Rentsch, D., Łodziana, Z., Battaglia, C. A lithium amide-borohydride solid-state electrolyte with lithium-ion conductivities comparable to liquid electrolytes. Adv. Energy Mater. 2017, 1700294.

[65] Burankova, T., Duchêne, L., Łodziana, Z., Frick, B., Yan, Y., Kühnel, R.S., Hagemann, H., Remhof, A., Embs, J.P. J. Phys. Chem. C 2017, 121, 17693–17702.

Christiane Stephan-Scherb

9 Crystallographic challenges in corrosion research

Abstract: High-temperature corrosion is a widespread problem in various industries. As soon as a hot and reactive gas (CO_2, O_2, H_2O, SO_2, NO_x, etc.) is in contact with a solid, physico-chemical processes at the surface and interfaces lead to material degradation. The processes are dynamic and controlled by thermodynamic and kinetic boundary conditions. Whether a reaction product is protective or not depends on various factors, such as chemical composition of the solid and the reactive media, surface treatment as well as diffusion and transport paths of cations and anions. Resulting chemical and structural inhomogeneities with the corrosion layers are characterized by off stoichiometry within cationic and anionic sub lattices. The competitive processes can be studied by various techniques of applied crystallography. This chapter gives an overview on the challenges of chemical-structural analysis of reaction products by crystallographic methods such as X-ray diffraction and X-ray near-edge structure spectroscopy and scanning electron microscopy electron backscatter diffraction (SEM-EBSD) for corrosion science.

Keywords: high-temperature corrosion, oxidation, diffraction, spectroscopy, oxides

9.1 Introduction

A variety of materials of technological interest change their properties through contact with reactive media. The types and properties of the respective reaction products are complex. Depending on the material (metal, ceramic, polymer) and the existing environment (gaseous, liquid, temperature, pressure), protective layers can form, or the material can be damaged more deeply by corrosive processes.

Damage caused by corrosion has a significant impact on the economy of an industry-based nation. Figure 9.1 shows an industrial turbine blade made of Ni-based alloys. The blade edges show significant damage caused by corrosion. The annual losses due to corrosion for the industry-based economy amount to 3–4% of the gross domestic product. If a component is damaged by corrosion, it must be replaced. Research into corrosion and protection mechanisms is essential to increase the lifetime of components that are susceptible to corrosion. By improving the corrosion properties of a material used, a direct impact on the resource efficiency of the manufacturing industry can be made [1].

Christiane Stephan-Scherb, Bundesanstalt für Materialforschung und -prüfung (BAM)

https://doi.org/10.1515/9783110674910-009

Fig. 9.1: Turbine blade of an industrial 96 MW gas turbine with inlet temperature of 990 °C after use. The enlarged leading edge shows distinct damage caused by high-temperature corrosion. Image source: BAM.

9.1.1 High-temperature corrosion phenomena

High-temperature gas corrosion of metallic alloys plays a crucial role for the safety of components in structural and high-temperature applications.

Due to their cost-effective use, medium alloyed Fe-Cr alloys (9–12% by weight chromium) are of strong technical relevance for various applications at medium to high temperatures (500–800 °C). In most applications, for instance as heat exchanger materials in biomass or co-fired power plants, the materials must resist high temperatures and harsh environments (CO_2, SO_2, O_2, H_2O, etc.). Figure 9.2 shows a light optical micrograph on a cross section of a ferritic model alloy with 9% in weight of chromium (Fe-9Cr) after the exposure to 0.5% SO_2 and 10% H_2O at 650 °C. Different contrasted areas are visible. Diffusion-reaction processes between incoming gas species and outgoing metallic species form the areas visible there. The white dashed line corresponds to the initial alloy surface. Everything above this line, toward the gas side, is called the external corrosion zone (ECZ). Everything below this line, the darker grey contrast within the alloy, is the inner corrosion zone (ICZ). The corrosion layers observed after gas exposure can be porous and non-protective. Whether a corrosion layer is protective or not especially depends on the composition and morphology of the layer phases, which in fact depends on the chemical potential of the metallic and gas components at the various interfaces. For instance, on pure iron or low-alloyed Fe-Cr steels, the stable phase in contact with SO_2 is Fe_2O_3 [2–4]. With decreasing chemical potential of the oxidizing species within the oxide layer, Fe_3O_4 is formed and

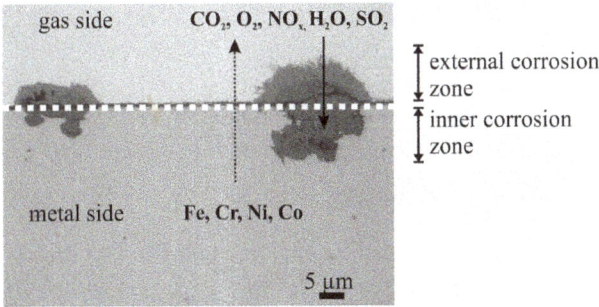

Fig. 9.2: Light optical micrograph of a Fe-Cr alloy with 9% chromium by weight after exposure for 5 h to 0.5% SO_2 and 10% H_2O at 650 °C.

known to be a fast growing oxide. By this, the production of voids and pores, leading to further outward diffusion of metal components and inward diffusion of anions, is favored. This supports fast material loss and spallation of the oxide layers until failure. The processes to form these corrosion layers are rather complex and must be understood on the interplay between external and internal interfaces and on the time scale.

9.1.2 From gas adsorption to layer thickening

The corrosion process is influenced by external and internal interfaces and various other factors. Thermodynamics and kinetics compete at different times and locations in the layer forming processes. On the one hand the composition of the solid and the gas species present have an impact. On the other hand, the temperature, the existing pressure of the gas, as well as the grain sizes and orientations and the presence of precipitates in the base material contribute to the overall process. One key parameter for the thermodynamic considerations, which oxide scale is formed, is the Gibbs free energy (G), defined as [5]:

$$G = H - TS = U + pV - TS \tag{9.1}$$

where H is the enthalpy, S the entropy, U the internal energy, V the volume of the respective system, p the pressure, and T the temperature. Since in alloys compositional changes through chemical reactions take place, a reversible internal energy change (dU) must be considered, defined as:

$$dU = TdS - pdV + \sum_i \mu_i dn_i \tag{9.2}$$

In an isothermal, isobaric system, the equilibrium is achieved when $dG = 0$ is observed. For these conditions, the molar free energy of a component is equivalent to its chemical potential μ_i [5].

$$\overline{G}_i = \left(\frac{\partial G}{\partial n_i}\right)_{T,p,n_{j\neq i}} = \mu_i \qquad (9.3)$$

Moreover, the chemical potential μ_i is a function of the temperature (T) and the chemical activity (a_c) of the respective oxide forming transition metal according to:

$$\mu_i = \mu_0 + RT ln a_c \qquad (9.4)$$

Figure 9.3 summarizes the development of chemical gradients schematically. Therein μ^M corresponds to the chemical potential of the oxide forming transition metal and chemical potential of oxygen. At the gas/oxide interface Fe_2O_3 is stable with M/O = 0.666, and at the alloy/oxide interface M_3O_4 is stable with M/O = 0.75. Thus, μ^M is high at the metal/oxide and low at the oxide/gas interface. μ^o behaves vice versa with μ^o is high at the oxide/gas and low at the metal/oxide interface. Experimental observation of oxide layers grown on Fe-9Cr show after 5 h of exposure strong gradients in metal(M)/oxygen(O) ratios from gas to alloy direction (X) (see Fig. 9.2).

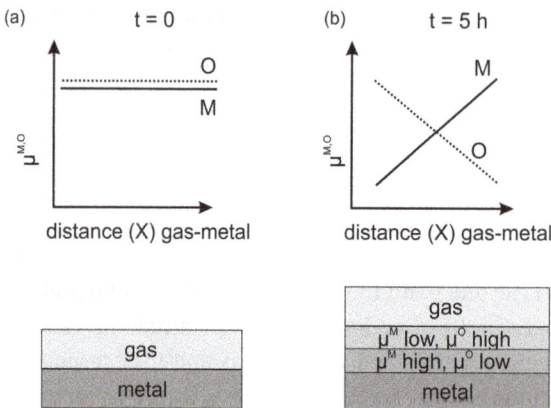

Fig. 9.3: Schematic representation of the chemical potential of M and O as a function of the gas-metal distance (X) for a) the initial situation at t = 0 and b) after an exemplary exposure time of 5 h.

As the corrosion process is highly dynamic, the chemical potential of the respective alloy elements like iron and chromium, and of the gas components (oxygen, sulfur, carbon, etc.), changes over exposure time and at different locations within the system "metal/oxide/gas." By this, the standard free energy for the formation of the respective reaction products is different for different locations within the corrosion layer.

Multiple, time-dependent reactions occur and each moment in time contributes to the outcome of the corrosion process. In the first step, adsorption and absorption of the gas specie dominate. As soon as an initial oxide layer is formed, Cabrera-Mott and Wagner theories are frequently applied to describe the growth kinetics of the oxide layers [6, 7]. Cabrera-Mott theory is valid for growth kinetics from an oxide layer thickness, ranging from 1 to 10 nm. For this early oxidation region, the growth is dominated by ionic transports through the layer, driven by an electric field between the oxide-metal and oxide-gas interfaces. At one point a limiting thickness is reached, and the electric field is diminished. From this limiting thickness on, Wagner theory applies, assuming a net charge neutral diffusion of ions and a field arising via ambipolar diffusion.

The kinetics of the oxide layer growth for high-temperature application can then mainly be expressed by a parabolic rate law, characteristic for diffusion-controlled processes.

$$\left(\frac{\Delta m}{A}\right)^2 = k_p \cdot t \tag{9.5}$$

The growth described by this law is strongly influenced by the outward diffusion of cations and inward transport of the gas components. Diffusion-dominated processes in this case can be described by Fick's second law, which demonstrated how diffusion causes the change in concentration over time:

$$\frac{\partial C}{\partial t} = D \frac{\partial^2 C}{\partial x^2} 0 \tag{9.6}$$

with D is the diffusion constant of the respective element.

As diffusion is a key aspect for oxide layer growth, the diffusion characteristics are strongly affected by the type of oxides grown on the alloy. When considering the oxidation of iron-chromium alloys, the Fe_3O_4 and $FeCr_2O_4$ formation plays thereby an important role. Due to the structural similarities the compositional range for a solid solution between them is wide. Ehlers et al. [8] demonstrated an almost continuous $(Fe,Cr)_3O_4$ layer on a Fe-9Cr steel after H_2O exposure at 650 °C. This layer slows the diffusivity of chromium down, due to the high crystal field preference energy (CFPE) of Cr^{3+} compared to the iron cations ($Fe^{3+} < Fe^{2+} < Cr^{3+}$). This causes a partitioning of the metal cations in the oxide scales [9]. Considering these aspects for diffusion, transport, and growth it is clearly evident that crystallographic aspects, such as the predominant structures and lattice matching capabilities, play a decisive role in the corrosion process.

9.2 Structural aspects of metal oxides

A number of metallic alloys can be traced back to the principle of close-packed lattices. A large part crystallizes in the face centered cubic (*fcc*), body centered cubic (*bcc*), and hexagonal close-packed (*hcp*) structures. The matrix of Ni-based and aluminum alloys, for instance, crystallizes in an *fcc* lattice. For Fe-based alloys, the most prominent phases are α-Fe (ferrite – *bcc*) and γ-Fe (austenite – *fcc*). The *fcc* structure of the austenite phase is stable only at temperatures > 912 °C or with Ni additions. In general, many properties depend on the crystal structure. Various elements are added to modify the properties such as corrosion resistance or increase in strength. These elements then either occupy a place in the lattice or form secondary precipitates in the overall structure. Chromium is often added to increase the corrosion resistance of Fe, Ni or Co alloys. This occupies the Fe, Ni or Co sites in the lattice in a statistically distributed manner. Another frequently added element is manganese. The variety of metallic elements in the alloy contributes to the formation of corrosion layers.

Considering only metal and oxygen, several reactions may take place, resulting in chemical and structural inhomogeneous oxide layers. In dependence on the aforementioned thermodynamic and kinetic aspects several reactions take place. It is possible to differentiate between the reactions of pure metal components with oxygen (I–III) and the reactions of already formed oxides with further incoming oxygen (IV–VI). A selection of possible reactions is summarized in the following:

Metal + Oxygen reactions:
I) $2M + O_2 \rightarrow 2MO$
II) $4/3\,M + O_2 \rightarrow 2/3M_2O_3$
III) $3/2\,M + O_2 \rightarrow 1/2M_3O_4$

Metal oxide + Oxygen reactions:
IV) $4MO + O_2 \rightarrow 2M_2O_3$
V) $6MO + O_2 \rightarrow 2M_3O_4$
VI) $M_3O_4 + O_2 \rightarrow 3M_2O_3$

In case of Fe-Cr alloys and their reaction with oxygen the most prominent oxides are: $Fe_{1-x}O$ (wüstite), Fe_2O_3 (hematite), Fe_3O_4 (magnetite), $FeCr_2O_4$ (chromite), and chromia (Cr_2O_3). The latter is the most protective oxide. In the following part, the structures of the most important oxides are revisited.

9.2.1 MO oxides

The mono-oxides, FeO (wüstite), MnO (manganosite), CoO, and NiO, crystallize, in a cubic sodium chloride (NaCl)-type structure (space group $Fm\bar{3}m$; see Fig. 9.4). The M^{2+} and O^{2-} form *fcc* cation and anion sub-lattices, which are translated by ½ 0 0. Metal and oxygen ions are mutually octahedrally surrounded by 6 neighbors each. This gives in theory the stoichiometric composition M:O = 1:1. However, it is well known that especially wüstite is highly defective and contains many defects on the cation sub-lattice [10, 11].

9.2.2 M₃O₄ oxides

The M_3O_4 oxides are given a special role in the corrosion layers. The multicomponent nature of engineering materials and the thermodynamic boundary conditions at the oxide-metal and metal-gas interfaces often lead to the formation of M_3O_4 mixed oxides with different protective properties. For instance, while Mn_3O_4 and Fe_3O_4 (magnetite) are known as fast growing, non-protective oxides, progressive material loss due to corrosion is minimized by the formation of $FeCr_2O_4$.

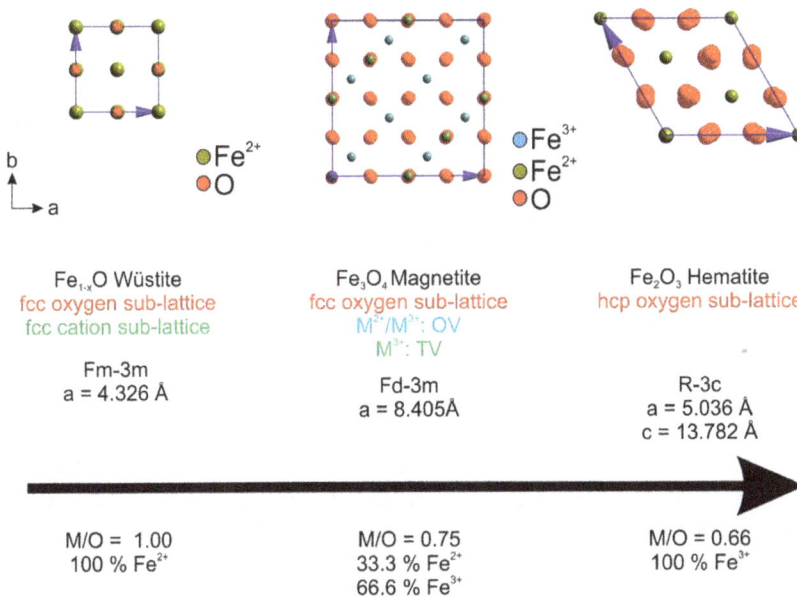

Fig. 9.4: Crystal structures of iron oxides frequently observed in surface layers produced by high-temperature corrosion phenomena.

Fe_3O_4 is known to be an inverse spinel with the tetrahedral sites ideally occupied by Fe^{3+} and the octahedral sites by Fe^{3+} and Fe^{2+} [12]. Chromite ($FeCr_2O_4$) is known to be a spinel with Fe^{2+} on the tetrahedral and Cr^{3+} on the octahedral sites [13]. Among the chromites frequently found in corrosion layers are also the Ni-chromite ($NiCr_2O_4$), the Mn-chromite ($MnCr_2O_4$), and the Co-chromite ($CoCr_2O_4$). The chromites with Ni^{2+}, Mn^{2+}, or Co^{2+} all crystallize, like $FeCr_2O_4$, in the spinel-type crystal structure. Due to the structural similarities, the compositional range for a solid solution between the different M_3O_4 oxides is wide. In the following, the remarkable characteristics in terms of high-temperature corrosion are explained in more detail.

As reported by Ehlers et al. [8] the exposure of Fe with 9% chromium by weight steel at 650 °C in H_2O atmosphere leads to an almost continuous $(Fe,Cr)_3O_4$ oxide layer. Considering the classical corrosion microstructure as presented in Chapter 9.1.1

(High-temperature corrosion phenomena; Fig. 9.2) into account, the co-formation of external and internal corrosion layers is frequently called duplex layer. These duplex layers are found frequently in CO_2 and H_2O environments [14, 15]. The external oxide layer constitutes magnetite, which can be topped with hematite, and the inner layer contains Fe-Cr spinel.

This spinel contains large amounts of voids and pores and is not protective against further oxidation. A detailed analysis of the inner oxide layer exposed to CO_2 and H_2O at 550 °C revealed a mean stoichiometry of $Fe_{2.3}Cr_{0.7}O_4$ [16] which corresponds more to a magnetite with chromium dissolved in the magnetite structure, than to chromite ($FeCr_2O_4$) phase. A similar observation for the characteristics of the inner oxide layer was presented by Weber et al. [17] after exposure of Fe-9Cr and Fe-13Cr to 0.5% SO_2 at 650 °C.

These observations are important when combining the coupling of diffusion characteristics and structural properties of the lattices. This correlation is demonstrated in Fig. 9.5. According to Azaroff the diffusion of cations in a close-packed oxygen sublattice, like it is the case for magnetite (inverse spinel – Fe_3O_4) or chromite (spinel – $FeCr_2O_4$), proceeds preferentially via alternate, adjacent octahedral, and tetrahedral positions (Azaroff mechanism) [18]. Regarding the difference in potential energy, it is easier for a cation to hop from a tetrahedral site to an octahedral site than vice versa. This theory directly links the importance of the predominant oxide phase and structure with the diffusion properties. The potential energy is expected to be related to the difference in the crystal-field preference energy (CFPE) for the different cations in the *fcc* oxygen sub-lattice. For the case of Fe-Cr alloys and the resulting oxides the CFPE for Fe^{2+}, Fe^{3+}, and Cr^{3+} in a close-packed oxide lattice follows $Fe^{3+} < Fe^{2+} < Cr^{3+}$. Cr^{3+} in the spinel-type crystal structure has the greatest and Fe^{3+} the lowest CFPE and the possible diffusion rate in the oxides and distance to be transported in relation to the original alloy surface should be larger for Fe^{3+} than for Cr^{3+}. This chemical partitioning of transition metal elements in oxide layers was first proposed by Cox et al. [9] and is illustrated in Fig. 9.5c.

9.2.3 M$_2$O$_3$ oxides

In this oxide class, Fe_2O_3 (hematite) and Cr_2O_3 (chromia – escolaite) and Mn_2O_3 (bixbyite) are prominent representatives, observed in corrosion layers. Hematite and chromia both crystallize in a rhombohedral corundum–type crystal structure (space group $R\bar{3}c$), see Fig. 9.4. This structure is characterized by a slightly distorted hexagonal close-packed (*hcp*) oxygen sub-lattice, with 2/3 of the voids between the octahedra occupied by Fe^{3+} or Cr^{3+}[19]. Hematite and chromia are isotypic and a continuous Fe_2O_3-Cr_2O_3 solid solution can form [20]. From a corrosion science perspective, chromia is the most protective oxide, and an intermixing with iron reduces the protective character of the oxide. In alloys in which Mn is added,

a)

b)

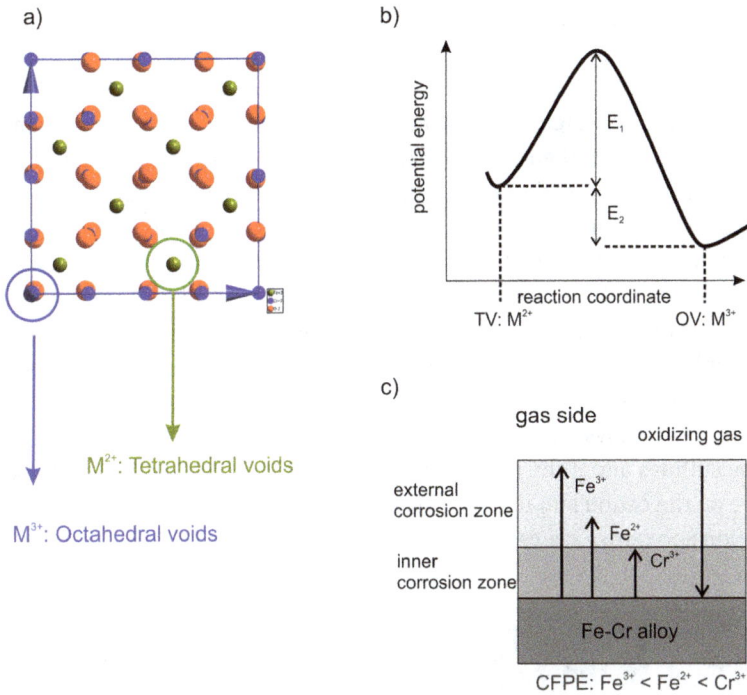

M^{2+}: Tetrahedral voids

M^{3+}: Octahedral voids

c)

Fig. 9.5: The relationship between crystal structure and diffusion mechanisms in spinel structures. a) spinel-type crystal structure. b) Energy barrier for hopping mechanism between octahedral (OV) and tetrahedral (TV) voids. c) Chemical partitioning of transition metal atoms in oxide layers.

additional formation of Mn-oxides often occurs. Mn$_2$O$_3$ (bixbyite) crystallizes in a cubic crystal structure with space group $I a\bar{3}$.

9.3 Sulfides

In fossil fuels, petrochemical processes and other combustion processes sulfur-containing gases such as SO$_2$, SO$_3$, or H$_2$S are often found in addition to oxygen and water. SO$_2$ in particular leads to catastrophic corrosion, because metal sulfides are very different from the corresponding oxides. These metal sulfides, which precipitate at the grain boundaries of the metals or within the oxide layers, thus minimize the corrosion protection by the oxide layers. In Fig. 9.6 a secondary electron image collected in a scanning electron microscope of the technical alloy VM12-SHC (12% chromium; ferritic-martensitic microstructure) is displayed. Below the inner corrosion zone small grey precipitates are clearly visible at the grain boundaries. These are metal sulfides that have a higher hardness than the base material. This leads to deep damage to the alloy, through grain boundary embrittlement

In the case of Fe-Cr alloys, $Fe_{1-x}S$, Cr_xS_y, or $(Fe,Cr)_xS_y$ occurs. For a model alloy made of iron with 13% by weight chromium, these precipitates were comprehensively characterized and identified as Cr_5S_6 [21]. The structure of the mono-sulfides differs strongly from the monoxide, due to the strong difference in anionic radii of oxygen (0.28 nm) and sulfur (0.368 nm). The mono-oxides are cubic, and oxygen forms an *fcc* sub-lattice. The transition metal sulfides commonly adopt the NiAs structure, in which the sulfur anions form a hexagonal close-packed (*hcp*) lattice. Here, the cations are located in the octahedral gaps formed by six anions in the hexagonally dense spherical packing, while the arsenic atoms are surrounded by six cations, which in turn form a trigonal prism. The Cr-sulfides all form defective NiAs structures [22]. In Fe-Cr alloys with low chromium concentration $Fe_{1-x}S$ is frequently observed. These sulfides adopt the NiAs structure type as well and are known to be highly defective on the cation sub-lattice. This supports accelerated cation diffusion through the cation sub-lattice via cation vacancies [23].

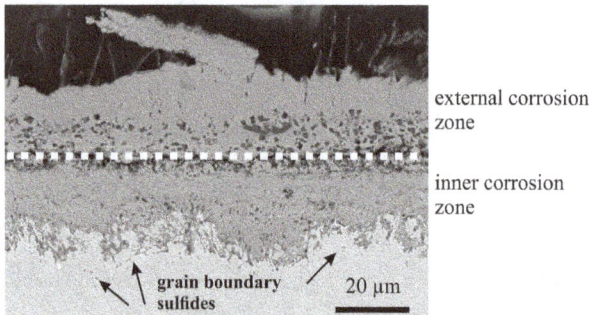

external corrosion zone

inner corrosion zone

grain boundary sulfides   20 μm

Fig. 9.6. Secondary electron microscopy image of a VM12 technical steel after 960 h exposure to an oxyfuel gas (27% H_2O; 60% CO_2; 1% SO_2; 10% N_2, 2% O_2) atmosphere at 600 °C. The dotted white line corresponds to the original sample surface [24].

9.4 Experimental approaches to access crystallographic challenges

In the previous sections, the basics of crystal structures of important alloys and their common corrosion products for high-temperature applications were discussed in more detail. In the following, the challenges for the analysis of corrosion processes from a crystallographic point of view will be discussed.

9.4.1 Microscopy

Different sophisticated microscopy techniques are used to visualize orientation relationships and initial corrosion mechanisms. In the initial stage of corrosion, adsorption and absorption processes of gas molecules at the alloy surfaces dominate and
bias the continuous growth of protective layers against further corrosion. The degree
of space filling of the lattice planes of the metals at the surface varies, depending on
the orientation of the crystal. For a *bcc* lattice the lattice plane with the highest atomic
density is the {110} plane, followed by the {100} and {111} planes. For the *fcc* lattice
the sequence is {111} > {100} > {110}. The notation of the {hkl} plane in curly brackets
{} contains all symmetry equivalent lattice planes to the defined plane. Accordingly,
for instance the $(\bar{h}k0)$, $(\bar{h}k0)$, and $(h\bar{k}0)$ planes belong to the {hk0}. Whereas (hkl) in
round brackets only corresponds to the notation of this specific lattice plane.

α-Fe and chromium both crystallize in *bcc* with a space filling of basic planes following the sequence {110} > {100} > {111}. The oxidation of Cr (100) at room temperature
starts with incorporation of oxygen atoms into the 4-fold symmetry. The oxidation of
Ni–Cr below 700 °C, with Cr < 20% Cr, begins with the formation of NiO (cubic-rock
salt type crystal structure) instead of Cr_2O_3 formation (rhombohedral corundum–type
crystal). The activation barrier for nucleation on a cube-on-cube epitaxy (cubic Ni
alloy – cubic NiO) is lower compared to a nucleation of a rhombohedral structure on a
cubic alloy. By this, NiO dominates the initial oxide layer, and Cr_2O_3 subsequently
forms at the alloy–oxide interface by internal oxidation [25, 26]. Blades et al. [27] cap-
tured these early stages of Ni-Cr (100) oxidation by scanning tunneling microscopy
and spectroscopy. The authors confirmed the cube-on-cube epitaxy by the observation
of NiO-Ni (6x7) and NiO-Ni (7x8) coincidence lattices. The subsequent statistical analysis of the geometric features of the surface oxide including step edge heights, and NiO
wedge angles illustrates the layer-by-layer growth mode of NiO and the restructuring
of the alloy–oxide interface during the early stages of the oxidation process. This cube-
on-cube epitaxy for NiO growth on Ni (001) was confirmed by a study presented by
Luoa et al. [28]. The authors studied the early stages of oxidation of Ni (001) thin films
with 10 or 20% chromium at 700 °C and directly visualized the process by in situ transmission electron microscopy (TEM). In the first stages, NiO (cubic rock salt–type crystal
structure) islands nucleate, which coalescence with further oxidation to a continuous
film.

For Fe-Cr alloys, the reduced misfit between two cubic phases' results in the
preferred formation of a further cubic phase compared to rhombohedral structures. An early study by Leygraf et al. [29] showed that the most probable adsorption site of O is the center of the unit cell of the (100) plane of ferrite. As soon as
all centers are occupied a monolayer of a film with *fcc* symmetry forms. The *fcc*
type oxide acts as a base for further spinel oxide formation during prolonged oxidation. Orientation relationships between Fe-Cr alloys and corrosion products formed
after high-temperature SO_2 exposure have been demonstrated in a scanning electron

microscope by electron backscatter diffraction (SEM-EBSD) [30]. On Fe with only 2% of chromium by weight a clear epitaxial relationship of magnetite (Fe_3O_4) and ferrite following $<110>_f || <111>_m$ was observed. A hematite (Fe_2O_3) layer on top of magnetite was not observed in this low-alloyed material, but on Fe with 9% by weight chromium. In this case, the hematite above magnetite has a clear [001] orientation. This is explained by the coincidence of the *fcc* and *hcp* oxygen sub-lattices for magnetite and hematite, respectively. Between the spinel and the magnetite, the dense oxygen packing only must change from *fcc* to *hcp*, respectively [31]. The coincidence of these two oxygen sub-lattices is illustrated in Fig. 9.7. Chromia (Cr_2O_3), crystallizing in the rhombohedral corundum–type crystal structure, was not observed in this study. It was substantiated that the high strain rates between cubic ferrite and chromia additionally limit the formation of a dense Cr_2O_3 oxide layer in the initial stages of high-temperature SO_2 corrosion for medium- and low-alloyed (Cr < 16%) ferritic alloys.

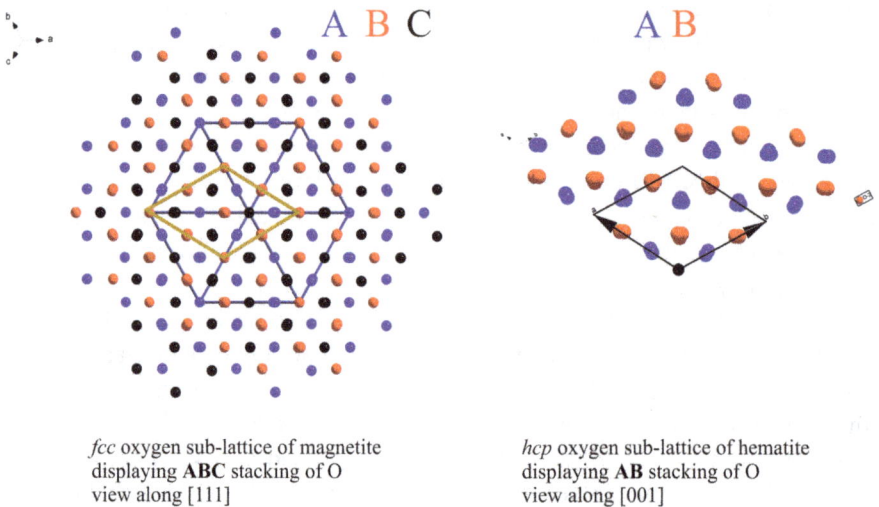

fcc oxygen sub-lattice of magnetite displaying **ABC** stacking of O view along [111]

hcp oxygen sub-lattice of hematite displaying **AB** stacking of O view along [001]

Fig. 9.7: Oxygen sub-lattices of cubic magnetite (Fe_3O_4) and rhombohedral hematite (Fe_2O_3). For magnetite the cubic (blue) and the trigonal (yellow) arrangement of the unit cell is displayed.

9.4.2 Diffraction: in situ

A large number of reaction products of high-temperature corrosion processes are crystalline and long-range order of lattice atoms is achieved. At a very early stage of corrosion, the nucleation of the reaction products in the inner corrosion zone can begin with the accumulation of the elements involved. This has been observed for internal sulfidation on Fe-Cr alloys with a Cr content of >9% [4]. The inner sulfidation started by sulfur accumulation in the inner corrosion zone. These phases are not necessarily crystalline and cannot be detected by diffraction at this stage in the internal corrosion zone. The growth happens mainly at the alloy surfaces and sub-surface interfaces. This enables X-ray diffraction as a powerful tool for the identification and characterization of reaction products and their growth behavior. The analysis of diffraction pattern of heterogeneous corrosion layers is challenging, due to strong structural similarities, for instance, between the MO, M_2O_3, and M_3O_4 oxides. Even strong similarities occur between the MO, M_2O_3, and M_3O_4 oxides; the XRD patterns of Fe_3O_4 and Fe_2O_3 are quite different and clearly distinguishable. However, the interdiffusion of metal ions like Fe and Cr into the oxides has an impact on the lattice dimensions and therefore on the position and the shape of the reflections. Thus, the separation of the contribution of, for instance, the structural similar magnetite (Fe_3O_4) and chromite ($FeCr_2O_4$) on a specific reflection corresponding to a spinel-type crystal structure can be challenging. Nevertheless, a careful inspection and comparison of simulated and collected diffraction pattern allow for the analysis of nucleation, growth, and dissolution of corrosion products.

To perform XRD in situ in a high-temperature corrosion environment, different requirements must be fulfilled on the analytical and on the corrosion experiment side. To expose the alloys to hot reactive gases, a high-temperature furnace with gas supply is required, which ensures additionally gas tightness. On the analytical side this furnace must be transparent for X-rays and the detection must enable the fast collection of enough intensities of the reaction's products even when only thin layers grow. The application of conventional lab XRD in θ-2θ or grazing incidence geometry using Cu-Kα radiation is challenging for in situ studies of corrosion phenomena, because often the given conditions of sample environment and X-ray intensity cannot be guaranteed with short data collection time. Kolarik et al. [32] studied the oxidation of ARMCO-iron in air at high temperatures (from 500 °C to 1,000 °C) by energy-dispersive X-ray diffraction (EDXRD) using a tungsten tube and a Si(Li) detector. The measurements took place under a constant diffractometer angle of $2\theta = 20°$. For the identification of Bragg-peaks search-match procedures, as applied for conventional diffraction pattern analysis, are inadequate since the relevant databases use the classical Bragg equation according to:

$$n\lambda = 2d_{hkl}\sin\theta \qquad (9.7)$$

with λ is the wavelength, d_{hkl} is the lattice spacing, and θ the Bragg angle.

For energy-dispersive X-ray diffraction high-energy white X-ray radiation covering a broad spectrum of energies or wavelengths is used instead of monochromatic radiation. The crystal phases occurring in the corrosion zones can be unambiguously assigned via their d_{hkl} values and can also be structurally characterized. The theoretic Bragg reflections of the reaction products can be calculated using the structural data of the expected compounds at room temperature and applying the energy-dispersive Bragg-equation:

$$E_{hkl} = \frac{hc}{2d_{hkl}\sin\theta}$$

(9.8)

where c is the velocity of light, h is Planck's constant, E_{hkl} is the energy of the radiation, and θ is the diffraction angle. The diffraction signal reveals a broad spectrum of structural information of the incorporating phases. The crystallization and the growth of the crystal phases can be correlated with their structural and chemical properties via the development of the diffraction signals over time, under isothermal conditions, applying a fixed gas partial pressure.

The following paragraph describes a case study of combined oxidation/sulfidation mechanisms studied by EDXRD during high-temperature corrosion. As first the experimental design must be considered. The analysis of corrosion in controlled gas mixtures is rather challenging compared to oxidation studies in air as reactive medium. There must be paid attention to the corrosivity and toxicity of gases such as SO_2 for the design of the sample environment for in situ EDXRD. The key part for the construction of such environmental chambers is the choice of the window material, sealing, and the temperature management within the reaction chamber as well as at the gas in- and outlets. For the analysis of high-temperature corrosion under SO_2 and SO_2/H_2O mixed atmosphere a tailor-made reaction chamber was used [2, 33].

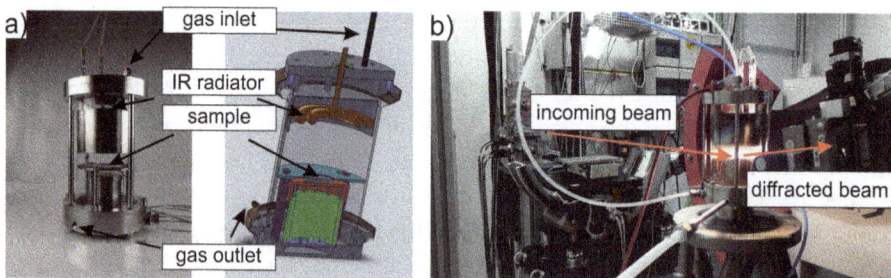

Fig. 9.8: a) Photograph and construction sketch of the tailor-made reaction chamber for high-temperature corrosion studies under aggressive atmosphere by X-ray diffraction. (b) Reaction chamber installed inside the experimental hutch on a HEXAPOD 6-axis positioning system.

This chamber consists of a quartz glass cylinder installed into stainless steel made covers on top and on the bottom. Both top and bottom covers are cooled by an integrated water cooling system. The top cover includes a high-power (2,200 W) infra-red radiator. The radiator allows for fast heating of the sample and the gas stream on an area of at least 35 mm in diameter. The gas stream enters and leaves the chamber via the integrated gas in- and outlets. Flow meters outside the chamber guarantee a continuous and controlled gas supply. The sample temperature and the temperature of the environment around are registered by two thermocouples. The first one is placed inside of the sample. The sample holder is made of an Inconel 725 plate and connected to springs to correct the thermal expansion of the sample and the sample holder by back pressure.

The X-rays necessary for data collection during the corrosion experiment are attenuated by the quartz glass walls of the chamber. To achieve high X-ray intensities during the corrosion experiment highly energetic synchrotron radiation was used. The high-energy radiation enables the in situ investigation of crystallization processes which take place on short time scales and have to be studied in gas-tight reaction chambers. By measuring Bragg signals of the crystalline material, their occurrence can be correlated with thermodynamic and kinetic parameters, such as sulfur and oxygen partial pressure, temperature, and reaction time. The EDXRD experiments took place at the energy-dispersive diffraction beamline EDDI at BESSY II synchrotron in an energy range of $E = 10–120$ keV. For the EDXRD corrosion experiment, the presented reaction chamber was tailor-made for the EDDI endstation (Fig. 9.8a) and mounted onto a 6-axis positioning system (HEXAPOD) within the EDDI cabin (Fig. 9.8b).

9.4.2.1 Ferritic model alloys exposed to 0.5% SO_2 at 650 °C

Figure 9.9a shows the development of the corrosion products as a function of time during the reaction of Fe with 2% Cr by weight (Fe-2Cr) with the gas mixture 0.5% SO_2 / 99.5% Ar at T = 650 °C for 5 h (300 min). Heating to 650 °C took place under argon atmosphere. The patterns are shown as a function of time. Red color refers to high and blue to low intensity in the diffraction pattern. The red line, for instance, at $E = 58$ keV corresponds to the α-Fe 110 reflection. It changes its intensity for Fe-2Cr and stays constant for Fe-9Cr. At the initial stage of the experiment, the oxygen partial pressure within the chamber is very low since SO_2 slowly floats the reaction chamber at a rate of 1 mL/min. At this low p_{O2}, FeO initially forms on Fe-2Cr as observed by the distinct reflection at $E = 54.5$ keV. Once sufficient SO_2 is present in the chamber, FeO is oxidized to form FeS and Fe_3O_4 (see reflections at 56 keV and 46 keV, respectively). In an early stage of corrosion an oxide is clearly formed and the sulfide growth then takes place parallel to the oxide formation. With proceeding reaction the layer thickens and the formation of an oxygen gradient within the corrosion layer is observed. This leads to a low oxygen partial pressure at the alloy/scale interface after 100 min of reaction, which promotes once again the formation of FeO. Parallel to the formation of oxides and sulfides, the intensity of the ferrite 110 reflection decreases, continuously. This loss of intensity indicates an increased absorption of the X-ray beam by the corrosion layer thickening. The mean linear attenuation coefficient at $E = 58$ keV (energy position of the Fe 110 reflection) is $\bar{\mu} = 4.84$ cm^2/s for a mixture of Fe_2O_3, Fe_3O_4 and FeS, as found in the corrosion layers. The quantitative phase analysis of a comparable sample was used to determine a mean value for the attenuation coefficient. In the analysis of a previous sample the fraction of the individual phases in the cross section was determined by light microscopy and analysis of the layer components with the help of image analysis software. Details about this procedure can be found in reference [4]. The percentage of Fe_2O_3, Fe_3O_4, and FeS, as well as the respective attenuation coefficients was used to obtain a mean value for the phase mixture.

Using the law of Lambert-Beer, the absorption of the X-rays by a mixture of Fe_2O_3, Fe_3O_4, and FeS results in a loss of intensity of 12% for a corrosion layer with a thickness of 30 μm. The absorption of the incident X-ray by the growing layer can therefore not be responsible solely for the loss of intensity of the ferrite reflection. The experiments took place under a constant focal plane. The loss of intensity of the ferrite reflection can also be attributed to a reduced diffracted volume of ferrite. The reaction of Fe-2Cr with the hot SO_2-containing gas produces oxides and sulfides at the interface between metal and gas. The oxide / sulfide layer grows into the direction of the gas atmosphere by progressive diffusion of the Fe atoms through the scale from the alloy. With proceeding growth more and more metal ions from the original alloy react to form oxides and sulfides. The degradation of the alloy and the loss of the original surface are reflected in the in situ EDXRD experiment by the intensity decrease of the ferrite 110 reflection with increasing reaction time.

The ferritic iron with 9% Cr by weight (Fe-9Cr) reacts with 0.5% SO_2 / 99.5% Ar at T = 650 °C and forms an oxide first (Fig. 9.9b)). The ferrite reflection does not lose intensity, which indicates a much thinner corrosion layer after 24 h of reaction, compared to iron with 2% Cr. This assumption was confirmed by metallographic analysis on cross sections after the EDXRD experiments. The energy position of the first occurring reflections index a mixed Fe-Cr oxide, such as $(Fe,Cr)_3O_4$, as initial phase. Sulfides and other iron oxides (Fe_3O_4, Fe_2O_3) occur as the reaction time proceeds. A reduction in the intensity of the ferrite 110 reflex with ongoing reaction is not observed. In the subsequent analysis of the metallographic cross section of the Fe-9Cr sample, it was shown that Cr-sulfides are located in the inner corrosion zone below the $Fe_2O_3/Fe_3O_4/(Fe,Cr)_3O_4$ layer stack at the interface to the alloy. The following mechanism is proposed after the analysis by in situ EDXRD data and the post-characterization of the corrosion layer cross section.

The Fe-9Cr alloy reacts with SO_2 in the beginning and forms at first a mixed Fe-Cr-oxide. This oxide is not completely dense and closed and sulfur is continuously transported via pores, cracks, and grain boundaries throughout the growing oxide layer towards the oxide/alloy interface. The alloy surface gets more and more iron depleted, since Fe-ions diffuse through $(Fe,Cr)_3O_4$ into the gas direction and react there to Fe_3O_4 and Fe_2O_3 at the gas/scale interface. Sulfur diffuses through the scale and reacts at the oxide/alloy interface with the Cr-enriched surface of the alloy to Cr_5S_6. The two processes, the external oxide growth and the inner sulfide growth, cause the typical scale structures as observed by previous studies [4] and by the in situ EDXRD experiments.

Both studied alloys showed oxides as first reaction products. The corrosion layers grown on Fe-2Cr exhibited a significant amount of FeO (wüstite) and FeS

a)

b)

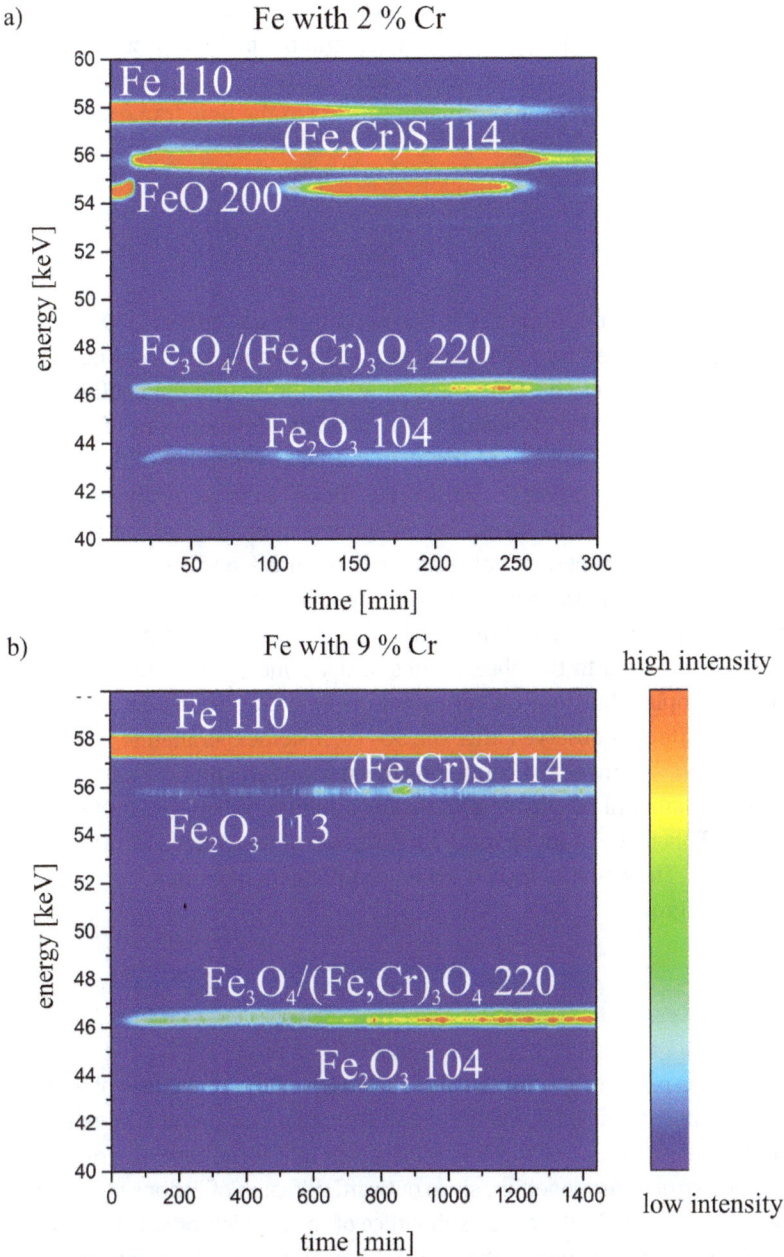

Fig. 9.9: a) EDXRD pattern (30 sec/pattern) in the energy region of 40–60 keV as a function of time of Fe with 2 % by weight Cr during the reaction with 0.5% SO_2/99.5% Ar at T = 650 °C for 5 h (300 min) b) EDXRD pattern (120sec/pattern) in the energy region of 40–60 keV as a function of time of Fe with 9 wt.% Cr during the reaction with 0.5% SO_2/99.5% Ar at T = 650 °C for 24 h (1,400 min).

(troilite). Both compounds are known to be highly defective on the cation sub-lattice, which accelerate the iron outward diffusion and the material loss by the corrosion process. On Fe-9Cr Cr-enriched oxides and sulfides are observed. Wüstite was not identified at any point of the corrosion process in the time frame studied by in situ EDXRD.

9.4.2.2 Ferritic model alloys exposed to 0.5% SO$_2$ and H$_2$O at 650 °C

In order to study the impact of humidity in SO$_2$ containing gas mixture, 30% of H$_2$O were additionally introduced into the gas stream. The heating and cooling procedures again took place under constant argon flow. All diffraction data collected by in situ EDXRD were analyzed with regard to the intensity development of the individual phases. The integral intensity of the characteristic reflections of the individual phases was extracted by fitting the single reflections of the respective phases to a Gaussian function. Figure 9.10 shows the integrated intensities of the corrosion products as a function of experimental time for Fe with 2% and 9% chromium as collected by in situ EDXRD in 0.5% SO$_2$/ 30% H$_2$O with argon as balancing gas.

In both corrosion layers, oxides are the first phases occurring immediately after experiment start. Compared to the observations in dry atmosphere, wüstite (FeO) is present on both samples, but to a higher extent on Fe with only 2% Cr. In the dry atmosphere no wüstite was observed after the SO$_2$ corrosion experiments. The water-containing atmosphere changes the conditions toward the formation of highly defective wüstite. A high amount of wüstite also means a high number of vacant sites on the cation lattice. This leaves a lot of space for diffusion of the cations from the alloy toward the gas side. The material thus loses its components more quickly and a porous corrosion layer grows.

9.4.3 X-ray absorption near-edge structure spectroscopy

The previous sections provided a comprehensive overview of the challenges of evaluating corrosion phenomena caused by chemical and structural changes at different interfaces. The protective properties of the corrosion layers that form depend strongly on their chemical and structural properties, since outward diffusion of cations mainly proceeds via vacant cation sites in the cation sub-lattice of the reaction products. To better understand diffusion and transport paths of metal ions and gas species via structural defects within the heterogeneous corrosion layers, it is beneficial to achieve a deeper knowledge about the valence states of reaction products. Analyses for the determination of not only the chemical compositions of the layers, extended to phase identification with sufficient lateral resolution, are rare. Several oxide phases within the corrosion layers demonstrate structural similarity. For instance, it is rather

a)

Fe with 2% Cr

b)

Fe with 9% Cr

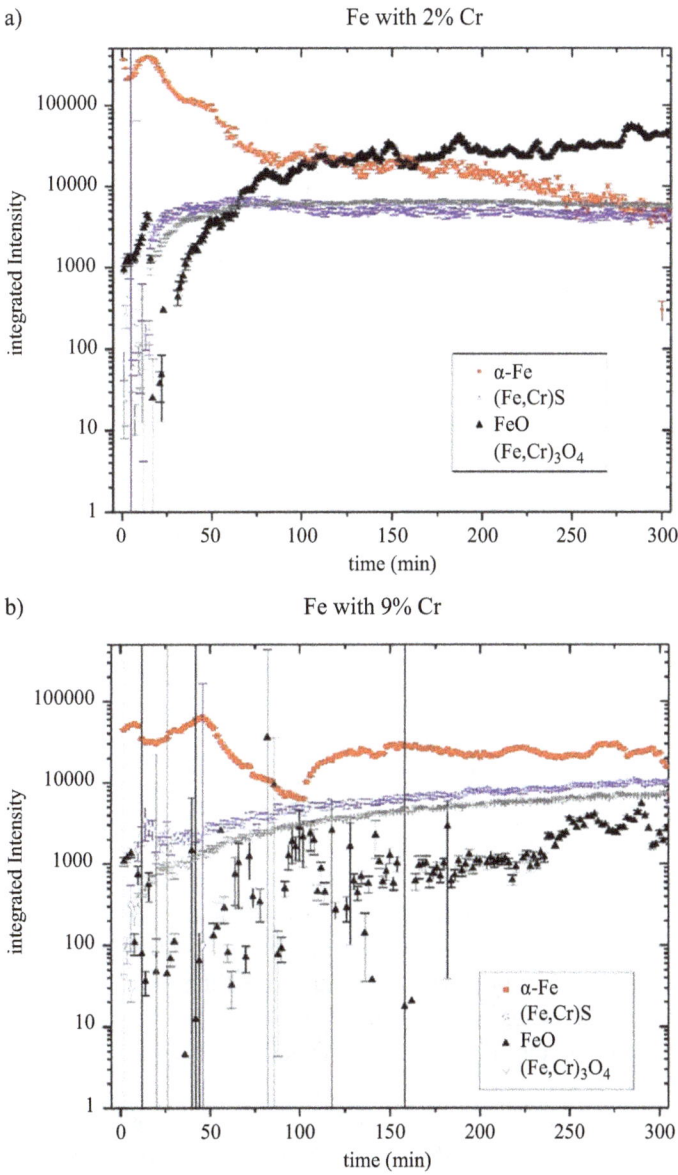

Fig. 9.10: Integrated intensities extracted from EDXRD pattern as a function of experiment time for reaction products grown on a) Fe-2Cr and b) Fe-9Cr exposed to 0.5% SO_2 and 30% H_2O at 650 °C.

challenging to differentiate the two oxides magnetite – Fe_3O_4 and chromite – $FeCr_2O_4$ (both space group $Fd\bar{3}m$), by conventional techniques such as XRD, SEM-EBSD, or TEM analysis in thin films with small grains and texture effects. The application of X-ray absorption near-edge structure spectroscopy (XANES) enables the structural and chemical information of the reaction products to be obtained. Based on the photoionization effect the X-ray absorption spectrum of an element of interest in a corrosion zone, such as an oxide, yields valuable insights into the chemistry and structure of the respective phase. The spectrum contains information about the local geometry around the absorbing atom, the oxidation state, bonding characteristics and interatomic distances and bond angles. On Fe-Cr corrosion layers, XANES spectroscopy at the Fe- and Cr-K-edges renders information on the various possible corrosion products based on the local atomic environment of the element of interest. The characteristic spectra for the different compounds are used as a fingerprinting method to identify the various corrosion products within the heterogeneous scale.

For μ-beam XANES analysis of corrosion zones the experiments are frequently applied on metallographic cross sections (see Fig. 9.11). In Fig. 9.11a the beam path as used for μ-beam is schematically illustrated. More details about the experimental setup can be found in Ref. [17]. The supplementary analysis by X-ray diffraction, light optical microscopy, and electron beam analyses using a SEM or an electron microprobe system is highly beneficial. A smart combination of supplementary analytics for XANES even enables quantitative measures about phase contents in very heterogeneous corrosion zones.

Addressing valence states and the local atomic environment of the metal within the respective oxides, μ-XANES was already shown to yield distinct insides into oxidation mechanisms. A study by Ja'baz [34] applied μ-beam XANES and synchrotron X-ray fluorescence mapping to analyze the spatial distribution of Cr-bearing species in corrosion layers grown after exposure of alloys used as heat exchanger tubes to an oxy-fuel environment (CO_2^{bal}; SO_2; O_2; H_2O; N_2). To qualitatively reveal the various species possibly present in corrosion zones, well-defined reference materials are necessary. For instance, the analysis of the corrosion zones after oxyfuel firing requires reference material of Cr_3C_2, Cr_2S_3, $FeCr_2O_4$, and Cr_2O_3. The valence state of chromium is in all compounds 3 +, but structurally they are different, leading to a different characteristics in the absorption spectra. The μ-beam XANES application led to a precise identification and localization of Cr_2C_3 and $FeCr_2O_4$ within the corrosion zones of the analyzed samples.

Another study by Couet et al. [35] used μ-beam XANES to investigate the evolution of oxidation states of Fe and Nb in the corrosion layers of various industrial zirconium alloys, which are used for nuclear fuel cladding. Several Fe and Nb references were used to determine by linear combination fitting (LCF) the oxidized fractions of Fe and Nb as a function of distance from the alloy/oxide interface. As a complementary technique, the authors applied transmission electron microscopy (TEM) to qualitatively confirm the results obtained by μ-XANES. In another work, Couet et al. [36] applied the knowledge of the Fe and Nb oxidation state evolution to build a new model for zirconium alloy oxidation explained by a coupled current charge oxidation model. For this purpose it was assumed the space charges of the

a)

0 X-ray beam
1 monochromator
2 lens system
3 ion-chamber (I_0)
4 samples
5 fluorescence X-rays
6 Si-drift detector

b)

0 X-ray beam
1 epoxy
2 corrosion layer
3 metal
4 fluorescence X-rays
5 Si-drift detector

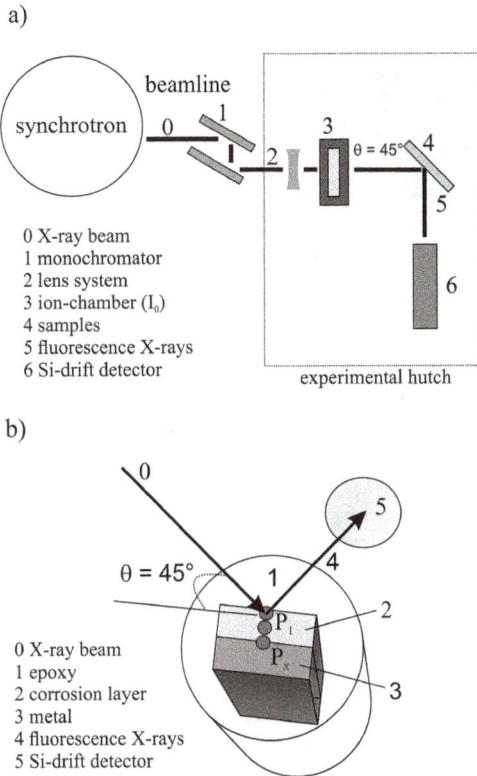

Fig. 9.11: General experimental setup for μ-beam X-ray absorption near-edge structure spectroscopy. a) Shows the beamline layout and b) the measurement strategy on metallographic cross sections.

oxide layers are compensated by the present Nb ions. In this context the XANES observations showed that there are enough aliovalent Nb ions in the oxide layers to verify oxide electroneutrality as predicted by the model.

The evolution of oxidation states of Fe as a function of the alloy/oxide interface on ferritic Fe-Cr alloys exposed at 650 °C to 0.5% SO_2 using μ-beam XANES was studied by Weber et al. [17]. Independent from the Cr content, a clear compositional gradient could be observed in the first step by electron microprobe analysis (see Fig. 9.12a). The XANES spectra show a clear shift in the absorption edge of Fe (E_0) as a function of the distance from the alloy/oxide interface. This shift is a clear indication of the change in oxidation states within the oxide layer (see Fig. 9.12b). The amount of Fe^{2+} and Fe^{3+} could be quantified by linear combination fitting (LCF) of previously recorded XANES spectra of reference compounds, as well as by complementary analysis using EMPA.

By μ-beam XANES a high concentration of Fe^{3+} cations at the gas interface and a high concentration of Fe^{2+} cations on the alloy interface could be observed. This gradient is independent of corrosion layer thickness and exposure time for iron

a)

gas side

metal side

b)

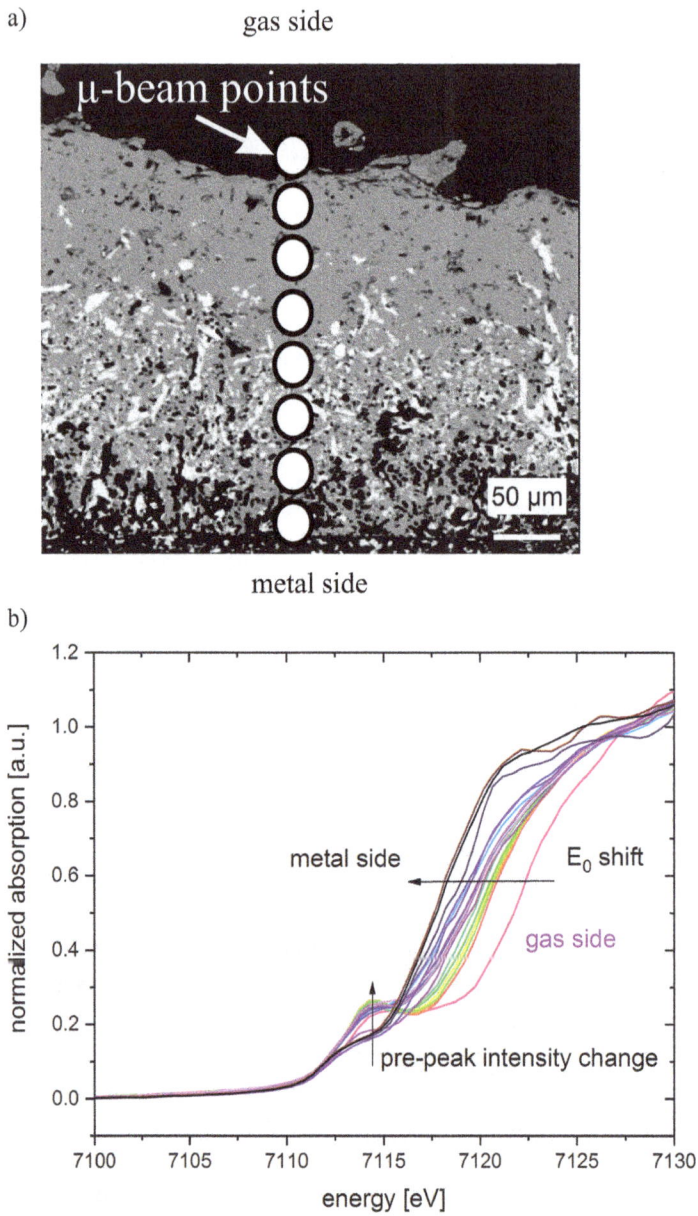

Fig. 9.12: Fe-2Cr exposed to 0.5% SO_2 at 650 °C after 250 of exposure. a) BSE image collected in an electron microprobe system and b) µ-beam XANES spectra collected at the iron K-absorption edge at different points of the corrosion layer.

with only 2% of chromium (Fe-2Cr) and more pronounced for the higher-alloyed materials (Fe-9Cr and Fe-13Cr). A higher concentration of Fe^{3+} within the scale grown on Fe-13Cr after 250 h of exposure is observed not only at the gas interface, but also within the center of the corrosion layer, too. Thus, a high amount of Fe_2O_3 was formed there and the external oxide layer consists of two types of oxides, Fe_2O_3 and Fe_3O_4. The inner corrosion zone shows localized areas of $FeCr_2O_4$, and a not negligible amount of chromium within this layer is dissolved within the magnetite phase. This was not intentionally expected from the EMPA results. The separation of Fe^{3+}, Fe^{2+} and Cr^{3+} within the corrosion layers is consistent with the chemical fractioning model according to Cox et al. [9] explained in a previous section here (see M_3O_4 oxides). This study demonstrated that the CPFE of the cations within a close-packed oxide lattice is shown to play an important role for the formation and stability of corrosion layers formed under combined oxidizing/sulfurizing atmospheres.

9.5 Concluding remarks

The interaction of hot and reactive gases with metallic alloys changes the chemical and structural nature of the surface and various interfaces. The corrosion process can be divided into material gain and loss forming different corrosion products. These products are frequently off stoichiometric and defects are present on the cationic and anionic sub-lattices. Defects on the atomic and microscopic level strongly influence the protective character of the corrosion products. The various oxides show structural similarities and relationships and it is in parts challenging to differentiate between them. These issues are important to understand and open a lot of old and new questions for applied crystallography as a tool for a better understanding of corrosion mechanisms.

References

[1] Raabe, D., Tasan, C.C., Olivetti, E.A. Strategies for improving the sustainability of structural metals. Nature 2019, 64–74.

[2] Falk, F., Menneken, M., Stephan-Scherb, C. Real-Time Oberservation of high-temperature gas corrosion in dry and Wet SO2-containing atmospheres. JOM 2019, 1560–1565.

[3] Olszewski, H.T., Schiek, M., Lutz, B., Holcomb, G., Shemet, V., Nowak, W., Meier, G., Singheiser, L., Quadakkers, W. Effect of SO2 on oxidation of metallic materials in CO2/H2O-rich gases relevant to oxyfuel environments. Mater. Corros. 2014, 121–131.

[4] Nützmann, K., Kranzmann, A., Stephan-Scherb, C. The influence of chromium on early high temperature corrosion of ferritic alloys under SO2 atmosphere. Mater. High Temp. 2018, 558–568.

[5] Young, D.J. High Temperature Oxidation and Corrosion of Metals, 2nd Edition. Elsevier, 2016.

[6]   Cabrera, N., Mott, N. Theory of the oxidation of metals. Rep. Prog. Phys. 1949, 163–184.

[7]   Wagner, C. Equations for transport in solid oxides and sulfides of transition metals. Prog. Solid State Chem. 1975, 3–16.

[8]   Ehlers, J., Young, D., Smaardijk, E., Tyagi, A., Penkalla, H., Singheiser, L., Quadakkers, W. Enhanced oxidation of the 9%Cr steel P91 in water vapour containing environments. Corros. Sci. 2006, 3428–3454.

[9]   Cox, M., Mcenaney, B., Scott, V. A chemical diffusion model for partitioning of transition elements in oxide scales on alloys. Philos. Mag.: J. Theor. Exp. Appl. Phys. 1972, 839–851.

[10]   Fjellvåg, H., Grønvold, F., Stølen, S., Hauback, B. On the crystallographic and magnetic structures of nearly stoichiometric iron monoxide. J Solid State Chem 1996, 52–57.

[11]   Koch, F., Cohen, J.B. The defect structure of $Fe_{1-x}O$. Acta Cryst. 1969, B25, 275–287.

[12]   Haavik, C., Stølen, S., Fjellvåg, H., Hanfland, M., Häusermann, D. Equation of state of magnetite and its high-pressure modification: Thermodynamics of the Fe-O system at high pressure. Am. Mineral. 2000, 514–523.

[13]   Lenaz, D., Skogby, H., Princivalle, F., Hålenius, U. Structural changes and valence states in the $MgCr_2O_4$–$FeCr_2O_4$ solid solution series. Phys. Chem. Miner. 2004, 633–642.

[14]   Rouillard, F., Moine, G., Martinelli, L., Ruiz, J.C. Corrosion of 9Cr steel in CO_2 at intermediate temperature I: mechanism of void-induced duplex oxide formation. Oxid. Met. 2012, 27–55.

[15]   Martinelli, L., Balbaud-Célérier, F., Terlain, A., Delpech, S., Santarini, G., Favergeon, J., Moulin, G., Tabarant, M., Picard, G. Oxidation mechanism of a Fe–9Cr–1Mo steel by liquid Pb–Bi eutectic alloy (Part I). Corros. Sci. 2008, 2523–2536.

[16]   Martinelli, L., Desgranges, C., Rouillard, F., Ginestar, K., Tabarant, M., Rousseau, K. Comparative oxidation behaviour of Fe-9Cr steel in CO_2 and H_2O at 550 °C: Detailed analysis of the inner oxide layer. Corros. Sci. 2015, 253–266.

[17]   Weber, K., Guilherme Buzanich, A., Radtke, M., Reinholz, U., Stephan-Scherb, C. A µ-XANES study of the combined oxidation/sulfidation of Fe–Cr model alloys. Mater. Corros. 2019, 1360–1370.

[18]   Azároff, L. Role of Crystal Structure in Diffusion. I. Diffusion Paths in Closest-Packed Crystals. J. Appl. Phys. 1961, 1658–1662.

[19]   Maslen, E., Streltsov, V.A., Streltsova, N.R., Ishizawa, N. Synchrotron X-ray study of the electron density in α-Fe_2O_3. Acta Cryst 1994, B50, 435–441.

[20]   Di Cerbo, R., Seybolt, A.U. Lattice parameters of the α-Fe_2O_3-Cr_2O_3 solid solution. J. Am. Ceram. Soc. 1959, 430–431.

[21]   Nützmann, K., Wollschläger, N., Rockenhäuser, C., Kranzmann, A., Stephan-Scherb, C. Identification and 3D reconstruction of Cr_5S_6 Precipitates along grain boundaries in Fe13Cr. J. Miner. Met. Mater. Soc. 2018, 1478–1483.

[22]   Jellinek, F. The structures of the chromium sulphides. Acta Crystallogr. 1957, 620–628.

[23]   Mrowec, S., Przybylski, K. Transport properties of sulfide scales and sulfidation of metals and alloys. Oxid. Met. 1985, 107–139.

[24]   Mosquera Feijoo, M., Influence of Surface Ash Layer on Dual Corrosion, Universida de Vigo: PhD Thesis, 2019.

[25]   Chattopadhyay, B., Wood, G. The transient oxidation of alloys. Oxid. Met. 1970, 373–399.

[26]   Calvarin, G., Molins, R., Huntz, A.M. Oxidation mechanism of Ni – 20Cr foils and Its relation to the oxide-scale microstructure. Oxid. Met. 2000, 25–48.

[27]   Blades, W., Reinke, P. From alloy to oxide: Capturing the early stages of oxidation on Ni–Cr(100) alloys. ACS Appl. Mater. Interfaces 2018, 43219–43229.

[28]   Luoa, L., Zou, L., Schreiber, D., Baer, D., Bruemmer, S.M., Zhou, G., Wang, C.-M. In-situ transmission electron microscopy study of surface oxidation for Ni–10Cr and Ni–20Cr alloys. Scr. Mater. 2016, 129–132.

[29] Leygraf, C., Hultquist, G. Initial oxidation stages on Fe-Cr(100) and Fe-Cr(110) surfaces. Surf. Sci. 1976, 69–84.

[30] Stephan-Scherb, C., Menneken, M., Weber, K., Agudo Jácome, L., Nolze, G., Elucidation of orientation relations between Fe-Cr alloys and corrosion products after high temperature SO2 corrosion. Corros. Sci. 2020, 108809.

[31] Nolze, G., Winkelmann, A., Exploring structural similarities between crystal phases using EBSD pattern comparison. Cryst. Res. Technol. 2014, 490–501.

[32] Kolarik, V., Juezlorenzo, M., Engel, W., Eisenreich, N., Application of a Fast X-Ray-Diffraction Method for Studies of High-Temperature Corrosion of Steel Surfaces. Fres. J. Anal. Chem. 1991, 436–438.

[33] Stephan-Scherb,C., Nützmann, K., Kranzmann, A., Klaus, M., Genzel, C. Real time observation of high temperature oxidation and sulfidation of Fe-Cr model alloys. Mater. Corros. 2018, 678–689.

[34] Ja'baz, I., Zhou, S., Estchmann, B., Paterson, D., Ninomiya, Y., Zhang, L. Spatial distribution of Cr-bearing species on the corroded tube surface characterised by synchrotron X-ray fluorescence (SXRF) mapping and micro-XANES: exposure of tubes in oxy-firing flue gas. J. Mater. Sci. (2018), 11791–11812.

[35] Couet, A., Motta, A.T., de Gabory, B., Cai, Z.H. Microbeam X-ray Absorption Near-Edge Spectroscopy study of the oxidation of Fe and Nb in zirconium alloy oxide layers. J. Nucl. Mater. 2014 614–627.

[36] Couet, A., Motta, A.T., Ambard, A. The coupled current charge compensation model for zirconium alloy fuel cladding oxidation: I. Parabolic oxidation of zirconium alloys: Corros. Sci. 2015, 73–84.

Daniel Fritsch

10 Crystallographic diffraction techniques and density functional theory: two sides of the same coin?

Abstract: Over the last decades, materials science has developed into an independent research area of science and engineering, thereby merging elements of disciplines such as solid-state physics, chemistry, and crystallography. With the advent of density functional theory and the widespread availability of high-performance computing facilities, computational materials science emerged as a particularly important subfield. Materials science pursues the improvement of already known and the design and discovery of new materials to be utilized in current and future technological applications, ideally taking into account their environmental impact and sustainability.

One prominent way to influence material properties uses substitutional and/or occupational disorder, as in solid solutions of different materials or controlled changes in defect concentrations. The accurate investigation of substitutional and/or occupational disorder with experimental and theoretical methods poses fundamentally different problems. On the one hand, experimental diffraction techniques usually employed in crystallography, provide only averaged information of material properties and require additional experimental techniques to explore local disorder. On the other hand, commonly applied periodic boundary conditions within density functional theory (DFT) require the application of supercells to describe substitutional and/or occupational disorder. In a way, diffraction techniques and DFT can be viewed as a top-down and bottom-up approach to materials science investigations.

This chapter presents recent advances in the investigation of structure–property relations from bottom-up approaches (DFT) with a particular focus on functional energy materials. Advantages, disadvantages, and limitations of this method will be discussed and how the investigation of material properties can be supplemented by top-down approaches (diffraction techniques) usually applied in crystallography.

Acknowledgment: The author is grateful to S. Levcenko for providing the spectroscopic ellipsometry data for $Cu_2ZnSnSe_4$ [1]. This work made use of computational resources provided by the North-German Supercomputing Alliance (HLRN, www.hlrn.de), and the ZEDAT and DIRAC high-performance computing facilities of the Freie Universität Berlin and the Helmholtz-Zentrum Berlin, respectively.

Daniel Fritsch, Department Structure and Dynamics of Energy Materials, Helmholtz-Zentrum Berlin für Materialien und Energie, Hahn-Meitner-Platz 1, 14109 Berlin, Germany, Daniel.Fritsch@helmholtz-berlin.de

https://doi.org/10.1515/9783110674910-010

Keywords: density functional theory, diffraction techniques, structure–property relations

10.1 Introduction

The last decades witnessed an impressing development in sample preparation techniques, allowing for the design and realization of complex functional materials with tailor-made material properties. However, in order to suitably influence material properties for certain applications, a detailed and accurate knowledge about their atomic scale structural properties is indispensable. It becomes apparent that sample characterization techniques have to be able to fulfil this requirement. One of the most common sample characterization techniques to date is single crystal or powder diffraction using X-ray, neutron, or synchrotron radiation. All of the mentioned radiation sources have advantages and disadvantages, and their usability partly depends on the specific experimental sample. One point to mention here are the atomic scattering factors, which make specific atomic species indistinguishable for different radiation sources. Another peculiarity of experimental diffraction data lies in their collection itself, meaning that always a data set averaged over the whole sample will be measured. Depending on the quality of the powder sample, this might include contributions from possible secondary phases, which ideally have to be known prior to any subsequent analysis. The details about the specific atomic scale structural properties can then be extracted by the application of numerical methods, such as Rietveld refinement or analyzing the pair distribution function. The latter one provides a measure of the distribution of interatomic distances within a given sample and can be supplemented by additional experimental techniques, for example, extended X-ray absorption fine structure (EXAFS) spectroscopy, to obtain more information about local atomic arrangements.

However, with the need to tailor specific material properties for certain applications, ever more complicated samples are grown, which have then to be analyzed reliably. Among others these include solid solutions between different materials, samples with intentionally high defect densities, local atomic disorder, or amorphous materials. In a way, they can be viewed as being derived from a parent crystal structure, superimposed by local disorder of various kinds.

The experimental implications on the crystal structure determination of a crystal structure and those derived from it by generalizations dates back to the late 40s of the twentieth century when Buerger coined the phrase of a *derivative crystal structure* [2]. To this end, Buerger distinguished between the original crystal structure (*basic crystal structure*) and those structures derived from it by generalizations. From a symmetry point of view, derivative crystal structures are generally defined as any crystal structures derived from a basic crystal structure by the suppression of one or more sets of operations of the space group of the basic crystal structure. As a special case, deriva-tive crystal structures contain the so-called *superstructures*, nowadays more com-monly known as supercells. In the simplest case, these superstructures are those crystal structures, where one or more edges of the superstructure are integer multi-ples of the corresponding edges of the underlying basic crystal structure. In terms of symmetry, superstructures can be obtained from a basic crystal structure by suppress-ing certain sets of translational operations in the space group of the basic crystal structure. Already back then it was recognized that it might be difficult to distinguish between experimental powder diffraction data obtained for the basic crystal structure and a possible derivative crystal structure, eventually leading to a wrongly deter-mined crystal structure of an unknown powder sample. The concept of derivative crystal structures has been further elaborated on in the seminal work by Megaw [3] on the ferroelectric perovskite $NaNbO_3$ [4]. Therein, the basic and derivative crystal structures have been renamed as *aristotype* and *hettotypes*, respectively.

As an example, Fig. 10.1 shows the possible pathways from a basic crystal struc-ture (aristotype), for example, zinc blende ZnSe, to various derivative crystal structures (hettotypes) for the quaternary $Cu_2ZnSnSe_4$, one of the key materials for thin-film solar cell applications. In more detail, the $Cu_2ZnSnSe_4$ crystal structure can be derived from the ZnSe crystal structure by doubling the zinc blende conventional unit cell along the crystallographic c axis and applying several cation substitutions. In order to obtain sta-ble cation substitutions, the bonding to adjacent cations has to fulfil the so-called octet rule, that is, the valence shell of each atom comprises eight electrons. Pairwise cation substitutions of the group-II element (Zn) by a group-I and group-III element, for exam-ple, Cu and Ga, can lead to the ternary $CuGaSe_2$ chalcopyrite and the CuAu-like crystal structures, respectively. A further pairwise cation substitution of the group-III element (Ga) in the chalcopyrite crystal structure by a group-II and group-IV element, for exam-ple, Zn and Sn, can lead to the $Cu_2ZnSnSe_4$ kesterite crystal structure, shown to the left in Fig. 10.1. However, performing the same replacement starting from the CuAu-like crystal structure can lead to the $Cu_2ZnSnSe_4$ stannite or a primitive-mixed CuAu-like crystal structure (PMCA), which has not been observed experimentally yet.

As will become apparent in the following sections, a successful and reliable de-termination of atomic scale structural properties relies on various factors, including among others the initial guesses for the numerical refinement procedures. Without any prior knowledge of a more complex sample, the refinement process can be quite tedious to nearly impossible. Luckily, with the advent of density functional theory (DFT) and the availability of high-performance computing (HPC) facilities, computa-tional materials science nowadays offers a lot of complementary tools to understand structure–property relations of unknown materials at the nanoscale. While experimen-tal diffraction techniques can be viewed as top-down approaches, that is, from a dif-fraction pattern averaged over the whole sample, the atomic scale structural properties have to be extracted, DFT investigations can be viewed as bottom-up approaches. One specifically can set up various different unit cells, that is allowing for changes in the

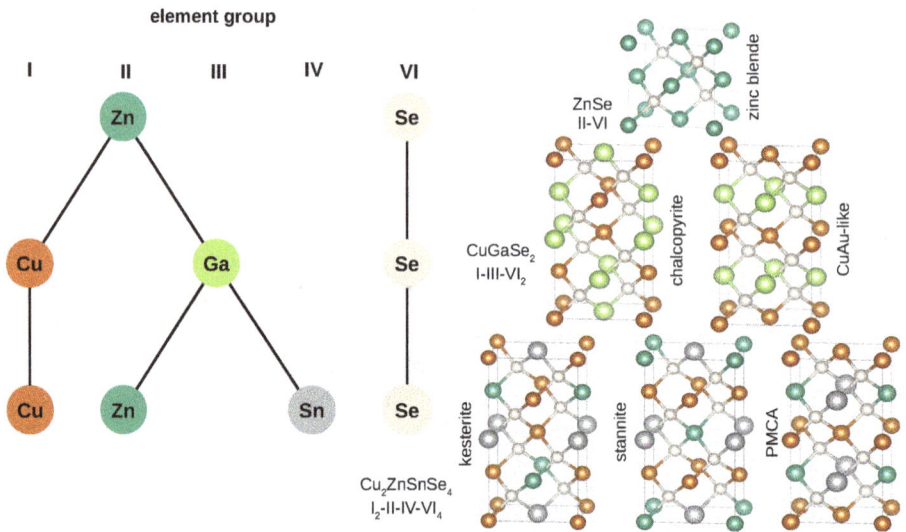

Fig. 10.1: Possible pathways from a basic crystal structure (aristotype) for zinc blende ZnSe toward various derivative crystal structures (hettotypes) for the quaternary $Cu_2ZnSnSe_4$ (kesterite, stannite, and primitive-mixed CuAu-like (PMCA) crystal structures), obtained via successive pairwise cation substitutions.

underlying symmetry, lattice parameters, or occupation of the Wyckoff positions, and derives macroscopic material properties. Among others, with the underlying known atomic scale structural properties, diffraction pattern for specific set ups can be generated and compared to the experimental ones. While this is also partly possible during the refinement process, the DFT calculations yield additional information about the energetic order of various possible crystal structures, that is how likely are they to occur in a real sample, and have immediate access to electronic and optical properties for comparison with additional experimental investigations.

This chapter will shed some light on the complementary character of crystallographic diffraction techniques and computational materials science investigations based on DFT methods. To this end, both approaches will shortly be introduced in Sections 10.2.1 and 10.2.2, respectively, before going into more detail about the possible treatment of local disorder in theoretical investigations in Section 10.2.3. As an illustrative case-study, the Cu–Zn disorder in kesterite-type $Cu_2ZnSnSe_4$ will be introduced in Section 10.3, and the results presented in Section 10.4.

10.2 Setting the stage

10.2.1 Heads: crystallographic diffraction techniques

Two of the most commonly used diffraction techniques to investigate the atomic scale structural properties of materials to date are based on single crystals and, where the former are not available due to various reasons, polycrystalline powder samples. The data collected by single crystal diffraction allows for a direct analysis of ground state material properties. However, this analysis is more complicated in case of powder diffraction data. One of the main reasons lies in the intrinsic nature of powder diffraction data being collected on a one-dimensional 2θ axis, contrary to the three-dimensional reciprocal space of single crystals. The most important information about the underlying atomic scale crystal structure lies hidden in the position and intensity of the Bragg peaks, and the collected background. All of these contributions to the powder diffraction pattern yield different information on the specific atomic microstructure of the sample under investigation. In a way, diffraction techniques can be viewed as a top-down approach in characterizing atomic scale structural properties.

Usually, in order to extract such information, the Rietveld method for refinement is applied. In short, it is an optimization method where an educated guess on the underlying lattice parameters and atomic positions of the sample under investigation is generated and, using a least-squares numerical analysis, one tries to optimize the agreement between the generated and measured diffraction pattern. There are several other parameters entering this optimization process, such as parameters influencing the width and shape of individual peaks, or to account for the experimentally collected machine-specific background. For more information, we refer to excellent standard textbooks on this topic [5, 6].

The quality and accuracy of the Rietveld refinement strongly depends on the starting parameters of the refinement process, which requires a good combination of experience and intuition. However, even with the best starting conditions, refinement processes can fail or deliver insufficient atomic scale structural properties. The main reason lies in the sample averaged nature of the experimental data, which has then to be extensively analyzed in order to extract the structural properties of interest. In case of more complicated powder samples, such as solid solutions, samples with intentionally high defect densities, local atomic disorder, or amorphous materials, one quickly approaches what has been coined the *nanostructure problem* [7], namely the need to determine atomic arrangements in nanostructured material, qualitatively and with high precision.

10.2.2 Tails: density functional theory

Since its inception in the seminal works of Hohenberg and Kohn [8] and Kohn and Sham [9] DFT has developed into an indispensable tool for theoretical materials science investigations. The foundations for its success were laid with the formulation of the Hohenberg–Kohn theorems [8], stating that the ground state properties of a many electron system can be accurately described by its electron density, and that the introduced energy functionals will be minimized by the correct ground state electron density. Further developments by Kohn and Sham [9] mapped the intractable problem of an interacting many electron system onto noninteracting electrons moving in an effective potential. Written down in a generalized Kohn–Sham (GKS) way, this effective potential $v_{GKS}(r, r')$ reads

$$v_{GKS}(r, r') = v_H(r) + v_{xc}(r, r') + v_{ext}(r). \tag{10.1}$$

Therein, the Hartree potential $v_H(r)$ takes into account the interaction between the electrons of the system, and the external potential $v_{ext}(r)$ describes the structure and elemental composition of the system. While these two terms, the Hartree $v_H(r)$ and the external potential $v_{ext}(r)$, are in principle known, the exchange and correlation potential $v_{xc}(r, r')$ in eq. (10.1) has to be suitably approximated [10].

Apart from its theoretical foundations, the enormous influence of DFT-based methods on theoretical materials science investigations has been further advanced by the growing availability of local, regional, and national HPC facilities, allowing for the necessary computations to be carried out at a pace not imaginable a decade ago. This increase in available resources has also triggered the development of ever more accurate approximations for the exchange and correlation functionals to be employed in theoretical materials science investigations, improving the agreement with experimental results on the structural, electronic, and optical properties, and ultimately leading to the stage where *predictive* materials science investigations became reality.

For the uninitiated, the sheer number of available exchange and correlation functionals makes it difficult to judge or compare the accuracy of reported theoretical results. In order to allow for some guidance, the so-called Jacob's ladder can provide a first hint [11], as adopted in Fig. 10.2.

Functionals of the local density approximation (LDA) can be found at the lowest rung and are based solely on the electron density, whereas the next two rungs take into account the gradient and the second derivative of the electron density to yield the generalized gradient approximation (GGA) functionals and the meta-GGAs, respectively. The hybrid functionals are placed one rung further up, where a pre-defined fraction of the underlying exchange and correlation energy is replaced by a Hartree–Fock exact exchange (EXX) term. In recent years, a lot of investigations focused on an improved adjustment of this pre-defined fraction of Hartree–Fock exact exchange, ultimately leading to a fully self-consistent procedure [12–15]. Of course, the expected increase in

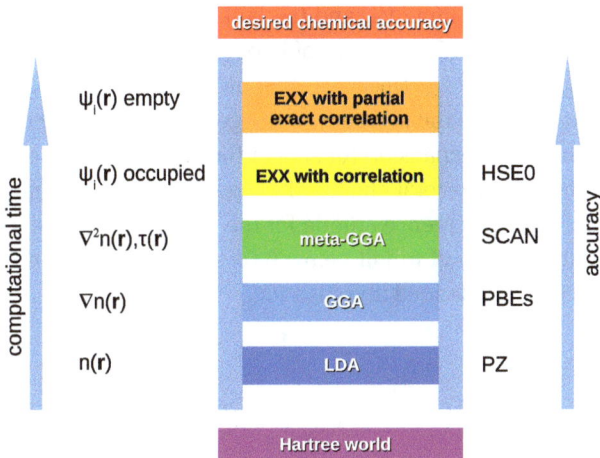

Fig. 10.2: Jacob's ladder, placing exchange and correlation functionals of similar capabilities at the same rung. Details about the displayed exchange and correlation functionals can be found in Section 10.3.2. After [11].

accuracy has to be paid for by an increase in computational time, as indicated by the two arrows in Fig. 10.2.

In principle, calculations based on the exchange and correlation functionals described so far, can serve as a starting point for subsequent quasiparticle calculations based on the GW approximation introduced by Hedin [16]. Here, for illustrative purposes, the investigation will be restricted to the numerically most simple implementation, namely the single-shot G_0W_0 method, with G_0 being the noninteracting Green's function of the system and W_0 its screened Coulomb interaction.

After having chosen an exchange and correlation functional according to the required accuracy and/or available computational resources, a typical DFT calculation for an unknown (or known) material is then performed in the following way: first, one has to set up the unit cell, ideally taking into account already known lattice parameters and other structural information, for example, Wyckoff positions of the various atoms. Second, for various volumes around the starting unit cell volume (typically ± a few per cent), the lattice parameters and internal atomic positions are allowed to relax, thereby keeping the respective unit cell volume constant. Finally, the obtained total energies with respect to the various unit cell volumes, that is, the total energy curve, is analyzed by an equation of state (EOS) to determine the ground state volume of that material. Commonly known is the Murnaghan EOS [17, 18]

$$E(V) = E(V_0) + \frac{B_0 V}{B'_0} \left[\frac{(V_0/V)^{B'_0}}{B'_0 - 1} + 1 \right] - \frac{V_0 B_0}{B'_0 - 1}, \qquad (10.2)$$

which gives access to the ground state volume V_0, and the bulk modulus B_0 and its pressure derivative B'_0, respectively. A final relaxation of lattice constants and internal atomic positions yields the ground state geometry of the material under investigation. Numerical parameters that influence the accuracy of the geometry optimization are the cut-off energy of the plane-wave expansion (or similar for nonplane-wave approaches), the convergence criteria for the total energy, the k-point grid for integrating over the Brillouin zone, and the convergence criteria for the forces on the atoms. The latter one typically has to be optimized to fall below 0.01 to 0.001 eV Å^{-1}, depending on the size of the unit cell and intended investigation.

10.2.3 Flipping the coin

As already mentioned in the introduction, *superstructures* or supercells are necessary for the accurate description of any disturbances of the crystal symmetry. This includes, among others, the description of solid solutions between different materials or the correct description of defects in the crystal structure. From a DFT perspective, one faces at least two different problems that impact on the required computational effort. The first one originates from the most commonly used numerical setup, employing periodic boundary conditions, where the investigated unit cell is repeated again after every lattice constant. For solid solutions and in particular for defect calculations, one has to make sure that the periodically repeated images are far enough apart in order to prevent any artificial self-interactions. The second problem arises from the use of only integer occupancies within the set up of the unit cells, thereby artificially lowering the symmetry of the intended material under investigation. Special treatments have been developed over the years to address these problems. In terms of efficiently setting up unit cells to investigate solid solutions or disorder in some materials, some of these approaches will be discussed now.

It has been realized relatively early, that in order to simulate more complicated cases, such as solid solutions, intentionally high defect densities, local atomic disorder, or amorphous materials, one has to resort to either very large unit cells or similarly, a large number of supercells with slightly different distributions of atoms over available sites. Accordingly, at a time when computational resources were sparse, approaches like the virtual crystal approximation (VCA) [19] or the coherent potential approximation (CPA) were the methods of choice. In them, the surroundings of particular atoms have been described as a uniform average medium, neglecting all local structural distortions and their subsequent influence on the structural, electronic, and optical properties of the materials. In a way, these early approaches were similar to the mentioned experimental crystal structure techniques, allowing only for an investigation of averaged local structures on the properties of materials.

Early attempts to tackle this problem have been made by Zunger et al. [20] with the introduction of so-called *special quasirandom structures*. Initially they investigated binary alloys of the form $A_{1-x}B_x$ and realized that, in order to properly take into account all possible atomic distributions of an N site alloy, they would have to calculate 2^N possible atomic configurations, which would have to be relaxed and suitably averaged over. Back in the day, the computational effort was too much. With their introduction of *special quasirandom structures*, constructed by a *selective* occupation of the N lattice sites, the computational demands could be reduced significantly, thereby allowing for binary alloys to be treated efficiently by DFT methods. This approach has been later extended to systematically generate superstructures for a given basic structure [21].

A more complicated example compared to binary alloys involves ferromagnetic spinel ferrites $CoFe_2O_4$ and $NiFe_2O_4$, which have been investigated as building blocks in artificial multiferroic heterostructures [22], for their dielectric response in strained thin films [23], and as possible spintronics devices [24]. Furthermore, the addition of nanosized $CoFe_2O_4$ enhances the superconducting properties of the host materials [25], encouraging the need for a proper modeling of superconducting properties [26–29].

The crystal structure of spinel (space group $Fd\bar{3}m$, no. 227) with the general formula AB_2O_4 contains two inequivalent cation sites, the tetrahedrally coordinated A site and the octahedrally coordinated B site. In the *normal* spinel structure, all A (B) sites are solely occupied by divalent (trivalent) cation species. However, in the *inverse* spinel structure the trivalent cations solely occupy the A sites, and the B sites see a mixture of di- and trivalent cations. If their distribution over the B sites is completely random, the overall cubic $Fd\bar{3}m$ symmetry is preserved. There also exist cases with incomplete inversion, where an inversion parameter x measures the degree of inversion ranging from $x = 0$ (normal spinel) to $x = 1$ (inverse spinel). While $NiFe_2O_4$ crystallizes in the inverse spinel structure, for $CoFe_2O_4$ the inversion is not complete, and lies around $x = 0.8 \ldots 0.9$, close to the inverse spinel structure.

From a DFT perspective, the inverse spinel structure has to be modeled with a random occupation of Co (Ni) and Fe cations on the B sites of the spinel structure, leading to an artificial symmetry lowering from the parent compound's $Fd\bar{3}m$ symmetry to *Imma* in the smallest possible unit cell, containing two spinel formula units [30]. Detailed first-principles calculations on the influence of exchange and correlation functionals belonging to different rungs of Jacob's ladder revealed their influence on the magnetostrictive properties [30, 31]. Choosing a slightly larger unit cell, now containing four spinel formula units, allowed for the identification of the influence of symmetry on the magnetostrictive properties [32]. In addition, this particular unit cell setup allowed for the identification of a long-range ordered ground state in case of $NiFe_2O_4$ [33], which has been independently confirmed by experimental Raman studies on single crystals [34] and thin films [35].

Over the years, the idea of special quasirandom structures served as inspiration and starting point for more elaborate approaches, taking into account the underlying symmetry of the basic crystal structure in Buerger's notation or the aristotype in Megaw's notation. Among the various available references, the work of Hart and Forcade [36] develops a universal algorithm for generating derivative structures, starting from

Buerger's derivative structures and taking into account the systematic generation of superstructures [21]. More recent approaches involve the introduction of available computer programs dealing with the generation of superstructures, for example, the symmetry-adapted configurational modeling of fractional site occupancies in solids by Grau-Crespo et al. [37], the use of symmetry in configurational analysis for the simulation of disordered solids by Mustapha et al. [38], or the combinatorial structure-generation approach for the local-level modeling of atomic substitutions and partial occupancies in crystals by the supercell program of Okhotnikov et al. [39], respectively. While on first sight all these available computer programs serve the same purpose, there are small differences in the targeted problems; in the present work, and after extensive testing, the supercell structure generation will be done by means of the supercell program of Okhotnikov et al. [39].

10.3 Picking a coin: illustrative case study

10.3.1 Cu–Zn disorder in kesterite-type $Cu_2ZnSnSe_4$

As a particular example serves Cu–Zn disorder in $Cu_2ZnSnSe_4$ [40]. At ambient conditions, $Cu_2ZnSnSe_4$ crystallizes in the kesterite crystal structure (space group $I\bar{4}$, no. 82), as shown in Fig. 10.3a. It belongs to the novel photovoltaic materials based on the kesterite crystal structure that are ingredients for third-generation thin-film solar cells incorporating only earth abundant and nontoxic elements. Together with Cu_2ZnSnS_4, their solid solution allows for a continuous tuning of the electronic band gap between around 1.0 eV for $Cu_2ZnSnSe_4$ [41, 42] and 1.53 to 1.67 eV for Cu_2ZnSnS_4 [43], respectively, thereby including the optimum band gap of 1.34 eV for single junction photovoltaic cells [44].

In order to better understand the kesterite crystal structure, one typically explores their similarities with the zinc blende crystal structure under the application of pairwise cation substitutions, as has been discussed in Section 10.1. One of these possible routes leads to the kesterite crystal structure, as depicted in Fig. 10.3a. However, a different possible route can lead to the stannite crystal structure (space group $I\bar{4}2m$, no. 121), depicted in Fig. 10.3c.

For the particular example of $Cu_2ZnSnSe_4$, a further complication arises. The identical Shannon ionic radii of Cu^+ and Zn^{2+} of 0.6 Å [46] give rise to a high probability of cation exchange in the Cu–Zn planes located at $z = 0.25$ and $z = 0.75$. A partial exchange of Cu and Zn cations leads to an increase in the Cu_{Zn} and Zn_{Cu} defect density, whereas a total exchange within the Cu–Zn planes leads to the introduction of additional symmetry elements and has therefore been named disordered kesterite structure, depicted in Fig. 10.3b. In terms of structural investigations this disordered kesterite structure crystallizes in the same space group as the stannite crystal structure, and has

(a) ordered kesterite ($I\bar{4}$, no. 82)   (b) disordered kesterite ($I\bar{4}2m$, no. 121)   (c) stannite ($I\bar{4}2m$, no. 121)

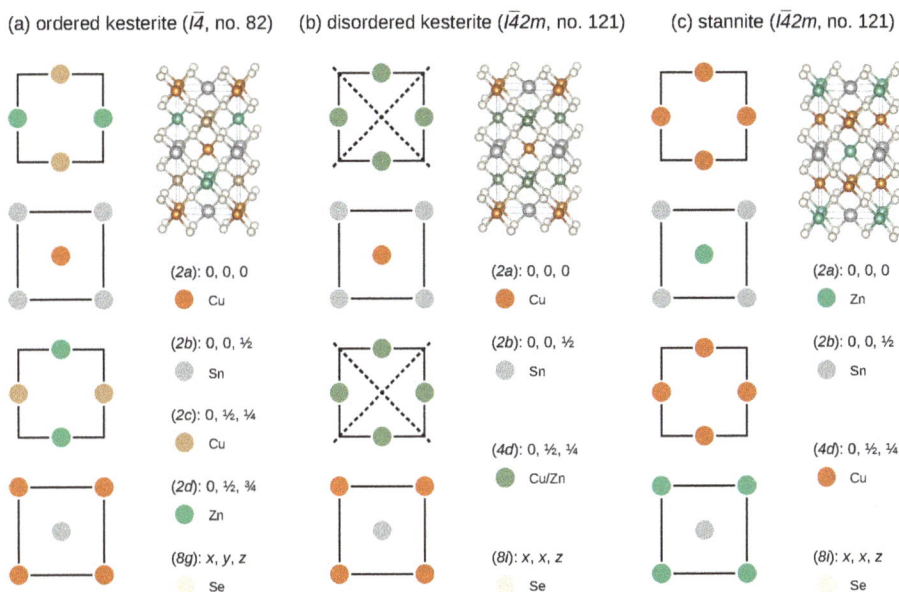

Fig. 10.3: Cu, Zn, and Sn cation planes with $z = 0.0, 0.25, 0.5$ and 0.75 (bottom to top) in the (a) ordered kesterite (space group $I\bar{4}$, no. 82), (b) disordered kesterite (space group $I\bar{4}2m$, no. 121), and (c) stannite (space group $I\bar{4}2m$, no. 121) crystal structures, respectively. The dashed lines in he disordered kesterite structure (b) denote additional symmetry elements due to the total Cu–Zn disorder in the $z = 0.25$ and $z = 0.75$ planes. Also given are the color-coded Wyckoff positions for each of the structural polymorphs. The crystal structure figures have been prepared using VESTA [45].

hindered the correct crystal structure determination of possible solar absorber materials for some time.

In order to investigate the Cu–Zn disorder in $Cu_2ZnSnSe_4$, DFT calculations have been performed on the kesterite based crystal structure including specific cation distributions over the Cu $2c$ and the Zn $2d$ Wyckoff positions. Here, a $2 \times 1 \times 1$ supercell has been generated based on the standard conventional unit cell of $Cu_2ZnSnSe_4$, optimized using the SCAN exchange and correlation functional [47]. This led to a total of eight different possible cation positions at the $2c$ and $2d$ Wyckoff positions, where Cu and Zn were distributed randomly utilizing the supercell program [39]. Special care has been taken to obey the 50/50 mixture of Cu/Zn cations over the $2c/2d$ Wyckoff positions. With these constraints, and the choice of a $2 \times 1 \times 1$ supercell, this resulted in 36 different cation distributions of Cu and Zn over the $2c$ and $2d$ Wyckoff positions of the parent kesterite crystal structure. However, several of the supercells with different cation distributions are related by symmetry. Taking these symmetry relations into account, the number of symmetry-inequivalent cation distributions is reduced to seven. The $2 \times 1 \times 1$ supercells including these symmetry-inequivalent cation distributions of

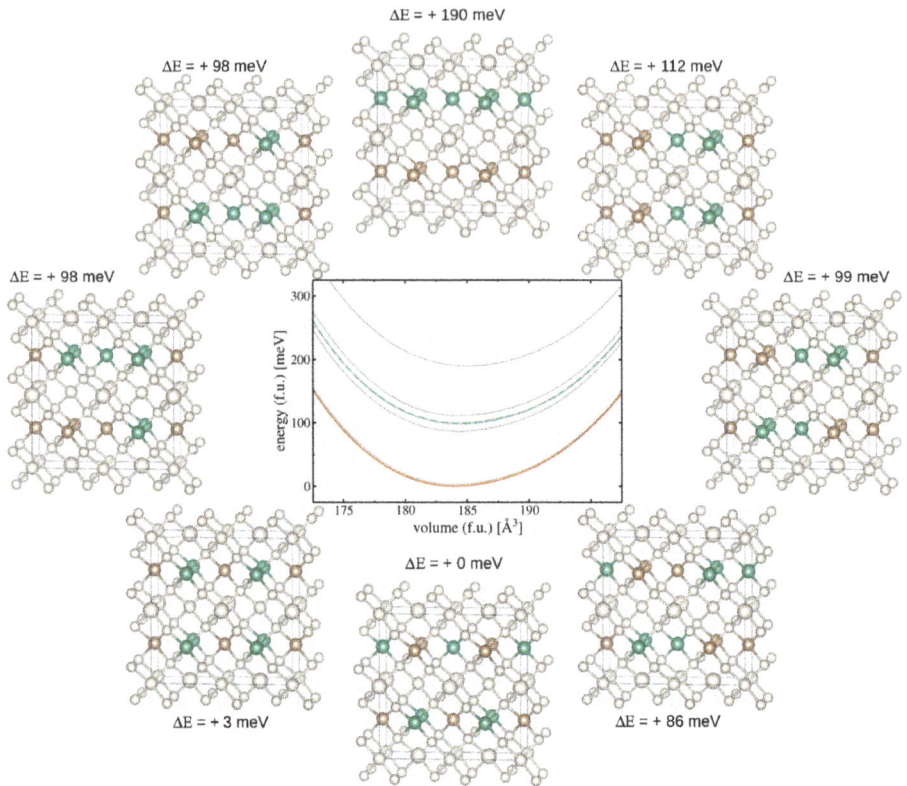

Fig. 10.4: 2 ×1×1 supercells including the seven symmetry-inequivalent cation distributions of Cu (orange) and Zn (green) over the 2c and 2d Wyckoff positions as generated by the supercell program [39]. The lowest supercell depicts the parent kesterite crystal structure. Structural optimization by means of the SCAN exchange and correlation functional of each of these supercells yielded their respective total energy curves, as shown in the middle figure, where the gray lines correspond to the seven symmetry inequivalent supercells, the dashed green line is their weighted average, and the orange line shows the results from the kesterite crystal structure for comparison. The crystal structure figures have been prepared using VESTA [45].

Cu (orange) and Zn (green) over the 2c and 2d Wyckoff positions of the original kesterite crystal structure are shown in Fig. 10.4.

The supercells with the seven symmetry-inequivalent cation distributions have been subjected to a geometry optimization utilizing the SCAN exchange and correlation functional, as outlined in Section 10.2.2. Thereby, cation positions have been held fixed to their respective ideal kesterite crystal structure Wyckoff positions, respectively.

It should be mentioned that a recent investigation based on high-resolution neutron diffraction experiments on the very similar material Cu_2ZnSnS_4 hints towards a Cu–Zn disorder including the Cu 2a Wyckoff position as well [48]. This has been confirmed theoretically [49, 50] using cluster expansion techniques and Monte Carlo simulations. The additional disorder at the Cu $2a$ Wyckoff position can be included in the calculations as well, but is beyond the scope of the present work.

10.3.2 Other computational details

All calculations presented in this work have been performed using the Vienna *ab initio* Simulation Package (VASP 5.4.4) [51–53]. The electron ion interaction is modelled using the projector-augmented wave formalism [54, 55], which provides a good compromise between plane-wave based and all-electron approaches. For the structural relaxations the recommended PAW potentials provided by VASP were employed, contributing 17, 12, 14, and 6 valence electrons for the Cu, Zn, Sn, and Se atoms, respectively. The electronic band structures, the optical properties, and the many-body perturbation theory calculations based on the GW approximation made use of the PAW potentials recommended for GW calculations, and provided 19, 20, 14, and 6 valence electrons for the Cu, Zn, Sn, and Se atoms instead.

In a separate work, the influence of different exchange and correlation functionals on the structural properties of $Cu_2ZnSnSe_4$ and $Ag_2ZnSnSe_4$ has already been investigated [56], choosing the Perdew and Zunger parametrization (PZ) for the LDA functional [57, 58], the Perdew, Burke, and Ernzerhof parametrization revised for solids (PBEsol) for the GGA functional [59], the newly introduced SCAN functional for the meta-GGA functional [47], and the Heyd, Scuseria, and Ernzerhof parametrization (HSE06) for the hybrid functional [60, 61], respectively. The mentioned SCAN functional satisfies all known possible exact constraints for the exact density functional, and has been claimed to match or improve on the accuracy of computationally more demanding hybrid functionals [62].

Structural relaxations of the kesterite and stannite crystal structures have been performed for the standard primitive unit cells, containing one formula unit for both structural polymorphs. The ground state structures have been optimized by analyzing the total energy curves, which have been obtained for several volumes around the experimentally known ground state volume. Keeping these volumes fixed, all internal coordinates have been allowed to relax until the forces on all atoms were below 0.001 eV $Å^{-1}$. The final ground state volumes have been obtained by a spline fit to the total energy curves.

The obtained relaxed ground state structures served as a starting point for subsequent calculations of the electronic band structures and the real and imaginary parts of the dielectric functions, which have been obtained by summing over empty states using Fermi's Golden Rule, transition matrix elements, and applying a Kramers–Kronig transformation [63]. In order to ensure converged results the number of empty bands in the calculations of the optical properties has been increased by a factor of four. The final real and imaginary parts of the dielectric functions have been obtained by diagonalizing the dielectric tensors for every energy point and averaging over the resulting main diagonal elements, as applied before to non-cubic oxide [4] and amorphous materials [64, 65].

Together with the other technical parameters, k-point grid of $6 \times 6 \times 6$, cut-off energy of the plane-wave expansion of 500 eV, and a convergence criteria for the total energy of 10^{-6} eV, this ensured well-converged results. Due to the increased numerical demand the k-point grid has been reduced to $4 \times 4 \times 4$ for the GW calculations.

10.4 Results and discussion

10.4.1 Kesterite and stannite crystal structures of $Cu_2ZnSnSe_4$

Starting with the ordered kesterite and stannite crystal structures of $Cu_2ZnSnSe_4$ their geometries have been optimized using various exchange and correlation functionals as outlined in Section 10.3.2. From the total energy curves shown in Fig. 10.5, it can be seen that the newly introduced SCAN exchange and correlation functional performs best compared to experimental results. The SCAN optimized lattice constants for kesterite type $Cu_2ZnSnSe_4$ of $a = 5.695(1)$ Å and $c = 11.340(1)$ Å are in favorable agreement with the experimental results of Siebentritt and Schorr [41] of $a = 5.695(2)$ Å and $c = 11.345(4)$ Å, Gong et al. [66] of $a = 5.692(3)$ Å and $c = 11.340(2)$ Å, and Gurieva et al. [42] of $a = 5.693(1)$ Å and $c = 11.347(2)$ Å, respectively.

Based on the optimized ground state geometry, electronic band structures and the real and imaginary parts of the dielectric functions can be calculated as well. The electronic band structures for the kesterite crystal structure of $Cu_2ZnSnSe_4$ are shown

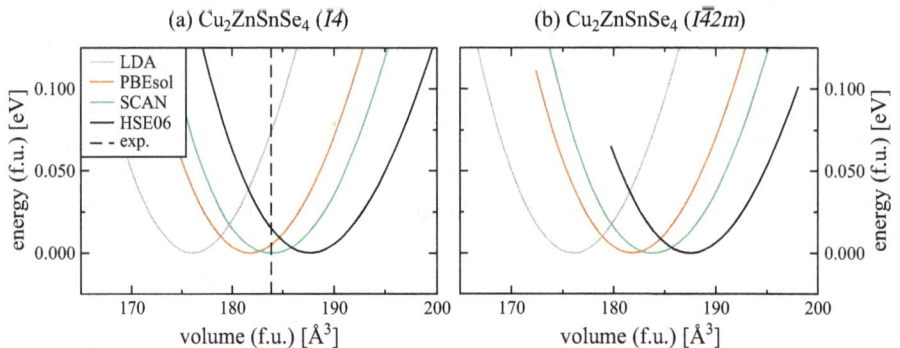

Fig. 10.5: Total energy curves for the (a) kesterite and (b) stannite crystal structures of $Cu_2ZnSnSe_4$, calculated using various exchange and correlation functionals. The dashed line in (a) indicates the experimental value of Gurieva et al. [42], and all energies are normalized to one functional unit (f.u.) and rescaled to zero energy, respectively.

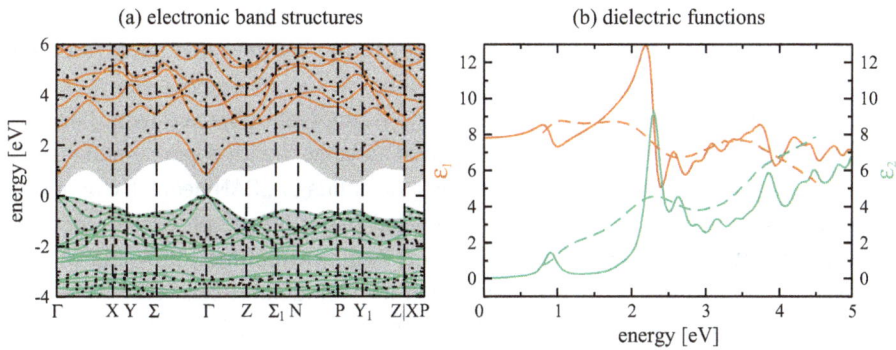

(a) electronic band structures (b) dielectric functions

Fig. 10.6: (a) Electronic band structures for the kesterite crystal structure of $Cu_2ZnSnSe_4$, with zero energy at the top of the valence bands. Shown are the valence (green) and conduction bands (orange), calculated using the hybrid HSE06 functional [60, 61]. The dotted lines and the shaded gray backgrounds show the results from the G_0W_0 and SCAN calculations, respectively. (b) Real (orange) and imaginary (green) parts of the dielectric functions for the kesterite crystal structure of $Cu_2ZnSnSe_4$, obtained by hybrid HSE06 calculations. The dashed lines in (b) present experimental results of León et al. [1], obtained via spectroscopic ellipsometry on bulk $Cu_2ZnSnSe_4$ crystals.

in Fig. 10.6a, with zero energy at the top of the valence bands. Shown are the valence (green) and conduction bands (orange), calculated using the hybrid HSE06 functional [60, 61]. The dotted lines and the shaded gray backgrounds show the results from the G_0W_0 and SCAN calculations, respectively. It can be seen that, although the SCAN exchange and correlation functional yields superior agreement in terms of the structural properties, for a more accurate description of the electronic band structure and a subsequent calculation of the real and imaginary parts of the dielectric functions, one has to resort to more advanced hybrid functional calculations [67]. The dielectric functions calculated with the latter ones are shown in Fig. 10.6b together with experimental results of León et al. [1], obtained via spectroscopic ellipsometry on bulk $Cu_2ZnSnSe_4$ crystals (dashed lines), showing overall good agreement.

10.4.2 Disorder results

In order to investigate in detail the Cu–Zn disorder on the Cu $2c$ and Zn $2d$ Wyckoff positions in $Cu_2ZnSnSe_4$, several $2\times1\times1$ supercells have been generated, as outlined in Section 10.3.1 and shown in Fig. 10.4, respectively. Therein, all possible distributions of Cu and Zn cations have been taken into account, subject to the condition of a 50/50 mixture of Cu/Zn cations over the $2c/2d$ Wyckoff positions. Considering symmetry relations between various of the generated supercells, this yielded seven symmetry-inequivalent cation distributions, as shown in Fig. 10.4. For the generation and symmetry analysis the supercell program of Okhotnikov et al. [39] has been employed.

The influence of several exchange and correlation functionals on the structural, electronic, and optical properties of $Cu_2ZnSnSe_4$ in the kesterite and stannite crystal structures served as a starting point for calculating the Cu–Zn disorder effects. As has been shown in the previous section, the structural properties obtained with the SCAN exchange and correlation functional yielded best agreement with experimental results. Therefore, and because of moderate numerical demands of SCAN geometry optimizations, the $2 \times 1 \times 1$ supercells have been generated based on the previously obtained ground state structure of kesterite $Cu_2ZnSnSe_4$. Again, as outlined in Section 10.2.2, total energy calculations have been performed for various unit cell volumes around the starting volume, where the volumes have been kept fixed and only the atoms were allowed to relax. For each of the seven supercells, this yielded the total energy curves, shown in the middle of Fig. 10.4, respectively. Therein, the unit cell volumes are given per formula unit and the total energy difference per formula unit is given with respect to the ground state kesterite crystal structure, shown in addition at the bottom of Fig. 10.4.

As can be seen, the different cation distributions have a huge influence on the energy difference compared to the kesterite ground state crystal structure. There is one particular cation distribution, shown in the lower left part of the figure, which has a similar ordering in the Cu–Zn planes compared to the ordered kesterite crystal structure, and is only 3 meV above it. On the other hand, occupying the Cu–Zn planes solely by one cation species, as shown in the upper figure, clearly is energetically most unfavorable.

The generation of the supercells with a specific cation distribution also yielded the weight of each supercell, that is, how often one specific setup appears before taking into account the symmetry relations between them. These weights range from two to eight in the present case and allow for the calculation of weighted averages over various properties. Exemplary, this is shown in the figure with the total energy curves, where the dashed green line represents the weighted average of all seven symmetry inequivalent supercells. This total energy curve has then been analyzed as outlined in Section 10.2.2 in order to determine the ground state volume of the disordered $Cu_2ZnSnSe_4$. In principle, one could then calculate the electronic band structures and dielectric functions, and take the weighted average in the same way. However, here the restriction is on the structural properties.

As already mentioned in the introduction and in more detail in Section 10.2.1, experimental diffraction techniques typically only yield averaged results over the whole sample. For the determination of local atomic arrangements, one has to resort to other techniques. Here, with the DFT calculations, it is basically the other way around. From the specific setup of the unit cells one already knows the local atomic arrangements, and can utilize the geometry optimized ground states of the various supercells to calculate X-ray or neutron diffraction pattern. In terms of the disordered $Cu_2ZnSnSe_4$, the weighted average approach is applied to the seven different supercell results. Exemplary, this is shown in Fig. 10.7, where the calculated

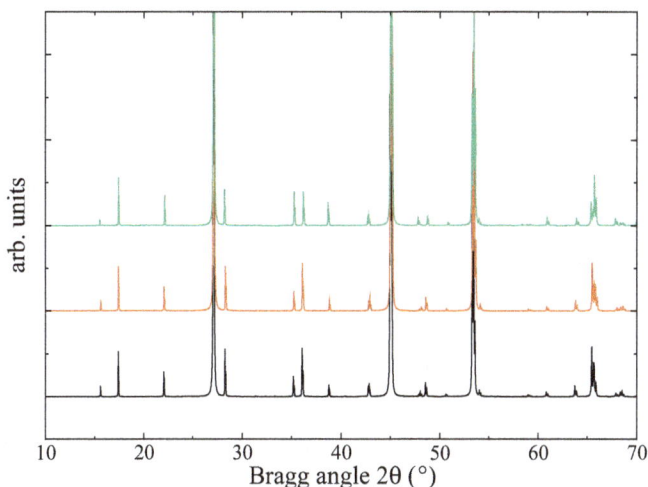

Fig. 10.7: $Cu_2ZnSnSe_4$ X-ray diffraction pattern calculated from the geometry optimized unit cells of the disordered kesterite (black), the kesterite (orange), and the stannite (green) crystal structures, respectively. In case of the disordered kesterite crystal structure, a weighted average over the seven different cation distributions is shown.

averaged X-ray diffraction pattern of the disordered $Cu_2ZnSnSe_4$ is shown in black, followed by the calculated X-ray diffraction pattern of the kesterite (orange) and stannite (green) crystal structures, respectively.

Even at the broad scale shown in Fig. 10.7, clear features unique to the parent kesterite and stannite crystal structures can be distinguished, as well as features solely due to the Cu–Zn disorder introduced at the 2c and 2d Wyckoff positions. At the moment this serves only illustrative purposes, however, for a real diffraction pattern analysis and a comparison with experimental data, a lot more detailed cal-culations would be required, which is beyond the scope of the present work.

10.5 Summary and outlook

Here, it has been shown that crystallographic diffraction techniques and DFT inves-tigations can be viewed as two sides of the same coin, namely the detailed under-standing of atomic scale structural properties of tailor-made materials. While the crystallographic diffraction techniques naturally only allow for the collection of an averaged data set over the whole sample, the atomic scale structural properties have to be extracted by extensive refinement methods, that is, it can be viewed as a top-down approach to materials science investigations. On the other hand, the specific atomic scale structural properties are the starting point for any DFT investigation, that is, it can be viewed as a complementary bottom-up approach to materials science investigations. From there, macroscopic material properties can be inferred and compared to experimental results, thereby allowing for a deeper understanding and more comprehensive analysis of experimental data sets.

As an illustrative example, the Cu–Zn disorder in kesterite-type $Cu_2ZnSnSe_4$ has been discussed, and how novel theoretical techniques can be applied to account for deviations from the perfect crystal lattice, that is, taking into account local disorder in derivative crystal structures. While the mandatory use of supercells are computationally quite demanding, the nowadays widespread availability of local, regional, and national HPC facilities provides access to the required resources.

What has not been elaborated on so far is the possible occurrence of secondary phases in the experimental samples, influencing refinement strategies and their possible outcome. Here, DFT based methods can be helpful as well with theoretical investigations of phase diagrams involving the atomic species of the experimental samples. The employed methods are quite different from the ones presented here, and a note to their existence might suffice [67].

Going beyond the structural analysis, by employing diffraction techniques one might be interested in looking at the pair distribution functions, holding a vast amount of information on local atomic structure within a given sample. While the specific atomic scale structural properties are input parameters for the DFT calculations, the pair distribution functions become easily accessible after geometry optimization. However, the experimental determination of pair distribution functions is still in its infancies, but remarkable progress has been made in the last decade. In terms of a comprehensive characterization of atomic scale properties of tailor-made materials, a closer collaboration between theory and experiment would be highly desirable. In the end it is the atomic scale structural properties of the materials that determine their macroscopic material properties, which we want to tailor for certain applications.

Bibliography

[1] León, M., et al. Spectroscopic ellipsometry study of $Cu_2ZnSnSe_4$ bulk crystals. Appl. Phys. Lett. 2014, 105, 061909.
[2] Buerger, M.J. Derivative crystal structures. J. Chem. Phys. 1947, 15, 1.
[3] Megaw, H.D. The seven phases of sodium niobate. Ferroelectrics 1974, 7, 87.
[4] Fritsch, D. Electronic and optical properties of sodium niobate: A density functional theory study. Adv. Mater. Sci. Eng. 2018, 2018, Article ID 6416057.
[5] Pecharsky, V.K., Zavalij, P.Y. Fundamentals of powder diffraction and structural characterization of materials. Springer, 2008.
[6] Dinnebier, R.E., Billinge, S.L.J. Powder Diffraction: Theory and Praxis. RSC Publishing, 2008.

[7] Billinge, S.J.L., Levin, I. The problem with determining atomic structure at the nanoscale. Science 2007, 316, 561.

[8] Hohenberg, P., Kohn, W. Inhomogeneous electron gas. Phys. Rev. 1964, 136, B864.

[9] Kohn, W., Sham, L.J. Self-consistent equations including exchange and correlation effects. Phys. Rev. 1965, 140, A1133.

[10] Bechstedt, F. Many-Body Approach To Electronic Excitations. Springer, 2015.

[11] Perdew, J.P., Schmidt, K. Jacob's ladder of density functional approximations for the exchange-correlation energy. AIP Conf. Proc. 2001, 577, 1.

[12] Shimazaki, T., Nakajima, T. First principles band structure calculations based on self-consistent screened Hartree-Fock exchange potential. J. Chem. Phys. 2009, 130, 164702.

[13] Skone, J.H., Govoni, M., Galli, G. Self-consistent hybrid functional for condensed systems. Phys. Rev. 2014, B 89, 195112.

[14] Fritsch, D., Morgan, B.J., Walsh, A. Self-consistent hybrid functional calculations: Implications for structural, electronic, and optical properties of oxide semiconductors. Nanoscale Res. Lett. 2017, 12, 19.

[15] Fritsch, D. Self-consistent hybrid functionals: What we've learned so far. In: Levchenko, E. V., Dappe, Y. L., Ori, G., (Ed. by), Theory and Simulation in Physics for Materials Applications: Cutting-Edge Techniques in Theoretical and Computational Materials Science. Cham: Springer International Publishing, 2020, 79.

[16] Hedin, L. New method for calculating the one-particle Green's function with application to the electron-gas problem. Phys. Rev. 1965, 139, A796.

[17] Murnaghan, F.D. Finite deformations of elastic solids. Amer. Jour. Math. 1937, 59, 235.

[18] Murnaghan, F.D. The compressibility of media under extreme pressure. PNAS 1944, 30, 244.

[19] Fritsch, D., Schmidt, H., Grundmann, M. Pseudopotential band structures of rocksalt MgO, ZnO, and $Mg_{1-x}Zn_xO$. Appl. Phys. Lett. 2006, 88, 134104.

[20] Zunger, A., et al. Special quasirandom structures. Phys. Rev. Lett. 1990, 65, 353.

[21] Ferreira, L.G., Wei, S.-H., Zunger, A. Stability, electronic structure, and phase diagrams of novel inter-semiconductor compounds. Int. J. Supercomput. Appl. 1991, 5, 34.

[22] Zheng, H., et al. Multiferroic $BaTiO_3$-$CoFe_2O_4$ nanostructures. Science 2004, 303, 661.

[23] Gutiérrez, D., et al. Dielectric response of epitaxially strained $CoFe_2O_4$ spinel thin films. Phys. Rev. B 2012, 86, 125309.

[24] Caffrey, N., et al. Spin-filtering efficiency of ferrimagnetic spinels $CoFe_2O_4$ and $NiFe_2O_4$. Phys. Rev. B 2013, 87, 024419.

[25] Abou Aly, A.I., et al. Determination of superconducting parameters of $GdBa_2Cu_3O_7$ added with nanosized ferrite $CoFe_2O_4$ from excess conductivity analysis. J. Supercond. Nov. Magn. 2012, 25, 2281.

[26] Fritsch, D., Annett, J.F. Proximity effect in superconductor/conical magnet/ferromagnet heterostructures. New J. Phys. 2014, 16, 055005.

[27] Fritsch, D., Annett, J.F. Proximity effect in superconductor/conical magnet heterostructures. J. Phys.: Condens. Matter 2014, 26, 274212.

[28] Fritsch, D., Annett, J.F. Spin-flipping with Holmium: Case study of proximity effect in superconductor/ferromagnet/superconductor heterostructures. Phil. Mag. 2015, 95, 441.

[29] Fritsch, D., Annett, J.F. Triplet superconductivity and proximity effect induced by Bloch and Néel domain walls. Supercond. Sci. Technol. 2015, 28, 085015.

[30] Fritsch, D., Ederer, C. Epitaxial strain effects in the spinel ferrites $CoFe_2O_4$ and $NiFe_2O_4$ from first principles. Phys. Rev. B 2010, 82, 104117.

[31] Fritsch, D., Ederer, C. Strain effects in spinel ferrite thin films from first principles calculations. J. Phys.: Conf. Ser. 2011, 292, 012014.

[32] Fritsch, D., Ederer, C. First-principles calculation of magnetoelastic coefficients and magnetostriction in the spinel ferrites $CoFe_2O_4$ and $NiFe_2O_4$. Phys. Rev. B 2012, 86, 014406.

[33] Fritsch, D., Ederer, C. Effect of epitaxial strain on the cation distribution in spinel ferrites $CoFe_2O_4$ and $NiFe_2O_4$: A density functional theory study. Appl. Phys. Lett. 2011, 99, 081916.

[34] Ivanov, V.G., et al. Short-range B-site ordering in the inverse spinel ferrite $NiFe_2O_4$. Phys. Rev. B 2010, 82, 024104.

[35] Iliev, M.N., et al. Monitoring B-site ordering and strain relaxation in $NiFe_2O_4$ epitaxial films by polarized Raman spectroscopy. Phys. Rev. B 2011, 83, 014108.

[36] Hart, G.L.W., Forcade, R.W. Algorithm for generating derivative structures. Phys. Rev. B 2008, 77, 224115.

[37] Grau-Crespo, R., et al. Symmetry-adapted configurational modelling of fractional site occupancy in solids. J. Phys.: Condens. Matter 2007, 19, 256201.

[38] Mustapha, S., et al. On the use of symmetry in configurational analysis for the simulation of disordered solids. J. Phys.: Condens. Matter 2013, 25, 105401.

[39] Okhotnikov, K., Charpentier, T., Cadars, S. Supercell program: A combinatorial structure-generation approach for the local-level modeling of atomic substitutions and partial occupancies in crystals. J. Cheminform 2016, 8, 17.

[40] Schorr, S. The crystal structure of kesterite type compounds: A neutron and x-ray diffraction study. Sol. Energy Mater. and Sol. Cells 2011, 95, 1482.

[41] Siebentritt, S., Schorr, S. Kesterites – a challenging material for solar cells. Prog. Photovolt: Res. Appl. 2012, 20, 512.

[42] Gurieva, G., et al. Effect of Ag incorporation on structure and optoelectronic properties of $(Ag_{1-x}Cu_x)_2ZnSnSe_4$ solid solutions. Phys. Rev. Mater. 2020, 4, 054602.

[43] Levcenko, S., et al. Spectroscopic ellipsometry study of Cu_2ZnSnS_4 bulk poly-crystals. Appl. Phys. Lett. 2018, 112, 161901.

[44] Rühle, S. Tabulated values of the Shockley-Queisser limit for single junction solar cells. Sol. Energy 2016, 130, 139.

[45] Momma, K., Izumi, F. VESTA 3 for three-dimensional visualization of crystal, volumetric and morphology data. J. Appl. Crystallogr. 2011, 44, 1272.

[46] Shannon, R.D. Revised effective ionic radii and systematic studies of interatomic distances in halides and chalcogenides. Acta Cryst. A 1976, 32, 751.

[47] Sun, J., Ruzsinszky, A., Perdew, J.P. Strongly constrained and appropriately normed semilocal density functional. Phys. Rev. Lett. 2015, 115, 036402.

[48] Bosson, C.J., et al. Cation disorder and phase transitions in the structurally complex solar cell material Cu_2ZnSnS. J. Mater. Chem. A 2017, 5, 16672.

[49] Ramkumar, S.P., et al. Insights into cation disorder and phase transitions in CZTS from a first-principles approach. Phys. Rev. Mater. 2018, 2, 085403.

[50] Zheng, Y.-F., Yang, J.-H., Gong, X.-G. Cu-Zn disorder in stoichiometric and nonstoichiometric $Cu_2ZnSnS_4/Cu_2ZnSnSe_4$. AIP Adv. 2019, 9, 035248.

[51] Kresse, G., Hafner, J. Ab initio molecular dynamics for liquid metals. Phys. Rev. B 1993, 47, 558.

[52] Kresse, G., Hafner, J. Ab initio molecular-dynamics simulation of the liquid-metal–amorphous-semiconductor transition in germanium. Phys. Rev. B 1994, 49, 14251.

[53] Kresse, G., Furthmüller, J. Efficiency of ab-initio total energy calculations for metals and semiconductors using a plane-wave basis set. Comp. Mat. Sci. 1996, 6, 15.

[54] Blöchl, P.E. Projector augmented-wave method. Phys. Rev. B 1994, 50, 17953.

[55] Kresse, G., Joubert, D. From ultrasoft pseudopotentials to the projector augmented-wave method. Phys. Rev. B 1999, 59, 1758.

[56] Fritsch, D., Schorr, S. Climbing Jacob's ladder: A density functional theory case study for $Ag_2ZnSnSe_4$ and $Cu_2ZnSnSe_4$. J. Phys. Energy 2021, 3, 015002.

[57] Ceperley, D.M., Alder, B.J. Ground state of the electron gas by a stochastic method. Phys. Rev. Lett. 1980, 45, 566.

[58] Perdew, J.P., Zunger, A. Self-interaction correction to density-functional approximations for many-electron systems. Phys. Rev. B 1981, 23, 5048.

[59] Perdew, J.P., et al. Restoring the density-gradient expansion for exchange in solids and surfaces. Phys. Rev. Lett. 2008, 100, 136406.

[60] Heyd, J., Scuseria, G.E., Ernzerhof, M. Hybrid functionals based on a screened Coulomb potential. J. Chem. Phys. 2003, 118, 8207.

[61] Heyd, J., Scuseria, G.E., Ernzerhof, M. Erratum: Hybrid functionals based on a screened Coulomb potential [J. Chem. Phys. 118, 8207 (2003)]. J. Chem. Phys. 2006, 124, 219906.

[62] Sun, J., et al. Accurate first-principles structures and energies of diversely bonded systems from an efficient density functional. Nat. Chem. 2016, 8, 831.

[63] Gajdoš, M., et al. Linear optical properties in the projector-augmented wave methodology. Phys. Rev. B 2006, 73, 045112.

[64] Fritsch, D. Amorphous Sn-Ti oxides: A combined molecular dynamics and density functional theory study. Phys. Status Solidi A 2018, 215, 1800071.

[65] Wahila, M.J., et al. Accelerated optimization of transparent, amorphous zinc-tin-oxide thin films for optoelectronic applications. APL Mater. 2019, 7, 022509.

[66] Gong, W., et al. Crystallographic and optical properties of $(Cu, Ag)_2ZnSnS_4$ and $(Cu, Ag)_2ZnSnSe_4$ solid solutions. Phys. Status Solidi C 2015, 12, 700.

[67] Hasnip, P.J., et al. Density functional theory in the solid state. Phil. Trans. R. Soc. A 2014, 372, 20130270.

Bernd Hinrichsen

11 Crystallographic deviants: modelling symmetry shirkers

Abstract: Crystalline perfection is a rarity. It is no wonder that the 2017 Nobel Prize in Chemistry was awarded to a method of protein structure elucidation that precluded the need for crystalline samples for diffraction experiments. Many technically useful materials such as polymers, layered hydroxides, and nanoscale catalysts show challenging diffraction patterns, which are immune to analysis using the basic crystallographic toolkit. Although in the past decades we would seek to improve the crystallinity of the sample thereby coaxing it into an interpretable motif, we now more often try to elucidate the structures as they are produced. In doing so we attempt to improve our understanding of real-world material. This chapter will cover some advances in structure modelling and experimental diffraction techniques particularly with regard to the available software, which can be used for the interpretation of diffuse scattering.

Keywords: disorder, diffuse scattering, Debye function, pair distribution function, global optimization

11.1 Introduction

Crystallography is the science of order and beauty in form and properties. These have enthralled both novices and scientists since antiquity. The term disorder seems to blemish its defining property. Many valuable materials show little regard for crystalline perfection [1–3]. Eons might not even suffice to make some materials order into perfect three-dimensional (3D) crystals as the various incommensurately modulated structures of Plagioclase suggest [4]. Technically useful materials such as polymers [2], layered hydroxides [5], and nanoscale catalysts [6] show challenging diffraction patterns, which are immune to analysis using the classical crystallographic toolkit. When confronted with a material that does not crystallize in a traditional space group with sensible translational symmetry, or which produces experimental intensities that prove difficult to match at first sight, completely plausible model, crystallographers can often be heard muttering "disorder." Where in the past decades we would seek to improve

Acknowledgments: I would like to thank my interview partners who have contributed most of what you have read in the pages above: Reinhard Neder, Martin Dove, Hans-Beat Bürgi, Arkadiy Simonov, Joe Paddison, Ray Osborn, Max Terban, Matteo Leoni, and Antonella Guagliardi.

Bernd Hinrichsen, BASF SE, Ludwigshafen am Rhein, Germany

https://doi.org/10.1515/9783110674910-011

the crystallinity of the sample, thereby coaxing it into an interpretable motif, an increase in inquisitiveness in the structures as they exist has formed. In doing so, we attempt to improve our understanding of structure and properties of real-world materials.

This chapter is the result of many interviews that have been performed with colleagues working at the forefront of the field. I confess to be personally far removed from expertise in this challenging field. The following should therefore be regarded as a journalistic excursion giving an overview of methods and code used in the field of disorder modelling. It should be noted that the scope has been limited to code that seeks to characterize materials which cannot be described by classical approaches or even elegant modulation functions.

11.2 Disorder and modelling

Static disorder contrasts to dynamic disorder which can be experimentally observed in many molecular structures. Dynamic disorder results from structural fragments that are actually moving at speeds so high that their electron density is smeared out over the observational period. As a rule, dynamical disorder increases at elevated temperatures [7, 8] but is well known in room temperature experiments [9, 10]. Disentangling the two can be as easy as cooling down the sample until the rotational fraction is "locked in" to position [11]. In addition, the reduction of the atomic displacement during cooling leads to higher intensity diffraction data at higher scattering angle. This increase in information allows the crystal structure to be described to a higher resolution than for higher temperature measurements.[1] Theory-based rotational energy calculations give good insight into the probability of certain orientations, as well as the extent to which energy barriers limit or allow rotational disorder [12]. What follows deals with disorder which remains unchanged during the experiment unless specifically stated otherwise.

Conventional crystallographic tools model static disorder making rather "non-atomistic" approximations, such as lower-than-unity or mixed occupations for point defects or statistically evenly distributed occupation by more than one element on a crystallographic position. Positional disorder can be forced into the classical description by extending atomic displacement parameters or by modelling atoms with lower occupations simulating their statistical distribution. These methods are useful to

---

[1] This is one of the many but arguably the most important reasons for routinely cooling samples in single-crystal structure determination experiments. For protein and "electron-density" crystallographers, the resolution of their data (typically given in Angstroms) is singularly important and a highly visible experimental criterion. In most structure publications, the experimental resolution is not given, probably because even in low-resolution data the average crystal structure can usually be determined reliably.

create a diffraction pattern matching experiment, and can deliver values that are pertinent to material performance. They are not as useful in creating an understanding of the structural chemistry in relation to material properties but are used when conventional "small box" models are more opportune.

The typical *ansatz* is to populate a supercell "big box" with enough atoms to represent a statistically relevant portion of the sample material. The size of this box is elementary to the extent to which disorder can be modeled. It is also the dominant cost factor in the computations. Some nifty methods reduce the required computations dramatically [13–16]; however, the orders of magnitude remain more expensive than classical structural refinements. Within these initially pristine supercells, defects are inserted based on crystal chemical considerations and – ideally – investigations using complementary analytical techniques (electron microscopy, electron diffraction and solid-state NMR). Even when the chemical idea is successfully postulated and the structure is cerebrally envisaged, putting this into a model costs experience and effort. Here, software tools or programming proficiency can make a huge difference in productivity.

Modelling a nanoparticle by describing the position of a few thousand points in space, which are manually added, deleted, or nudged is a completely different task to populating a space from an average structure file and then pruning to a set of lattice planes or generic shapes and possibly adding a surface relaxation function. Surface modifications – either by manually adding molecules or varying the atomic configuration – in a chemically sound manner is a well-nigh impossible task. Code incorporating molecule-orientation tools with chemically sensible values are of great help. On the other hand, the interoperability with various related methods, be they experimental, such as high-resolution tomography, or theoretical, such as quantum chemical [17] or crystal growth simulation code [18], can offer synergistic effects. The experimental technique most often used to model these is powder diffraction. The number of parameters required to describe the model are limited, and powder diffraction can offer enough data to describe these sufficiently.

A great part of the work published on disordered materials is on layered structures. Many naturally occurring and industrially relevant materials show interlayer faulting [5, 19–27]. Describing a model of such a material starts with the description of the individual layers unblemished in their crystallographic integrity. The disorder aspect is introduced in the predominantly in-plane shift from one layer to the next. A set of vectors describes the positions of the layers relative to one another. A transition probability matrix is used to populate big boxes. These contain hundreds or thousands of layers depending on how low the probabilities of the faults are that are being simulated. Again, powder diffraction is the dominant experimental technique used to analyze this type of disorder. Laboratory data are often sufficient for such experiments. Complex structures can profit from multiple experimental sources such as single crystal electron diffraction, high resolution transmission electron microscopy, NMR, or theoretical calculations [28, 29].

Complex disorder that spans 3D space [30] often requires more detailed experimental data that is accessible from single crystal type experiments. In reality, the causality is most of the time reversed. Single-crystal diffraction data show signs of disorder through Bragg peak intensities and then if the material properties merit the effort, it is investigated. The intensities of diffuse scattering which are to be interpreted by a disorder model are generally far weaker than the Bragg intensities. Therefore, the average structure derived from the latter is taken as a starting model for the study of the deviations from it. An ingenious method to give further clues as to the origin of the disorder is the 3D-ΔPDF method devised by Thomas Weber and Arkadiy Simonov [31, 32]. Here, direct space aspects (vector maps) of the disorder are calculated after subtracting the dominant effects of the average structure.

Finally, all models have to face the experimental data and attempt to describe them as best as possible. The optimization routines are to a large degree established methods that have proven their capabilities over decades using relentlessly slow workstations and eons of processor time. Monte Carlo, reverse Monte Carlo, simulated annealing and genetic algorithms [33] such as differential evolution are all flavors of routines which sample parameter space efficiently, avoid local minima, and find the most reliable solution. The author has experimented with Bayesian optimization for a layer disorder problem involving a small number of parameters, and has found this to be a rather efficient method. It has found its way into a quality control routine for a battery material. Whatever the optimization method, the more chemical sanity that can be introduced into the model, the more reliable the answer is likely to be. Using such constraints very heavily can bog down the model into a local minima, so practising caution and ideally an experienced colleague can be worthwhile.

11.3 Software

11.3.1 Discus

DISCUS had its origins on a MicroVAX. Reinhard Neder developed it as an educational tool to visualize the relationship between a crystal structure and its diffraction signal. It has developed into a powerful package to model disordered structures and nanoparticles. The experimental data can be derived from X-ray, neutron, or electron, be it diffraction of either single crystals or powders. A model can be optimized against the experimental reciprocal space data or the converted direct space pair distribution function. Modelling of single-crystal diffuse scattering data can be performed as well. The package is also capable of single-model refinement from combined datasets. The software is actively developed by its main author Reinhard Neder.

A useful feature in the construction of faulted structures is that of dummy or pseudo atoms. They can be used as proxies for any structural fragment. To build a model, you start with an asymmetric unit which is expanded to the unit cell, and from there to a supercell. Modulations of the atomic parameters can be applied to the construct using various wave functions directed through the structure. These can be applied to the density, displacement, or – if applied to a molecule – rotation. Domains can be created to describe nanoparticulate structures. The surfaces of these can be modeled using either crystallographic planes, a sphere, an ellipsoid, or a cylinder. Intriguingly, this enables the formation of inner surfaces, that is, voids within a crystal structure. The surfaces can be relaxed, emulating experimentally observed bond distance lengthening. DISCUS contains a tool for decorating surfaces with previously defined atoms or molecules. Six different automated methods based on the structural chemistry of both the molecule and the surface enable an automated decoration, ensuring chemically sound configurations. In addition to the standard Fourier transform for diffraction signal calculation, DISCUS can utilize the Debye equation, allowing the reliable modelling of discrete nanoparticles.

Layered structures can be defined by creating the various layers proposed for the material and then defining the probability matrix defining the relative probabilities of the transition from one layer type to the other. A complementary matrix of translation vectors defining the relation of the different layers to one another is also required. Uncertainties for these translations can be provided, enabling a Gaussian distribution of translations. Interestingly, when calculating the faulted layers, DISCUS does not generate the total set of atomic positions, but only the origin distribution of the displaced layers. This enables a faster computation, as the Fourier transform is only performed once per unique layer. It is then multiplied with the Fourier transform of the origin distribution to achieve the final result.

To optimize the model parameters against experimental data, DISCUS uses classical least squares. Least least-squares computations of parameters for which analytical differentials cannot be determined are difficult, but can be replaced by, an alternative global optimization algorithm: differential evolution. This method allows the determination of all model descriptors with a high probability of locating the global minimum. Monte Carlo methods can also be used to optimize structure descriptors against experimental data. Lennard–Jones potentials help keep atoms at a chemically sound bonding and nonbonding distances.

Diffraction simulations are limited to the kinematical theory. Magnetic structures and glasses cannot be modeled by this package, though the author has announced magnetic scattering capabilities in a future release.

Yearly workshops on the program suite ensure that the community is kept up-to-date with the latest features and capabilities of the software package. Neder counts around 300 citations for the software. A mailing list and YouTube channel ensure that the users remain informed and have access to the community [34–36].

11.3.2 Single crystal

11.3.2.1 Yell

Starting out as a PhD project of Arkadiy Simonov in the group of Thomas Weber at the ETHZ, Yell has developed into a powerful code for the interpretation of 3D pair distribution function (3D-PDF) data [31]. The software starts off with the reciprocal space reconstruction of 3D single-crystal diffraction data. This is Fourier transformed to a 3D-PDF. The 3D-PDF of the average structure (the Patterson function) is subtracted from the 3D-PDF resulting in a map of the deviations from the average structure, called the 3D-ΔPDF map. The code then searches for correlations linking substitutional or displacement disorder or size effects to the diffuse signals. Yell has mainly been used on synchrotron data; however, Arkadiy Simonov is optimistic that currently available high brilliance laboratory X-ray sources paired with improved area detectors could suffice to collect reliable and accurate diffuse scattering data. For the application of the method, 100% coverage of reciprocal space is necessary, therefore nontrivial symmetries are helpful in reducing the required angular range of data collection or in improving the quality of the diffuse data. It has to be mentioned that the diffuse scattering can reduce the symmetry of the average structure, possibly leading to changing data collection requirements. Arkadiy Simonov describes the data preparation as extensive and suggests scientific cooperation with an experienced user before tackling a solo analysis.

11.3.2.2 NeXpy/CCTW

An active group in the Materials Science Division of Argonne National Laboratory has been utilizing the advanced photon source (APS) to investigate crystalline disorder. Ray Osborn and colleagues have developed a Python framework for reducing single-crystal diffuse scattering data from high-energy beamlines at the APS, NeXpy [37]. It was developed as a graphical user interface (GUI) for handling and manipulating NeXus files [38]. NeXus brings the modern hierarchical data format HDF5 into the diffraction world and enables the storage of large 3D diffraction data with all the relevant metadata so that retrieval and analysis can ideally be performed on the server without the necessity of moving large amounts data through the networks. The data are transformed from raw detector images into 3D reciprocal space maps by the crystal coordinate transformation workflow (CCTW), a C++ program, written by Guy Jennings [39] at the APS. As the data sets are over 30 GB in size, this is performed in chunks that are ~100 × 100 × 100 elements in size, limiting the required memory to 10 GB per calculation, which is orders of magnitude smaller than the memory required for the entire transformation in a single step. This parallelization enables a substantial speed gain when projecting the data into reciprocal space. The group at Argonne employ simulations of 3D diffuse scattering data, using a combination of DISCUS [13] and Joe Paddison's FFT code [15], as well as their own implementation of the 3D-ΔPDF method discussed in the previous section, in order to interpret these large data sets.

11.3.3 Powder diffraction

11.3.3.1 DiffPY

DiffPY is a project that has developed out of Simon Billinge's group at Columbia University with Pavol Juhás [40] being the main contributor. It shares some code with PDFgui [41], likely one of the most important drivers of the methodological expansion that pair distribution function analysis has enjoyed over the last decade. Although DiffPY is its successor, many PDF refinements are still performed using the intuitive and fast interface of PDFgui. DiffPY is essentially a collection of python libraries that can be used to perform structure modelling and pair distribution function analysis. Fitting routines allow the fitting of multiple data to a single model. It has the advantage of being written in Python, the lingua franca of modern scientific programming and can therefore make use of a large ecosystem of code ranging from fundamental crystallography [42] to optimization routines [43, 44]. Due to the popularity of Python, many of these libraries have achieved a maturity rarely encountered before in scientific computing. The most developed components in DiffPY are the structural calculators. Building a radial distribution function from a structure file applying boundary conditions, or as a discrete nanoparticle or single molecule, is straightforward. Pair distribution functions are calculated either directly or via the Debye function and Fourier transformation. The access to libraries for structure model manipulation is a key advantage. The additional capability of analyzing small-angle scattering data offers more detail in nanoparticle characterization. Combined refinements including both experiments, are a feature of the SrFit library within DiffPY [45]. The fact that DiffPY is typically used in a development environment type interface [46, 47] does call for some coding affinity by the user. However, this allows for a large productivity gain, especially the automation of repetitive tasks or the setting up of optimization runs. Recent studies [2, 48] on industrially relevant polymers using PDF have been performed using this tool. Extending features or adding new capabilities is a task that is possible with some experience. Adding features from other code is also possible, as calling up functions using dynamic libraries is a relatively simple task within Python. An interesting development is the PDF search/match routine which allows a fast and simple identification of the structure portrayed by the PDF data [49].

The growing user base and open source nature of the code are indicators of a bright future in the PDF world.

11.3.3.2 DIFFaX

DIFFaX [50] is the most established disorder simulation tool for powder diffraction of materials containing planar defects. It has not garnered a large user base as it has the aura of being a nonintuitive and complex tool. It requires a text file input written in a determined format as well as scattering factors. The creation of a file that be interpreted by DIFFaX may be time consuming. It requires the planes containing the defects to lie normal to the c-axis. DIFFaX can not only output the simulation of a powder diffraction pattern, but also deliver electron diffraction data [51] or linear profiles along streaks seen in single crystal datasets. Instrumental resolution and crystallite size effects can be accounted for within DIFFaX. Users with a low frustration threshold can resort to the interface to DIFFaX that GSAS-II [52] provides. Great effort has been made to make all of the functions available via an intuitive user interface. A couple of default settings for various instruments ease the way into the use of DIFFaX.

11.3.3.3 FAULTS

FAULTS [53, 54] is a software distributed with the FullProf Suite [55] and represents another package which aims to ease the use of DIFFaX via an established Rietveld refinement package. FAULTS offers the capability of refining parameters. This is a feature that DIFFaX does not offer. Users should be aware that the software has only recently been published. The integration into the FullProf Suite does not seem quite as tight as one might hope.

11.3.3.4 DIFFaX+

DIFFaX+ [56, 57] was the first code to encase DIFFaX into a structure refinement suite including microstructural modelling as well as the refinement of the transition matrix elements. Matteo Leoni is the author of the software and is planning various features into a new version. These include a faster implementation of the recursion method and a new core allowing the modelling of composite faulted lattices. Another goal is to implement multiphase analysis.

11.3.3.5 BGMN

In 2012, BGMN, a Rietveld refinement programme that had become well known for its use in quantitative analysis, adopted an approach similar to DIFFaX to model layered structures [20, 21]. This made BGMN the first software package to allow quantitative phase analysis of materials containing layered disordered structures. These are prevalent in layered silicates which represent the most challenging geological samples to quantify using XRD. Ufer used BGMN to quantify illitic–smectitic intergrowths. One advantage of this implementation is the possibility of refining the transition matrix elements. These converged reliably to the values of the DIFFaX simulated pattern [20].

11.3.3.6 RMCProfile

The Reverse Monte Carlo algorithm was introduced with RMC in 1988, modelling the structure of liquid Argon against PDF data [58]. RMCProfile [59] is a strongly collaborative effort between Martin Dove, Andrew Goodwin, Matt Tucker, Dave Keen, and with various other collaborators. It has profited from research funding, which has enabled the modelling of magnetism and the introduction of chemically realistic constraints. It is a big box disorder modelling software for neutron and X-ray diffraction data. Total scattering and PDF data can be fitted to a model. An often-noted issue with this algorithm is the fact that it drives toward the solution with the highest entropy. This is due to the method being governed by statistical mechanics. The complex bond and bond angle restraining potentials that can be set should hinder such unrealistic solutions. A fascinating feature was added by Igor Levin: EXAFS modelling. This capability is unique among the discussed programmes. Optimization runs require substantial computing time, typically on the order of one day. GSAS-II contains a beautifully made user interface to RMCProfile [60], giving visual feedback on results and a useful integration to the initial Rietveld refinement.

11.3.3.7 TOPAS

A relatively new addition to the field of modelling disordered structures is the software TOPAS [61, 62]. Developed as a Rietveld software with a powerful algebraic language, it has achieved great success in the field of powder diffraction, despite the fact that it is commercial software. Since the advent of version 6, it has the capability of fitting PDF data and simulating layer disorder in DIFFaX style. This is an interesting development, as TOPAS is respected for the high level of optimization as well as robustness that are characteristic for most of its algorithms. To describe a layered disorder, the appropriate layers are defined, and the vectors connecting one to the other as well as the probability matrix for the described transitions need to be given [63]. TOPAS then iterates the layers a given number of times. This separation into multiple smaller boxes instead of one big one serves to reduce the calculation times. This shortcut has a side effect of producing less peak positions than a larger box, possibly causing unintended ripples in the simulated pattern. An ingenious selective peak broadening applied to faulted peaks, and leaving unfaulted peaks sharp, leads to patterns similar to boxes 10 to 20 times larger [14]. Further enhancements such as utilizing efficient peak buffers and efficient stacking algorithms allow an acceleration of four orders of magnitude compared to the performance of the traditional code operating on a disordered big box structure. In version 7, TOPAS [71] includes cloud-based computation enabling the parallel optimization of potentially thousands of structures.

11.3.3.8 DIANNA

DIANNA is a Debye scattering package written by researchers at the Boreskov Institute of Catalysis [64]. This is software keyed to the characterization of nanoparticles and sports an intuitive interface that allows quick initial refinements. It has been used primarily in the field of catalysis and is especially useful in characterizing the shape and size of nanoparticles using laboratory diffraction data. The software package also allows the calculation of the pair distribution function.

Tab. 11.1: Overview of some features differentiating the discussed software packages.

Software	Sample		Radiation		Model building	Chemical constraints	Optimization	
	Single crystal	Powder	X-ray	Neutron			Local	Global
Discus	x	x	x	x	xx	x	x	x
Yell	x		x	x				
CCTW	x		x	x				
DiffPY		x	x	x	x	x	x	x
DIFFaX		x	x	x	x			
FAULTS		x	x	x	x		x	
DIFFaX+		x	x	x	x		x	
BGMN		x	x		x		x	
RMCProfile		x	x	x	x	x		x
TOPAS	(x)	x	x	x	x	x	x	x
DIANNA		x	x		x		x	
DebUsSy		x	x	x	x		x	x

11.3.3.9 DebUsSy

This is another Debye scattering software authored by Antonella Guagliardi and Antonio Cervellino [65] who have been joined by Federica Bertolotti and Ruggero Frison for the current version [66]. It contains a package (Claude) to create nanoparticles or populations of nanoparticles of variable shapes and sizes and another (Debussy) to refine these against experimental data. A useful size distribution complete with graphics is provided within the user interface. The software can be used in a text input/output mode as well as via a GUI. The possibility of calculating a PDF has been added. The software has shown encouraging progress since its debut and is currently carrying the version number 2.2 with the latest update from 2018.

11.4 Conclusion

Should you wish to discover the reasons for the disorder in your material and if you are in possession of single-crystal diffuse scattering data, then the only choice lies between Discus and Yell/CCTW. The different approaches of the two will be the deciding factor. Where Discus models the diffraction signal, the 3Δ-PDF gives you a direct space conception of the source of disorder. Yell will be able to quantify this for you.

In the powder diffraction field, the options are varied. If you would like to simulate nanostructure effects on the diffraction pattern, the Debye function refinements offered by DIANNA and DebUsSy will be your preferred choice. They can model various shapes of nanoparticles with intuitive user interfaces. If you are trying to solve a layer disorder effect then the choice is wide: Discus, FAULTS, DIFFaX, DIFFaX+, BGMN, and TOPAS offer very similar descriptors to build faulted layers into a large superstructure required to simulate a diffraction pattern reliably. DIFFaX+ and BGMN offer the ability to refine the transition probabilities, easing the search for the best model. If your disordered material shows a differing dimensionality, then you need to look at Discus, DiffPY, or RMCProfile. These are the most advanced programmes with regard to features, model building, and versatility.

In spite of the number of tools available, the study of disorder remains a challenging task. The effort is easily comparable to the solving of crystal structures half a century ago. Many developments point to a strong growth and a bright future for this field of study. The first is related to computing environments: high speed detectors paired with brilliant X-ray sources produce huge amounts of data. These demand a new software infrastructure capable of parallel processing in server- or cloud-based infrastructures [67, 68]. Such infrastructure has to meet modern data handling principles [69] as well as to allow efficient processing. Developments in this direction will undoubtedly lead to much more efficient ways to analyze disordered materials. Most of the computations used to solve big box structures belong to the embarrassingly parallel [70] variety and can profit immensely from modern cloud based implementations [71]. These have the capability of reducing day-long calculations to a few minutes. Second, the broad acceptance within the solid-state research community that disorder of materials plays a fundamental role in their properties will increase the need for more studies. It is encouraging to see the amount of collaborative efforts that are being published and the high quality of the work being produced.

References

[1] Stampfl, C., Van De Walle, C.G. Energetics and electronic structure of stacking faults in AlN, GaN, and InN. Phys. Rev. B: Condens. Matter Mater. Phys. 1998, 57, R15052. Doi: 10.1103/physrevb.57.r15052.

[2] Terban, M.W., Dabbous, R., Debellis, A.D., Pöselt, E., Billinge, S.J. Structures of hard phases in thermoplastic polyurethanes. Macromolecules 2016, 10, 49, 7350–7358. Doi: 10.1021/acs.macromol.6b00889.

[3] Bette, S., Hinrichsen, B., Pfister, D., Dinnebier, E. A routine for the determination of the microstructure of stacking-faulted nickel cobalt aluminium hydroxide precursors for lithium nickel cobalt aluminium oxide battery materials. J. Appl. Crystallogr. 2020, 2, 53, 76–87. Doi: 10.1107/S1600576719016212.

[4] Jin, S., Xu, H., Wang, X., Jacobs, R., Morgan, D. The incommensurately modulated structures of low-temperature labradorite feldspars: A single-crystal X-ray and neutron diffraction study. Acta Crystallogr., Sect. B: Struct. Sci 2020, 2, 76, 93–107. Doi: 10.1107/S2052520619017128.

[5] Radha, A.V., Kamath, P.V., Shivakumara, C. Conservation of order, disorder, and "crystallinity" during anion-exchange reactions among Layered Double Hydroxides (LDHs) of Zn with Al. J. Phys. Chem. B 2007, 111, 3411. Doi: 10.1021/jp0684170.

[6] Nguyen, P., Sleight, A., Roberts, N., Warren, W. Modeling of extended defects in the vanadium phosphate catalyst for butane oxidation, (VO)2P2O7. J. Solid State Chem. 1996, 122, 259. Doi: 10.1006/jssc.1996.0111.

[7] Derollez, P., Correia, N.T., Danède, F., Capet, F., Affouard, F., Lefebvre, J., Descamps, M. Ab initio structure determination of the high-temperature phase of anhydrous caffeine by X-ray powder diffraction. Acta Crystallogr., B 2005, 61, 329–334.

[8] Dinnebier, R., Sofina, N., Jansen, M. The structure of the high temperature modification of lithium triflate (γ-LiSO3CF3). Zeitschrift für anorganische und allgemeine Chemie 2004, 630, 1613–1616. Doi: 10.1002/zaac.200400224.

[9] Hildebrandt, L., Dinnebier, R., Jansen, M. Crystal structure and ionic conductivity of cesium trifluoromethyl sulfonate, CsSO3CF3. Zeitschrift für anorganische und allgemeine Chemie 2005, 7, 631, 1660–1666. Doi: 10.1002/zaac.200500097.

[10] Bertolotti, F., Protesescu, L., Kovalenko, M.V., Yakunin, S., Cervellino, A., Billinge, S.J., . . . Guagliardi, A. Coherent nanotwins and dynamic disorder in cesium lead halide perovskite nanocrystals. ACS Nano 2017, 11, 3819–3831. Doi: 10.1021/acsnano.7b00017.

[11] Kamarás, K., Klupp, G., Tanner, D.B., Hebard, A.F., Nemes, N.M., Fischer, J.E. Ordered low-temperature structure in K 4 C 60 detected by infrared spectroscopy. Phys. Rev. B 2002, 65, 052103.

[12] Zachariou, A., Hawkins, A.P., Collier, P., Howe, R.F., Lennon, D., Parker, S.F. The methyl torsion in unsaturated compounds. ACS Omega 2020, 2, 5, 2755–2765. Doi: 10.1021/acsomega.9b03351.

[13] Proffen, T., Neder, R.B. (2008). Diffuse Scattering and Defect Structure Simulations: A Cook Book Using the Program DISCUS.

[14] Coelho, A.A., Evans, J.S., Lewis, J.W. Averaging the intensity of many-layered structures for accurate stacking-fault analysis using Rietveld refinement. J. Appl. Crystallogr. 2016, 10, 49, 1740–1749. Doi: 10.1107/S1600576716013066.

[15] Paddison, J.A. Ultrafast calculation of diffuse scattering from atomistic models. Acta Crystallogr., Sect. A: Found. Crystallogr. 2019, 75, 14–24. Doi: 10.1107/s2053273318015632.

[16] Coelho, A.A., Chater, P.A., Kern, A. Fast synthesis and refinement of the atomic pair distribution function. J. Appl. Crystallogr. 2015, 6, 48, 869–875. Doi: 10.1107/S1600576715007487.

[17] Anderson, M.W., Terasaki, O., Ohsuna, T., Philippou, A., Mackay, S.P., Ferreira, A., . . . Lidin, S. Structure of the microporous titanosilicate ETS-10. Nature 1994, 367, 347. Doi: 10.1038/367347a0.

[18] Anderson, M.W., Gebbie-Rayet, J.T., Hill, A.R., Farida, N., Attfield, M.P., Cubillas, P., . . . Gale, J.D. Predicting crystal growth via a unified kinetic three-dimensional partition model. Nature 2017, 544, 456–459. Doi: 10.1038/nature21684.

[19] Bette, S., Dinnebier, R.E., Freyer, D. Structure solution and refinement of stacking-faulted NiCl(OH). J. Appl. Crystallogr. 2015, 48, 1706. Doi: 10.1107/s1600576715017719.

[20] Ufer, K., Kleeberg, R., Bergmann, J., Dohrmann, R. Rietveld refinement of disordered illite-smectite mixed-layer structures by a recursive algorithm. I: One-dimensional patterns. Clays Clay Miner. 2012, 60, 507. Doi: 10.1346/ccmn.2012.0600507.

[21] Ufer, K., Kleeberg, R., Bergmann, J., Dohrmann, R. Rietveld refinement of disordered illite-smectite mixed-layer structures by a recursive algorithm. II: Powder-pattern refinement and quantitative phase analysis. Clays Clay Miner. 2012, 60, 535. Doi: 10.1346/ccmn.2012.0600508.

[22] Ufer, K., Roth, G., Kleeberg, R., Stanjek, H., Dohrmann, R., Bergmann, J. Description of X-Ray powder pattern of turbostratically disordered layer structures with a rietveld compatible approach. Z. Kristallogr. – Cryst. Mater. 2004, 219, 519. Doi: 10.1524/zkri.219.9.519.44039.

[23] Sun, X., Bonnick, P., Nazar, L.F. Layered TiS2 positive electrode for Mg batteries, ACS Energy Lett., 1. 2016. Doi: 10.1021/acsenergylett.6b00145.

[24] Tepavcevic, S. Nanostructured layered cathode for rechargeable Mg-ion batteries. ACS Nano 2015, 9. Doi: 10.1021/acsnano.5b02450.

[25] Shao-Horn, Y., Levasseur, S., Weill, F., Delmas, C. Probing lithium and vacancy ordering in O3 layered LixCoO2 (x ≈ 0.5). J. Electrochem. Soc. 2003, 150. Doi: 10.1149/1.1553787.

[26] Krogstad, M.J., Rosenkranz, S., Wozniak, J.M., Jennings, G., Ruff, J.P., Vaughey, J.T., Osborn, R. Reciprocal space imaging of ionic correlations in intercalation compounds. Nat. Mater 2020, 19, 63–68. Doi: 10.1038/s41563-019-0500-7.

[27] Toumar, A.J., Ong, S.P., Richards, W.D., Dacek, S., Ceder, G. Vacancy ordering in O3-type layered metal oxide sodium-ion battery cathodes. Phys. Rev. Appl. 2015, 4. Doi: 10.1103/PhysRevApplied.4.064002.

[28] Bréger, J., Jiang, M., Dupré, N., Meng, Y.S., Shao-Horn, Y., Ceder, G., Grey, C.P. High-Resolution X-Ray Diffraction, DIFFaX, NMR and First Principles Study of Disorder in the Li2MnO3-Li[Ni1/2Mn1/2]O2 Solid Solution. J. Solid State Chem. 2005, 178, 2575. Doi: 10.1016/j.jssc.2005.05.027.

[29] Xu, H., Chen, W., Wu, Q., Lei, C., Zhang, J., Han, S., . . . Xiao, F.-S. Transformation synthesis of aluminosilicate SSZ-39 zeolite from ZSM-5 and beta zeolite. J. Mater. Chem. A 2019, 7(9), 4420–4425, Doi: 10.1039/C9TA00174C.

[30] Welberry, T.R., Goossens, D.J. Diffuse scattering and partial disorder in complex structures. IUCrJ 2014, 1, 550–562.

[31] Weber, T., Simonov, A. The three-dimensional pair distribution function analysis of disordered single crystals: basic concepts. Z. Kristallogr. 2012, 227. Doi: 10.1524/zkri.2012.1504.

[32] Simonov, A., De Baerdemaeker, T., Boström, H.L., Ríos Gómez, M.L., Gray, H.J., Chernyshov, D., . . . Goodwin, A.L. Hidden diversity of vacancy networks in Prussian blue analogues. Nature 2020, 578, 256–260. Retrieved from: https://doi.org/10.1038/s41586-020-1980-y.

[33] Weber, T. Cooperative Evolution – a new algorithm for the investigation of disordered structures via Monte Carlo modelling. Zeitschrift für Kristallographie – Crystalline Materials 2005 Dec, 01, 220, 1099–1107. Doi: https://doi.org/10.1524/zkri.2005.220.12.1099.

[34] Neder, R. (n.d.). Discus Facebook. Retrieved from: https://www.facebook.com/programDISCUS/

[35] Neder, R. (n.d.). Discus Youtube channel. Retrieved from: https://www.youtube.com/channel/UC6nA4_mMy__j6GYpbIKAz8w/

[36] Proffen, T., Neder, R. (n.d.). Discus Code. Retrieved from: http://tproffen.github.io/DiffuseCode/

[37] Osborne, R. (n.d.). NeXpy: A Python GUI to analyze NeXus data. Retrieved from NeXpy: http://nexpy.github.io/nexpy/

[38] Könnecke, M. The NeXus data format. J. Appl. Crystallogr. 2015, 48. Doi: 10.1107/S1600576714027575.

[39] Jennings, G. (2019). Crystal Coordinate Transformation Workflow (CCTW). Retrieved from Sourceforge: https://sourceforge.net/projects/cctw/

[40] Juhás, P., Farrow, C.L., Yang, X., Knox, K.R., Billinge, S.J. Complex modeling: A strategy and software program for combining multiple information sources to solve ill posed structure and nanostructure inverse problems. Acta Crystallogr., Sect. A: Found. Crystallogr. Adv. 2015, 11, 71(6), 562–568. Doi: 10.1107/S2053273315014473.

[41] Farrow, C.L., Juhás, P., Liu, J.W., Bryndin, D., Božin, E.S., Bloch, J., . . . Billinge, S.J. PDFfit2 and PDFgui: computer programs for studying nanostructure in crystals. J. Phys.: Condens. Matter 2007, 7, 19, 335219. Doi: 10.1088/0953-8984/19/33/335219.

[42] Crystals. (n.d.). Retrieved from pypi.org: https://pypi.org/project/crystals/

[43] scipy. (n.d.). Retrieved from pypi: https://pypi.org/project/scipy/

[44] Optimize. (n.d.). Retrieved from scipy.org: https://docs.scipy.org/doc/scipy/reference/tutorial/optimize.html#id20

[45] Farrow, C.L., Shi, C., Juhás, P., Peng, X., Billinge, S.J. Robust structure and morphology parameters for CdS nanoparticles by combining small-angle X-ray scattering and atomic pair distribution function data in a complex modeling framework. J. Appl. Crystallogr. 2014, 4, 47, 561–565. Doi: 10.1107/S1600576713034055.

[46] Jupyter. (n.d.). Retrieved from https://jupyter.org

[47] ipython. (n.d.). Retrieved from: https://ipython.org

[48] Terban, M.W., Pütz, A.M., Savasci, G., Heinemeyer, U., Hinrichsen, B., Desbois, P., Dinnebier, R.E. Improving the picture of atomic structure in nonoriented polymer domains using the pair distribution function: A study of polyamide 6. J. Polym. Sci. 2020, 58, 1843–1866. Doi: 10.1002/pol.20190272.

[49] Yang, L., Juhás, P., Terban, M.W., Tucker, M.G., Billinge, S.J. Structure-mining: Screening structure models by automated fitting to the atomic pair distribution function over large

numbers of models. Acta Crystallogr., Sect. A: Found. Crystallogr. 2020, 5, 76, 395–409. Doi: 10.1107/S2053273320002028.

[50] Treacy, M.M., Newsam, J.M., Deem, M.W. (1991, 6). A general recursion method for calculating diffracted intensities from crystals containing planar faults. Proceedings of the Royal Society of London Series A, 433, 499–520. Doi: 10.1098/rspa.1991.0062

[51] Treacy, M.M., Newsam, J.M., Deem, M.W. Simulation of electron diffraction patterns from partially ordered layer lattices. Ultramicroscopy 1993, 52, 512–522. Doi: 10.1016/0304-3991 (93)90068-9.

[52] Brian Toby, R.V. (n.d.). GSAS-II DIFFaX. Retrieved from: https://subversion.xray.aps.anl.gov/ pyGSAS/Tutorials/StackingFaults-I/Stacking%20Faults-I.htm

[53] Casas-Cabanas, M., Reynaud, M., Rikarte, J., Horbach, P., Rodríguez-Carvajal, J. A program for refinement of structures with extended defects. J. Appl. Crystallogr. 2016, 12, 49, 2259–2269. Doi: 10.1107/S1600576716014473.

[54] Casas-Cabanas, M., Rodríguez-Carvajal, J., Canales-Vázquez, J., Laligant, Y., Lacorre, P., Palacín, M.R. Microstructural characterisation of battery materials using powder diffraction data: DIFFaX, FAULTS and SH-FullProf approaches. J. Power Sources 2007, 1, 174, 414–420. Doi: 10.1016/j.jpowsour.2007.06.216.

[55] Rodríguez-Carvajal, J. Recent advances in magnetic structure determination by neutron powder diffraction. Phys. B Condens Matter 1993, 192, 55–69. Doi: https://doi.org/10.1016/ 0921-4526(93)90108-I.

[56] Leoni, M. Diffraction analysis of layer disorder. Z. Kristallogr. 2008, 223, 561. Doi: 10.1524/ zkri.2008.1214.

[57] Leoni, M., Gualtieri, A.F., Roveri, N. Simultaneous refinement of structure and microstructure of layered materials. J. Appl. Crystallogr. 2004, 37, 166. Doi: 10.1107/s0021889803022787.

[58] McGreevy, R.L., Pusztai, L. Reverse Monte Carlo Simulation: A New Technique for the Determination of Disordered Structures. Mol. Simul 1988, 1, 359–367. Doi: 10.1080/ 08927028808080958.

[59] Tucker, M.G., Keen, D.A., Dove, M.T., Goodwin, A.L., Hui, Q. RMCProfile: Reverse Monte Carlo for polycrystalline materials. J. Phys.: Condens. Matter 2007, 7, 19, 335218. Doi: 10.1088/ 0953-8984/19/33/335218.

[60] Toby, B., Von Dreele, R. (n.d.). GSAS RMCProfile Tutorial. Retrieved from: https://subversion. xray.aps.anl.gov/pyGSAS/Tutorials/RMCProfile-I/RMCProfile-I.htm

[61] Coelho, A.A. (2012). TOPAS Academic: General Profile and Structure Analysis Software for Powder Diffraction Data. (5, Ed.)

[62] Coelho, A.A., Evans, J.S., Evans, I.R., Kern, A., Parsons, S. The TOPAS symbolic computation system. Powder Diffr. 2011, 26, S22. Doi: 10.1154/1.3661087.

[63] Ainsworth, C.M., Lewis, J.W., Wang, C.-H., Coelho, A.A., Johnston, H.E., Brand, H.E., Evans, J.S. 3D transition metal ordering and rietveld stacking fault quantification in the new oxychalcogenides La2O2Cu2−4xCd2xSe2. Chem. Mater. 2016, 28, 3184–3195. Doi: 10.1021/ acs.chemmater.6b00924.

[64] Yatsenko, D., Tsybulya, S. DIANNA (diffraction analysis of nanopowders) – a software for structural analysis of nanosized powders. Zeitschrift für Kristallographie – Crystalline Materials 2018 Jan, 01, 233, 61–66. Doi: https://doi.org/10.1515/zkri-2017-2056.

[65] Cervellino, A., Giannini, C., Guagliardi, A. A Debye user system for nanocrystalline materials. J. Appl. Crystallogr. 2010, 12, 43, 1543–1547. Doi: 10.1107/S0021889810041889.

[66] Cervellino, A., Frison, R., Bertolotti, F., Guagliardi, A. DEBUSSY 2.0: The new release of a Debye user system for nanocrystalline and/or disordered materials. J. Appl. Crystallogr. 2015, 48, 2026–2032. Doi: 10.1107/S1600576715020488.

[67] Michels-Clark, T.M., Lynch, V.E., Hoffmann, C.M., Hauser, J., Weber, T., Harrison, R., Bürgi, H.B. Analyzing diffuse scattering with supercomputers. J. Appl. Crystallogr. 2013, 12, 46, 1616–1625. Doi: 10.1107/S0021889813025399.

[68] Billinge, S., Tucker, M., Jensen, K. (2020). PDF in the Could. (Columbia University & NSF) Retrieved from: https://pdfitc.org.

[69] Wilkinson, M.D., Dumontier, M., Aalbersberg, I.J., Appleton, G., Axton, M., Baak, A. others. The FAIR guiding principles for scientific data management and stewardship. Sci. Data 2016, 3, 1–9.

[70] Embarrasingly parallel. (n.d.). Retrieved from Wikipedia: https://en.wikipedia.org/wiki/Embarrassingly_parallel

[71] Coelho, A. (n.d.). TOPAS Academic V7. Retrieved from: http://www.topas-academic.net/

Index

https://doi.org/10.1515/9783110674910-012